1

INTRODUCTION TO STATICS

CHAPTER OUTLINE

1/1 MECHANICS

Mechanics is the physical science which deals with the effects of forces on objects. No other subject plays a greater role in engineering analysis than mechanics. Although the principles of mechanics are few, they have wide application in engineering. The principles of mechanics are central to research and development in the fields of vibrations, stability and strength of structures and machines, robotics, rocket and spacecraft design, automatic control, engine performance, fluid flow, electrical machines and apparatus, and molecular, atomic, and subatomic behavior. A thorough understanding of this subject is an essential prerequisite for work in these and many other fields.

Mechanics is the oldest of the physical sciences. The early history of this subject is synonymous with the very beginnings of engineering. The earliest recorded writings in mechanics are those of Archimedes (287–212 B.C.) on the principle of the lever and the principle of buoyancy. Substantial progress came later with the formulation of the laws of vector combination of forces by Stevinus (1548–1620), who also formulated most of the principles of statics. The first investigation of a dynamics problem is credited to Galileo (1564–1642) for his experiments with falling stones. The accurate formulation of the laws of motion, as well as the law of gravitation, was made by Newton (1642–1727), who also conceived the idea of the infinitesimal in mathematical analysis. Substantial contributions to the development of mechanics were also made by da Vinci, Varignon, Euler, D'Alembert, Lagrange, Laplace, and others.

In this book we will be concerned with both the development of the principles of mechanics and their application. The principles of mechanics as a science are rigorously expressed by mathematics, and thus mathematics plays an important role in the application of these principles to the solution of practical problems.

The subject of mechanics is logically divided into two parts: ***statics***, which concerns the equilibrium of bodies under the action of forces, and ***dynamics***, which concerns the motion of bodies. *Engineering Mechanics* is divided into these two parts, *Vol. 1 Statics* and *Vol. 2 Dynamics*.

1/2 Basic Concepts

The following concepts and definitions are basic to the study of mechanics, and they should be understood at the outset.

Space is the geometric region occupied by bodies whose positions are described by linear and angular measurements relative to a coordinate system. For three-dimensional problems, three independent coordinates are needed. For two-dimensional problems, only two coordinates are required.

Time is the measure of the succession of events and is a basic quantity in dynamics. Time is not directly involved in the analysis of statics problems.

Mass is a measure of the inertia of a body, which is its resistance to a change of velocity. Mass can also be thought of as the quantity of matter in a body. The mass of a body affects the gravitational attraction force between it and other bodies. This force appears in many applications in statics.

Force is the action of one body on another. A force tends to move a body in the direction of its action. The action of a force is characterized by its *magnitude*, by the *direction* of its action, and by its *point of application*. Thus force is a vector quantity, and its properties are discussed in detail in Chapter 2.

A ***particle*** is a body of negligible dimensions. In the mathematical sense, a particle is a body whose dimensions are considered to be near zero so that we may analyze it as a mass concentrated at a point. We often choose a particle as a differential element of a body. We may treat a body as a particle when its dimensions are irrelevant to the description of its position or the action of forces applied to it.

Rigid body. A body is considered rigid when the change in distance between any two of its points is negligible for the purpose at hand. For instance, the calculation of the tension in the cable which supports the boom of a mobile crane under load is essentially unaffected by the small internal deformations in the structural members of the boom. For the purpose, then, of determining the external forces which act on the boom, we may treat it as a rigid body. Statics deals primarily with the calculation of external forces which act on rigid bodies in equilibrium. Determination of the internal deformations belongs to the study of the mechanics of deformable bodies, which normally follows statics in the curriculum.

1/3 Scalars and Vectors

We use two kinds of quantities in mechanics—scalars and vectors. *Scalar quantities* are those with which only a magnitude is associated. Examples of scalar quantities are time, volume, density, speed, energy, and mass. *Vector quantities*, on the other hand, possess direction as well as magnitude, and must obey the parallelogram law of addition as described later in this article. Examples of vector quantities are displacement, velocity, acceleration, force, moment, and momentum. Speed is a scalar. It is the magnitude of velocity, which is a vector. Thus velocity is specified by a direction as well as a speed.

Vectors representing physical quantities can be classified as free, sliding, or fixed.

A ***free vector*** is one whose action is not confined to or associated with a unique line in space. For example, if a body moves without rotation, then the movement or displacement of any point in the body may be taken as a vector. This vector describes equally well the direction and magnitude of the displacement of every point in the body. Thus, we may represent the displacement of such a body by a free vector.

A ***sliding vector*** has a unique line of action in space but not a unique point of application. For example, when an external force acts on a rigid body, the force can be applied at any point along its line of action without changing its effect on the body as a whole,* and thus it is a sliding vector.

A ***fixed vector*** is one for which a unique point of application is specified. The action of a force on a deformable or nonrigid body must be specified by a fixed vector at the point of application of the force. In this instance the forces and deformations within the body depend on the point of application of the force, as well as on its magnitude and line of action.

Conventions for Equations and Diagrams

A vector quantity **V** is represented by a line segment, Fig. 1/1, having the direction of the vector and having an arrowhead to indicate the sense. The length of the directed line segment represents to some convenient scale the magnitude $|\mathbf{V}|$ of the vector, which is printed with lightface italic type V. For example, we may choose a scale such that an arrow one centimeter long represents a force of twenty newtons.

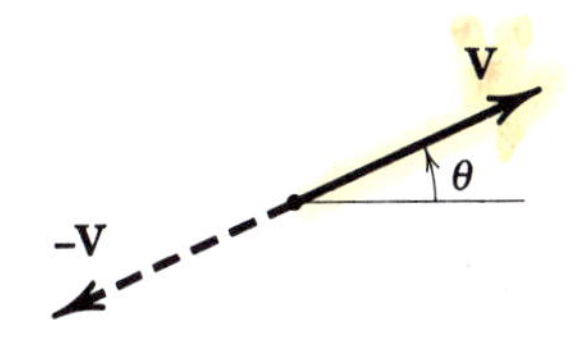

Figure 1/1

In scalar equations, and frequently on diagrams where only the magnitude of a vector is labeled, the symbol will appear in lightface italic type. Boldface type is used for vector quantities whenever the directional aspect of the vector is a part of its mathematical representation. When writing vector equations, *always* be certain to preserve the mathematical distinction between vectors and scalars. In handwritten work, use a dis-

*This is the *principle of transmissibility*, which is discussed in Art. 2/2.

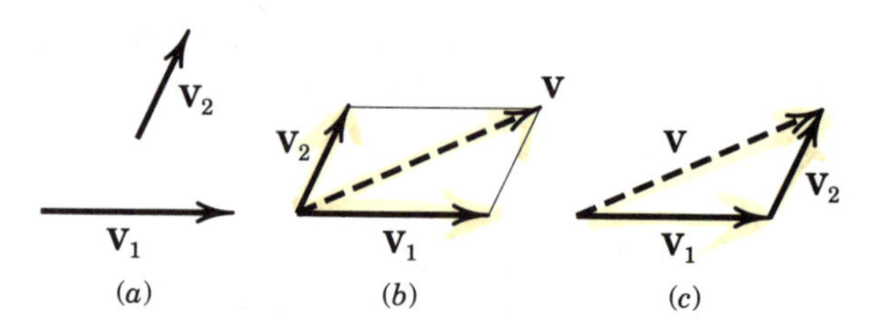

Figure 1/2

tinguishing mark for each vector quantity, such as an underline, $\underline{V}$, or an arrow over the symbol, $\vec{V}$, to take the place of boldface type in print.

Working with Vectors

The direction of the vector $\mathbf{V}$ may be measured by an angle θ from some known reference direction as shown in Fig. 1/1. The negative of $\mathbf{V}$ is a vector $-\mathbf{V}$ having the same magnitude as $\mathbf{V}$ but directed in the sense opposite to $\mathbf{V}$, as shown in Fig. 1/1.

Vectors must obey the parallelogram law of combination. This law states that two vectors $\mathbf{V}_1$ and $\mathbf{V}_2$, treated as free vectors, Fig. 1/2*a*, may be replaced by their equivalent vector $\mathbf{V}$, which is the diagonal of the parallelogram formed by $\mathbf{V}_1$ and $\mathbf{V}_2$ as its two sides, as shown in Fig. 1/2*b*. This combination is called the *vector sum,* and is represented by the vector equation

$$\mathbf{V} = \mathbf{V}_1 + \mathbf{V}_2$$

where the plus sign, when used with the vector quantities (in boldface type), means *vector* and not *scalar* addition. The scalar sum of the magnitudes of the two vectors is written in the usual way as $V_1 + V_2$. The geometry of the parallelogram shows that $V \neq V_1 + V_2$.

The two vectors $\mathbf{V}_1$ and $\mathbf{V}_2$, again treated as free vectors, may also be added head-to-tail by the triangle law, as shown in Fig. 1/2*c*, to obtain the identical vector sum $\mathbf{V}$. We see from the diagram that the order of addition of the vectors does not affect their sum, so that $\mathbf{V}_1 + \mathbf{V}_2 = \mathbf{V}_2 + \mathbf{V}_1$.

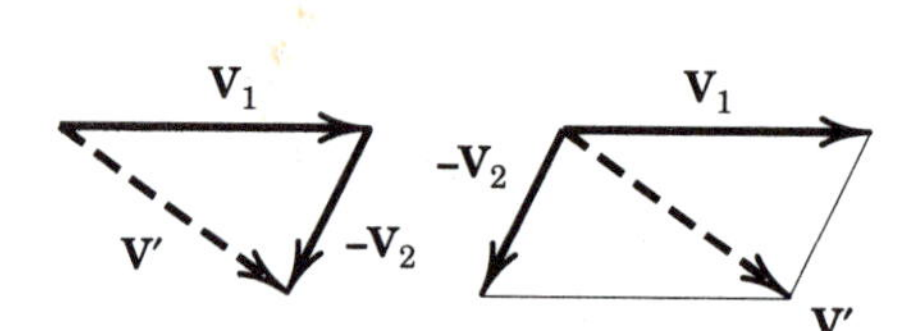

Figure 1/3

The difference $\mathbf{V}_1 - \mathbf{V}_2$ between the two vectors is easily obtained by adding $-\mathbf{V}_2$ to $\mathbf{V}_1$ as shown in Fig. 1/3, where either the triangle or parallelogram procedure may be used. The difference $\mathbf{V}'$ between the two vectors is expressed by the vector equation

$$\mathbf{V}' = \mathbf{V}_1 - \mathbf{V}_2$$

where the minus sign denotes *vector subtraction*.

Any two or more vectors whose sum equals a certain vector $\mathbf{V}$ are said to be the *components* of that vector. Thus, the vectors $\mathbf{V}_1$ and $\mathbf{V}_2$ in Fig. 1/4*a* are the components of $\mathbf{V}$ in the directions 1 and 2, respectively. It is usually most convenient to deal with vector components which are mutually perpendicular; these are called *rectangular components*. The vectors $\mathbf{V}_x$ and $\mathbf{V}_y$ in Fig. 1/4*b* are the x- and y-components,

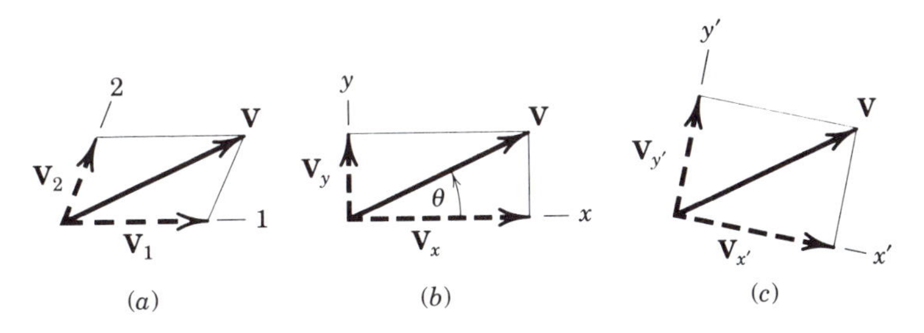

Figure 1/4

respectively, of $\mathbf{V}$. Likewise, in Fig. 1/4c, $\mathbf{V}_{x'}$ and $\mathbf{V}_{y'}$ are the x'- and y'-components of $\mathbf{V}$. When expressed in rectangular components, the direction of the vector with respect to, say, the x-axis is clearly specified by the angle θ, where

$$\theta = \tan^{-1} \frac{V_y}{V_x}$$

A vector $\mathbf{V}$ may be expressed mathematically by multiplying its magnitude V by a vector $\mathbf{n}$ whose magnitude is one and whose direction coincides with that of $\mathbf{V}$. The vector $\mathbf{n}$ is called a *unit vector*. Thus,

$$\mathbf{V} = V\mathbf{n}$$

In this way both the magnitude and direction of the vector are conveniently contained in one mathematical expression. In many problems, particularly three-dimensional ones, it is convenient to express the rectangular components of $\mathbf{V}$, Fig. 1/5, in terms of unit vectors $\mathbf{i}$, $\mathbf{j}$, and $\mathbf{k}$, which are vectors in the x-, y-, and z-directions, respectively, with unit magnitudes. Because the vector $\mathbf{V}$ is the vector sum of the components in the x-, y-, and z-directions, we can express $\mathbf{V}$ as follows:

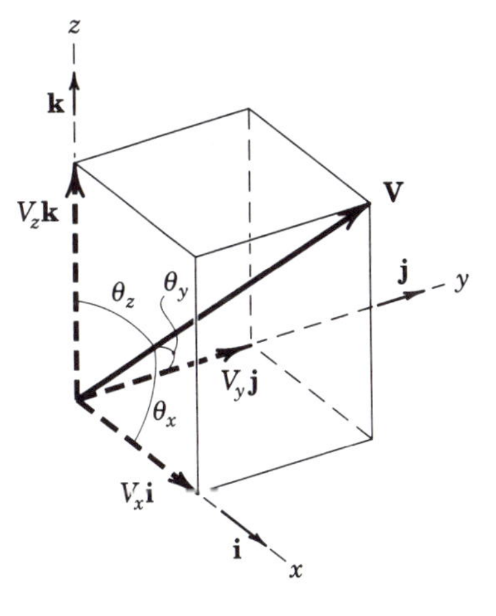

Figure 1/5

$$\mathbf{V} = V_x\mathbf{i} + V_y\mathbf{j} + V_z\mathbf{k}$$

We now make use of the *direction cosines* l, m, and n of $\mathbf{V}$, which are defined by

$$l = \cos\theta_x \qquad m = \cos\theta_y \qquad n = \cos\theta_z$$

Thus, we may write the magnitudes of the components of $\mathbf{V}$ as

$$V_x = lV \qquad V_y = mV \qquad V_z = nV$$

where, from the Pythagorean theorem,

$$V^2 = V_x^{\,2} + V_y^{\,2} + V_z^{\,2}$$

Note that this relation implies that $l^2 + m^2 + n^2 = 1$.

1/4 Newton's Laws

Sir Isaac Newton was the first to state correctly the basic laws governing the motion of a particle and to demonstrate their validity.* Slightly reworded with modern terminology, these laws are:

Law I. A particle remains at rest or continues to move with *uniform velocity* (in a straight line with a constant speed) if there is no unbalanced force acting on it.

Law II. The acceleration of a particle is proportional to the vector sum of forces acting on it, and is in the direction of this vector sum.

Law III. The forces of action and reaction between interacting bodies are equal in magnitude, opposite in direction, and *collinear* (they lie on the same line).

The correctness of these laws has been verified by innumerable accurate physical measurements. Newton's second law forms the basis for most of the analysis in dynamics. As applied to a particle of mass m, it may be stated as

$$\mathbf{F} = m\mathbf{a} \tag{1/1}$$

where $\mathbf{F}$ is the vector sum of forces acting on the particle and $\mathbf{a}$ is the resulting acceleration. This equation is a *vector* equation because the direction of $\mathbf{F}$ must agree with the direction of $\mathbf{a}$, and the magnitudes of $\mathbf{F}$ and $m\mathbf{a}$ must be equal.

Newton's first law contains the principle of the equilibrium of forces, which is the main topic of concern in statics. This law is actually a consequence of the second law, since there is no acceleration when the force is zero, and the particle either is at rest or is moving with a uniform velocity. The first law adds nothing new to the description of motion but is included here because it was part of Newton's classical statements.

The third law is basic to our understanding of force. It states that forces always occur in pairs of equal and opposite forces. Thus, the downward force exerted on the desk by the pencil is accompanied by an upward force of equal magnitude exerted on the pencil by the desk. This principle holds for all forces, variable or constant, regardless of their source, and holds at every instant of time during which the forces are applied. Lack of careful attention to this basic law is the cause of frequent error by the beginner.

In the analysis of bodies under the action of forces, it is absolutely necessary to be clear about which force of each action-reaction pair is being considered. It is necessary first of all to *isolate* the body under consideration and then to consider only the one force of the pair which acts *on* the body in question.

*Newton's original formulations may be found in the translation of his *Principia* (1687) revised by F. Cajori, University of California Press, 1934.

1/5 Units

In mechanics we use four fundamental quantities called *dimensions*. These are length, mass, force, and time. The units used to measure these quantities cannot all be chosen independently because they must be consistent with Newton's second law, Eq. 1/1. Although there are a number of different systems of units, only the two systems most commonly used in science and technology will be used in this text. The four fundamental dimensions and their units and symbols in the two systems are summarized in the following table.

QUANTITY	DIMENSIONAL SYMBOL	SI UNITS			U.S. CUSTOMARY UNITS		
			UNIT	SYMBOL		UNIT	SYMBOL
Mass	M	Base units	kilogram	kg		slug	—
Length	L	Base units	meter	m	Base units	foot	ft
Time	T	Base units	second	s	Base units	second	sec
Force	F		newton	N	Base units	pound	lb

SI Units

The International System of Units, abbreviated SI (from the French, Système International d'Unités), is accepted in the United States and throughout the world, and is a modern version of the metric system. By international agreement, SI units will in time replace other systems. As shown in the table, in SI, the units kilogram (kg) for mass, meter (m) for length, and second (s) for time are selected as the base units, and the newton (N) for force is derived from the preceding three by Eq. 1/1. Thus, force (N) = mass (kg) × acceleration (m/s^2) or

$$N = kg \cdot m/s^2$$

Thus, 1 newton is the force required to give a mass of 1 kg an acceleration of 1 m/s^2.

Consider a body of mass m which is allowed to fall freely near the surface of the earth. With only the force of gravitation acting on the body, it falls with an acceleration g toward the center of the earth. This gravitational force is the *weight* W of the body, and is found from Eq. 1/1:

$$W(\text{N}) = m\,(\text{kg}) \times g\,(\text{m/s}^2)$$

U.S. Customary Units

The U.S. customary, or British system of units, also called the foot-pound-second (FPS) system, has been the common system in business and industry in English-speaking countries. Although this system will in time be replaced by SI units, for many more years engineers must be able to work with both SI units and FPS units.

As shown in the table, in the U.S. or FPS system, the units of feet (ft) for length, seconds (sec) for time, and pounds (lb) for force are selected as base units, and the slug for mass is derived from Eq. 1/1. Thus, force (lb) = mass (slugs) × acceleration (ft/sec^2), or

$$\text{slug} = \frac{\text{lb-sec}^2}{\text{ft}}$$

Therefore, 1 slug is the mass which is given an acceleration of 1 ft/sec^2 when acted on by a force of 1 lb. If W is the gravitational force or weight and g is the acceleration due to gravity, Eq. 1/1 gives

$$m\,(\text{slugs}) = \frac{W\,(\text{lb})}{g\,(\text{ft/sec}^2)}$$

Note that seconds is abbreviated as *s* in SI units, and as *sec* in FPS units.

In U.S. units the pound is also used on occasion as a unit of mass, especially to specify thermal properties of liquids and gases. When distinction between the two units is necessary, the force unit is frequently written as lbf and the mass unit as lbm. In this book we use almost exclusively the force unit, which is written simply as lb. Other common units of force in the U.S. system are the *kilopound* (kip), which equals 1000 lb, and the *ton*, which equals 2000 lb.

The International System of Units (SI) is termed an *absolute* system because the measurement of the base quantity mass is independent of its environment. On the other hand, the U.S. system (FPS) is termed a *gravitational* system because its base quantity force is defined as the gravitational attraction (weight) acting on a standard mass under specified conditions (sea level and 45° latitude). A standard pound is also the force required to give a one-pound mass an acceleration of 32.1740 ft/sec^2.

In SI units the kilogram is used *exclusively* as a unit of mass—*never* force. In the MKS (meter, kilogram, second) gravitational system, which has been used for many years in non-English-speaking countries, the kilogram, like the pound, has been used both as a unit of force and as a unit of mass.

Primary Standards

Primary standards for the measurements of mass, length, and time have been established by international agreement and are as follows:

Mass. The kilogram is defined as the mass of a specific platinum–iridium cylinder which is kept at the International Bureau of Weights and Measures near Paris, France. An accurate copy of this cylinder is kept in the United States at the National Institute of Standards and Technology (NIST), formerly the National Bureau of Standards, and serves as the standard of mass for the United States.

Length. The meter, originally defined as one ten-millionth of the distance from the pole to the equator along the meridian through Paris, was later defined as the length of a specific platinum–iridium bar kept at the International Bureau of Weights and Measures. The difficulty of accessing the bar and reproducing accurate measurements prompted the adoption of a more accurate and reproducible standard of length for the meter, which is now defined as 1 650 763.73 wavelengths of a specific radiation of the krypton-86 atom.

Time. The second was originally defined as the fraction 1/(86 400) of the mean solar day. However, irregularities in the earth's rotation led to difficulties with this definition, and a more accurate and reproducible standard has been adopted. The second is now defined as the duration of 9 192 631 770 periods of the radiation of a specific state of the cesium-133 atom.

For most engineering work, and for our purpose in studying mechanics, the accuracy of these standards is considerably beyond our needs. The standard value for gravitational acceleration g is its value at sea level and at a 45° latitude. In the two systems these values are

$$\text{SI units} \qquad g = 9.806\ 65\ \text{m/s}^2$$

$$\text{U.S. units} \qquad g = 32.1740\ \text{ft/sec}^2$$

The approximate values of 9.81 m/s^2 and 32.2 ft/sec^2, respectively, are sufficiently accurate for the vast majority of engineering calculations.

Unit Conversions

The characteristics of SI units are shown inside the front cover of this book, along with the numerical conversions between U.S. customary and SI units. In addition, charts giving the approximate conversions

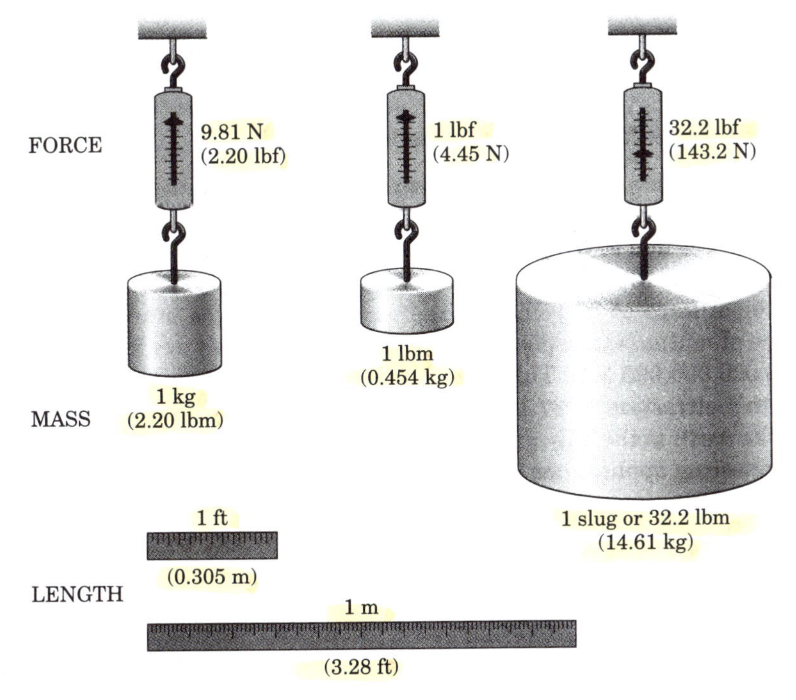

Figure 1/6

between selected quantities in the two systems appear inside the back cover for convenient reference. Although these charts are useful for obtaining a feel for the relative size of SI and U.S. units, in time engineers will find it essential to think directly in terms of SI units without converting from U.S. units. In statics we are primarily concerned with the units of length and force, with mass needed only when we compute gravitational force, as explained previously.

Figure 1/6 depicts examples of force, mass, and length in the two systems of units, to aid in visualizing their relative magnitudes.

1/6 Law of Gravitation

In statics as well as dynamics we often need to compute the weight of a body, which is the gravitational force acting on it. This computation depends on the *law of gravitation*, which was also formulated by Newton. The law of gravitation is expressed by the equation

$$F = G\frac{m_1 m_2}{r^2} \tag{1/2}$$

where F = the mutual force of attraction between two particles

G = a universal constant known as the *constant of gravitation*

m_1, m_2 = the masses of the two particles

r = the distance between the centers of the particles

The mutual forces F obey the law of action and reaction, since they are equal and opposite and are directed along the line joining the centers of the particles, as shown in Fig. 1/7. By experiment the gravitational constant is found to be $G = 6.673(10^{-11})\ \text{m}^3/(\text{kg}\cdot\text{s}^2)$.

Gravitational Attraction of the Earth

Gravitational forces exist between every pair of bodies. On the surface of the earth the only gravitational force of appreciable magnitude is the force due to the attraction of the earth. For example, each of two iron spheres 100 mm in diameter is attracted to the earth with a gravitational force of 37.1 N, which is its weight. On the other hand, the force of mutual attraction between the spheres if they are just touching is 0.000 000 095 1 N. This force is clearly negligible compared with the earth's attraction of 37.1 N. Consequently the gravitational attraction of the earth is the only gravitational force we need to consider for most engineering applications on the earth's surface.

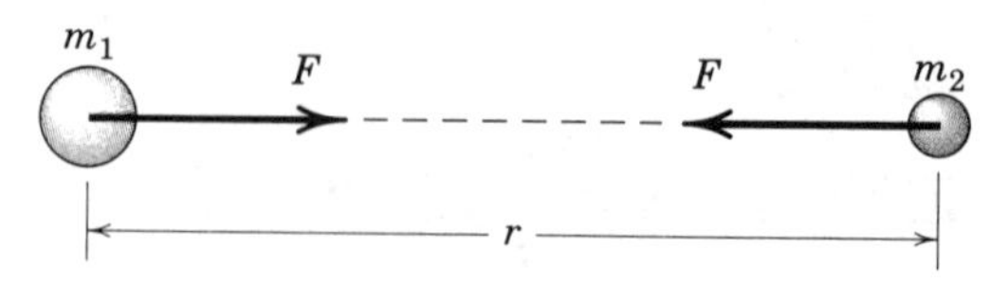

Figure 1/7

The gravitational attraction of the earth on a body (its weight) exists whether the body is at rest or in motion. Because this attraction is a force, the weight of a body should be expressed in newtons (N) in SI units and in pounds (lb) in U.S. customary units. Unfortunately in common practice the mass unit kilogram (kg) has been frequently used as a measure of weight. This usage should disappear in time as SI units become more widely used, because in SI units the kilogram is used exclusively for mass and the newton is used for force, including weight.

For a body of mass m near the surface of the earth, the gravitational attraction F on the body is specified by Eq. 1/2. We usually denote the magnitude of this gravitational force or weight with the symbol W. Because the body falls with an acceleration g, Eq. 1/1 gives

$$W = mg \tag{1/3}$$

The weight W will be in newtons (N) when the mass m is in kilograms (kg) and the acceleration of gravity g is in meters per second squared (m/s^2). In U.S. customary units, the weight W will be in pounds (lb) when m is in slugs and g is in feet per second squared. The standard values for g of 9.81 m/s^2 and 32.2 ft/sec^2 will be sufficiently accurate for our calculations in statics.

The true weight (gravitational attraction) and the apparent weight (as measured by a spring scale) are slightly different. The difference, which is due to the rotation of the earth, is quite small and will be neglected. This effect will be discussed in *Vol. 2 Dynamics*.

1/7 Accuracy, Limits, and Approximations

The number of significant figures in an answer should be no greater than the number of figures justified by the accuracy of the given data. For example, suppose the 24-mm side of a square bar was measured to the nearest millimeter, so we know the side length to two significant figures. Squaring the side length gives an area of 576 mm^2. However, according to our rule, we should write the area as 580 mm^2, using only two significant figures.

When calculations involve small differences in large quantities, greater accuracy in the data is required to achieve a given accuracy in the results. Thus, for example, it is necessary to know the numbers 4.2503 and 4.2391 to an accuracy of five significant figures to express their difference 0.0112 to three-figure accuracy. It is often difficult in lengthy computations to know at the outset how many significant figures are needed in the original data to ensure a certain accuracy in the answer. Accuracy to three significant figures is considered satisfactory for most engineering calculations.

In this text, answers will generally be shown to three significant figures unless the answer begins with the digit 1, in which case the answer will be shown to four significant figures. For purposes of calculation, consider all data given in this book to be exact.

Differentials

The *order* of differential quantities frequently causes misunderstanding in the derivation of equations. Higher-order differentials may always be neglected compared with lower-order differentials when the mathematical limit is approached. For example, the element of volume ΔV of a right circular cone of altitude h and base radius r may be taken to be a circular slice a distance x from the vertex and of thickness Δx. The expression for the volume of the element is

$$\Delta V = \frac{\pi r^2}{h^2} [x^2 \, \Delta x + x(\Delta x)^2 + \tfrac{1}{3}(\Delta x)^3]$$

Note that, when passing to the limit in going from ΔV to dV and from Δx to dx, the terms containing $(\Delta x)^2$ and $(\Delta x)^3$ drop out, leaving merely

$$dV = \frac{\pi r^2}{h^2} x^2 \, dx$$

which gives an exact expression when integrated.

Small-Angle Approximations

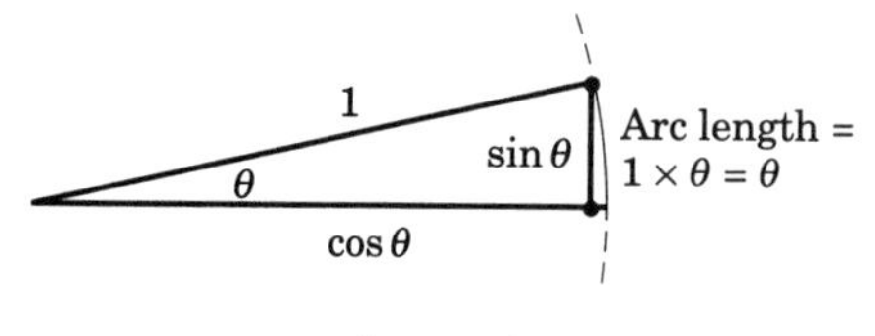

Figure 1/8

When dealing with small angles, we can usually make use of simplifying approximations. Consider the right triangle of Fig. 1/8 where the angle θ, expressed in radians, is relatively small. If the hypotenuse is unity, we see from the geometry of the figure that the arc length $1 \times \theta$ and $\sin \theta$ are very nearly the same. Also $\cos \theta$ is close to unity. Furthermore, $\sin \theta$ and $\tan \theta$ have almost the same values. Thus, for small angles we may write

$$\sin \theta \cong \tan \theta \cong \theta \qquad \cos \theta \cong 1$$

provided that the angles are expressed in radians. These approximations may be obtained by retaining only the first terms in the series expansions for these three functions. As an example of these approximations, for an angle of 1°

$$1° = 0.017\ 453 \text{ rad} \qquad \tan 1° = 0.017\ 455$$

$$\sin 1° = 0.017\ 452 \qquad \cos 1° = 0.999\ 848$$

If a more accurate approximation is desired, the first two terms may be retained, and they are

$$\sin \theta \cong \theta - \theta^3/6 \qquad \tan \theta \cong \theta + \theta^3/3 \qquad \cos \theta \cong 1 - \theta^2/2$$

where the angles must be expressed in radians. (To convert degrees to radians, multiply the angle in degrees by $\pi/180°$.) The error in replacing the sine by the angle for 1° (0.0175 rad) is only 0.005 percent. For 5° (0.0873 rad) the error is 0.13 percent, and for 10° (0.1745 rad), the error is still only 0.51 percent. As the angle θ approaches zero, the following relations are true in the mathematical limit:

$$\sin d\theta = \tan d\theta = d\theta \qquad \cos d\theta = 1$$

where the differential angle $d\theta$ must be expressed in radians.

1/8 Problem Solving in Statics

We study statics to obtain a quantitative description of forces which act on engineering structures in equilibrium. Mathematics establishes the relations between the various quantities involved and enables us to predict effects from these relations. We use a dual thought process in solving statics problems: We think about both the physical situation and the corresponding mathematical description. In the analysis of every problem, we make a transition between the physical and the mathematical. One of the most important goals for the student is to develop the ability to make this transition freely.

Making Appropriate Assumptions

We should recognize that the mathematical formulation of a physical problem represents an ideal description, or *model*, which approximates but never quite matches the actual physical situation. When we construct an idealized mathematical model for a given engineering problem, certain approximations will always be involved. Some of these approximations may be mathematical, whereas others will be physical.

For instance, it is often necessary to neglect small distances, angles, or forces compared with large distances, angles, or forces. Suppose a force is distributed over a small area of the body on which it acts. We may consider it to be a concentrated force if the dimensions of the area involved are small compared with other pertinent dimensions.

We may neglect the weight of a steel cable if the tension in the cable is many times greater than its total weight. However, if we must calculate the deflection or sag of a suspended cable under the action of its weight, we may not ignore the cable weight.

Thus, what we may assume depends on what information is desired and on the accuracy required. We must be constantly alert to the various assumptions called for in the formulation of real problems. The ability to understand and make use of the appropriate assumptions in the formulation and solution of engineering problems is certainly one of the most important characteristics of a successful engineer. One of the major aims of this book is to provide many opportunities to develop this ability through the formulation and analysis of many practical problems involving the principles of statics.

Using Graphics

Graphics is an important analytical tool for three reasons:

1. We use graphics to represent a physical system on paper with a sketch or diagram. Representing a problem geometrically helps us

with its physical interpretation, especially when we must visualize three-dimensional problems.

2. We can often obtain a graphical solution to problems more easily than with a direct mathematical solution. Graphical solutions are both a practical way to obtain results, and an aid in our thought processes. Because graphics represents the physical situation and its mathematical expression simultaneously, graphics helps us make the transition between the two.
3. Charts or graphs are valuable aids for representing results in a form which is easy to understand.

Formulating Problems and Obtaining Solutions

In statics, as in all engineering problems, we need to use a precise and logical method for formulating problems and obtaining their solutions. We formulate each problem and develop its solution through the following sequence of steps.

1. Formulate the problem:
   (*a*) State the given data.
   (*b*) State the desired result.
   (*c*) State your assumptions and approximations.
2. Develop the solution:
   (*a*) Draw any diagrams you need to understand the relationships.
   (*b*) State the governing principles to be applied to your solution.
   (*c*) Make your calculations.
   (*d*) Ensure that your calculations are consistent with the accuracy justified by the data.
   (*e*) Be sure that you have used consistent units throughout your calculations.
   (*f*) Ensure that your answers are reasonable in terms of magnitudes, directions, common sense, etc.
   (*g*) Draw conclusions.

Keeping your work neat and orderly will help your thought process and enable others to understand your work. The discipline of doing orderly work will help you develop skill in formulation and analysis. Problems which seem complicated at first often become clear when you approach them with logic and discipline.

The Free-Body Diagram

The subject of statics is based on surprisingly few fundamental concepts and involves mainly the application of these basic relations to a variety of situations. In this application the *method* of analysis is all-important. In solving a problem, it is essential that the laws which apply be carefully fixed in mind and that we apply these principles literally and exactly. In applying the principles of mechanics to analyze forces acting on a body, it is essential that we *isolate* the body in question from all other bodies so that a complete and accurate account of all forces acting on this body can be taken. This *isolation* should exist mentally and should be represented on paper. The diagram of such an isolated body with the representation of *all* external forces acting *on* it is called a *free-body diagram*.

The free-body-diagram method is the key to the understanding of mechanics. This is so because the *isolation* of a body is the tool by which *cause* and *effect* are clearly separated, and by which our attention is clearly focused on the literal application of a principle of mechanics. The technique of drawing free-body diagrams is covered in Chapter 3, where they are first used.

Numerical Values versus Symbols

In applying the laws of statics, we may use numerical values to represent quantities, or we may use algebraic symbols, and leave the answer as a formula. When numerical values are used, the magnitude of each quantity expressed in its particular units is evident at each stage of the calculation. This is useful when we need to know the magnitude of each term.

The symbolic solution, however, has several advantages over the numerical solution. First, the use of symbols helps to focus our attention on the connection between the physical situation and its related mathematical description. Second, we can use a symbolic solution repeatedly for obtaining answers to the same type of problem, but having different units or numerical values. Third, a symbolic solution enables us to make a dimensional check at every step, which is more difficult to do when numerical values are used. In any equation representing a physical situation, the dimensions of every term on both sides of the equation must be the same. This property is called *dimensional homogeneity*.

Thus, facility with both numerical and symbolic forms of solution is essential.

Solution Methods

Solutions to the problems of statics may be obtained in one or more of the following ways.

1. Obtain mathematical solutions by hand, using either algebraic symbols or numerical values. We can solve most problems this way.
2. Obtain graphical solutions for certain problems.
3. Solve problems by computer. This is useful when a large number of equations must be solved, when a parameter variation must be studied, or when an intractable equation must be solved.

Many problems can be solved with two or more of these methods. The method utilized depends partly on the engineer's preference and partly on the type of problem to be solved. The choice of the most expedient method of solution is an important aspect of the experience to be gained from the problem work. There are a number of problems in *Vol. 1 Statics* which are designated as *Computer-Oriented Problems*. These problems appear at the end of the Review Problem sets and are selected to illustrate the type of problem for which solution by computer offers a distinct advantage.

CHAPTER REVIEW

This chapter has introduced the concepts, definitions, and units used in statics, and has given an overview of the procedure used to formulate and solve problems in statics. Now that you have finished this chapter, you should be able to do the following:

1. **Express vectors in terms of unit vectors and perpendicular components, and perform vector addition and subtraction.**
2. **State Newton's laws of motion.**
3. **Perform calculations using SI and U.S. units, using appropriate accuracy.**
4. **Express the law of gravitation and calculate the weight of an object.**
5. **Apply simplifications based on differential and small-angle approximations.**
6. **Describe the methodology used to formulate and solve statics problems.**

Sample Problem 1/1

Determine the weight in newtons of a car whose mass is 1400 kg. Convert the mass of the car to slugs and then determine its weight in pounds.

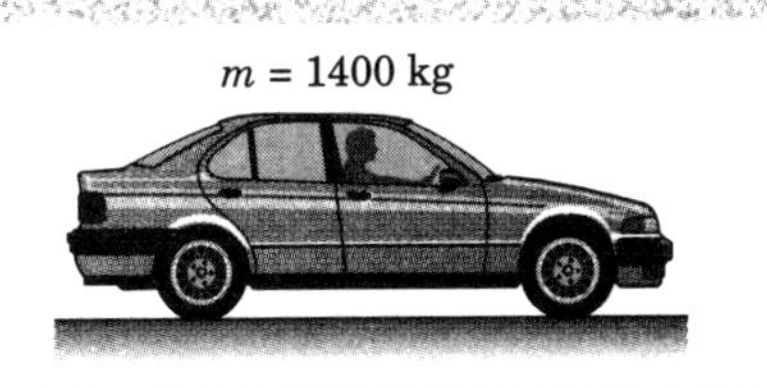

Solution. From relationship 1/3, we have

① $$W = mg = 1400(9.81) = 13\ 730\ \text{N} \qquad \textit{Ans.}$$

From the table of conversion factors inside the front cover of the textbook, we see that 1 slug is equal to 14.594 kg. Thus, the mass of the car in slugs is

② $$m = 1400\ \text{kg}\left[\frac{1\ \text{slug}}{14.594\ \text{kg}}\right] = 95.9\ \text{slugs} \qquad \textit{Ans.}$$

Finally, its weight in pounds is

③ $$W = mg = (95.9)(32.2) = 3090\ \text{lb} \qquad \textit{Ans.}$$

As another route to the last result, we can convert from kg to lbm. Again using the table inside the front cover, we have

$$m = 1400\ \text{kg}\left[\frac{1\ \text{lbm}}{0.45359\ \text{kg}}\right] = 3090\ \text{lbm}$$

The weight in pounds associated with the mass of 3090 lbm is 3090 lb, as calculated above. We recall that 1 lbm is the amount of mass which under standard conditions has a weight of 1 lb of force. We rarely refer to the U.S. mass unit lbm in this textbook series, but rather use the slug for mass. The sole use of slug, rather than the unnecessary use of two units for mass, will prove to be powerful and simple—especially in dynamics.

Helpful Hints

① Our calculator indicates a result of 13 734 N. Using the rules of significant-figure display used in this textbook, we round the written result to four significant figures, or 13 730 N. Had the number begun with any digit other than 1, we would have rounded to three significant figures.

② A good practice with unit conversion is to multiply by a factor such as $\left[\frac{1\ \text{slug}}{14.594\ \text{kg}}\right]$, which has a value of 1, because the numerator and the denominator are equivalent. Make sure that cancellation of the units leaves the units desired; here the units of kg cancel, leaving the desired units of slug.

③ Note that we are using a previously calculated result (95.9 slugs). We must be sure that when a calculated number is needed in subsequent calculations, it is retained in the calculator to its full accuracy (95.929834 · · ·) until it is needed. This may require storing it in a register upon its initial calculation and recalling it later. We must not merely punch 95.9 into our calculator and proceed to multiply by 32.2—this practice will result in loss of numerical accuracy. Some individuals like to place a small indication of the storage register used in the right margin of the work paper, directly beside the number stored.

Sample Problem 1/2

Use Newton's law of universal gravitation to calculate the weight of a 70-kg person standing on the surface of the earth. Then repeat the calculation by using $W = mg$ and compare your two results. Use Table D/2 as needed.

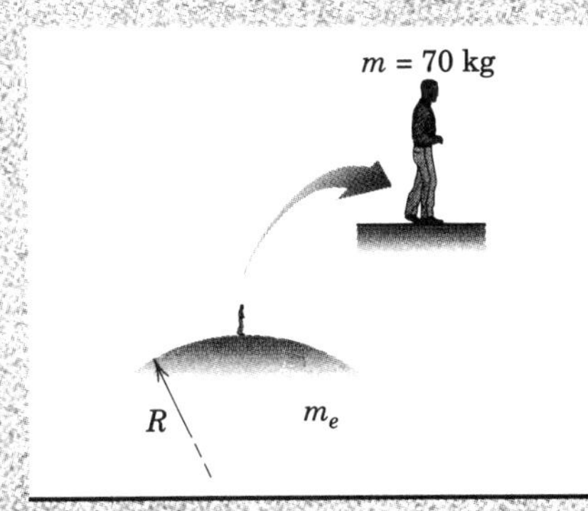

Solution. The two results are

① $$W = \frac{Gm_e m}{R^2} = \frac{(6.673 \cdot 10^{-11})(5.976 \cdot 10^{24})(70)}{[6371 \cdot 10^3]^2} = 688\ \text{N} \qquad \textit{Ans.}$$

$$W = mg = 70(9.81) = 687\ \text{N} \qquad \textit{Ans.}$$

The discrepancy is due to the fact that Newton's universal gravitational law does not take into account the rotation of the earth. On the other hand, the value $g = 9.81\ \text{m/s}^2$ used in the second equation does account for the earth's rotation. Note that had we used the more accurate value $g = 9.80665\ \text{m/s}^2$ (which likewise accounts for the earth's rotation) in the second equation, the discrepancy would have been larger (686 N would have been the result).

Helpful Hint

① The effective distance between the mass centers of the two bodies involved is the radius of the earth.

Sample Problem 1/3

For the vectors $\mathbf{V}_1$ and $\mathbf{V}_2$ shown in the figure,

(*a*) determine the magnitude S of their vector sum $\mathbf{S} = \mathbf{V}_1 + \mathbf{V}_2$

(*b*) determine the angle α between $\mathbf{S}$ and the positive x-axis

(*c*) write $\mathbf{S}$ as a vector in terms of the unit vectors $\mathbf{i}$ and $\mathbf{j}$ and then write a unit vector $\mathbf{n}$ along the vector sum $\mathbf{S}$

(*d*) determine the vector difference $\mathbf{D} = \mathbf{V}_1 - \mathbf{V}_2$

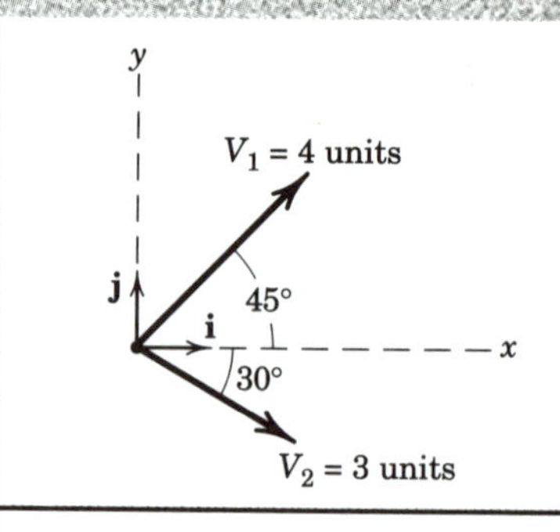

Solution (*a*) We construct to scale the parallelogram shown in Fig. *a* for adding $\mathbf{V}_1$ and $\mathbf{V}_2$. Using the law of cosines, we have

$$S^2 = 3^2 + 4^2 - 2(3)(4)\cos 105^\circ$$

$$S = 5.59 \text{ units} \qquad \textit{Ans.}$$

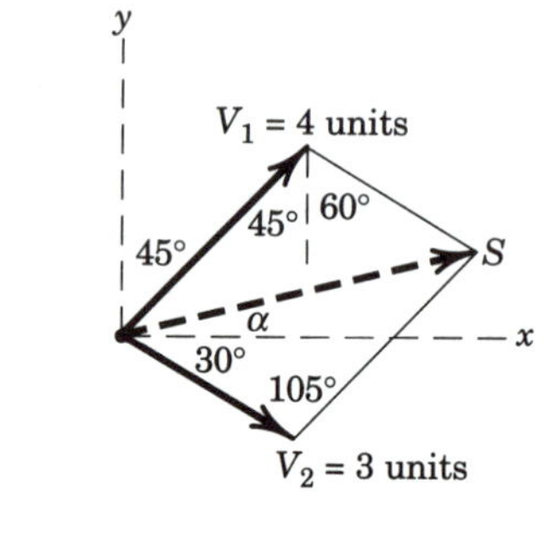

(*a*)

① (*b*) Using the law of sines for the lower triangle, we have

$$\frac{\sin 105^\circ}{5.59} = \frac{\sin(\alpha + 30^\circ)}{4}$$

$$\sin(\alpha + 30^\circ) = 0.692$$

$$(\alpha + 30^\circ) = 43.8^\circ \qquad \alpha = 13.76^\circ \qquad \textit{Ans.}$$

(*c*) With knowledge of both S and α, we can write the vector $\mathbf{S}$ as

$$\mathbf{S} = S[\mathbf{i}\cos\alpha + \mathbf{j}\sin\alpha]$$

$$= S[\mathbf{i}\cos 13.76^\circ + \mathbf{j}\sin 13.76^\circ] = 5.43\mathbf{i} + 1.328\mathbf{j} \text{ units} \qquad \textit{Ans.}$$

② Then $$\mathbf{n} = \frac{\mathbf{S}}{S} = \frac{5.43\mathbf{i} + 1.328\mathbf{j}}{5.59} = 0.971\mathbf{i} + 0.238\mathbf{j} \qquad \textit{Ans.}$$

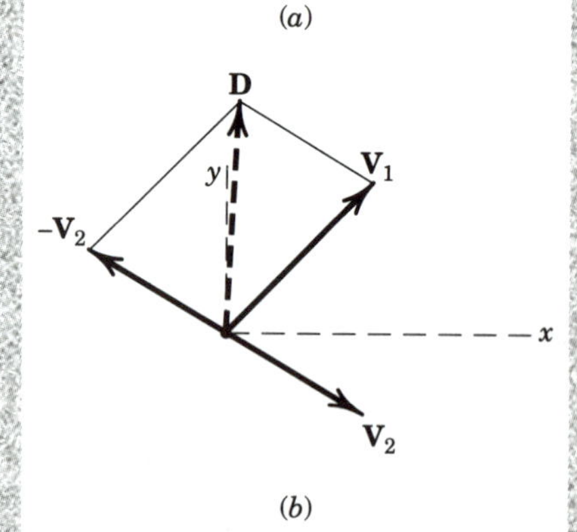

(*b*)

(*d*) The vector difference $\mathbf{D}$ is

$$\mathbf{D} = \mathbf{V}_1 - \mathbf{V}_2 = 4(\mathbf{i}\cos 45^\circ + \mathbf{j}\sin 45^\circ) - 3(\mathbf{i}\cos 30^\circ - \mathbf{j}\sin 30^\circ)$$

$$= 0.230\mathbf{i} + 4.33\mathbf{j} \text{ units} \qquad \textit{Ans.}$$

The vector $\mathbf{D}$ is shown in Fig. *b* as $\mathbf{D} = \mathbf{V}_1 + (-\mathbf{V}_2)$.

Helpful Hints

① You will frequently use the laws of cosines and sines in mechanics. See Art. C/6 of Appendix C for a review of these important geometric principles.

② A unit vector may always be formed by dividing a vector by its magnitude. Note that a unit vector is dimensionless.

PROBLEMS

1/1 Determine the angle made by the vector $\mathbf{V} = -10\mathbf{i} + 24\mathbf{j}$ with the positive x-axis. Write the unit vector $\mathbf{n}$ in the direction of $\mathbf{V}$.

Ans. $\theta_x = 112.6°$, $\mathbf{n} = -0.385\mathbf{i} + 0.923\mathbf{j}$

1/2 Determine the magnitude of the vector sum $\mathbf{V} = \mathbf{V}_1 + \mathbf{V}_2$ and the angle θ_x which $\mathbf{V}$ makes with the positive x-axis. Complete both graphical and algebraic solutions.

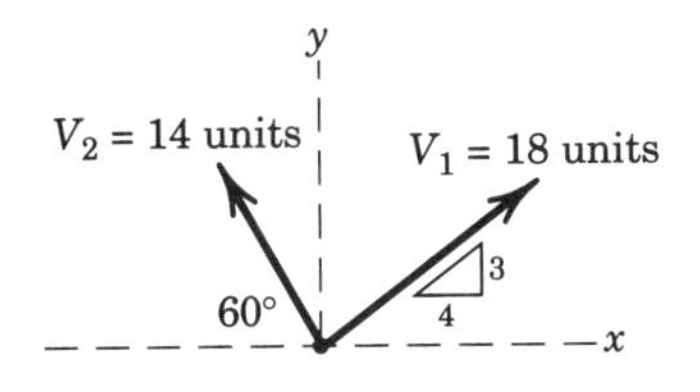

Problem 1/2

1/3 For the given vectors $\mathbf{V}_1$ and $\mathbf{V}_2$ of Prob. 1/2, determine the magnitude of the vector difference $\mathbf{V}' = \mathbf{V}_2 - \mathbf{V}_1$ and the angle θ_x which $\mathbf{V}'$ makes with the positive x-axis. Complete both graphical and algebraic solutions.

Ans. $V' = 21.4$ units, $\theta_x = 176.5°$

1/4 A force is specified by the vector $\mathbf{F} = 80\mathbf{i} - 40\mathbf{j} + 60\mathbf{k}$ N. Calculate the angles made by $\mathbf{F}$ with the x-, y-, and z-axes.

1/5 What is the weight in both newtons and pounds of a 75-kg beam?

Ans. $W = 736$ N, $W = 165.4$ lb

1/6 From the gravitational law calculate the weight W (gravitational force with respect to the earth) of an 80-kg man in a spacecraft traveling in a circular orbit 250 km above the earth's surface. Express W in both newtons and pounds.

1/7 Determine the weight in newtons of a woman whose weight in pounds is 130. Also, find her mass in slugs and in kilograms. Determine your own weight in newtons.

Ans. $W = 578$ N
$m = 4.04$ slugs, $m = 58.9$ kg

1/8 Suppose that two nondimensional quantities are given as $A = 8.69$ and $B = 1.427$. Using the rules for significant figures as stated in this chapter, determine the four quantities $(A + B)$, $(A - B)$, (AB), and (A/B).

1/9 Compute the magnitude F of the force which the earth exerts on the moon. Perform the calculation first in newtons and then convert your result to pounds. Refer to Table D/2 for necessary physical quantities.

Ans. $F = 1.984(10^{20})$ N, $F = 4.46(10^{19})$ lb

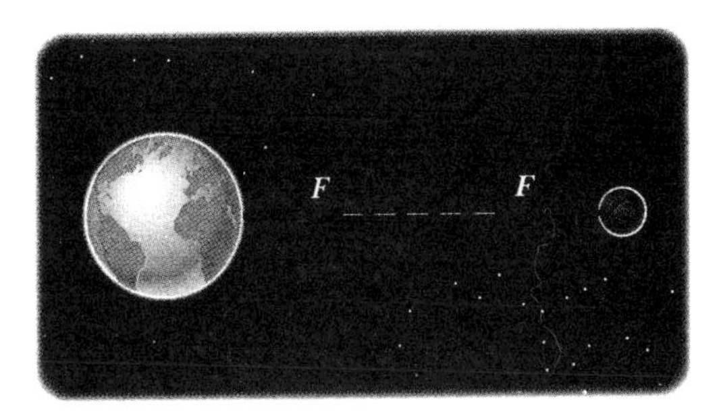

Problem 1/9

1/10 What is the percent error in replacing the sine of 20° by the value of the angle in radians? Repeat for the tangent of 20°, and explain the qualitative difference in the two error percentages.

Chapter

2

FORCE SYSTEMS

CHAPTER OUTLINE

2/1 INTRODUCTION

In this and the following chapters, we study the effects of forces which act on engineering structures and mechanisms. The experience gained here will help you in the study of mechanics and in other subjects such as stress analysis, design of structures and machines, and fluid flow. This chapter lays the foundation for a basic understanding not only of statics but also of the entire subject of mechanics, and you should master this material thoroughly.

2/2 FORCE

Before dealing with a group or *system* of forces, it is necessary to examine the properties of a single force in some detail. A force has been defined in Chapter 1 as an action of one body on another. In dynamics we will see that a force is defined as an action which tends to cause acceleration of a body. A force is a *vector quantity*, because its effect depends on the direction as well as on the magnitude of the action. Thus,

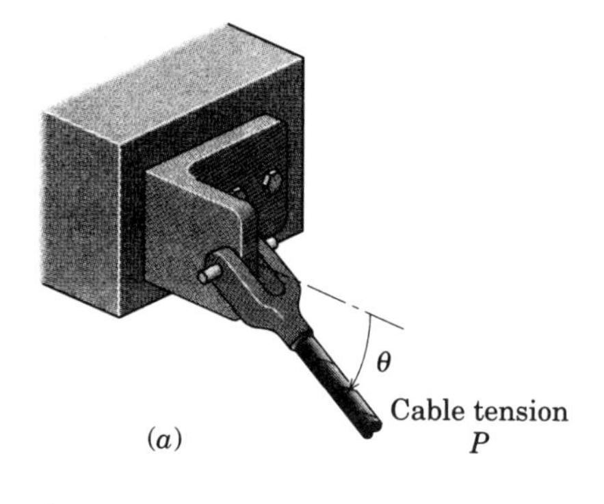

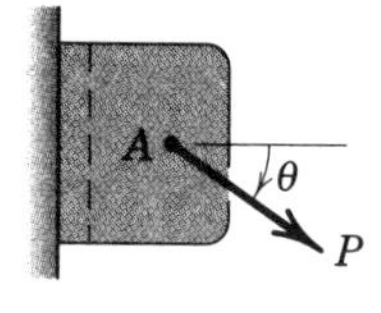

Figure 2/1

forces may be combined according to the parallelogram law of vector addition.

The action of the cable tension on the bracket in Fig. 2/1*a* is represented in the side view, Fig. 2/1*b*, by the force vector **P** of magnitude P. The effect of this action on the bracket depends on P, the angle θ, and the location of the point of application A. Changing any one of these three specifications will alter the effect on the bracket, such as the force in one of the bolts which secure the bracket to the base, or the internal force and deformation in the material of the bracket at any point. Thus, the complete specification of the action of a force must include its *magnitude*, *direction*, and *point of application*, and therefore we must treat it as a fixed vector.

External and Internal Effects

We can separate the action of a force on a body into two effects, *external* and *internal*. For the bracket of Fig. 2/1 the effects of **P** external to the bracket are the reactive forces (not shown) exerted on the bracket by the foundation and bolts because of the action of **P**. Forces external to a body can be either *applied* forces or *reactive* forces. The effects of **P** internal to the bracket are the resulting internal forces and deformations distributed throughout the material of the bracket. The relation between internal forces and internal deformations depends on the material properties of the body and is studied in strength of materials, elasticity, and plasticity.

Principle of Transmissibility

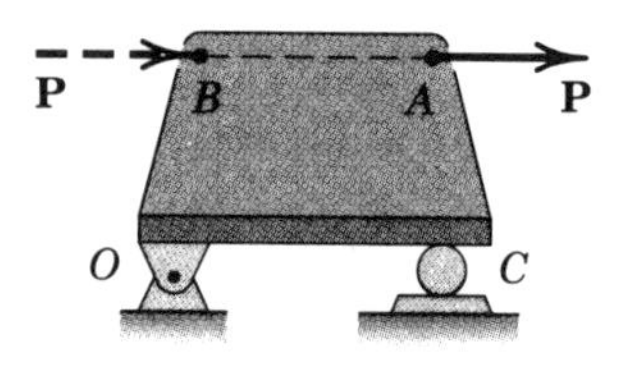

Figure 2/2

When dealing with the mechanics of a rigid body, we ignore deformations in the body and concern ourselves with only the net external effects of external forces. In such cases, experience shows us that it is not necessary to restrict the action of an applied force to a given point. For example, the force **P** acting on the rigid plate in Fig. 2/2 may be applied at A or at B or at any other point on its line of action, and the net external effects of **P** on the bracket will not change. The external effects are the force exerted on the plate by the bearing support at O and the force exerted on the plate by the roller support at C.

This conclusion is summarized by the *principle of transmissibility*, which states that a force may be applied at any point on its given line of action without altering the resultant effects of the force *external* to the *rigid* body on which it acts. Thus, whenever we are interested in only the resultant external effects of a force, the force may be treated as a *sliding* vector, and we need specify only the *magnitude*, *direction*, and *line of action* of the force, and not its *point of application*. Because this book deals essentially with the mechanics of rigid bodies, we will treat almost all forces as sliding vectors for the rigid body on which they act.

Force Classification

Forces are classified as either *contact* or *body* forces. A contact force is produced by direct physical contact; an example is the force exerted on a body by a supporting surface. On the other hand, a body force is generated by virtue of the position of a body within a force field such as a gravitational, electric, or magnetic field. An example of a body force is your weight.

Forces may be further classified as either *concentrated* or *distributed*. Every contact force is actually applied over a finite area and is therefore really a distributed force. However, when the dimensions of the area are very small compared with the other dimensions of the body, we may consider the force to be concentrated at a point with negligible loss of accuracy. Force can be distributed over an *area*, as in the case of mechanical contact, over a *volume* when a body force such as weight is acting, or over a *line*, as in the case of the weight of a suspended cable.

The *weight* of a body is the force of gravitational attraction distributed over its volume and may be taken as a concentrated force acting through the center of gravity. The position of the center of gravity is frequently obvious if the body is symmetric. If the position is not obvious, then a separate calculation, explained in Chapter 5, will be necessary to locate the center of gravity.

We can measure a force either by comparison with other known forces, using a mechanical balance, or by the calibrated movement of an elastic element. All such comparisons or calibrations have as their basis a primary standard. The standard unit of force in SI units is the newton (N) and in the U.S. customary system is the pound (lb), as defined in Art. 1/5.

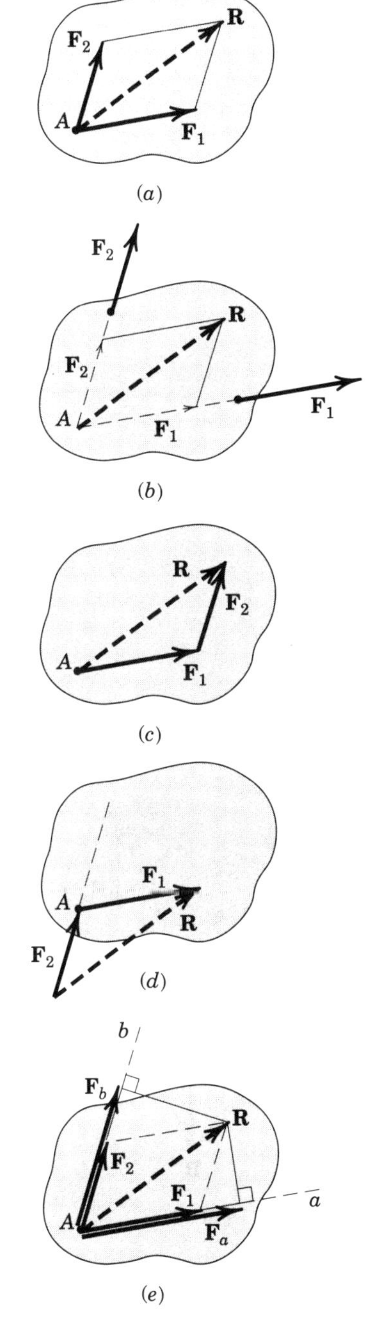

Figure 2/3

Action and Reaction

According to Newton's third law, the *action* of a force is always accompanied by an *equal* and *opposite reaction*. It is essential to distinguish between the action and the reaction in a pair of forces. To do so, we first *isolate* the body in question and then identify the force exerted *on* that body (not the force exerted *by* the body). It is very easy to mistakenly use the wrong force of the pair unless we distinguish carefully between action and reaction.

Concurrent Forces

Two or more forces are said to be *concurrent at a point* if their lines of action intersect at that point. The forces $\mathbf{F}_1$ and $\mathbf{F}_2$ shown in Fig. 2/3*a* have a common point of application and are concurrent at the point *A*. Thus, they can be added using the parallelogram law in their common plane to obtain their sum or *resultant* **R**, as shown in Fig. 2/3*a*. The resultant lies in the same plane as $\mathbf{F}_1$ and $\mathbf{F}_2$.

Suppose the two concurrent forces lie in the same plane but are applied at two different points as in Fig. 2/3*b*. By the principle of transmissibility, we may move them along their lines of action and complete their vector sum **R** at the point of concurrency *A*, as shown in Fig. 2/3*b*. We can replace $\mathbf{F}_1$ and $\mathbf{F}_2$ with the resultant **R** without altering the external effects on the body upon which they act.

We can also use the triangle law to obtain **R**, but we need to move the line of action of one of the forces, as shown in Fig. 2/3*c*. If we add the same two forces, as shown in Fig. 2/3*d*, we correctly preserve the magnitude and direction of **R**, but we lose the correct line of action, because **R** obtained in this way does not pass through *A*. Therefore this type of combination should be avoided.

We can express the sum of the two forces mathematically by the vector equation

$$\mathbf{R} = \mathbf{F}_1 + \mathbf{F}_2$$

Vector Components

In addition to combining forces to obtain their resultant, we often need to replace a force by its *vector components* in directions which are convenient for a given application. The vector sum of the components must equal the original vector. Thus, the force **R** in Fig. 2/3*a* may be replaced by, or *resolved* into, two vector components $\mathbf{F}_1$ and $\mathbf{F}_2$ with the specified directions by completing the parallelogram as shown to obtain the magnitudes of $\mathbf{F}_1$ and $\mathbf{F}_2$.

The relationship between a force and its vector components along given axes must not be confused with the relationship between a force and its perpendicular* projections onto the same axes. Figure 2/3*e* shows the perpendicular projections $\mathbf{F}_a$ and $\mathbf{F}_b$ of the given force **R** onto axes *a* and *b*, which are parallel to the vector components $\mathbf{F}_1$ and $\mathbf{F}_2$ of Fig. 2/3*a*. Figure 2/3*e* shows that the components of a vector are not necessarily equal to the projections of the vector onto the same axes. Furthermore, the vector sum of the projections $\mathbf{F}_a$ and $\mathbf{F}_b$ is not the vector **R**, because the parallelogram law of vector addition must be used to form the sum. The components and projections of **R** are equal only when the axes *a* and *b* are perpendicular.

A Special Case of Vector Addition

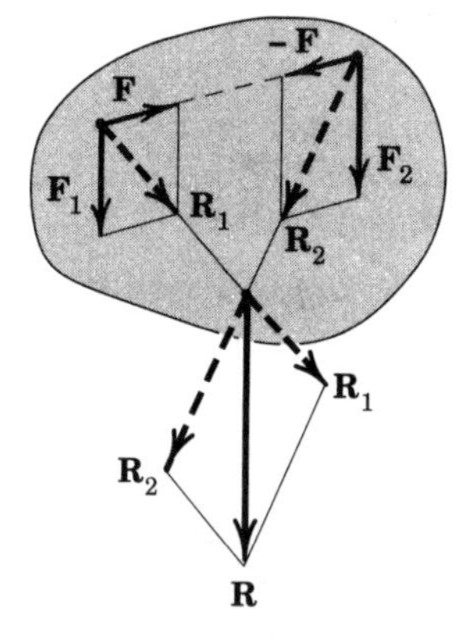

Figure 2/4

To obtain the resultant when the two forces $\mathbf{F}_1$ and $\mathbf{F}_2$ are parallel as in Fig. 2/4, we use a special case of addition. The two vectors are combined by first adding two equal, opposite, and collinear forces **F** and $-\mathbf{F}$ of convenient magnitude, which taken together produce no external effect on the body. Adding $\mathbf{F}_1$ and **F** to produce $\mathbf{R}_1$, and combining with the sum $\mathbf{R}_2$ of $\mathbf{F}_2$ and $-\mathbf{F}$ yield the resultant **R**, which is correct in magnitude, direction, and line of action. This procedure is also useful for graphically combining two forces which have a remote and inconvenient point of concurrency because they are almost parallel.

It is usually helpful to master the analysis of force systems in two dimensions before undertaking three-dimensional analysis. Thus the remainder of Chapter 2 is subdivided into these two categories.

*Perpendicular projections are also called *orthogonal* projections.

SECTION A. TWO-DIMENSIONAL FORCE SYSTEMS

2/3 RECTANGULAR COMPONENTS

The most common two-dimensional resolution of a force vector is into rectangular components. It follows from the parallelogram rule that the vector **F** of Fig. 2/5 may be written as

$$\mathbf{F} = \mathbf{F}_x + \mathbf{F}_y \tag{2/1}$$

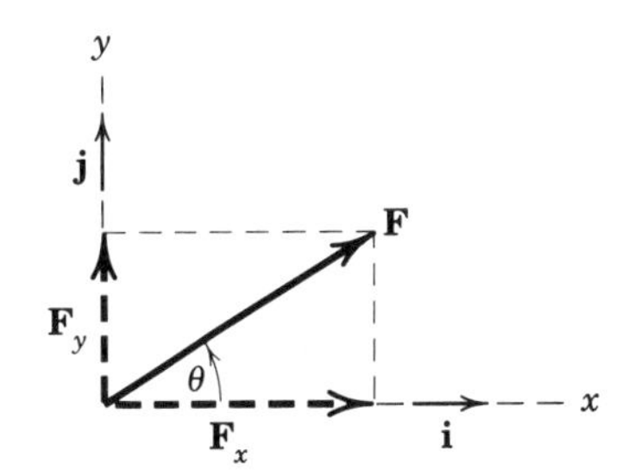

Figure 2/5

where $\mathbf{F}_x$ and $\mathbf{F}_y$ are *vector components* of **F** in the x- and y-directions. Each of the two vector components may be written as a scalar times the appropriate unit vector. In terms of the unit vectors **i** and **j** of Fig. 2/5, $\mathbf{F}_x = F_x\mathbf{i}$ and $\mathbf{F}_y = F_y\mathbf{j}$, and thus we may write

$$\mathbf{F} = F_x\mathbf{i} + F_y\mathbf{j} \tag{2/2}$$

where the scalars F_x and F_y are the x and y *scalar components* of the vector **F**.

The scalar components can be positive or negative, depending on the quadrant into which **F** points. For the force vector of Fig. 2/5, the x and y scalar components are both positive and are related to the magnitude and direction of **F** by

$$\begin{aligned} F_x &= F\cos\theta \qquad & F &= \sqrt{F_x^{\,2} + F_y^{\,2}} \\ F_y &= F\sin\theta \qquad & \theta &= \tan^{-1}\frac{F_y}{F_x} \end{aligned} \tag{2/3}$$

Conventions for Describing Vector Components

We express the magnitude of a vector with lightface italic type in print, that is, $|\mathbf{F}|$ is indicated by F, a quantity which is always *nonnegative*. However, the scalar components, also denoted by lightface italic type, will include sign information. See Sample Problems 2/1 and 2/3 for numerical examples which involve both positive and negative scalar components.

When both a force and its vector components appear in a diagram, it is desirable to show the vector components of the force with dashed lines, as in Fig. 2/5, and show the force with a solid line, or vice versa. With either of these conventions it will always be clear that a force and its components are being represented, and not three separate forces, as would be implied by three solid-line vectors.

Actual problems do not come with reference axes, so their assignment is a matter of arbitrary convenience, and the choice is frequently up to the student. The logical choice is usually indicated by the way in which the geometry of the problem is specified. When the principal dimensions of a body are given in the horizontal and vertical directions,

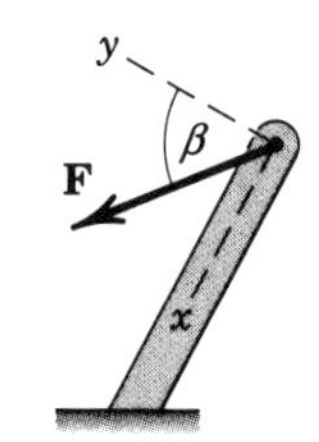

$F_x = F \sin \beta$
$F_y = F \cos \beta$

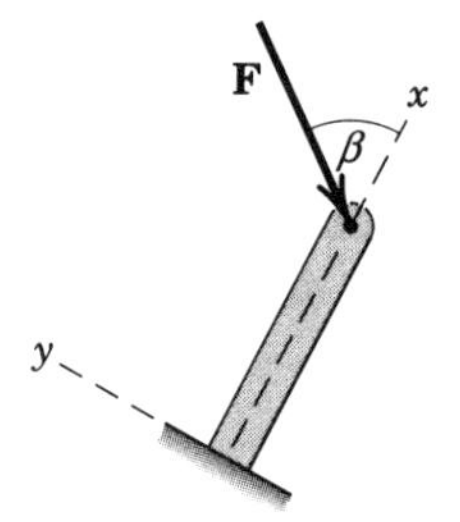

$F_x = -F \cos \beta$
$F_y = -F \sin \beta$

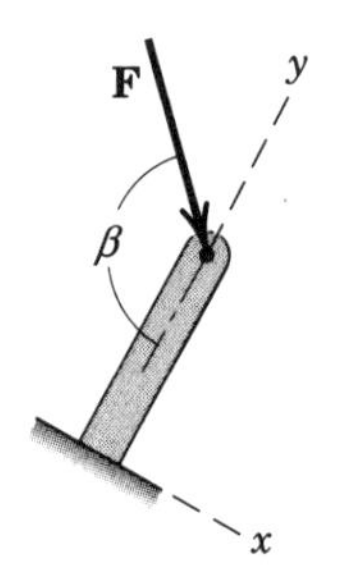

$F_x = F \sin(\pi - \beta)$
$F_y = -F \cos(\pi - \beta)$

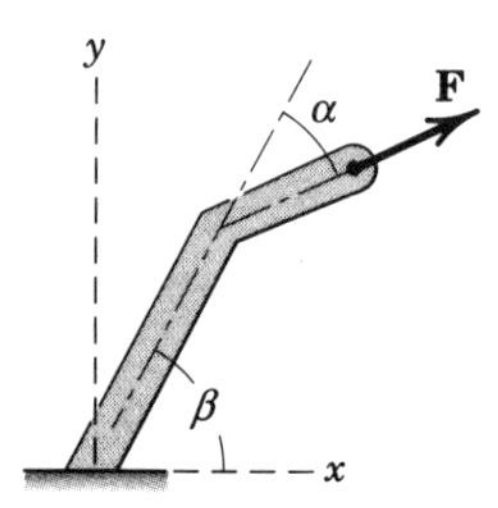

$F_x = F \cos(\beta - \alpha)$
$F_y = F \sin(\beta - \alpha)$

Figure 2/6

for example, you would typically assign reference axes in these directions.

Determining the Components of a Force

Dimensions are not always given in horizontal and vertical directions, angles need not be measured counterclockwise from the x-axis, and the origin of coordinates need not be on the line of action of a force. Therefore, it is essential that we be able to determine the correct components of a force no matter how the axes are oriented or how the angles are measured. Figure 2/6 suggests a few typical examples of vector resolution in two dimensions.

Memorization of Eqs. 2/3 is not a substitute for understanding the parallelogram law and for correctly projecting a vector onto a reference axis. A neatly drawn sketch always helps to clarify the geometry and avoid error.

Rectangular components are convenient for finding the sum or resultant $\mathbf{R}$ of two forces which are concurrent. Consider two forces $\mathbf{F}_1$ and $\mathbf{F}_2$ which are originally concurrent at a point O. Figure 2/7 shows the line of action of $\mathbf{F}_2$ shifted from O to the tip of $\mathbf{F}_1$ according to the triangle rule of Fig. 2/3. In adding the force vectors $\mathbf{F}_1$ and $\mathbf{F}_2$, we may write

$$\mathbf{R} = \mathbf{F}_1 + \mathbf{F}_2 = (F_{1_x}\mathbf{i} + F_{1_y}\mathbf{j}) + (F_{2_x}\mathbf{i} + F_{2_y}\mathbf{j})$$

or

$$R_x\mathbf{i} + R_y\mathbf{j} = (F_{1_x} + F_{2_x})\mathbf{i} + (F_{1_y} + F_{2_y})\mathbf{j}$$

from which we conclude that

$$\begin{aligned} R_x &= F_{1_x} + F_{2_x} = \Sigma F_x \\ R_y &= F_{1_y} + F_{2_y} = \Sigma F_y \end{aligned} \qquad \textbf{(2/4)}$$

The term ΣF_x means "the algebraic sum of the x scalar components". For the example shown in Fig. 2/7, note that the scalar component F_{2_y} would be negative.

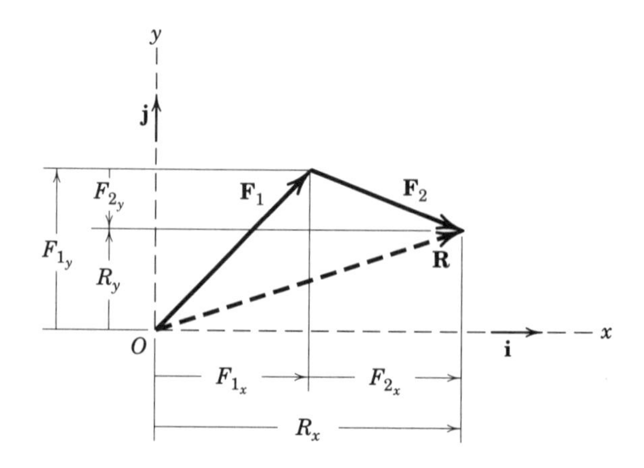

Figure 2/7

Sample Problem 2/1

The forces $\mathbf{F}_1$, $\mathbf{F}_2$, and $\mathbf{F}_3$, all of which act on point A of the bracket, are specified in three different ways. Determine the x and y scalar components of each of the three forces.

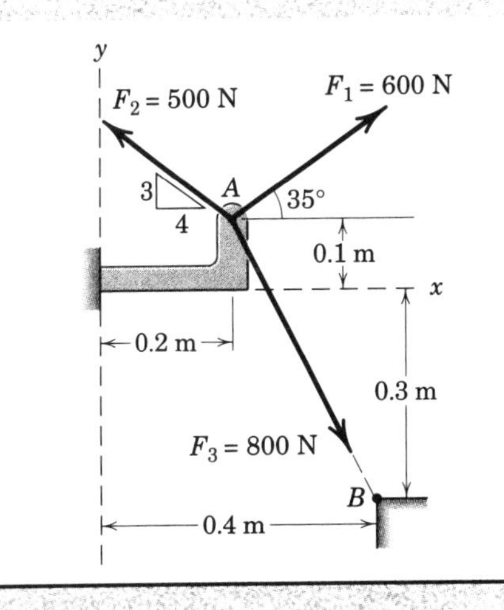

Solution. The scalar components of $\mathbf{F}_1$, from Fig. a, are

$$F_{1_x} = 600 \cos 35° = 491 \text{ N} \qquad \textit{Ans.}$$

$$F_{1_y} = 600 \sin 35° = 344 \text{ N} \qquad \textit{Ans.}$$

The scalar components of $\mathbf{F}_2$, from Fig. b, are

$$F_{2_x} = -500(\tfrac{4}{5}) = -400 \text{ N} \qquad \textit{Ans.}$$

$$F_{2_y} = 500(\tfrac{3}{5}) = 300 \text{ N} \qquad \textit{Ans.}$$

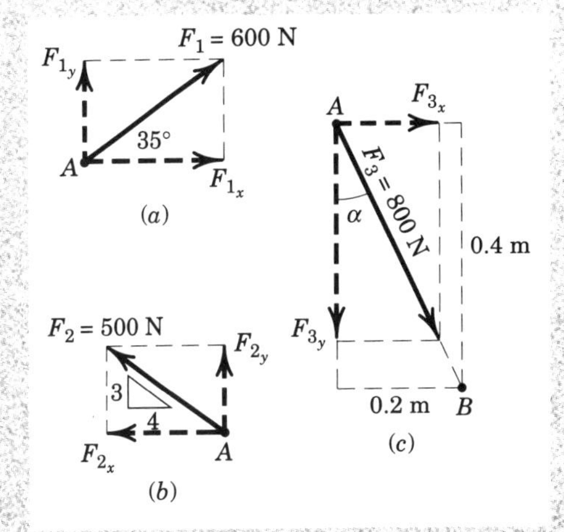

Note that the angle which orients $\mathbf{F}_2$ to the x-axis is never calculated. The cosine and sine of the angle are available by inspection of the 3-4-5 triangle. Also note that the x scalar component of $\mathbf{F}_2$ is negative by inspection.

The scalar components of $\mathbf{F}_3$ can be obtained by first computing the angle α of Fig. c.

$$\alpha = \tan^{-1}\left[\frac{0.2}{0.4}\right] = 26.6°$$

① Then $F_{3_x} = F_3 \sin \alpha = 800 \sin 26.6° = 358 \text{ N}$ *Ans.*

$$F_{3_y} = -F_3 \cos \alpha = -800 \cos 26.6° = -716 \text{ N} \qquad \textit{Ans.}$$

Alternatively, the scalar components of $\mathbf{F}_3$ can be obtained by writing $\mathbf{F}_3$ as a magnitude times a unit vector $\mathbf{n}_{AB}$ in the direction of the line segment AB. Thus,

② $$\mathbf{F}_3 = F_3\mathbf{n}_{AB} = F_3\frac{\overrightarrow{AB}}{\overline{AB}} = 800\left[\frac{0.2\mathbf{i} - 0.4\mathbf{j}}{\sqrt{(0.2)^2 + (-0.4)^2}}\right]$$

$$= 800[0.447\mathbf{i} - 0.894\mathbf{j}]$$

$$= 358\mathbf{i} - 716\mathbf{j} \text{ N}$$

The required scalar components are then

$$F_{3_x} = 358 \text{ N} \qquad \textit{Ans.}$$

$$F_{3_y} = -716 \text{ N} \qquad \textit{Ans.}$$

which agree with our previous results.

Helpful Hints

① You should carefully examine the geometry of each component-determination problem and not rely on the blind use of such formulas as $F_x = F \cos \theta$ and $F_y = F \sin \theta$.

② A unit vector can be formed by dividing *any* vector, such as the geometric position vector $\overrightarrow{AB}$, by its length or magnitude. Here we use the overarrow to denote the vector which runs from A to B and the overbar to denote the distance between A and B.

Sample Problem 2/2

Combine the two forces **P** and **T**, which act on the fixed structure at B, into a single equivalent force **R**.

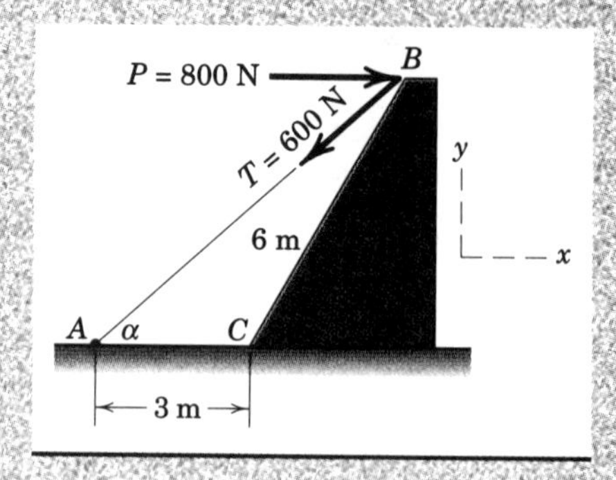

Graphical solution. The parallelogram for the vector addition of forces **T** and ① **P** is constructed as shown in Fig. a. The approximate scale used here is 1 cm = 400 N; a scale of 1 cm = 100 N would be more suitable for regular-size paper and would give greater accuracy. Note that the angle α must be determined prior to construction of the parallelogram. From the given figure

$$\tan\alpha = \frac{\overline{BD}}{\overline{AD}} = \frac{6\sin 60°}{3 + 6\cos 60°} = 0.866 \qquad \alpha = 40.9°$$

Measurement of the length R and direction θ of the resultant force **R** yields the approximate results

$$R = 525\text{ N} \qquad \theta = 49° \qquad \textit{Ans.}$$

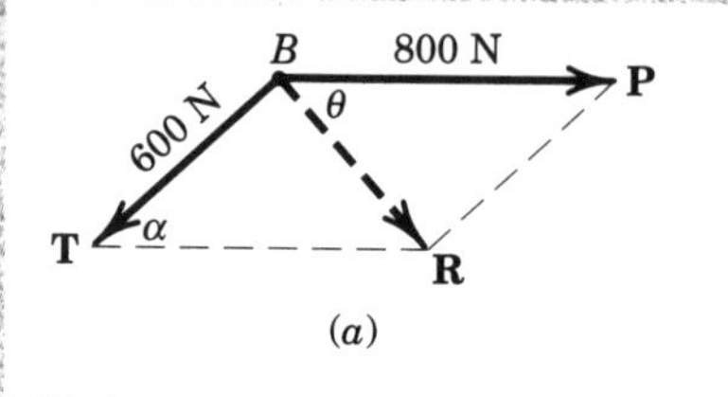

(a)

Geometric solution. The triangle for the vector addition of **T** and **P** is shown ② in Fig. b. The angle α is calculated as above. The law of cosines gives

$$R^2 = (600)^2 + (800)^2 - 2(600)(800)\cos 40.9° = 274{,}300$$

$$R = 524\text{ N} \qquad \textit{Ans.}$$

From the law of sines, we may determine the angle θ which orients **R**. Thus,

$$\frac{600}{\sin\theta} = \frac{524}{\sin 40.9°} \qquad \sin\theta = 0.750 \qquad \theta = 48.6° \qquad \textit{Ans.}$$

Helpful Hints

① Note the repositioning of **P** to permit parallelogram addition at B.

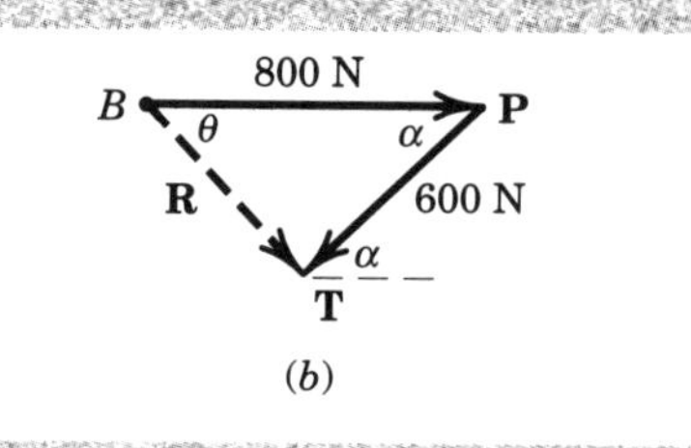

(b)

② Note the repositioning of **T** so as to preserve the correct line of action of the resultant **R**.

Algebraic solution. By using the x-y coordinate system on the given figure, we may write

$$R_x = \Sigma F_x = 800 - 600\cos 40.9° = 346\text{ N}$$

$$R_y = \Sigma F_y = -600\sin 40.9° = -393\text{ N}$$

The magnitude and direction of the resultant force **R** as shown in Fig. c are then

$$R = \sqrt{{R_x}^2 + {R_y}^2} = \sqrt{(346)^2 + (-393)^2} = 524\text{ N} \qquad \textit{Ans.}$$

$$\theta = \tan^{-1}\frac{|R_y|}{|R_x|} = \tan^{-1}\frac{393}{346} = 48.6° \qquad \textit{Ans.}$$

The resultant **R** may also be written in vector notation as

$$\mathbf{R} = R_x\mathbf{i} + R_y\mathbf{j} = 346\mathbf{i} - 393\mathbf{j}\text{ N} \qquad \textit{Ans.}$$

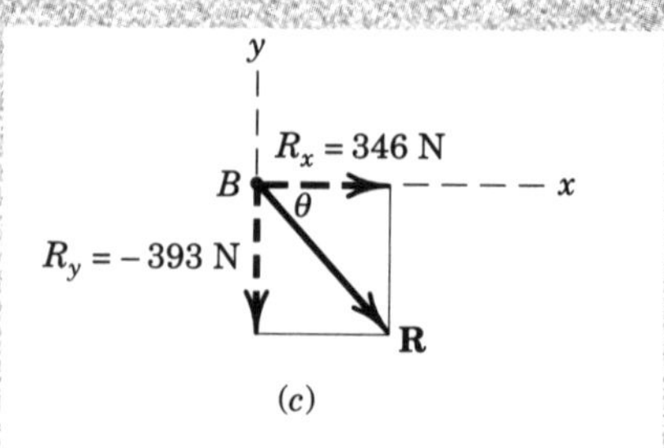

(c)

Sample Problem 2/3

The 500-N force **F** is applied to the vertical pole as shown. (1) Write **F** in terms of the unit vectors **i** and **j** and identify both its vector and scalar components. (2) Determine the scalar components of the force vector **F** along the x'- and y'-axes. (3) Determine the scalar components of **F** along the x- and y'-axes.

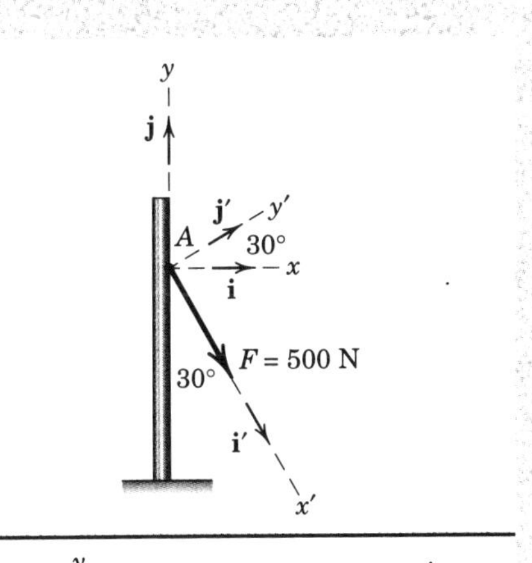

Solution. ***Part (1).*** From Fig. *a* we may write **F** as

$$\mathbf{F} = (F\cos\theta)\mathbf{i} - (F\sin\theta)\mathbf{j}$$
$$= (500\cos 60^\circ)\mathbf{i} - (500\sin 60^\circ)\mathbf{j}$$
$$= (250\mathbf{i} - 433\mathbf{j})\text{ N} \qquad \textit{Ans.}$$

The scalar components are $F_x = 250$ N and $F_y = -433$ N. The vector components are $\mathbf{F}_x = 250\mathbf{i}$ N and $\mathbf{F}_y = -433\mathbf{j}$ N.

Part (2). From Fig. *b* we may write **F** as $\mathbf{F} = 500\mathbf{i}'$ N, so that the required scalar components are

$$F_{x'} = 500 \text{ N} \qquad F_{y'} = 0 \qquad \textit{Ans.}$$

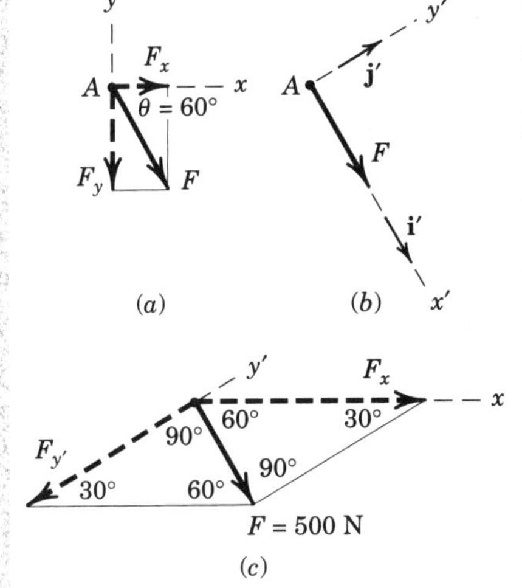

Part (3). The components of **F** in the x- and y'-directions are nonrectangular and are obtained by completing the parallelogram as shown in Fig. *c*. The magnitudes of the components may be calculated by the law of sines. Thus,

① $$\frac{|F_x|}{\sin 90^\circ} = \frac{500}{\sin 30^\circ} \qquad |F_x| = 1000 \text{ N}$$

$$\frac{|F_{y'}|}{\sin 60^\circ} = \frac{500}{\sin 30^\circ} \qquad |F_{y'}| = 866 \text{ N}$$

The required scalar components are then

$$F_x = 1000 \text{ N} \qquad F_{y'} = -866 \text{ N} \qquad \textit{Ans.}$$

Helpful Hint

① Obtain F_x and $F_{y'}$ graphically and compare your results with the calculated values.

Sample Problem 2/4

Forces $\mathbf{F}_1$ and $\mathbf{F}_2$ act on the bracket as shown. Determine the projection F_b of their resultant **R** onto the b-axis.

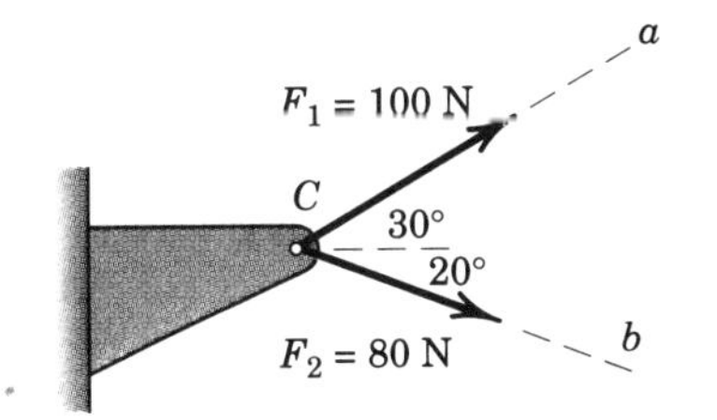

Solution. The parallelogram addition of $\mathbf{F}_1$ and $\mathbf{F}_2$ is shown in the figure. Using the law of cosines gives us

$$R^2 = (80)^2 + (100)^2 - 2(80)(100)\cos 130^\circ \qquad R = 163.4 \text{ N}$$

The figure also shows the orthogonal projection $\mathbf{F}_b$ of **R** onto the b-axis. Its length is

$$F_b = 80 + 100\cos 50^\circ = 144.3 \text{ N} \qquad \textit{Ans.}$$

Note that the components of a vector are in general not equal to the projections of the vector onto the same axes. If the a-axis had been perpendicular to the b-axis, then the projections and components of **R** would have been equal.

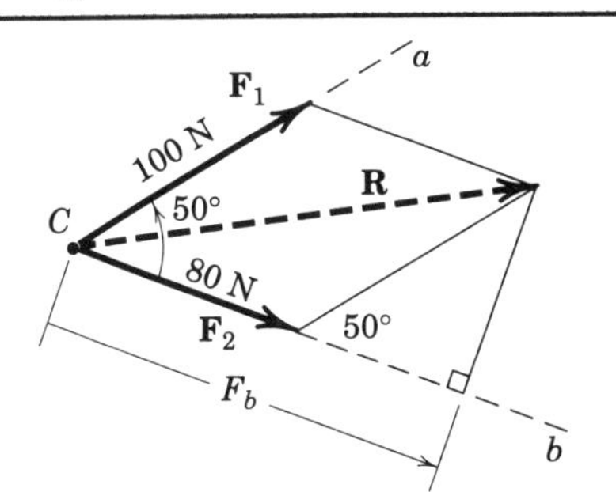

PROBLEMS

Introductory Problems

2/1 The force **F** has a magnitude of 500 N. Express **F** as a vector in terms of the unit vectors **i** and **j**. Identify the x and y scalar components of **F**.

Ans. $\mathbf{F} = 383\mathbf{i} - 321\mathbf{j}$ N, $F_x = 383$ N, $F_y = -321$ N

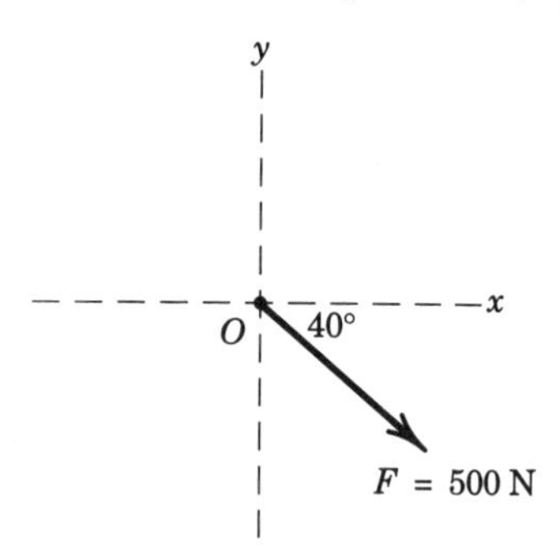

Problem 2/1

2/2 The magnitude of the force **F** is 400 N. Express **F** as a vector in terms of the unit vectors **i** and **j**. Identify both the scalar and vector components of **F**.

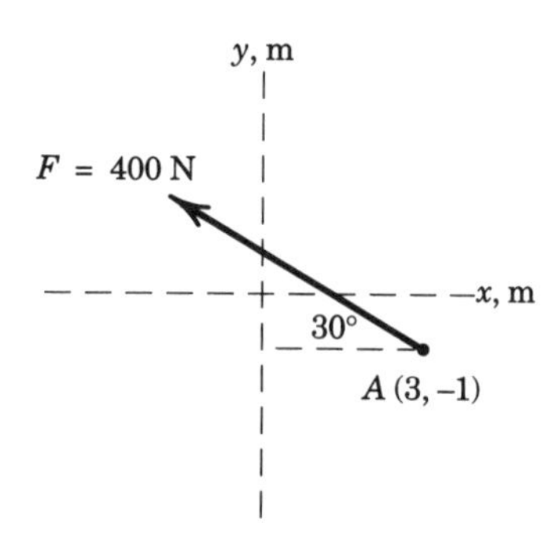

Problem 2/2

2/3 The slope of the 5.2-kN force **F** is specified as shown in the figure. Express **F** as a vector in terms of the unit vectors **i** and **j**.

Ans. $\mathbf{F} = -4.8\mathbf{i} - 2\mathbf{j}$ kN

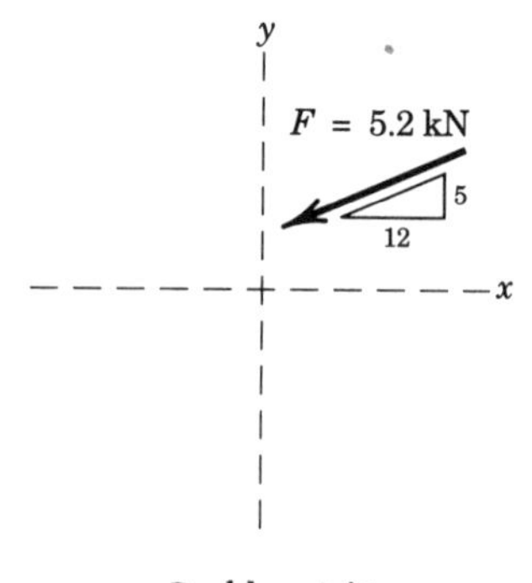

Problem 2/3

2/4 The line of action of the 34-kN force runs through the points A and B as shown in the figure. Determine the x and y scalar components of **F**.

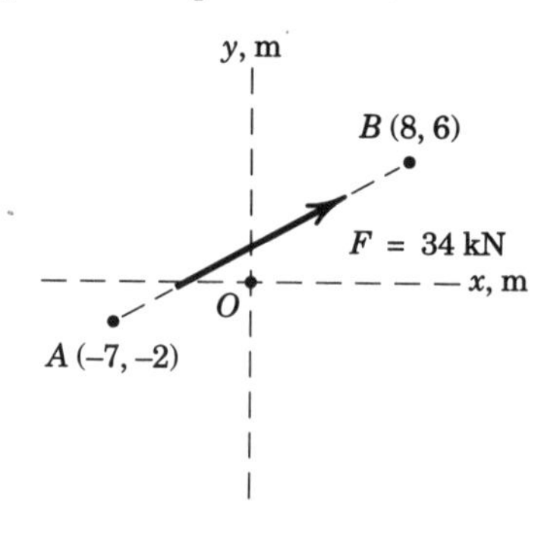

Problem 2/4

2/5 The 1800-N force **F** is applied to the end of the I-beam. Express **F** as a vector using the unit vectors **i** and **j**.

Ans. $\mathbf{F} = -1080\mathbf{i} - 1440\mathbf{j}$ N

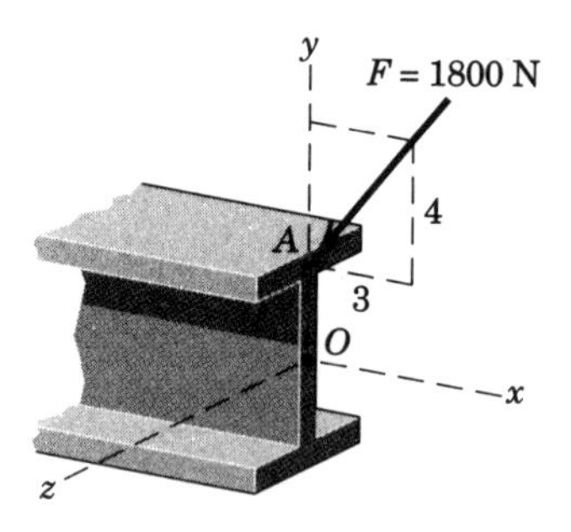

Problem 2/5

2/6 The two structural members, one of which is in tension and the other in compression, exert the indicated forces on joint O. Determine the magnitude of the resultant **R** of the two forces and the angle θ which **R** makes with the positive x-axis.

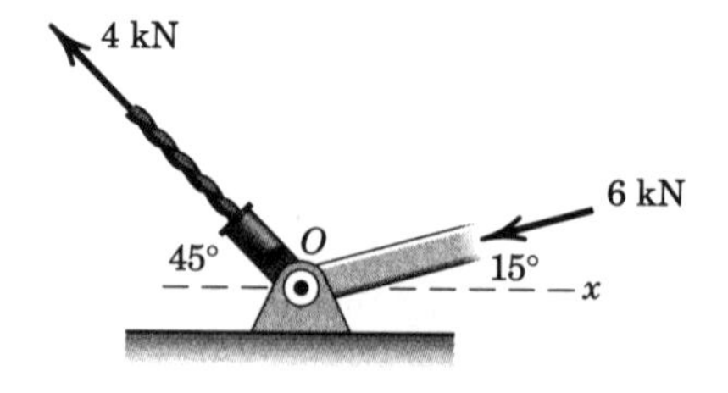

Problem 2/6

2/7 The y-component of the force $\mathbf{F}$ which a person exerts on the handle of the box wrench is known to be 320 N. Determine the x-component and the magnitude of $\mathbf{F}$.

Ans. $F_x = 133.3$ N, $F = 347$ N

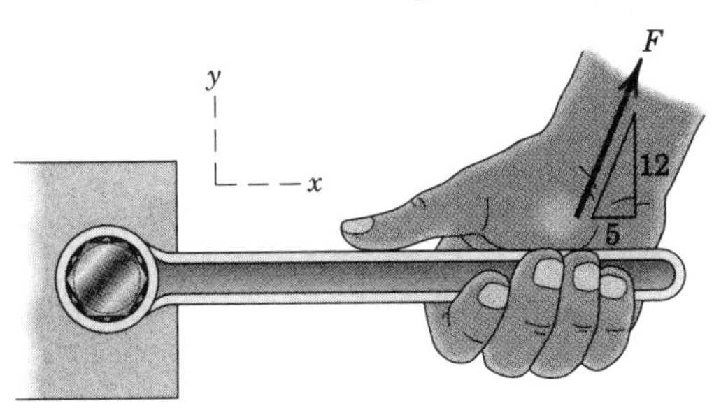

Problem 2/7

2/8 Determine the resultant $\mathbf{R}$ of the two forces shown by (*a*) applying the parallelogram rule for vector addition and (*b*) summing scalar components.

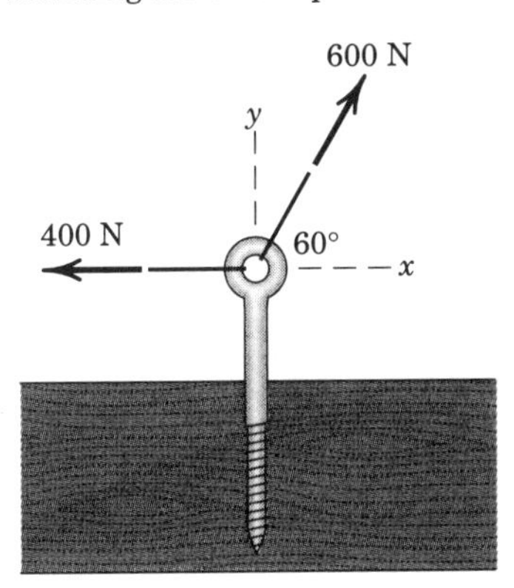

Problem 2/8

2/9 To satisfy design limitations it is necessary to determine the effect of the 2-kN tension in the cable on the shear, tension, and bending of the fixed I-beam. For this purpose replace this force by its equivalent of two forces at A, F_t parallel and F_n perpendicular to the beam. Determine F_t and F_n.

Ans. $F_t = 1.286$ kN, $F_n = 1.532$ kN

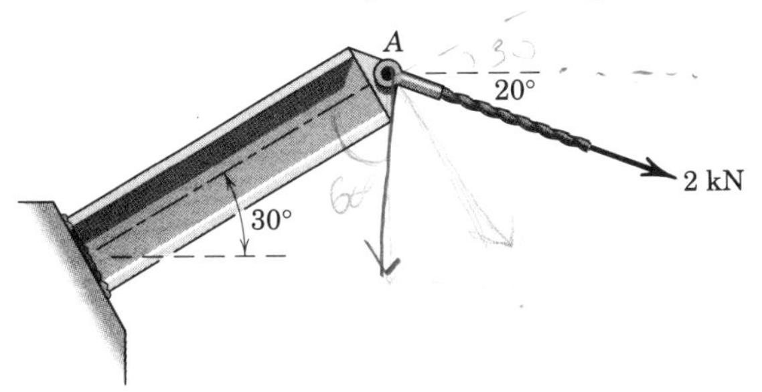

Problem 2/9

Representative Problems

2/10 Determine the magnitude F_s of the tensile spring force in order that the resultant of $\mathbf{F}_s$ and $\mathbf{F}$ is a vertical force. Determine the magnitude R of this vertical resultant force.

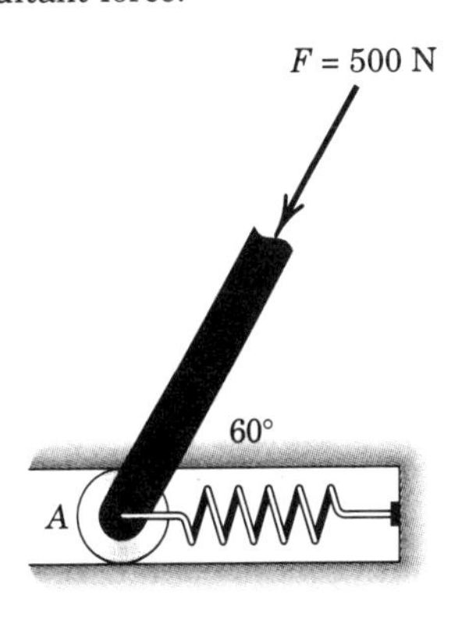

Problem 2/10

2/11 In the design of a control mechanism, it is determined that rod AB transmits a 260-N force $\mathbf{P}$ to the crank BC. Determine the x and y scalar components of $\mathbf{P}$.

Ans. $P_x = -240$ N
$P_y = -100$ N

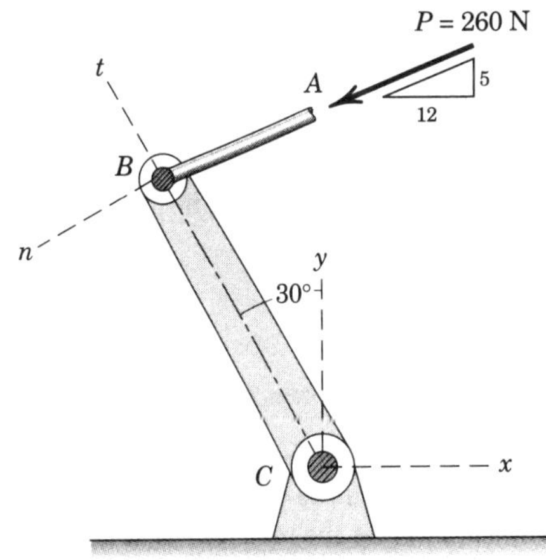

Problem 2/11

2/12 For the mechanism of Prob. 2/11, determine the scalar components P_t and P_n of $\mathbf{P}$ which are tangent and normal, respectively, to crank BC.

2/13 If the equal tensions T in the pulley cable are 400 N, express in vector notation the force **R** exerted on the pulley by the two tensions. Determine the magnitude of **R**.

Ans. $\mathbf{R} = 600\mathbf{i} + 346\mathbf{j}$ N, $R = 693$ N

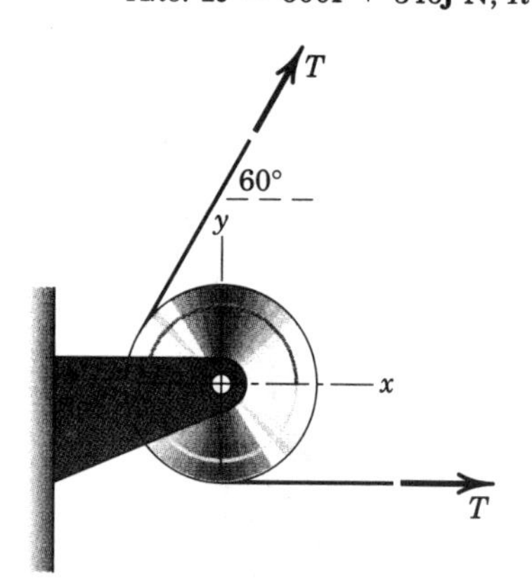

Problem 2/13

2/14 While steadily pushing the machine up an incline, a person exerts a 180-N force **P** as shown. Determine the components of **P** which are parallel and perpendicular to the incline.

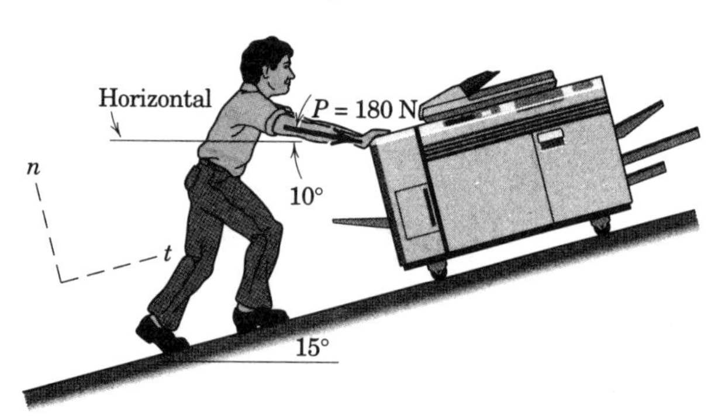

Problem 2/14

2/15 The normal reaction force N and the tangential friction force F act on the tire of a front-wheel-drive car as shown. Express the resultant **R** of these two forces in terms of the unit vectors (*a*) **i** and **j** along the *x-y* axes and (*b*) $\mathbf{e}_t$ and $\mathbf{e}_n$ along the *n-t* axes shown.

Ans. (*a*) $\mathbf{R} = 0.614\mathbf{i} + 3.89\mathbf{j}$ kN
(*b*) $\mathbf{R} = 1.6\mathbf{e}_t + 3.6\mathbf{e}_n$ kN

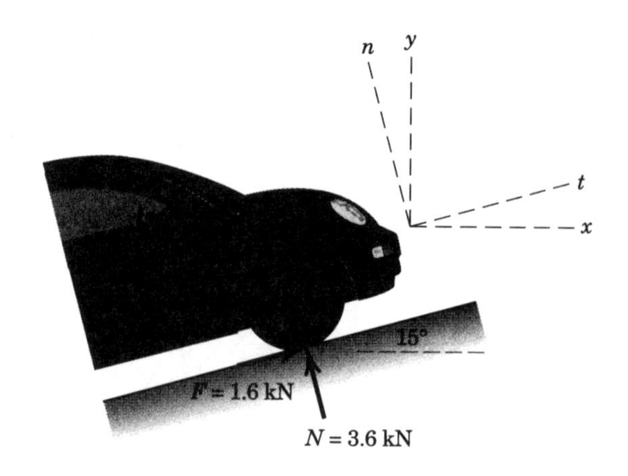

Problem 2/15

2/16 Determine the resultant **R** of the two forces applied to the bracket. Write **R** in terms of unit vectors along the x- and y-axes shown.

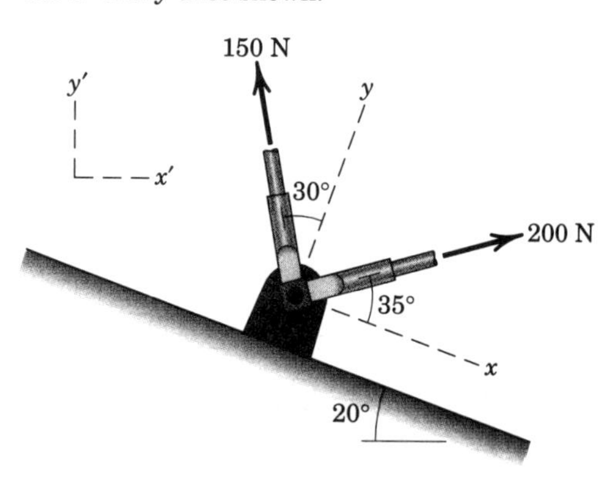

Problem 2/16

2/17 The ratio of the lift force L to the drag force D for the simple airfoil is $L/D = 10$. If the lift force on a short section of the airfoil is 200 N, compute the magnitude of the resultant force **R** and the angle θ which it makes with the horizontal.

Ans. $R = 201$ N, $\theta = 84.3°$

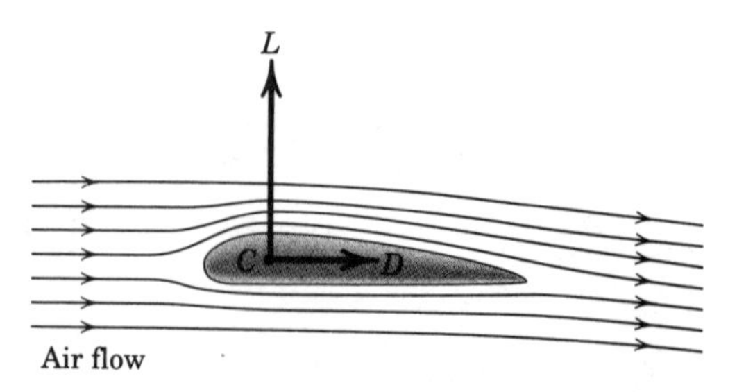

Problem 2/17

2/18 Determine the components of the 2-kN force along the oblique axes a and b. Determine the projections of $\mathbf{F}$ onto the a- and b-axes.

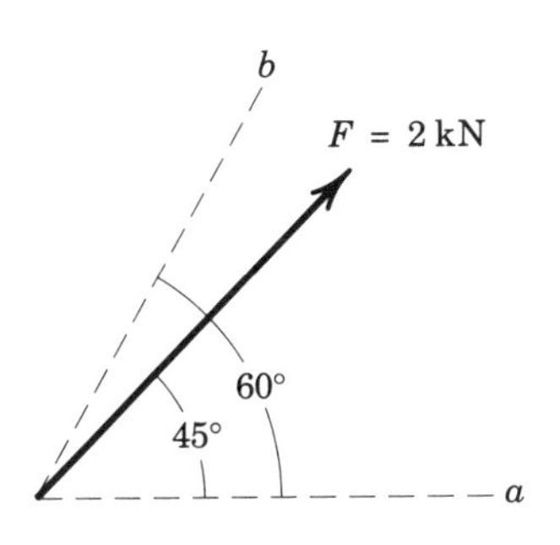

Problem 2/18

2/19 Determine the components of the 800-N force $\mathbf{F}$ along the oblique axes a and b. Also, determine the projections of $\mathbf{F}$ onto the a- and b-axes.

Ans. Components: $F_a = 1093$ N, $F_b = 980$ N
Projections: $F_a = 400$ N, $F_b = 207$ N

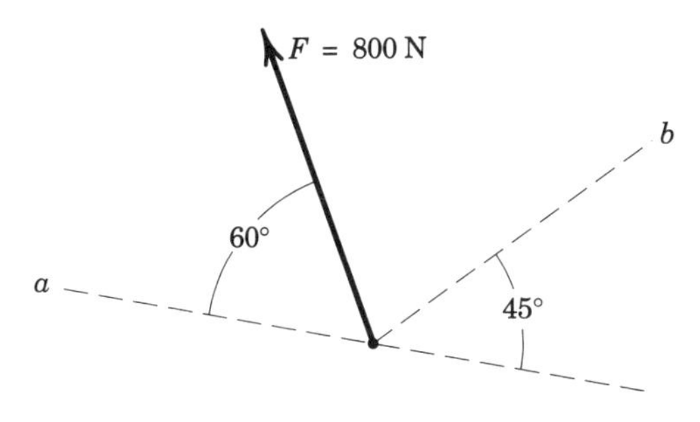

Problem 2/19

2/20 The 600-N force applied to the bracket at A is to be replaced by two forces, F_a in the a-a direction and F_b in the b-b direction, which together produce the same effect on the bracket as that of the 600-N force. Determine F_a and F_b.

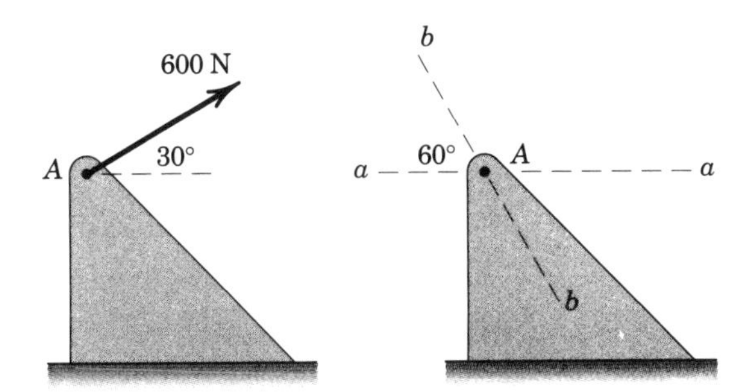

Problem 2/20

2/21 It is desired to remove the spike from the timber by applying force along its horizontal axis. An obstruction A prevents direct access, so that two forces, one 1.6 kN and the other $\mathbf{P}$, are applied by cables as shown. Compute the magnitude of $\mathbf{P}$ necessary to ensure a resultant $\mathbf{T}$ directed along the spike. Also find T.

Ans. $P = 2.15$ kN
$T = 3.20$ kN

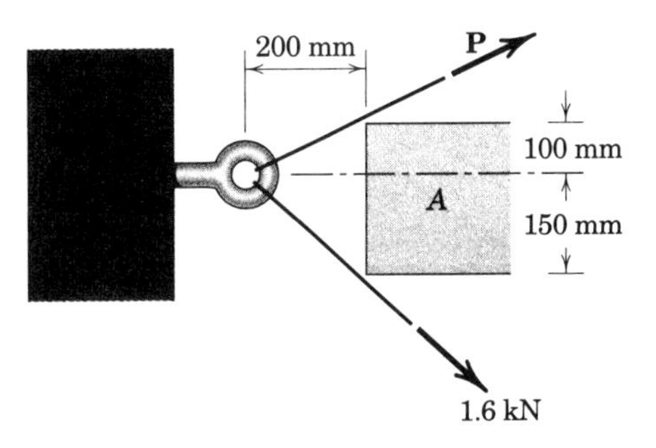

Problem 2/21

2/22 At what angle θ must the 800-N force be applied in order that the resultant $\mathbf{R}$ of the two forces has a magnitude of 2000 N? For this condition, determine the angle β between $\mathbf{R}$ and the vertical.

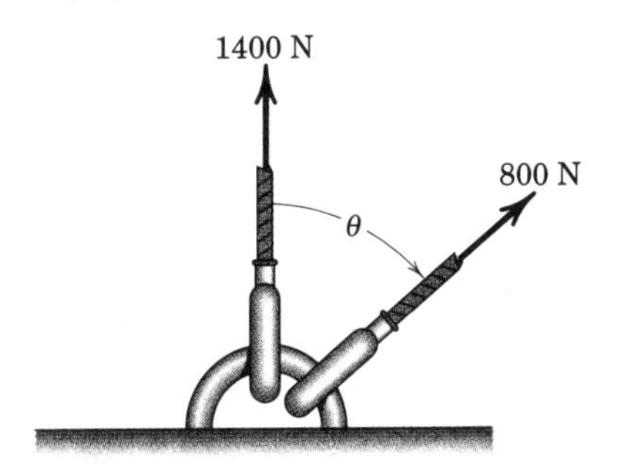

Problem 2/22

2/23 The cable AB prevents bar OA from rotating clockwise about the pivot O. If the cable tension is 750 N, determine the n- and t-components of this force acting on point A of the bar.

Ans. $T_n = 333$ N, $T_t = -672$ N

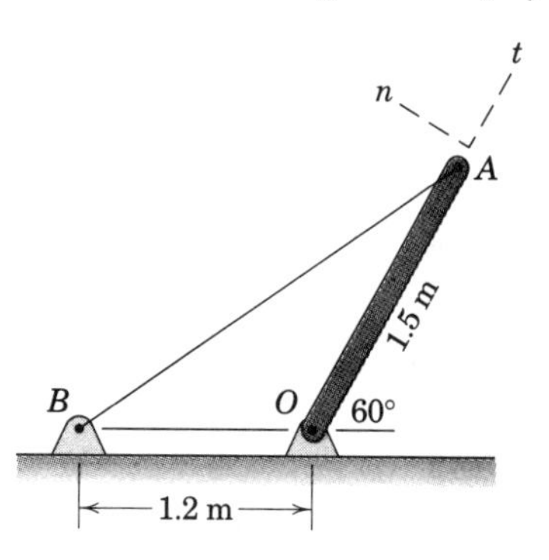

Problem 2/23

2/24 In the design of the robot to insert the small cylindrical part into a close-fitting circular hole, the robot arm must exert a 90-N force P on the part parallel to the axis of the hole as shown. Determine the components of the force which the part exerts *on* the robot along axes (*a*) parallel and perpendicular to the arm AB, and (*b*) parallel and perpendicular to the arm BC.

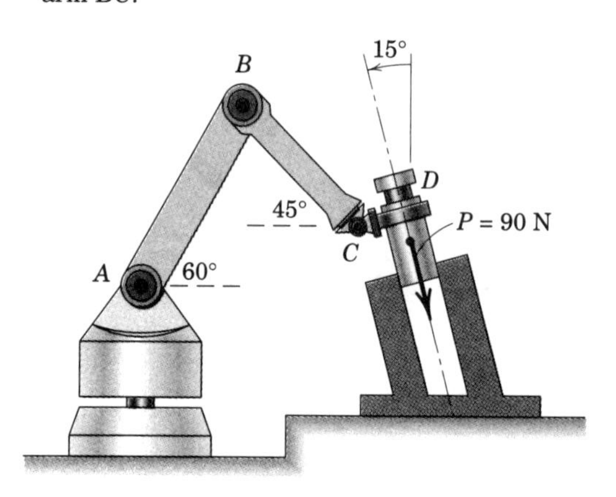

Problem 2/24

2/25 The guy cables AB and AC are attached to the top of the transmission tower. The tension in cable AC is 8 kN. Determine the required tension T in cable AB such that the net effect of the two cable tensions is a downward force at point A. Determine the magnitude R of this downward force.

Ans. $T = 5.68$ kN, $R = 10.21$ kN

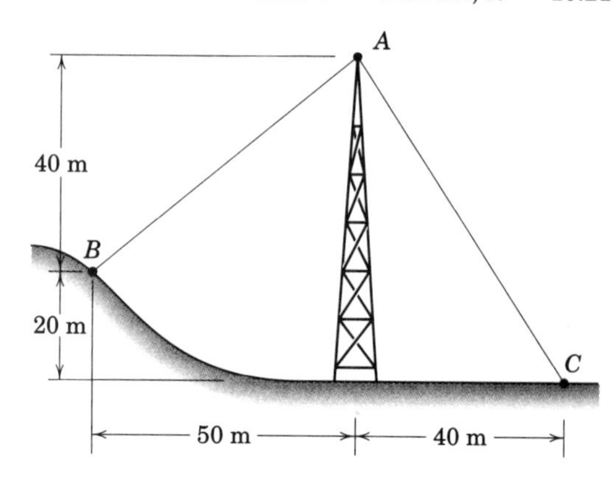

Problem 2/25

2/26 The gusset plate is subjected to the two forces shown. Replace them by two equivalent forces, F_x in the x-direction and F_a in the a-direction. Determine the magnitudes of F_x and F_a. Solve geometrically or graphically.

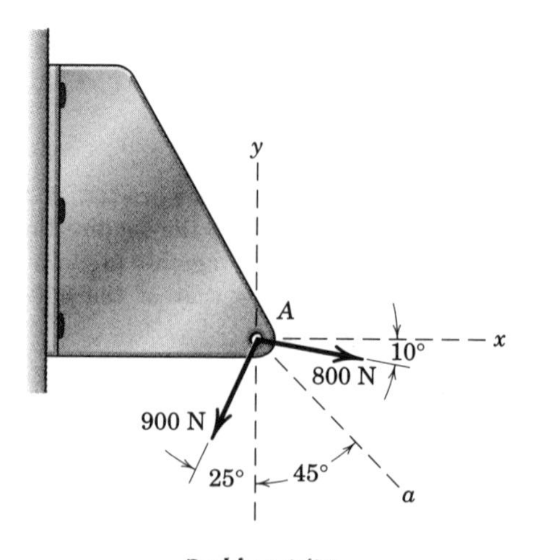

Problem 2/26

2/4 Moment

In addition to the tendency to move a body in the direction of its application, a force can also tend to rotate a body about an axis. The axis may be any line which neither intersects nor is parallel to the line of action of the force. This rotational tendency is known as the *moment* **M** of the force. Moment is also referred to as *torque*.

As a familiar example of the concept of moment, consider the pipe wrench of Fig. 2/8*a*. One effect of the force applied perpendicular to the handle of the wrench is the tendency to rotate the pipe about its vertical axis. The magnitude of this tendency depends on both the magnitude F of the force and the effective length d of the wrench handle. Common experience shows that a pull which is not perpendicular to the wrench handle is less effective than the right-angle pull shown.

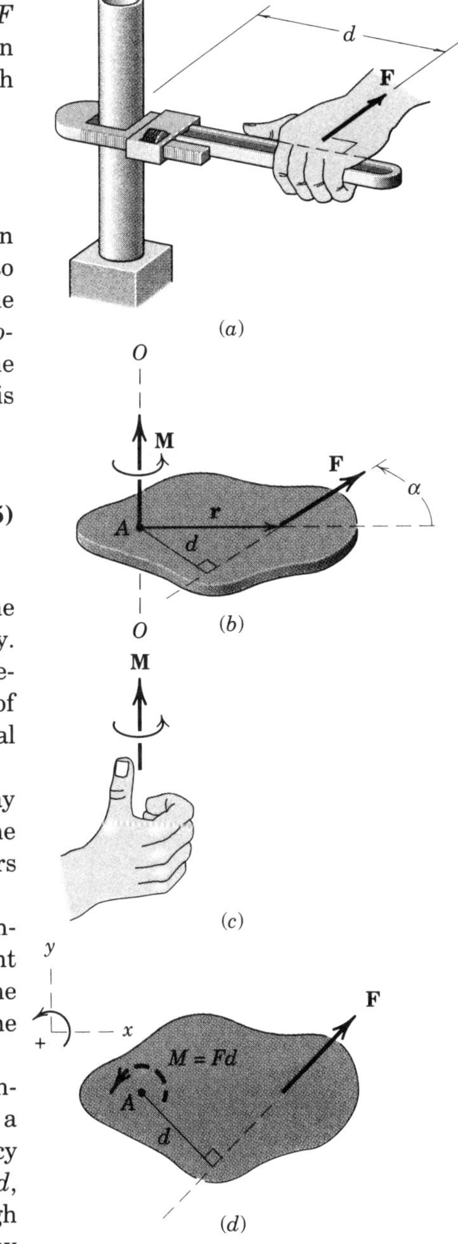

Figure 2/8

Moment about a Point

Figure 2/8*b* shows a two-dimensional body acted on by a force **F** in its plane. The magnitude of the moment or tendency of the force to rotate the body about the axis *O-O* perpendicular to the plane of the body is proportional both to the magnitude of the force and to the *moment arm d*, which is the perpendicular distance from the axis to the line of action of the force. Therefore, the magnitude of the moment is defined as

$$M = Fd \tag{2/5}$$

The moment is a vector **M** perpendicular to the plane of the body. The sense of **M** depends on the direction in which **F** tends to rotate the body. The right-hand rule, Fig. 2/8*c*, is used to identify this sense. We represent the moment of **F** about *O-O* as a vector pointing in the direction of the thumb, with the fingers curled in the direction of the rotational tendency.

The moment **M** obeys all the rules of vector combination and may be considered a sliding vector with a line of action coinciding with the moment axis. The basic units of moment in SI units are newton-meters (N·m), and in the U.S. customary system are pound-feet (lb-ft).

When dealing with forces which all act in a given plane, we customarily speak of the moment *about a point*. By this we mean the moment with respect to an axis normal to the plane and passing through the point. Thus, the moment of force **F** about point A in Fig. 2/8*d* has the magnitude $M = Fd$ and is counterclockwise.

Moment directions may be accounted for by using a stated sign convention, such as a plus sign (+) for counterclockwise moments and a minus sign (−) for clockwise moments, or vice versa. Sign consistency within a given problem is essential. For the sign convention of Fig. 2/8*d*, the moment of **F** about point A (or about the z-axis passing through point A) is positive. The curved arrow of the figure is a convenient way to represent moments in two-dimensional analysis.

The Cross Product

In some two-dimensional and many of the three-dimensional problems to follow, it is convenient to use a vector approach for moment calculations. The moment of **F** about point A of Fig. 2/8*b* may be represented by the cross-product expression

$$\mathbf{M} = \mathbf{r} \times \mathbf{F} \tag{2/6}$$

where **r** is a position vector which runs from the moment reference point A to *any* point on the line of action of **F**. The magnitude of this expression is given by*

$$M = Fr \sin \alpha = Fd \tag{2/7}$$

which agrees with the moment magnitude as given by Eq. 2/5. Note that the moment arm $d = r \sin \alpha$ does not depend on the particular point on the line of action of **F** to which the vector **r** is directed. We establish the direction and sense of **M** by applying the right-hand rule to the sequence $\mathbf{r} \times \mathbf{F}$. If the fingers of the right hand are curled in the direction of rotation from the positive sense of **r** to the positive sense of **F**, then the thumb points in the positive sense of **M**.

We must maintain the sequence $\mathbf{r} \times \mathbf{F}$, because the sequence $\mathbf{F} \times \mathbf{r}$ would produce a vector with a sense opposite to that of the correct moment. As was the case with the scalar approach, the moment **M** may be thought of as the moment about point A or as the moment about the line O-O which passes through point A and is perpendicular to the plane containing the vectors **r** and **F**. When we evaluate the moment of a force about a given point, the choice between using the vector cross product or the scalar expression depends on how the geometry of the problem is specified. If we know or can easily determine the perpendicular distance between the line of action of the force and the moment center, then the scalar approach is generally simpler. If, however, **F** and **r** are not perpendicular and are easily expressible in vector notation, then the cross-product expression is often preferable.

In Section B of this chapter, we will see how the vector formulation of the moment of a force is especially useful for determining the moment of a force about a point in three-dimensional situations.

Varignon's Theorem

One of the most useful principles of mechanics is *Varignon's theorem*, which states that the moment of a force about any point is equal to the sum of the moments of the components of the force about the same point.

*See item 7 in Art. C/7 of Appendix C for additional information concerning the cross product.

To prove this theorem, consider the force **R** acting in the plane of the body shown in Fig. 2/9*a*. The forces **P** and **Q** represent any two nonrectangular components of **R**. The moment of **R** about point O is

$$\mathbf{M}_O = \mathbf{r} \times \mathbf{R}$$

Because $\mathbf{R} = \mathbf{P} + \mathbf{Q}$, we may write

$$\mathbf{r} \times \mathbf{R} = \mathbf{r} \times (\mathbf{P} + \mathbf{Q})$$

Using the distributive law for cross products, we have

$$\mathbf{M}_O = \mathbf{r} \times \mathbf{R} = \mathbf{r} \times \mathbf{P} + \mathbf{r} \times \mathbf{Q} \qquad (2/8)$$

which says that the moment of **R** about O equals the sum of the moments about O of its components **P** and **Q.** This proves the theorem.

Varignon's theorem need not be restricted to the case of two components, but it applies equally well to three or more. Thus we could have used any number of concurrent components of **R** in the foregoing proof.*

Figure 2/9*b* illustrates the usefulness of Varignon's theorem. The moment of **R** about point O is Rd. However, if d is more difficult to determine than p and q, we can resolve **R** into the components **P** and **Q**, and compute the moment as

$$M_O = Rd = -pP + qQ$$

where we take the clockwise moment sense to be positive.

Sample Problem 2/5 shows how Varignon's theorem can help us to calculate moments.

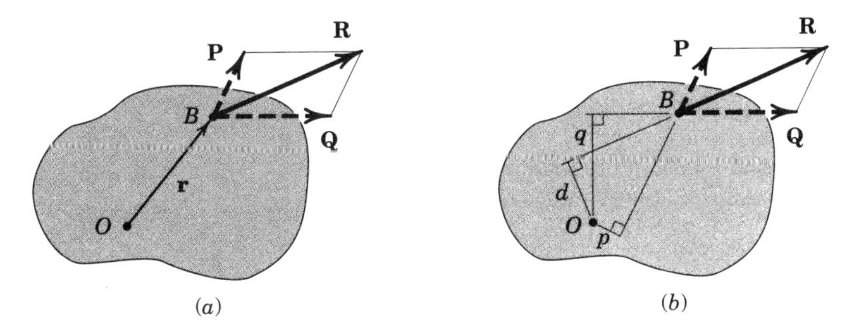

Figure 2/9

*As originally stated, Varignon's theorem was limited to the case of two concurrent components of a given force. See *The Science of Mechanics*, by Ernst Mach, originally published in 1883.

Sample Problem 2/5

Calculate the magnitude of the moment about the base point O of the 600-N force in five different ways.

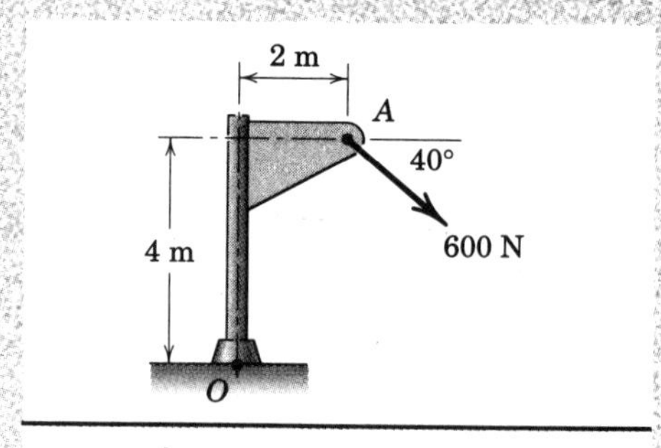

Solution. (*I*) The moment arm to the 600-N force is

$$d = 4 \cos 40^\circ + 2 \sin 40^\circ = 4.35 \text{ m}$$

① By $M = Fd$ the moment is clockwise and has the magnitude

$$M_O = 600(4.35) = 2610 \text{ N·m} \qquad \textit{Ans.}$$

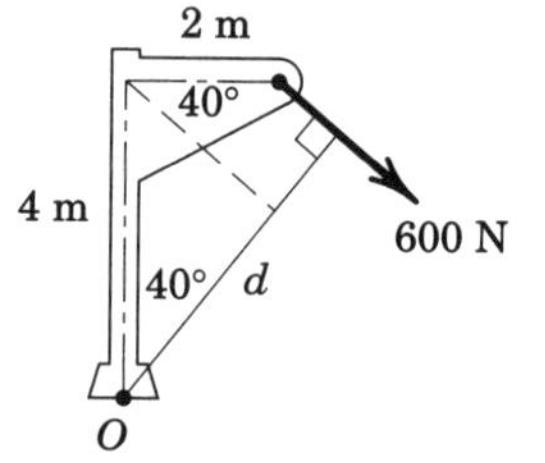

(*II*) Replace the force by its rectangular components at A

$$F_1 = 600 \cos 40^\circ = 460 \text{ N}, \qquad F_2 = 600 \sin 40^\circ = 386 \text{ N}$$

By Varignon's theorem, the moment becomes

② $$M_O = 460(4) + 386(2) = 2610 \text{ N·m} \qquad \textit{Ans.}$$

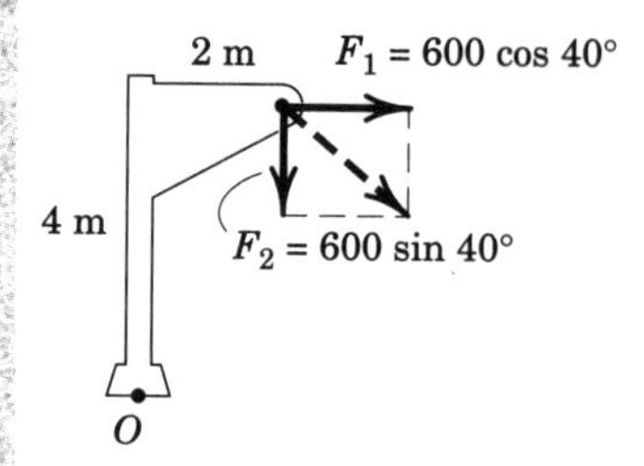

(*III*) By the principle of transmissibility, move the 600-N force along its line of action to point B, which eliminates the moment of the component F_2. The moment arm of F_1 becomes

$$d_1 = 4 + 2 \tan 40^\circ = 5.68 \text{ m}$$

and the moment is

$$M_O = 460(5.68) = 2610 \text{ N·m} \qquad \textit{Ans.}$$

③ (*IV*) Moving the force to point C eliminates the moment of the component F_1. The moment arm of F_2 becomes

$$d_2 = 2 + 4 \cot 40^\circ = 6.77 \text{ m}$$

and the moment is

$$M_O = 386(6.77) = 2610 \text{ N·m} \qquad \textit{Ans.}$$

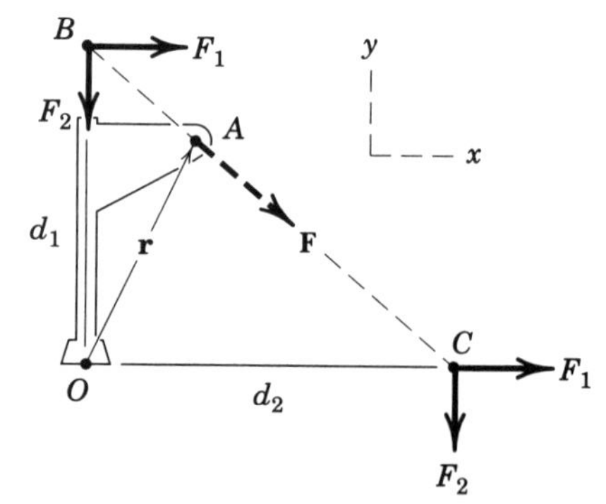

(*V*) By the vector expression for a moment, and by using the coordinate system indicated on the figure together with the procedures for evaluating cross products, we have

④ $$\mathbf{M}_O = \mathbf{r} \times \mathbf{F} = (2\mathbf{i} + 4\mathbf{j}) \times 600(\mathbf{i} \cos 40^\circ - \mathbf{j} \sin 40^\circ)$$
$$= -2610\mathbf{k} \text{ N·m}$$

The minus sign indicates that the vector is in the negative z-direction. The magnitude of the vector expression is

$$M_O = 2610 \text{ N·m} \qquad \textit{Ans.}$$

Helpful Hints

① The required geometry here and in similar problems should not cause difficulty if the sketch is carefully drawn.

② This procedure is frequently the shortest approach.

③ The fact that points B and C are not on the body proper should not cause concern, as the mathematical calculation of the moment of a force does not require that the force be on the body.

④ Alternative choices for the position vector $\mathbf{r}$ are $\mathbf{r} = d_1\mathbf{j} = 5.68\mathbf{j}$ m and $\mathbf{r} = d_2\mathbf{i} = 6.77\mathbf{i}$ m.

PROBLEMS

Introductory Problems

2/27 The 4-kN force **F** is applied at point A. Compute the moment of **F** about point O, expressing it both as a scalar and as a vector quantity. Determine the coordinates of the points on the x- and y-axes about which the moment of **F** is zero.

Ans. $M_O = 2.68$ kN·m CCW, $\mathbf{M}_O = 2.68\mathbf{k}$ kN·m
$(x, y) = (-1.3, 0)$ and $(0, 0.78)$ m

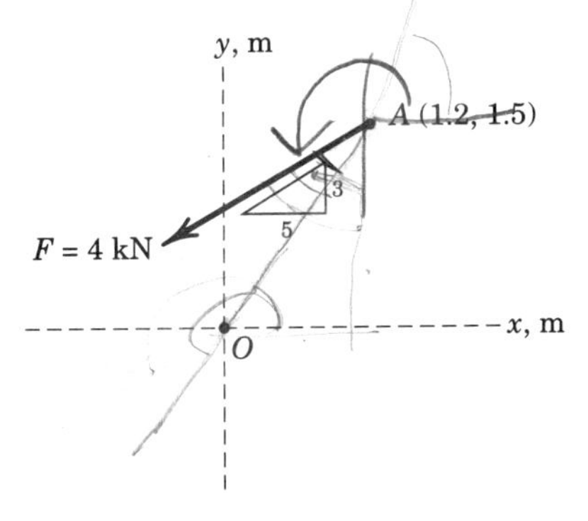

Problem 2/27

2/28 The rectangular plate is made up of 1-m squares as shown. A 75-N force is applied at point A in the direction shown. Determine the moment of this force about point B and about point C.

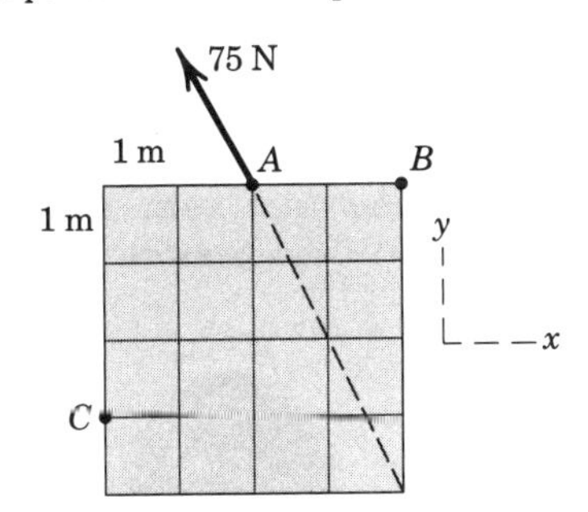

Problem 2/28

2/29 The throttle-control sector pivots freely at O. If an internal torsional spring exerts a return moment $M = 2$ N·m on the sector when in the position shown, for design purposes determine the necessary throttle-cable tension T so that the net moment about O is zero. Note that when T is zero, the sector rests against the idle-control adjustment screw at R.

Ans. $T = 40$ N

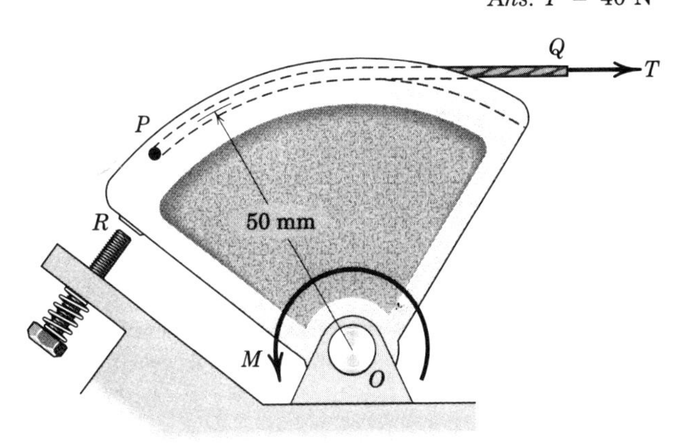

Problem 2/29

2/30 The entire branch OA has a mass of 180 kg with mass center at G. Determine the moment of the weight of this branch about point O.

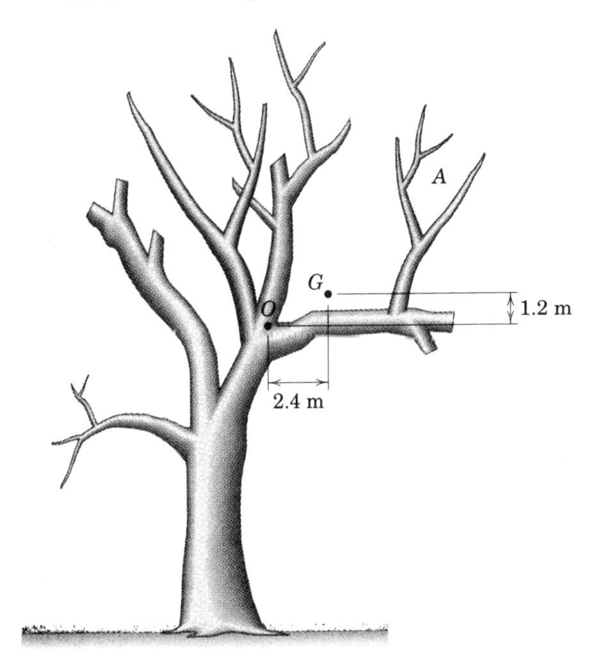

Problem 2/30

2/31 A force **F** of magnitude 60 N is applied to the gear. Determine the moment of **F** about point O.

Ans. $M_O = 5.64 \text{ N}\cdot\text{m CW}$

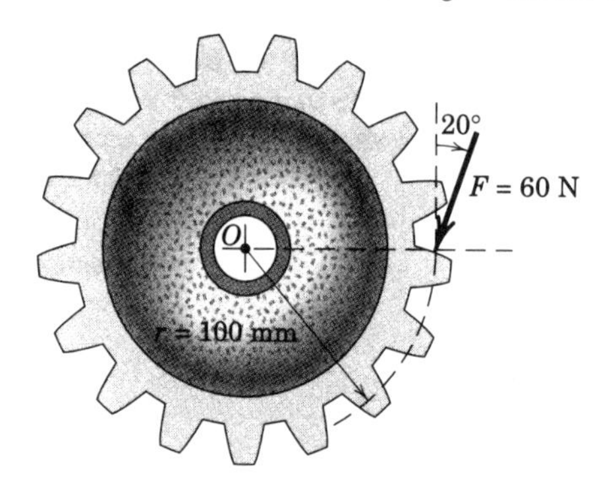

Problem 2/31

2/32 Calculate the moment of the 250-N force on the handle of the monkey wrench about the center of the bolt.

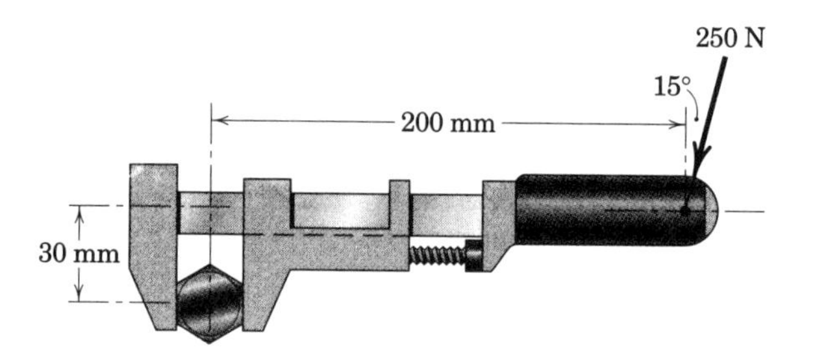

Problem 2/32

2/33 A prybar is used to remove a nail as shown. Determine the moment of the 240-N force about the point O of contact between the prybar and the small support block.

Ans. $M_O = 84.0 \text{ N}\cdot\text{m CW}$

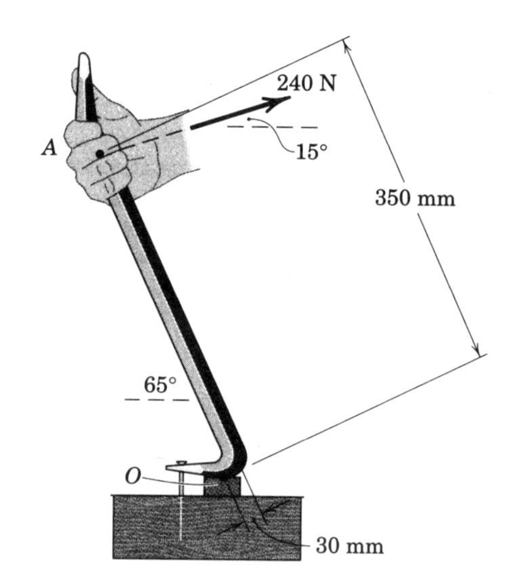

Problem 2/33

Representative Problems

2/34 A portion of a mechanical coin sorter works as follows: Pennies and dimes roll down the 20° incline, the last triangular portion of which pivots freely about a horizontal axis through O. Dimes are light enough (2.28 grams each) so that the triangular portion remains stationary, and the dimes roll into the right collection column. Pennies, on the other hand, are heavy enough (3.06 grams each) so that the triangular portion pivots clockwise, and the pennies roll into the left collection column. Determine the moment about O of the weight of the penny in terms of the slant distance s in millimeters.

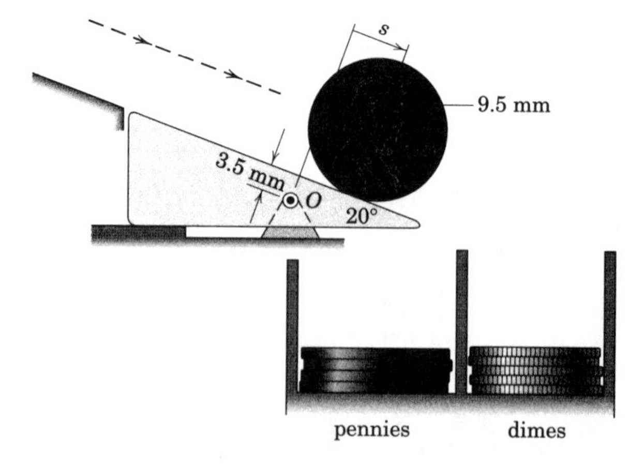

Problem 2/34

2/35 Elements of the lower arm are shown in the figure. The mass of the forearm is 2.3 kg with mass center at G. Determine the combined moment about the elbow pivot O of the weights of the forearm and the 3.6-kg homogeneous sphere. What must the biceps tension force T be so that the overall moment about O is zero?

Ans. $M_O = 14.25$ N·m CW, $T = 285$ N

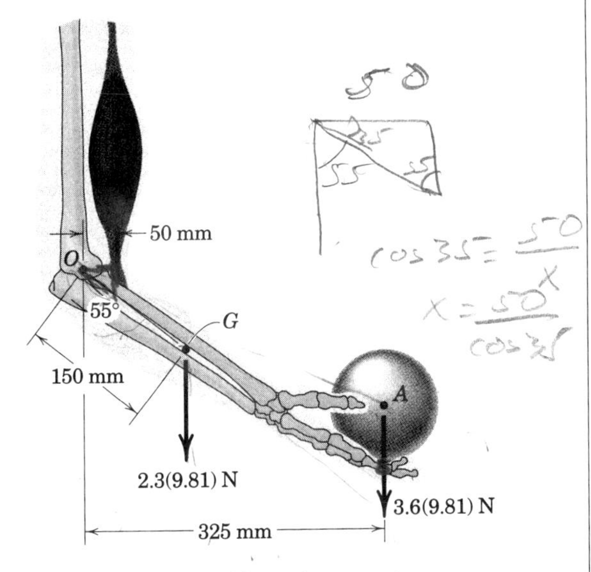

Problem 2/35

2/36 The 30-N force **P** is applied perpendicular to the portion BC of the bent bar. Determine the moment of **P** about point B and about point A.

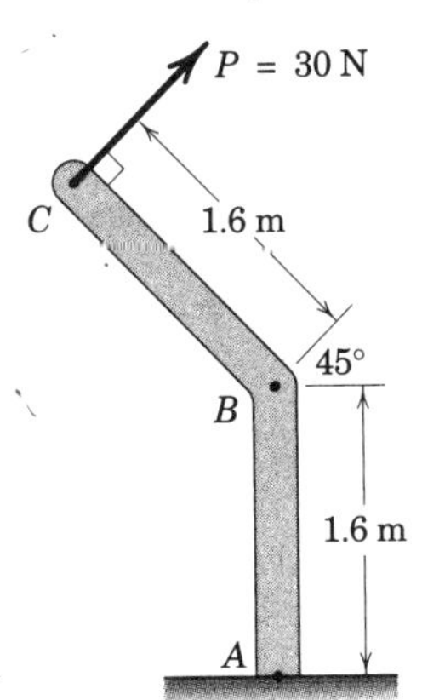

Problem 2/36

2/37 In order to raise the flagpole OC, a light frame OAB is attached to the pole and a tension of 3.2 kN is developed in the hoisting cable by the power winch D. Calculate the moment M_O of this tension about the hinge point O.

Ans. $M_O = 6.17$ kN·m CCW

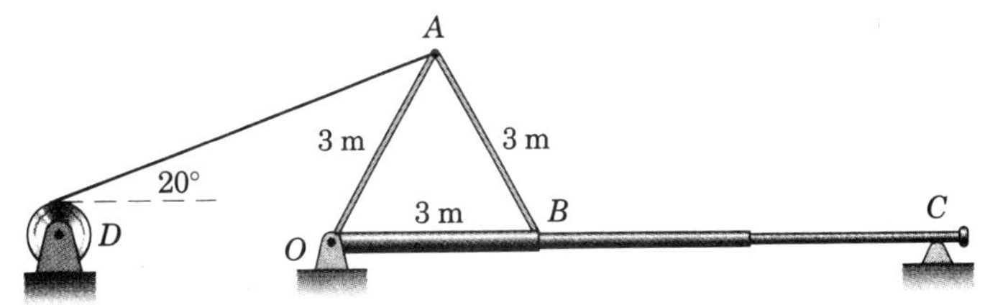

Problem 2/37

2/38 Compute the moment of the 1.6-N force about the pivot O of the wall-switch toggle.

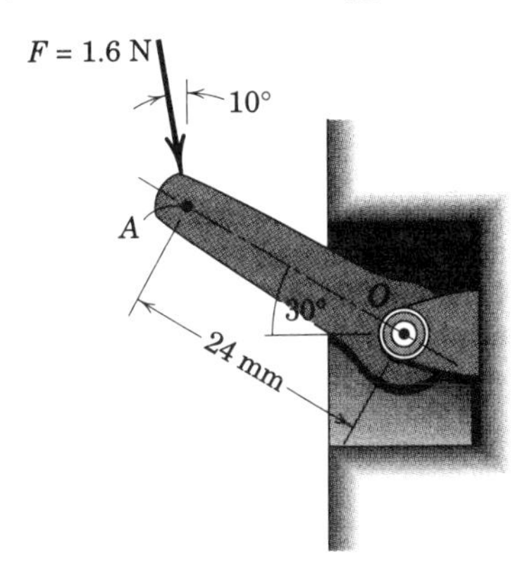

Problem 2/38

2/39 A force of 200 N is applied to the end of the wrench to tighten a flange bolt which holds the wheel to the axle. Determine the moment M produced by this force about the center O of the wheel for the position of the wrench shown.

Ans. $M = 78.3$ N·m CW

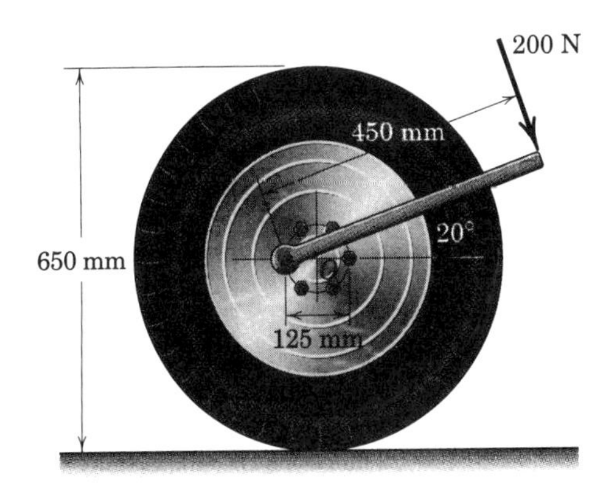

Problem 2/39

2/40 The lower lumbar region A of the spine is the part of the spinal column most susceptible to abuse while resisting excessive bending caused by the moment about A of a force F. For given values of F, b, and h, determine the angle θ which causes the most severe bending strain.

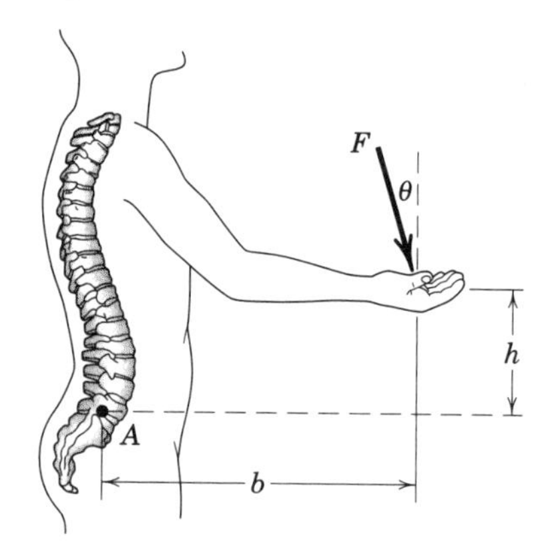

Problem 2/40

2/41 Determine the combined moment about O due to the weight of the mailbox and the cross member AB. The mass of the mailbox is 2 kg and that of the uniform cross member is 5 kg. The weights act at the geometric centers of the respective items.

Ans. $M_O = 15.94\ \text{N}\cdot\text{m CCW}$

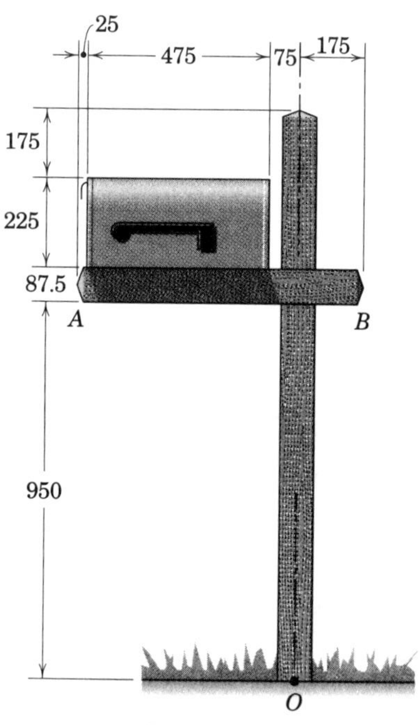

Problem 2/41

2/42 The force exerted by the plunger of cylinder AB on the door is 40 N directed along the line AB, and this force tends to keep the door closed. Compute the moment of this force about the hinge O. What force F_C normal to the plane of the door must the door stop at C exert on the door so that the combined moment about O of the two forces is zero?

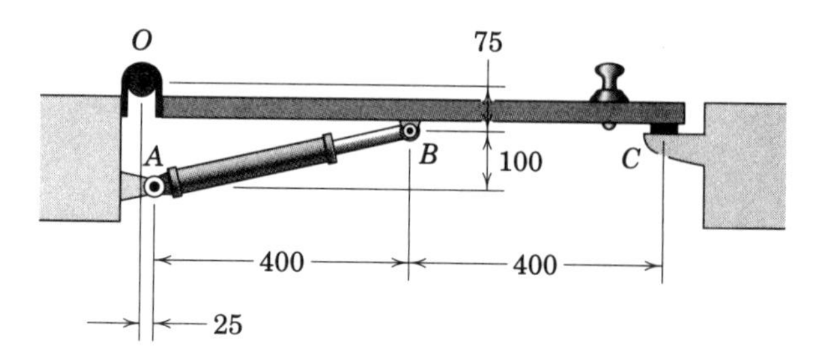

Problem 2/42

2/43 (*a*) Calculate the moment of the 90-N force about point O for the condition $\theta = 15°$. Also, determine the value of θ for which the moment about O is (*b*) zero and (*c*) a maximum.

Ans. (*a*) $M_O = 33.5\ \text{N}\cdot\text{m CCW}$
(*b*) $\theta = 36.9°$ (or 217°)
(*c*) $\theta = 126.9°$ (or 307°)

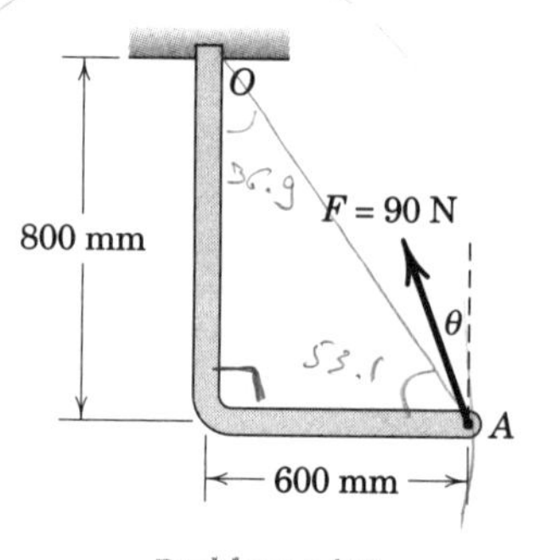

Problem 2/43

2/44 Determine the angle θ which will maximize the moment M_O of the 200-N force about the shaft axis at O. Also compute M_O.

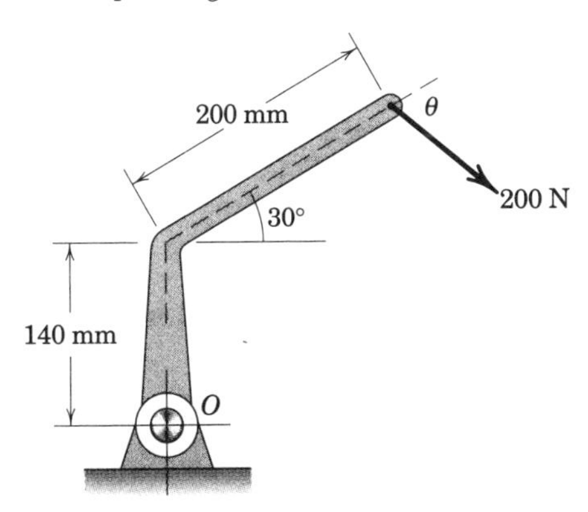

Problem 2/44

2/45 The spring-loaded follower A bears against the circular portion of the cam until the lobe of the cam lifts the plunger. The force required to lift the plunger is proportional to its vertical movement h from its lowest position. For design purposes determine the angle θ for which the moment of the contact force on the cam about the bearing O is a maximum. In the enlarged view of the contact, neglect the small distance between the actual contact point B and the end C of the lobe.

Ans. $\theta = 57.5°$

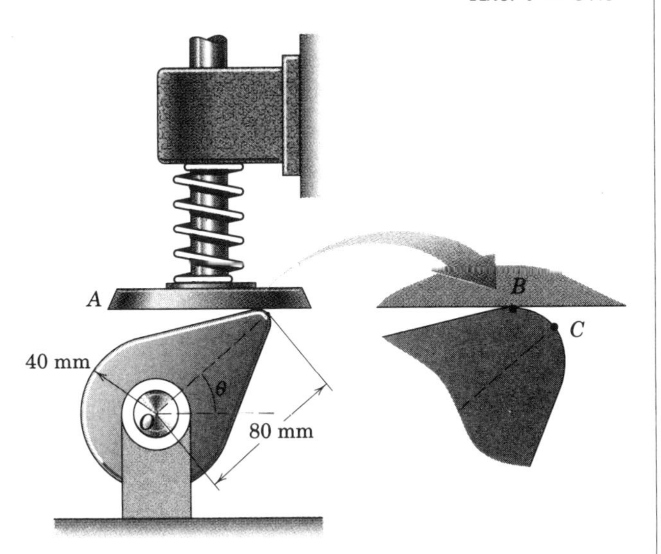

Problem 2/45

2/46 The small crane is mounted along the side of a pickup bed and facilitates the handling of heavy loads. When the boom elevation angle is $\theta = 40°$, the force in the hydraulic cylinder BC is 4.5 kN, and this force applied at point C is in the direction from B to C (the cylinder is in compression). Determine the moment of this 4.5-kN force about the boom pivot point O.

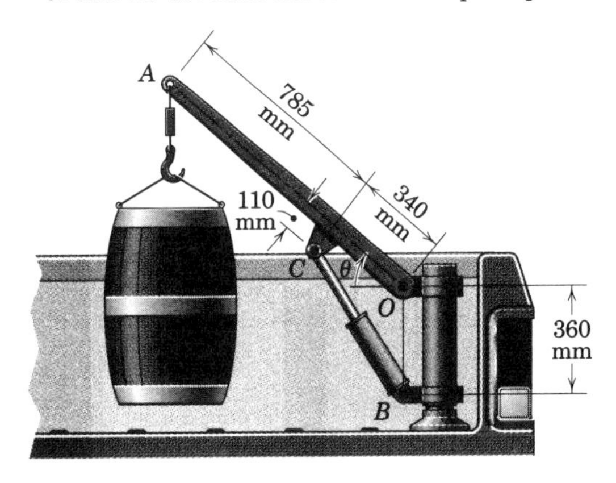

Problem 2/46

2/47 Design criteria require that the robot exert the 90-N force on the part as shown while inserting a cylindrical part into the circular hole. Determine the moment about points A, B, and C of the force which the part exerts on the robot.

Ans. $M_A = 68.8$ N·m, $M_B = 33.8$ N·m
$M_C = 13.50$ N·m (all CCW)

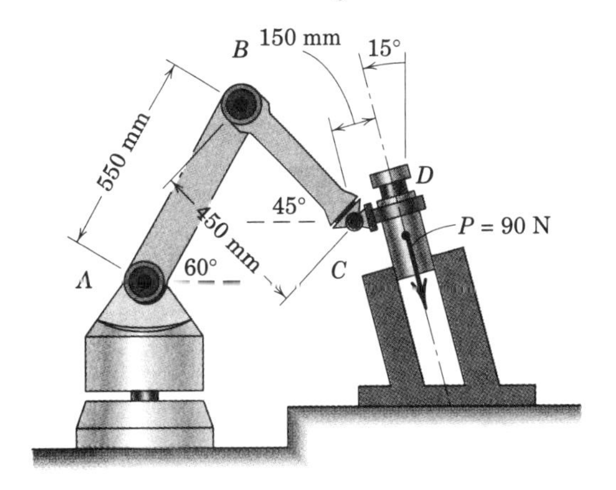

Problem 2/47

2/48 As the result of a wind blowing normal to the plane of the rectangular sign, a uniform pressure of 175 N/m^2 is exerted in the direction shown in the figure. Determine the moment of the resulting force about point O. Express your result as a vector using the coordinates shown.

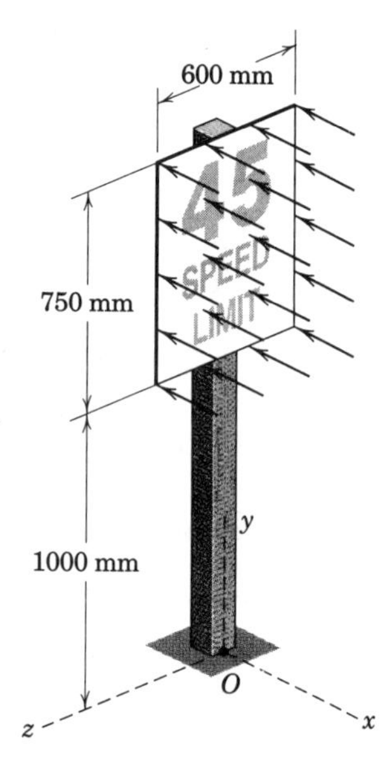

Problem 2/48

2/49 The masthead fitting supports the two forces shown. Determine the magnitude of **T** which will cause no bending of the mast (zero moment) at point O.

Ans. $T = 4.04$ kN

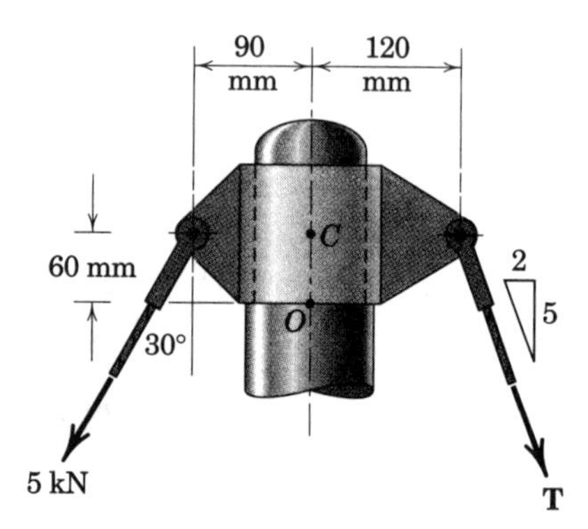

Problem 2/49

2/50 The rocker arm BD of an automobile engine is supported by a nonrotating shaft at C. If the design value of the force exerted by the pushrod AB on the rocker arm is 360 N, determine the force which the valve stem DE must exert at D in order for the combined moment about point C to be zero. Compute the resultant of these two forces exerted on the rocker arm. Note that the points B, C, and D lie on a horizontal line and that both the pushrod and valve stem exert forces along their axes.

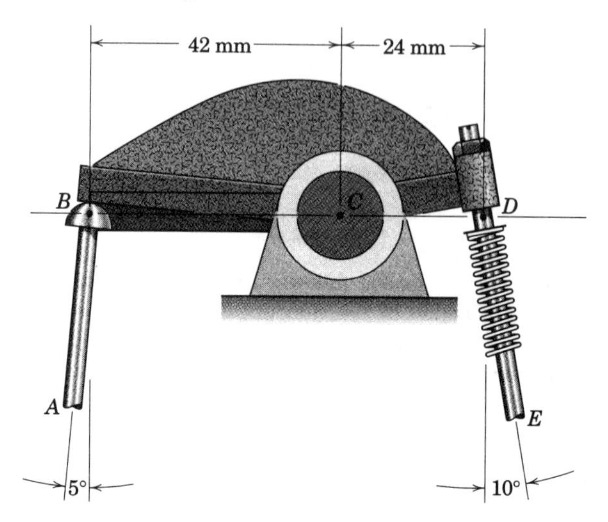

Problem 2/50

2/51 The 120-N force is applied as shown to one end of the curved wrench. If $\alpha = 30°$, calculate the moment of F about the center O of the bolt. Determine the value of α which would maximize the moment about O; state the value of this maximum moment.

Ans. $M_O = 41.5 \text{ N·m CW}$
$\alpha = 33.2°$, $(M_O)_{max} = 41.6 \text{ N·m CW}$

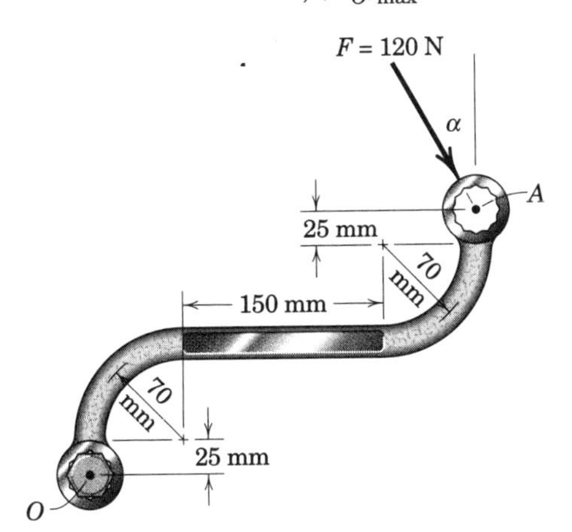

Problem 2/51

2/52 The piston, connecting rod, and crankshaft of a diesel engine are shown in the figure. The crank throw OA is half the stroke of 200 mm, and the length AB of the rod is 350 mm. For the position indicated, the rod is under a compression along AB of 16 kN. Determine the moment M of this force about the crankshaft axis O.

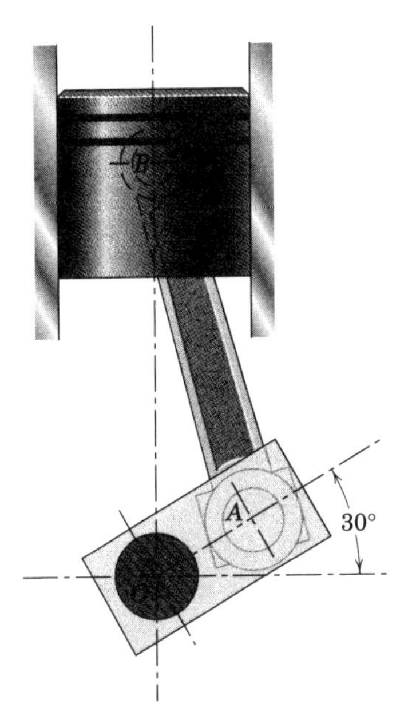

Problem 2/52

Chapter

3

EQUILIBRIUM

CHAPTER OUTLINE

3/1 INTRODUCTION

Statics deals primarily with the description of the force conditions necessary and sufficient to maintain the equilibrium of engineering structures. This chapter on equilibrium, therefore, constitutes the most important part of statics, and the procedures developed here form the basis for solving problems in both statics and dynamics. We will make continual use of the concepts developed in Chapter 2 involving forces, moments, couples, and resultants as we apply the principles of equilibrium.

When a body is in equilibrium, the resultant of *all* forces acting on it is zero. Thus, the resultant force **R** and the resultant couple **M** are both zero, and we have the equilibrium equations

$$\mathbf{R} = \Sigma \mathbf{F} = \mathbf{0} \qquad \mathbf{M} = \Sigma \mathbf{M} = \mathbf{0} \tag{3/1}$$

These requirements are both necessary and sufficient conditions for equilibrium.

All physical bodies are three-dimensional, but we can treat many of them as two-dimensional when the forces to which they are subjected act in a single plane or can be projected onto a single plane. When this simplification is not possible, the problem must be treated as three-dimensional. We will follow the arrangement used in Chapter 2, and discuss in Section A the equilibrium of bodies subjected to two-dimensional force systems and in Section B the equilibrium of bodies subjected to three-dimensional force systems.

SECTION A. EQUILIBRIUM IN TWO DIMENSIONS

3/2 System Isolation and the Free-Body Diagram

Before we apply Eqs. 3/1, we must define unambiguously the particular body or mechanical system to be analyzed and represent clearly and completely *all* forces acting *on* the body. Omission of a force which acts *on* the body in question, or inclusion of a force which does not act *on* the body, will give erroneous results.

A *mechanical system* is defined as a body or group of bodies which can be conceptually isolated from all other bodies. A system may be a single body or a combination of connected bodies. The bodies may be rigid or nonrigid. The system may also be an identifiable fluid mass, either liquid or gas, or a combination of fluids and solids. In statics we study primarily forces which act on rigid bodies at rest, although we also study forces acting on fluids in equilibrium.

Once we decide which body or combination of bodies to analyze, we then treat this body or combination as a single body *isolated* from all surrounding bodies. This isolation is accomplished by means of the ***free-body diagram***, which is a diagrammatic representation of the isolated system treated as a single body. The diagram shows all forces applied to the system by mechanical contact with other bodies, which are imagined to be removed. If appreciable body forces are present, such as gravitational or magnetic attraction, then these forces must also be shown on the free-body diagram of the isolated system. Only after such a diagram has been carefully drawn should the equilibrium equations be written. Because of its critical importance, we emphasize here that

the free-body diagram is the most important single step in the solution of problems in mechanics.

Before attempting to draw a free-body diagram, we must recall the basic characteristics of force. These characteristics were described in Art. 2/2, with primary attention focused on the vector properties of force. Forces can be applied either by direct physical contact or by remote action. Forces can be either internal or external to the system under consideration. Application of force is accompanied by reactive force, and both applied and reactive forces may be either concentrated or distributed. The principle of transmissibility permits the treatment of force as a sliding vector as far as its external effects on a rigid body are concerned.

We will now use these force characteristics to develop conceptual models of isolated mechanical systems. These models enable us to write the appropriate equations of equilibrium, which can then be analyzed.

Modeling the Action of Forces

Figure 3/1 shows the common types of force application on mechanical systems for analysis in two dimensions. Each example shows the force exerted *on* the body to be *isolated, by* the body to be *removed.* Newton's third law, which notes the existence of an equal and opposite reaction to every action, must be carefully observed. The force exerted *on* the body in question *by* a contacting or supporting member is always in the sense to oppose the movement of the isolated body which would occur if the contacting or supporting body were removed.

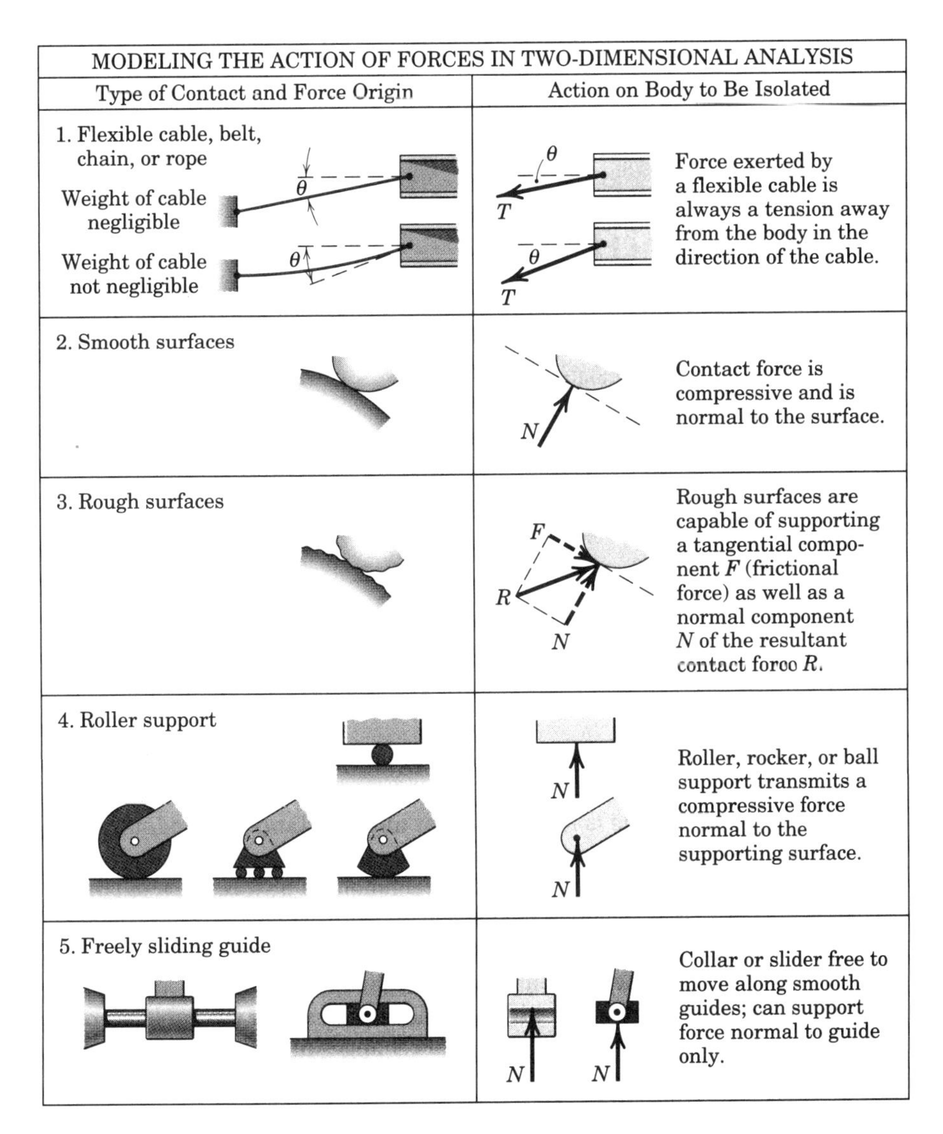

Figure 3/1

MODELING THE ACTION OF FORCES IN TWO-DIMENSIONAL ANALYSIS (*cont.*)	
Type of Contact and Force Origin	Action on Body to Be Isolated
6. Pin connection 	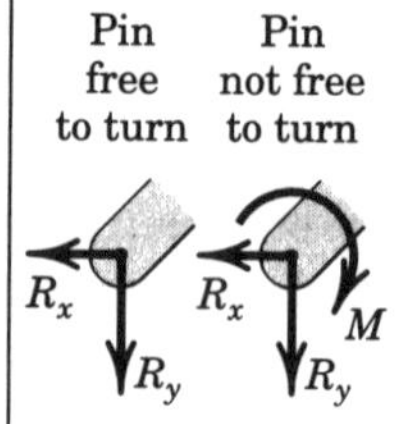 Pin free to turn; Pin not free to turn. A freely hinged pin connection is capable of supporting a force in any direction in the plane normal to the axis; usually shown as two components R_x and R_y. A pin not free to turn may also support a couple M.
7. Built-in or fixed support 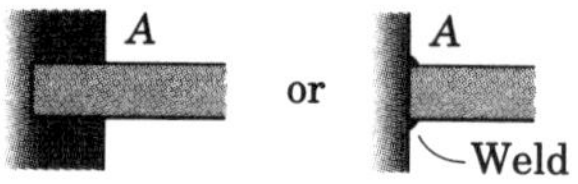 (A or A, Weld)	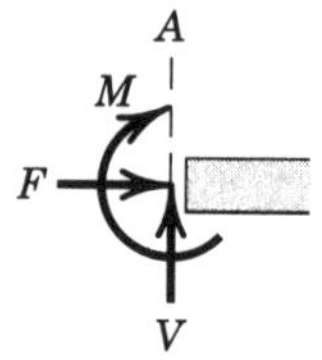 A built-in or fixed support is capable of supporting an axial force F, a transverse force V (shear force), and a couple M (bending moment) to prevent rotation.
8. Gravitational attraction 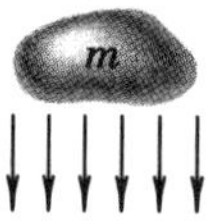	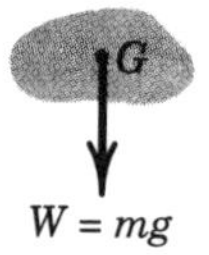 $W = mg$. The resultant of gravitational attraction on all elements of a body of mass m is the weight $W = mg$ and acts toward the center of the earth through the center mass G.
9. Spring action 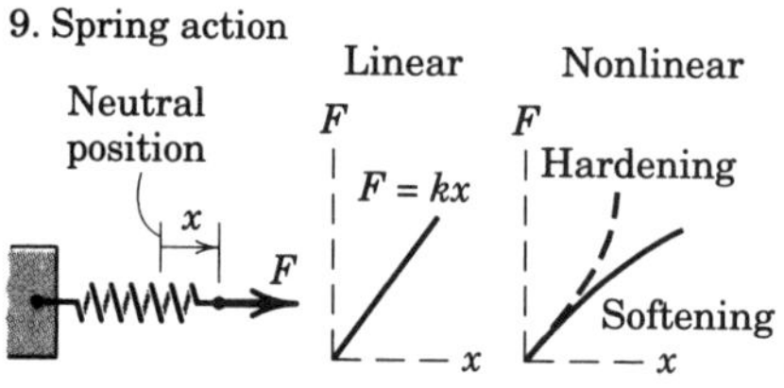 (Neutral position; Linear: $F = kx$; Nonlinear: Hardening, Softening)	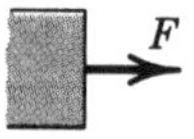 Spring force is tensile if spring is stretched and compressive if compressed. For a linearly elastic spring the stiffness k is the force required to deform the spring a unit distance.

Figure 3/1, continued

In Fig. 3/1, Example 1 depicts the action of a flexible cable, belt, rope, or chain on the body to which it is attached. Because of its flexibility, a rope or cable is unable to offer any resistance to bending, shear, or compression and therefore exerts only a tension force in a direction tangent to the cable at its point of attachment. The force exerted *by* the cable *on* the body to which it is attached is always *away* from the body. When the tension T is large compared with the weight of the cable, we may assume that the cable forms a straight line. When the cable weight is not negligible compared with its tension, the sag of the cable becomes important, and the tension in the cable changes direction and magnitude along its length.

When the smooth surfaces of two bodies are in contact, as in Example 2, the force exerted by one on the other is *normal* to the tangent to the surfaces and is compressive. Although no actual surfaces are perfectly smooth, we can assume this to be so for practical purposes in many instances.

When mating surfaces of contacting bodies are rough, as in Example 3, the force of contact is not necessarily normal to the tangent to the surfaces, but may be resolved into a *tangential* or *frictional component* F and a *normal component* N.

Example 4 illustrates a number of forms of mechanical support which effectively eliminate tangential friction forces. In these cases the net reaction is normal to the supporting surface.

Example 5 shows the action of a smooth guide on the body it supports. There cannot be any resistance parallel to the guide.

Example 6 illustrates the action of a pin connection. Such a connection can support force in any direction normal to the axis of the pin. We usually represent this action in terms of two rectangular components. The correct sense of these components in a specific problem depends on how the member is loaded. When not otherwise initially known, the sense is arbitrarily assigned and the equilibrium equations are then written. If the solution of these equations yields a positive algebraic sign for the force component, the assigned sense is correct. A negative sign indicates the sense is opposite to that initially assigned.

If the joint is free to turn about the pin, the connection can support only the force R. If the joint is not free to turn, the connection can also support a resisting couple M. The sense of M is arbitrarily shown here, but the true sense depends on how the member is loaded.

Example 7 shows the resultants of the rather complex distribution of force over the cross section of a slender bar or beam at a built-in or fixed support. The sense of the reactions F and V and the bending couple M in a given problem depends, of course, on how the member is loaded.

One of the most common forces is that due to gravitational attraction, Example 8. This force affects all elements of mass in a body and is, therefore, distributed throughout it. The resultant of the gravitational forces on all elements is the weight $W = mg$ of the body, which passes through the center of mass G and is directed toward the center of the earth for earthbound structures. The location of G is frequently obvious from the geometry of the body, particularly where there is symmetry. When the location is not readily apparent, it must be determined by experiment or calculations.

Similar remarks apply to the remote action of magnetic and electric forces. These forces of remote action have the same overall effect on a rigid body as forces of equal magnitude and direction applied by direct external contact.

Example 9 illustrates the action of a *linear* elastic spring and of a *nonlinear* spring with either hardening or softening characteristics. The force exerted by a linear spring, in tension or compression, is given by $F = kx$, where k is the *stiffness* of the spring and x is its deformation measured from the neutral or undeformed position.

The representations in Fig. 3/1 are *not* free-body diagrams, but are merely elements used to construct free-body diagrams. Study these nine conditions and identify them in the problem work so that you can draw the correct free-body diagrams.

Construction of Free-Body Diagrams

The full procedure for drawing a free-body diagram which isolates a body or system consists of the following steps.

Step 1. Decide which system to isolate. The system chosen should usually involve one or more of the desired unknown quantities.

Step 2. Next isolate the chosen system by drawing a diagram which represents its *complete external boundary*. This boundary defines the isolation of the system from *all* other attracting or contacting bodies, which are considered removed. This step is often the most crucial of all. Make certain that you have *completely isolated* the system before proceeding with the next step.

Step 3. Identify all forces which act *on* the isolated system as applied *by* the removed contacting and attracting bodies, and represent them in their proper positions on the diagram of the isolated system. Make a systematic traverse of the entire boundary to identify all contact forces. Include body forces such as weights, where appreciable. Represent all known forces by vector arrows, each with its proper magnitude, direction, and sense indicated. Each unknown force should be represented by a vector arrow with the unknown magnitude or direction indicated by symbol. If the sense of the vector is also unknown, you must arbitrarily assign a sense. The subsequent calculations with the equilibrium equations will yield a positive quantity if the correct sense was assumed and a negative quantity if the incorrect sense was assumed. It is necessary to be *consistent* with the assigned characteristics of unknown forces throughout all of the calculations. If you are consistent, the solution of the equilibrium equations will reveal the correct senses.

Step 4. Show the choice of coordinate axes directly on the diagram. Pertinent dimensions may also be represented for convenience. Note, however, that the free-body diagram serves the purpose of focusing attention on the action of the external forces, and therefore the diagram should not be cluttered with excessive extraneous information. Clearly distinguish force arrows from arrows representing quantities other than forces. For this purpose a colored pencil may be used.

Completion of the foregoing four steps will produce a correct free-body diagram to use in applying the governing equations, both in statics and in dynamics. Be careful not to omit from the free-body diagram certain forces which may not appear at first glance to be needed in the calculations. It is only through *complete* isolation and a systematic representation of *all* external forces that a reliable accounting of the effects of all applied and reactive forces can be made. Very often a force which at first glance may not appear to influence a desired result does indeed have an influence. Thus, the only safe procedure is to include on the free-body diagram all forces whose magnitudes are not obviously negligible.

The free-body method is extremely important in mechanics because it ensures an accurate definition of a mechanical system and focuses attention on the exact meaning and application of the force laws of statics and dynamics. Review the foregoing four steps for constructing a free-body diagram while studying the sample free-body diagrams shown in Fig. 3/2 and the Sample Problems which appear at the end of the next article.

Examples of Free-Body Diagrams

Figure 3/2 gives four examples of mechanisms and structures together with their correct free-body diagrams. Dimensions and magnitudes are omitted for clarity. In each case we treat the entire system as a single body, so that the internal forces are not shown. The characteristics of the various types of contact forces illustrated in Fig. 3/1 are used in the four examples as they apply.

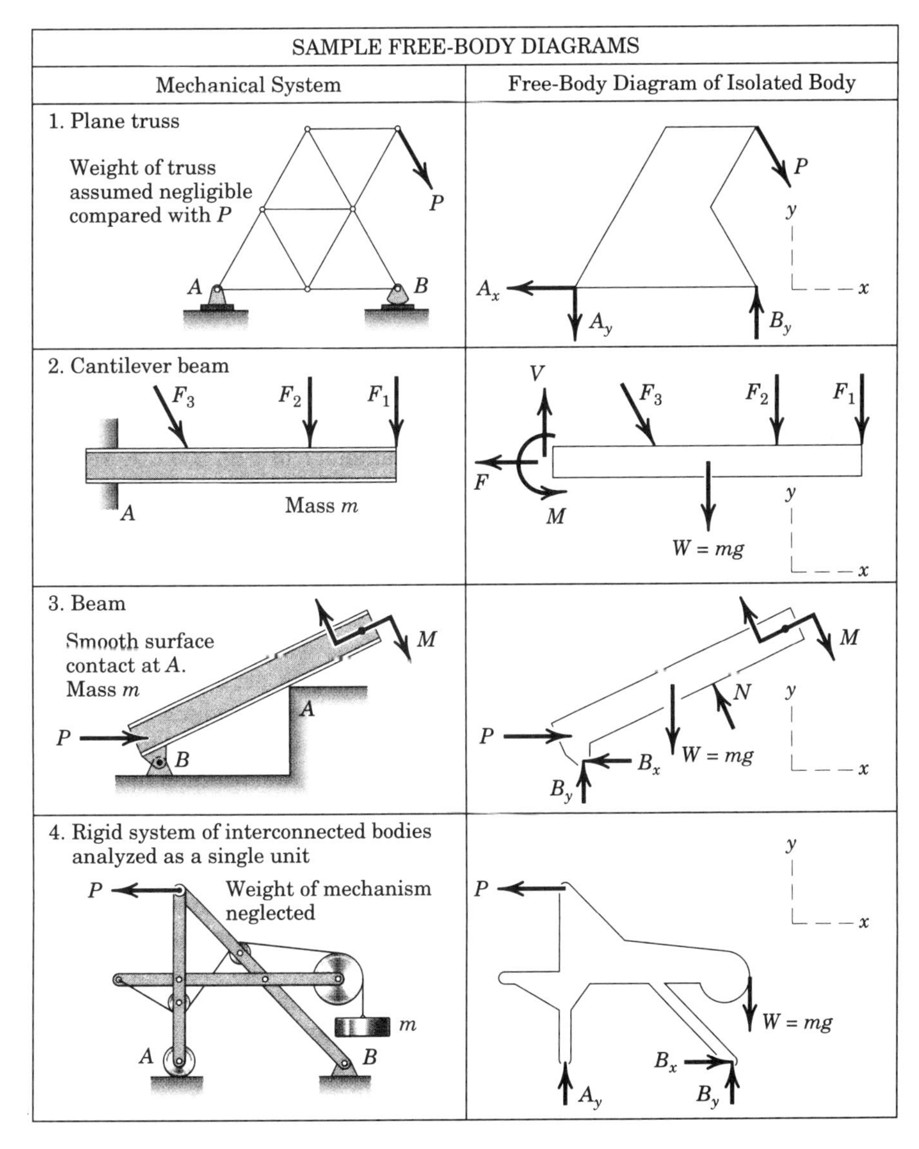

Figure 3/2

In Example 1 the truss is composed of structural elements which, taken all together, constitute a rigid framework. Thus, we may remove the entire truss from its supporting foundation and treat it as a single rigid body. In addition to the applied external load P, the free-body diagram must include the reactions on the truss at A and B. The rocker at B can support a vertical force only, and this force is transmitted to the structure at B (Example 4 of Fig. 3/1). The pin connection at A (Example 6 of Fig. 3/1) is capable of supplying both a horizontal and a vertical force component to the truss. If the total weight of the truss members is appreciable compared with P and the forces at A and B, then the weights of the members must be included on the free-body diagram as external forces.

In this relatively simple example it is clear that the vertical component A_y must be directed down to prevent the truss from rotating clockwise about B. Also, the horizontal component A_x will be to the left to keep the truss from moving to the right under the influence of the horizontal component of P. Thus, in constructing the free-body diagram for this simple truss, we can easily perceive the correct sense of each of the components of force exerted *on* the truss *by* the foundation at A and can, therefore, represent its correct physical sense on the diagram. When the correct physical sense of a force or its component is not easily recognized by direct observation, it must be assigned arbitrarily, and the correctness of or error in the assignment is determined by the algebraic sign of its calculated value.

In Example 2 the cantilever beam is secured to the wall and subjected to three applied loads. When we isolate that part of the beam to the right of the section at A, we must include the reactive forces applied *to* the beam *by* the wall. The resultants of these reactive forces are shown acting on the section of the beam (Example 7 of Fig. 3/1). A vertical force V to counteract the excess of downward applied force is shown, and a tension F to balance the excess of applied force to the right must also be included. Then, to prevent the beam from rotating about A, a counterclockwise couple M is also required. The weight mg of the beam must be represented through the mass center (Example 8 of Fig. 3/1).

In the free-body diagram of Example 2, we have represented the somewhat complex system of forces which actually act on the cut section of the beam by the equivalent force–couple system in which the force is broken down into its vertical component V (shear force) and its horizontal component F (tensile force). The couple M is the bending moment in the beam. The free-body diagram is now complete and shows the beam in equilibrium under the action of six forces and one couple.

In Example 3 the weight $W = mg$ is shown acting through the center of mass of the beam, whose location is assumed known (Example 8 of Fig. 3/1). The force exerted by the corner A on the beam is normal to the smooth surface of the beam (Example 2 of Fig. 3/1). To perceive this action more clearly, visualize an enlargement of the contact point A, which would appear somewhat rounded, and consider the force exerted by this rounded corner on the straight surface of the beam, which is assumed to be smooth. If the contacting surfaces at the corner were not smooth, a tangential frictional component of force could exist. In addition to the applied force P and couple M, there is the pin connection at B, which exerts both an x- and a y-component of force on the beam. The positive senses of these components are assigned arbitrarily.

In Example 4 the free-body diagram of the entire isolated mechanism contains three unknown forces if the loads mg and P are known. Any one of many internal configurations for securing the cable leading from the mass m would be possible without affecting the external response of the mechanism as a whole, and this fact is brought out by the free-body diagram. This hypothetical example is used to show that the forces internal to a rigid assembly of members do not influence the values of the external reactions.

We use the free-body diagram in writing the equilibrium equations, which are discussed in the next article. When these equations are solved, some of the calculated force magnitudes may be zero. This would indicate that the assumed force does not exist. In Example 1 of Fig. 3/2, any of the reactions A_x, A_y, or B_y can be zero for specific values of the truss geometry and of the magnitude, direction, and sense of the applied load P. A zero reaction force is often difficult to identify by inspection, but can be determined by solving the equilibrium equations.

Similar comments apply to calculated force magnitudes which are negative. Such a result indicates that the actual sense is the opposite of the assumed sense. The assumed positive senses of B_x and B_y in Example 3 and B_y in Example 4 are shown on the free-body diagrams. The correctness of these assumptions is proved or disproved according to whether the algebraic signs of the computed forces are plus or minus when the calculations are carried out in an actual problem.

The isolation of the mechanical system under consideration is a crucial step in the formulation of the mathematical model. The most important aspect to the correct construction of the all-important free-body diagram is the clear-cut and unambiguous decision as to what is included and what is excluded. This decision becomes unambiguous only when the boundary of the free-body diagram represents a complete traverse of the body or system of bodies to be isolated, starting at some arbitrary point on the boundary and returning to that same point. The system within this closed boundary is the isolated free body, and all contact forces and all body forces transmitted to the system across the boundary must be accounted for.

The following exercises provide practice with drawing free-body diagrams. This practice is helpful before using such diagrams in the application of the principles of force equilibrium in the next article.

FREE-BODY DIAGRAM EXERCISES

3/A In each of the five following examples, the body to be isolated is shown in the left-hand diagram, and an *incomplete* free-body diagram (FBD) of the isolated body is shown on the right. Add whatever forces are necessary in each case to form a complete free-body diagram. The weights of the bodies are negligible unless otherwise indicated. Dimensions and numerical values are omitted for simplicity.

	Body	Incomplete *FBD*
1. Bell crank supporting mass *m* with pin support at *A*.	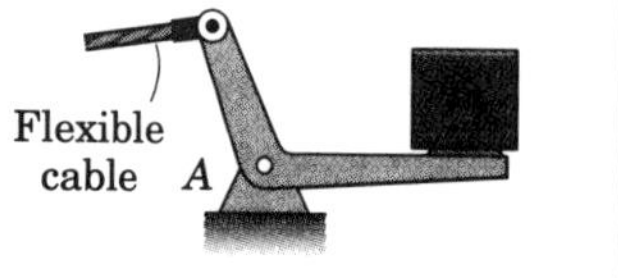 Flexible cable *A*	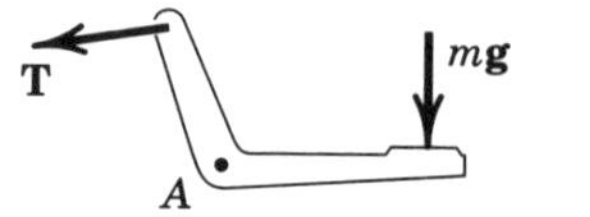 **T** *m***g** *A*
2. Control lever applying torque to shaft at *O*.	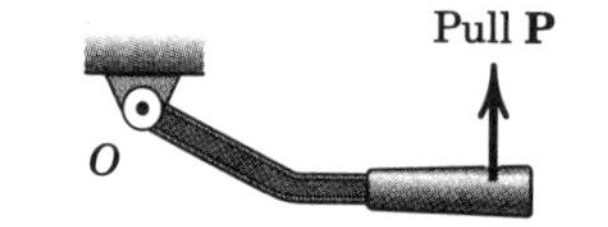 Pull **P** *O*	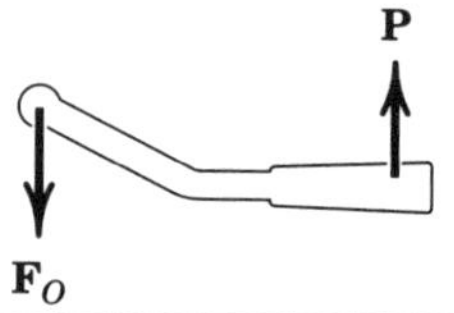 **P** $\mathbf{F}_O$
3. Boom *OA*, of negligible mass compared with mass *m*. Boom hinged at *O* and supported by hoisting cable at *B*.	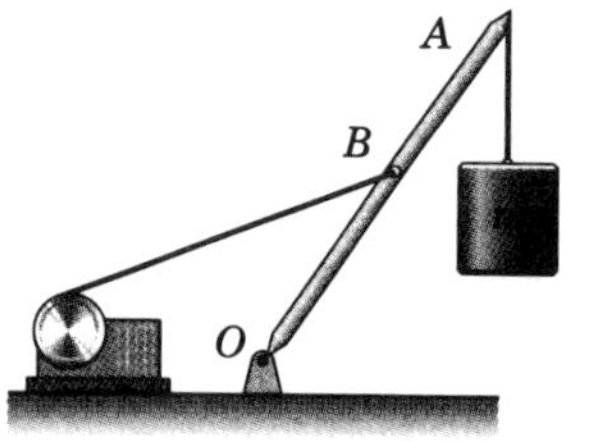 *A* *B* *O*	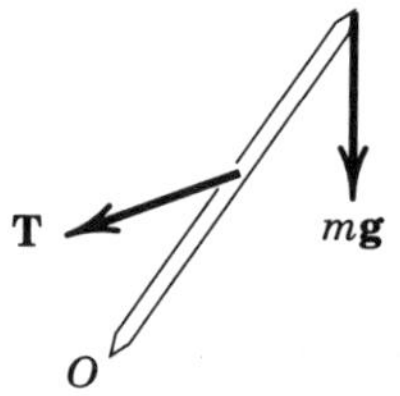 **T** *m***g** *O*
4. Uniform crate of mass *m* leaning against smooth vertical wall and supported on a rough horizontal surface.	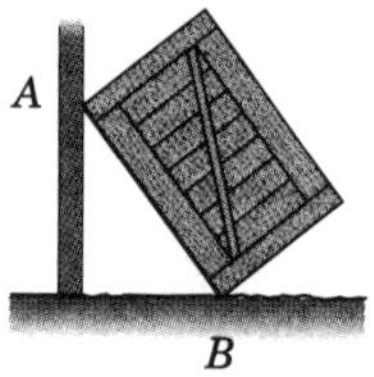 *A* *B*	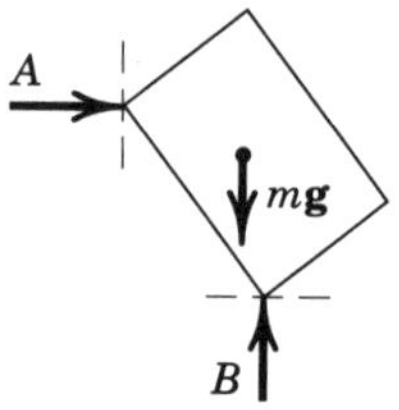 *A* *m***g** *B*
5. Loaded bracket supported by pin connection at *A* and fixed pin in smooth slot at *B*.	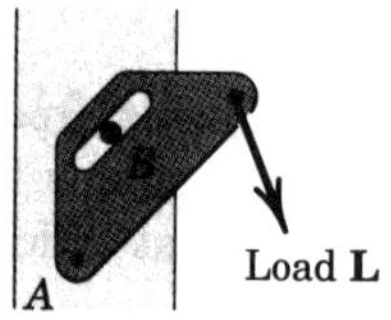 *B* *A* Load **L**	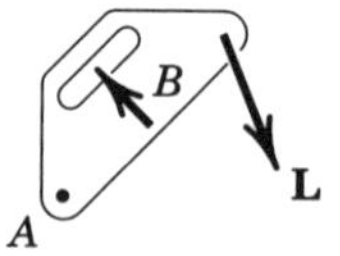 *B* **L** *A*

Figure 3/A

3/B In each of the five following examples, the body to be isolated is shown in the left-hand diagram, and either a *wrong* or an *incomplete* free-body diagram (FBD) is shown on the right. Make whatever changes or additions are necessary in each case to form a correct and complete free-body diagram. The weights of the bodies are negligible unless otherwise indicated. Dimensions and numerical values are omitted for simplicity.

	Body	Wrong or Incomplete *FBD*
1. Lawn roller of mass m being pushed up incline θ.	 $\mathbf{P}$, θ	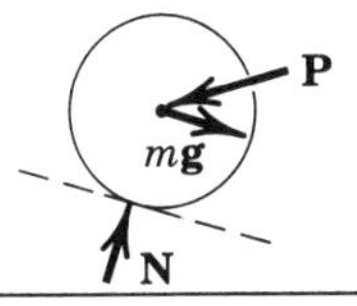 $\mathbf{P}$, $m\mathbf{g}$, $\mathbf{N}$
2. Prybar lifting body A having smooth horizontal surface. Bar rests on horizontal rough surface.	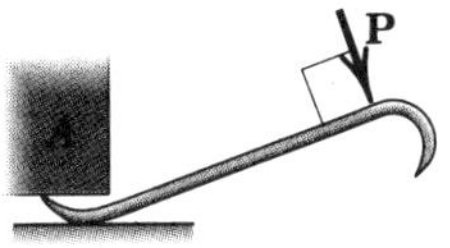 A, $\mathbf{P}$	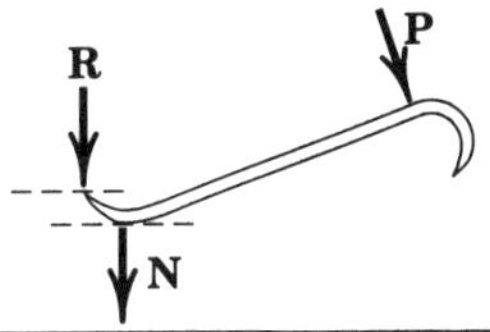 $\mathbf{R}$, $\mathbf{P}$, $\mathbf{N}$
3. Uniform pole of mass m being hoisted into position by winch. Horizontal supporting surface notched to prevent slipping of pole.	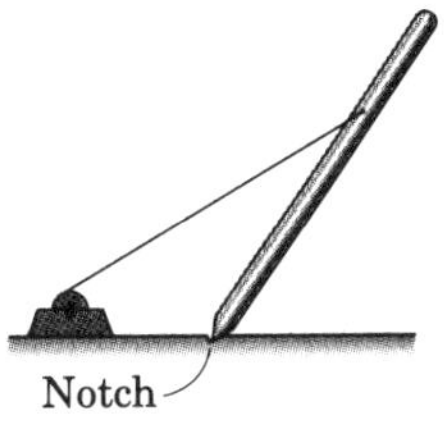 Notch	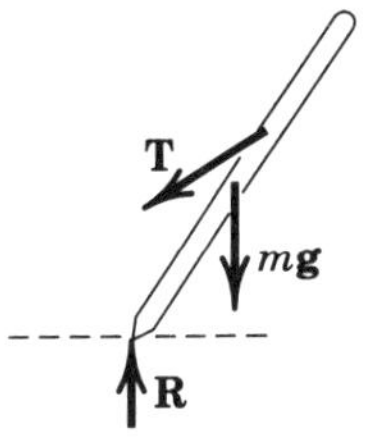 $\mathbf{T}$, $m\mathbf{g}$, $\mathbf{R}$
4. Supporting angle bracket for frame; Pin joints.	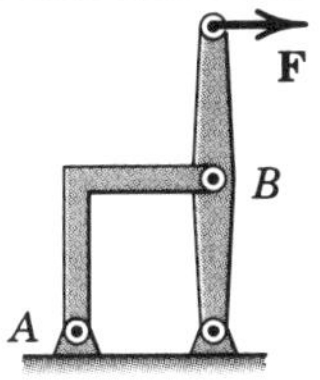 $\mathbf{F}$, B, A	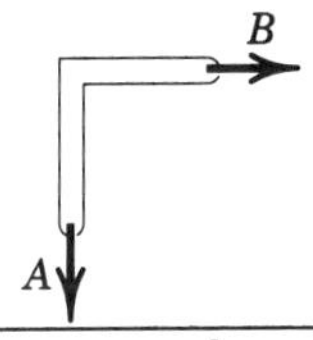 B, A
5. Bent rod welded to support at A and subjected to two forces and couple.	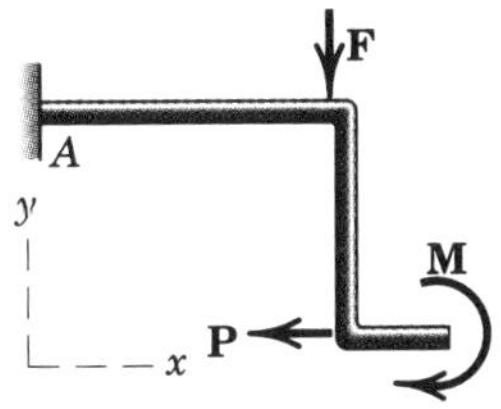 $\mathbf{F}$, A, y, $\mathbf{M}$, x, $\mathbf{P}$	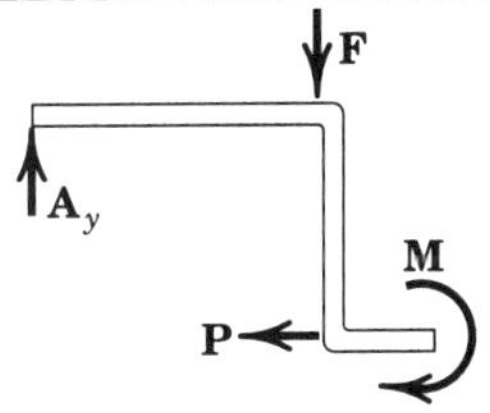 $\mathbf{F}$, $\mathbf{A}_y$, $\mathbf{M}$, $\mathbf{P}$

Figure 3/B

3/C Draw a complete and correct free-body diagram of each of the bodies designated in the statements. The weights of the bodies are significant only if the mass is stated. All forces, known and unknown, should be labeled. (*Note*: The sense of some reaction components cannot always be determined without numerical calculation.)

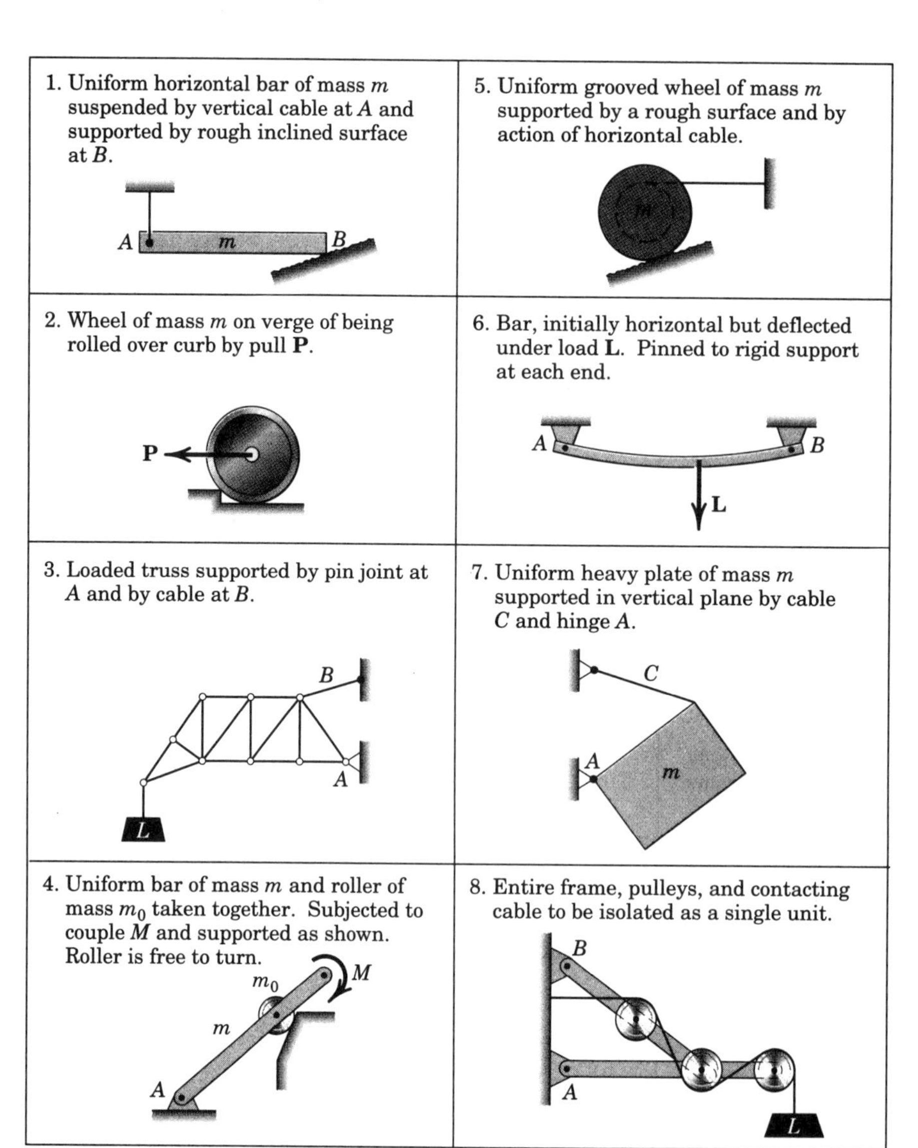

Figure 3/C

3/3 Equilibrium Conditions

In Art. 3/1 we defined equilibrium as the condition in which the resultant of all forces and moments acting on a body is zero. Stated in another way, a body is in equilibrium if all forces and moments applied to it are in balance. These requirements are contained in the vector equations of equilibrium, Eqs. 3/1, which in two dimensions may be written in scalar form as

$$\Sigma F_x = 0 \qquad \Sigma F_y = 0 \qquad \Sigma M_O = 0 \tag{3/2}$$

The third equation represents the zero sum of the moments of all forces about any point O on or off the body. Equations 3/2 are the necessary and sufficient conditions for complete equilibrium in two dimensions. They are necessary conditions because, if they are not satisfied, there can be no force or moment balance. They are sufficient because once they are satisfied, there can be no imbalance, and equilibrium is assured.

The equations relating force and acceleration for rigid-body motion are developed in *Vol. 2 Dynamics* from Newton's second law of motion. These equations show that the acceleration of the mass center of a body is proportional to the resultant force $\Sigma\mathbf{F}$ acting on the body. Consequently, if a body moves with constant velocity (zero acceleration), the resultant force on it must be zero, and the body may be treated as in a state of translational equilibrium.

For complete equilibrium in two dimensions, all three of Eqs. 3/2 must hold. However, these conditions are independent requirements, and one may hold without another. Take, for example, a body which slides along a horizontal surface with increasing velocity under the action of applied forces. The force–equilibrium equations will be satisfied in the vertical direction where the acceleration is zero, but not in the horizontal direction. Also, a body, such as a flywheel, which rotates about its fixed mass center with increasing angular speed is not in rotational equilibrium, but the two force–equilibrium equations will be satisfied.

Categories of Equilibrium

Applications of Eqs. 3/2 fall naturally into a number of categories which are easily identified. The categories of force systems acting on bodies in two-dimensional equilibrium are summarized in Fig. 3/3 and are explained further as follows.

Category 1, equilibrium of collinear forces, clearly requires only the one force equation in the direction of the forces (x-direction), since all other equations are automatically satisfied.

Category 2, equilibrium of forces which lie in a plane (x-y plane) and are concurrent at a point O, requires the two force equations only, since the moment sum about O, that is, about a z-axis through O, is necessarily zero. Included in this category is the case of the equilibrium of a particle.

Category 3, equilibrium of parallel forces in a plane, requires the one force equation in the direction of the forces (x-direction) and one moment equation about an axis (z-axis) normal to the plane of the forces.

CATEGORIES OF EQUILIBRIUM IN TWO DIMENSIONS		
Force System	Free-Body Diagram	Independent Equations
1. Collinear		$\Sigma F_x = 0$
2. Concurrent at a point		$\Sigma F_x = 0$ $\Sigma F_y = 0$
3. Parallel		$\Sigma F_x = 0 \quad \Sigma M_z = 0$
4. General		$\Sigma F_x = 0 \quad \Sigma M_z = 0$ $\Sigma F_y = 0$

Figure 3/3

Category 4, equilibrium of a general system of forces in a plane (x-y), requires the two force equations in the plane and one moment equation about an axis (z-axis) normal to the plane.

Two- and Three-Force Members

You should be alert to two frequently occurring equilibrium situations. The first situation is the equilibrium of a body under the action of two forces only. Two examples are shown in Fig. 3/4, and we see that for such a *two-force member* to be in equilibrium, the forces must be *equal, opposite,* and *collinear*. The shape of the member does not affect this simple requirement. In the illustrations cited, we consider the weights of the members to be negligible compared with the applied forces.

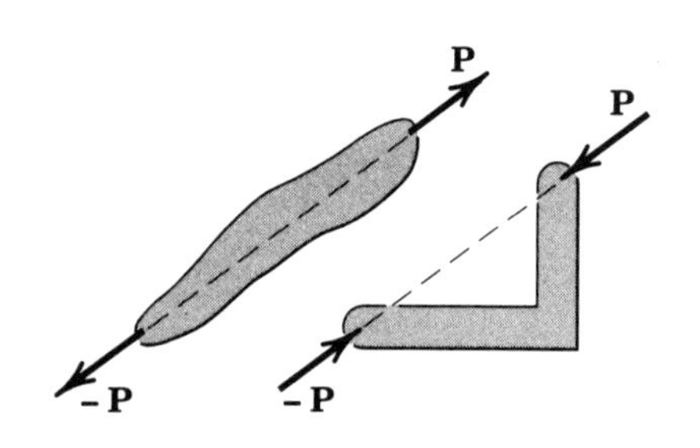

Two-force members

Figure 3/4

The second situation is a *three-force member*, which is a body under the action of three forces, Fig. 3/5a. We see that equilibrium requires the lines of action of the three forces to be *concurrent*. If they were not concurrent, then one of the forces would exert a resultant moment about the point of intersection of the other two, which would violate the requirement of zero moment about every point. The only exception occurs when the three forces are parallel. In this case we may consider the point of concurrency to be at infinity.

The principle of the concurrency of three forces in equilibrium is of considerable use in carrying out a graphical solution of the force equations. In this case the polygon of forces is drawn and made to close, as shown in Fig. 3/5*b*. Frequently, a body in equilibrium under the action of more than three forces may be reduced to a three-force member by a combination of two or more of the known forces.

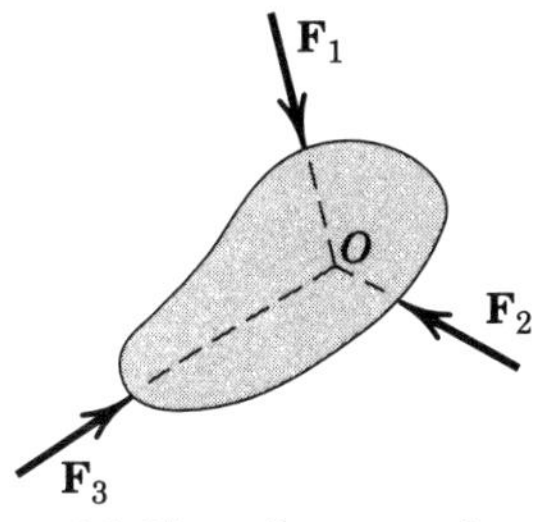

(*a*) Three-force member

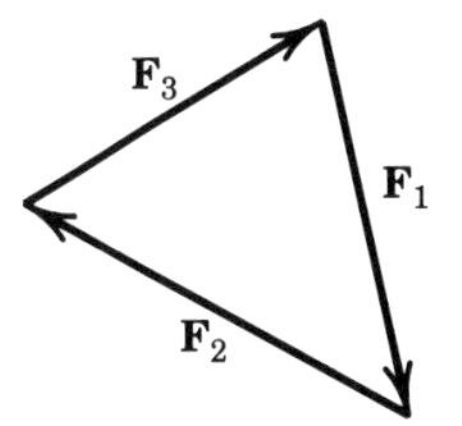

(*b*) Closed polygon satisfies $\Sigma\mathbf{F} = \mathbf{0}$

Figure 3/5

Alternative Equilibrium Equations

In addition to Eqs. 3/2, there are two other ways to express the general conditions for the equilibrium of forces in two dimensions. The first way is illustrated in Fig. 3/6, parts (*a*) and (*b*). For the body shown in Fig. 3/6*a*, if $\Sigma M_A = 0$, then the resultant, if it still exists, cannot be a couple, but must be a force **R** passing through A. If now the equation $\Sigma F_x = 0$ holds, where the x-direction is arbitrary, it follows from Fig. 3/6*b* that the resultant force **R**, if it still exists, not only must pass through A, but also must be perpendicular to the x-direction as shown. Now, if $\Sigma M_B = 0$, where B is any point such that the line AB is not perpendicular to the x-direction, we see that **R** must be zero, and thus the body is in equilibrium. Therefore, an alternative set of equilibrium equations is

$$\Sigma F_x = 0 \qquad \Sigma M_A = 0 \qquad \Sigma M_B = 0$$

where the two points A and B must not lie on a line perpendicular to the x-direction.

A third formulation of the equilibrium conditions may be made for a coplanar force system. This is illustrated in Fig. 3/6, parts (*c*) and (*d*). Again, if $\Sigma M_A = 0$ for any body such as that shown in Fig. 3/6*c*, the resultant, if any, must be a force **R** through A. In addition, if $\Sigma M_B = 0$, the resultant, if one still exists, must pass through B as shown in Fig. 3/6*d*. Such a force cannot exist, however, if $\Sigma M_C = 0$, where C is not

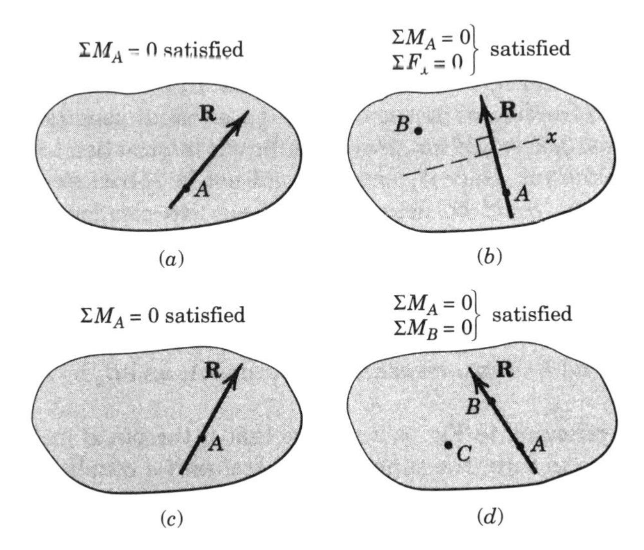

Figure 3/6

collinear with A and B. Thus, we may write the equations of equilibrium as

$$\Sigma M_A = 0 \qquad \Sigma M_B = 0 \qquad \Sigma M_C = 0$$

where A, B, and C are any three points not on the same straight line.

When equilibrium equations are written which are not independent, redundant information is obtained, and a correct solution of the equations will yield $0 = 0$. For example, for a general problem in two dimensions with three unknowns, three moment equations written about three points which lie on the same straight line are not independent. Such equations will contain duplicated information, and solution of two of them can at best determine two of the unknowns, with the third equation merely verifying the identity $0 = 0$.

Constraints and Statical Determinacy

The equilibrium equations developed in this article are both necessary and sufficient conditions to establish the equilibrium of a body. However, they do not necessarily provide all the information required to calculate all the unknown forces which may act on a body in equilibrium. Whether the equations are adequate to determine all the unknowns depends on the characteristics of the constraints against possible movement of the body provided by its supports. By *constraint* we mean the restriction of movement.

In Example 4 of Fig. 3/1 the roller, ball, and rocker provide constraint normal to the surface of contact, but none tangent to the surface. Thus, a tangential force cannot be supported. For the collar and slider of Example 5, constraint exists only normal to the guide. In Example 6 the fixed-pin connection provides constraint in both directions, but offers no resistance to rotation about the pin unless the pin is not free to turn. The fixed support of Example 7, however, offers constraint against rotation as well as lateral movement.

If the rocker which supports the truss of Example 1 in Fig. 3/2 were replaced by a pin joint, as at A, there would be one additional constraint beyond those required to support an equilibrium configuration with no freedom of movement. The three scalar conditions of equilibrium, Eqs. 3/2, would not provide sufficient information to determine all four unknowns, since A_x and B_x could not be solved for separately; only their sum could be determined. These two components of force would be dependent on the deformation of the members of the truss as influenced by their corresponding stiffness properties. The horizontal reactions A_x and B_x would also depend on any initial deformation required to fit the dimensions of the structure to those of the foundation between A and B. Thus, we cannot determine A_x and B_x by a rigid-body analysis.

Again referring to Fig. 3/2, we see that if the pin B in Example 3 were not free to turn, the support could transmit a couple to the beam through the pin. Therefore, there would be four unknown supporting reactions acting on the beam, namely, the force at A, the two components of force at B, and the couple at B. Consequently the three independent scalar equations of equilibrium would not provide enough information to compute all four unknowns.

A rigid body, or rigid combination of elements treated as a single body, which possesses more external supports or constraints than are necessary to maintain an equilibrium position is called *statically indeterminate*. Supports which can be removed without destroying the equilibrium condition of the body are said to be *redundant*. The number of redundant supporting elements present corresponds to the *degree of statical indeterminacy* and equals the total number of unknown external forces, minus the number of available independent equations of equilibrium. On the other hand, bodies which are supported by the minimum number of constraints necessary to ensure an equilibrium configuration are called *statically determinate*, and for such bodies the equilibrium equations are sufficient to determine the unknown external forces.

The problems on equilibrium in this article and throughout *Vol. 1 Statics* are generally restricted to statically determinate bodies where the constraints are just sufficient to ensure a stable equilibrium configuration and where the unknown supporting forces can be completely determined by the available independent equations of equilibrium.

We must be aware of the nature of the constraints before we attempt to solve an equilibrium problem. A body can be recognized as statically indeterminate when there are more unknown external reactions than there are available independent equilibrium equations for the force system involved. It is always well to count the number of unknown variables on a given body and to be certain that an equal number of independent equations can be written; otherwise, effort might be wasted in attempting an impossible solution with the aid of the equilibrium equations only. The unknown variables may be forces, couples, distances, or angles.

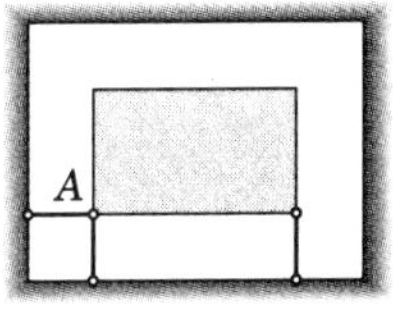

(*a*) Complete fixity
Adequate constraints

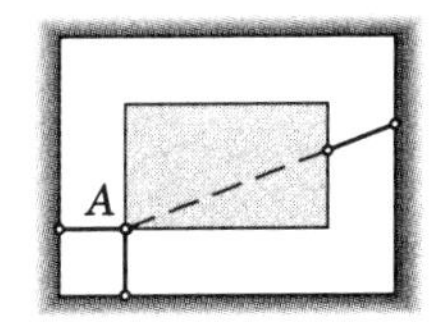

(*b*) Incomplete fixity
Partial constraints

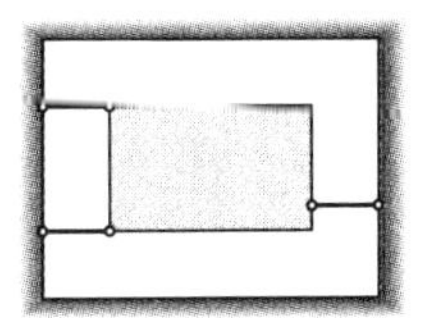
(*c*) Incomplete fixity
Partial constraints

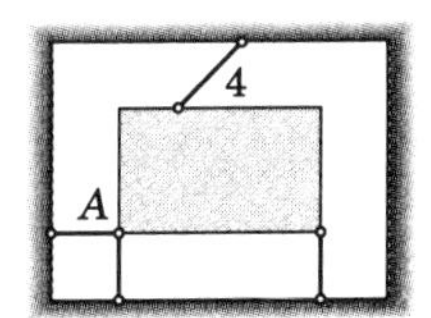

(*d*) Excessive fixity
Redundant constraint

Figure 3/7

Adequacy of Constraints

In discussing the relationship between constraints and equilibrium, we should look further at the question of the adequacy of constraints. The existence of three constraints for a two-dimensional problem does not always guarantee a stable equilibrium configuration. Figure 3/7 shows four different types of constraints. In part *a* of the figure, point *A* of the rigid body is fixed by the two links and cannot move, and the third link prevents any rotation about *A*. Thus, this body is *completely fixed* with three *adequate (proper) constraints*.

In part *b* of the figure, the third link is positioned so that the force transmitted by it passes through point *A* where the other two constraint forces act. Thus, this configuration of constraints can offer no initial resistance to rotation about *A*, which would occur when external loads were applied to the body. We conclude, therefore, that this body is *incompletely fixed* under *partial constraints*.

The configuration in part *c* of the figure gives us a similar condition of incomplete fixity because the three parallel links could offer no initial resistance to a small vertical movement of the body as a result of external loads applied to it in this direction. The constraints in these two examples are often termed *improper*.

In part d of Fig. 3/7 we have a condition of complete fixity, with link 4 acting as a fourth constraint which is unnecessary to maintain a fixed position. Link 4, then, is a *redundant constraint*, and the body is statically indeterminate.

As in the four examples of Fig. 3/7, it is generally possible by direct observation to conclude whether the constraints on a body in two-dimensional equilibrium are adequate (proper), partial (improper), or redundant. As indicated previously, the vast majority of problems in this book are statically determinate with adequate (proper) constraints.

Approach to Solving Problems

The sample problems at the end of this article illustrate the application of free-body diagrams and the equations of equilibrium to typical statics problems. These solutions should be studied thoroughly. In the problem work of this chapter and throughout mechanics, it is important to develop a logical and systematic approach which includes the following steps:

1. Identify clearly the quantities which are known and unknown.
2. Make an unambiguous choice of the body (or system of connected bodies treated as a single body) to be isolated and draw its complete free-body diagram, labeling all external known and unknown but identifiable forces and couples which act on it.
3. Choose a convenient set of reference axes, always using right-handed axes when vector cross products are employed. Choose moment centers with a view to simplifying the calculations. Generally the best choice is one through which as many unknown forces pass as possible. Simultaneous solutions of equilibrium equations are frequently necessary, but can be minimized or avoided by a careful choice of reference axes and moment centers.
4. Identify and state the applicable force and moment principles or equations which govern the equilibrium conditions of the problem. In the following sample problems these relations are shown in brackets and precede each major calculation.
5. Match the number of independent equations with the number of unknowns in each problem.
6. Carry out the solution and check the results. In many problems engineering judgment can be developed by first making a reasonable guess or estimate of the result prior to the calculation and then comparing the estimate with the calculated value.

Sample Problem 3/1

Determine the magnitudes of the forces **C** and **T**, which, along with the other three forces shown, act on the bridge-truss joint.

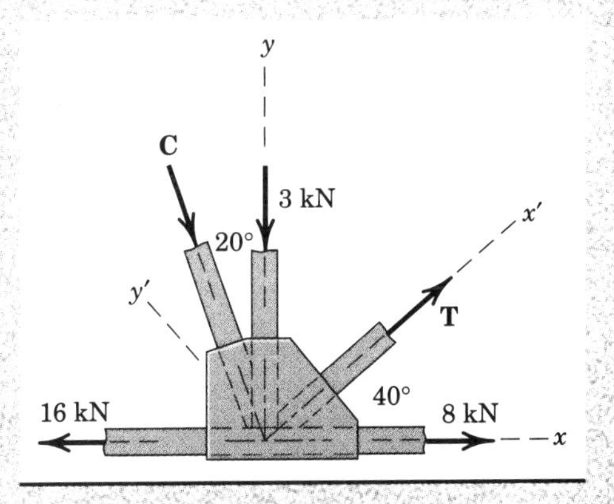

Solution. The given sketch constitutes the free-body diagram of the isolated ① section of the joint in question and shows the five forces which are in equilibrium.

Solution I (scalar algebra). For the x-y axes as shown we have

$[\Sigma F_x = 0]$ $$8 + T \cos 40° + C \sin 20° - 16 = 0$$

$$0.766T + 0.342C = 8 \qquad (a)$$

$[\Sigma F_y = 0]$ $$T \sin 40° - C \cos 20° - 3 = 0$$

$$0.643T - 0.940C = 3 \qquad (b)$$

Simultaneous solution of Eqs. (a) and (b) produces

$$T = 9.09 \text{ kN} \qquad C = 3.03 \text{ kN} \qquad \textit{Ans.}$$

Solution II (scalar algebra). To avoid a simultaneous solution, we may use axes x'-y' with the first summation in the y'-direction to eliminate reference to ② T. Thus,

$[\Sigma F_{y'} = 0]$ $$-C \cos 20° - 3 \cos 40° - 8 \sin 40° + 16 \sin 40° = 0$$

$$C = 3.03 \text{ kN} \qquad \textit{Ans.}$$

$[\Sigma F_{x'} = 0]$ $$T + 8 \cos 40° - 16 \cos 40° - 3 \sin 40° - 3.03 \sin 20° = 0$$

$$T = 9.09 \text{ kN} \qquad \textit{Ans.}$$

Solution III (vector algebra). With unit vectors **i** and **j** in the x- and y-directions, the zero summation of forces for equilibrium yields the vector equation

$[\Sigma \mathbf{F} = \mathbf{0}]$ $$8\mathbf{i} + (T \cos 40°)\mathbf{i} + (T \sin 40°)\mathbf{j} - 3\mathbf{j} + (C \sin 20°)\mathbf{i} - (C \cos 20°)\mathbf{j} - 16\mathbf{i} = \mathbf{0}$$

Equating the coefficients of the **i**- and **j**-terms to zero gives

$$8 + T \cos 40° + C \sin 20° - 16 = 0$$

$$T \sin 40° - 3 - C \cos 20° = 0$$

which are the same, of course, as Eqs. (a) and (b), which we solved above.

Solution IV (geometric). The polygon representing the zero vector sum of the five forces is shown. Equations (a) and (b) are seen immediately to give the projections of the vectors onto the x- and y-directions. Similarly, projections onto the x'- and y'-directions give the alternative equations in Solution II.

A graphical solution is easily obtained. The known vectors are laid off head- ③ to-tail to some convenient scale, and the directions of **T** and **C** are then drawn to close the polygon. The resulting intersection at point P completes the solution, thus enabling us to measure the magnitudes of **T** and **C** directly from the drawing to whatever degree of accuracy we incorporate in the construction.

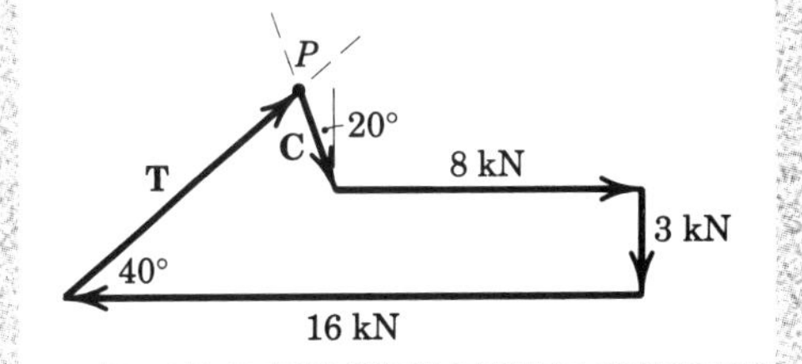

Helpful Hints

① Since this is a problem of concurrent forces, no moment equation is necessary.

② The selection of reference axes to facilitate computation is always an important consideration. Alternatively in this example we could take a set of axes along and normal to the direction of **C** and employ a force summation normal to **C** to eliminate it.

③ The known vectors may be added in any order desired, but they must be added before the unknown vectors.

Sample Problem 3/2

Calculate the tension T in the cable which supports the 500-kg mass with the pulley arrangement shown. Each pulley is free to rotate about its bearing, and the weights of all parts are small compared with the load. Find the magnitude of the total force on the bearing of pulley C.

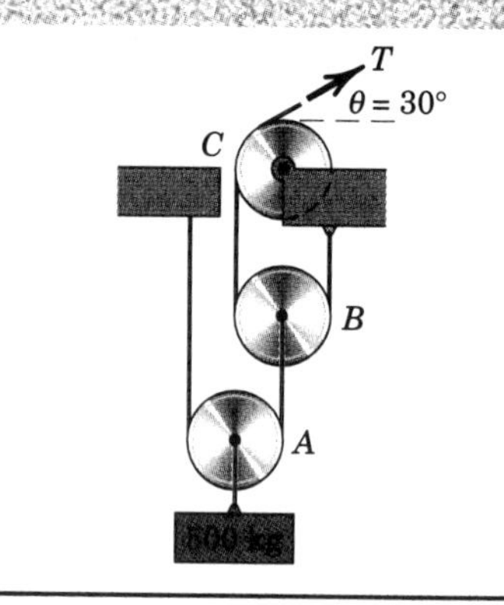

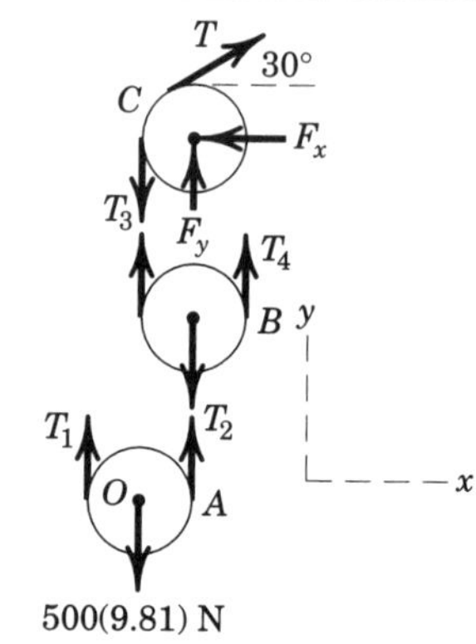

Solution. The free-body diagram of each pulley is drawn in its relative position to the others. We begin with pulley A, which includes the only known force. With the unspecified pulley radius designated by r, the equilibrium of moments about its center O and the equilibrium of forces in the vertical direction require

① $[\Sigma M_O = 0] \qquad T_1 r - T_2 r = 0 \qquad T_1 = T_2$

$[\Sigma F_y = 0] \qquad T_1 + T_2 - 500(9.81) = 0 \qquad 2T_1 = 500(9.81) \qquad T_1 = T_2 = 2450 \text{ N}$

From the example of pulley A we may write the equilibrium of forces on pulley B by inspection as

$$T_3 = T_4 = T_2/2 = 1226 \text{ N}$$

For pulley C the angle $\theta = 30°$ in no way affects the moment of T about the center of the pulley, so that moment equilibrium requires

$$T = T_3 \quad \text{or} \quad T = 1226 \text{ N} \qquad \textit{Ans.}$$

Equilibrium of the pulley in the x- and y-directions requires

$[\Sigma F_x = 0] \qquad 1226 \cos 30° - F_x = 0 \qquad F_x = 1062 \text{ N}$

$[\Sigma F_y = 0] \qquad F_y + 1226 \sin 30° - 1226 = 0 \qquad F_y = 613 \text{ N}$

$[F = \sqrt{F_x{}^2 + F_y{}^2}] \qquad F = \sqrt{1062^2 + 613^2} = 1226 \text{ N} \qquad \textit{Ans.}$

Helpful Hint

① Clearly the radius r does not influence the results. Once we have analyzed a simple pulley, the results should be perfectly clear by inspection.

Sample Problem 3/3

The uniform 100-kg I-beam is supported initially by its end rollers on the horizontal surface at A and B. By means of the cable at C it is desired to elevate end B to a position 3 m above end A. Determine the required tension P, the reaction at A, and the angle θ made by the beam with the horizontal in the elevated position.

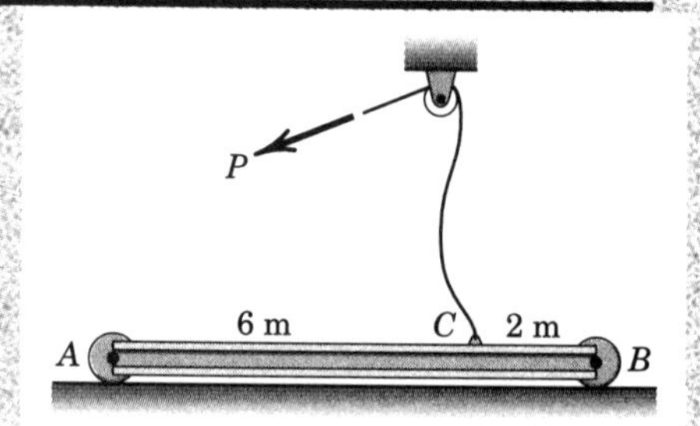

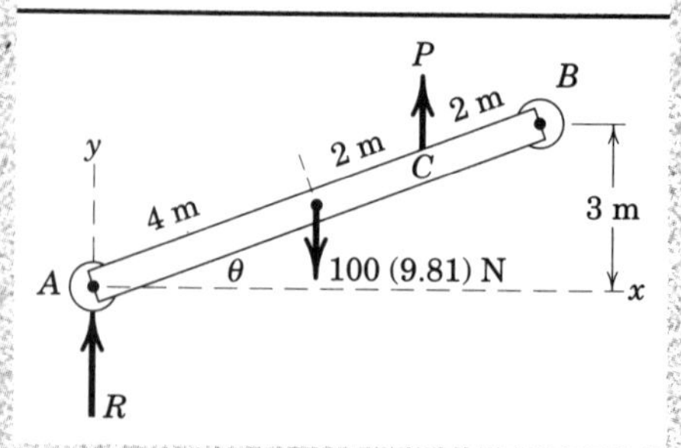

Solution. In constructing the free-body diagram, we note that the reaction on the roller at A and the weight are vertical forces. Consequently, in the absence of other horizontal forces, P must also be vertical. From Sample Problem 3/2 we see immediately that the tension P in the cable equals the tension P applied to the beam at C.

Moment equilibrium about A eliminates force R and gives

① $[\Sigma M_A = 0] \qquad P(6 \cos \theta) - 981(4 \cos \theta) = 0 \qquad P = 654 \text{ N} \qquad \textit{Ans.}$

Equilibrium of vertical forces requires

$[\Sigma F_y = 0] \qquad 654 + R - 981 = 0 \qquad R = 327 \text{ N} \qquad \textit{Ans.}$

The angle θ depends only on the specified geometry and is

$$\sin \theta = 3/8 \qquad \theta = 22.0° \qquad \textit{Ans.}$$

Helpful Hint

① Clearly the equilibrium of this parallel force system is independent of θ.

Sample Problem 3/4

Determine the magnitude T of the tension in the supporting cable and the magnitude of the force on the pin at A for the jib crane shown. The beam AB is a standard 0.5-m I-beam with a mass of 95 kg per meter of length.

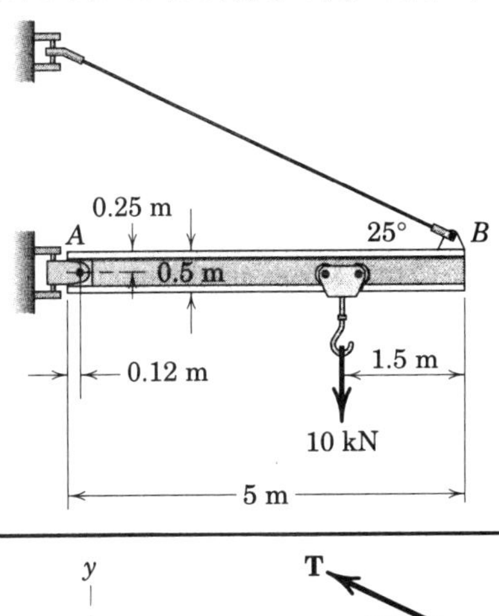

Algebraic solution. The system is symmetrical about the vertical x-y plane through the center of the beam, so the problem may be analyzed as the equilibrium of a coplanar force system. The free-body diagram of the beam is shown in the figure with the pin reaction at A represented in terms of its two rectangular components. The weight of the beam is $95(10^{-3})(5)9.81 = 4.66$ kN and acts through its center. Note that there are three unknowns A_x, A_y, and T which may be found from the three equations of equilibrium. We begin with a moment equation about A, which eliminates two of the three unknowns from the equation. ① In applying the moment equation about A, it is simpler to consider the moments of the x- and y-components of $\mathbf{T}$ than it is to compute the perpendicular distance from $\mathbf{T}$ to A. Hence, with the counterclockwise sense as positive we write

② $[\Sigma M_A = 0]$ $\quad (T \cos 25°)0.25 + (T \sin 25°)(5 - 0.12)$
$$- 10(5 - 1.5 - 0.12) - 4.66(2.5 - 0.12) = 0$$

from which $\quad T = 19.61 \text{ kN}$ *Ans.*

Equating the sums of forces in the x- and y-directions to zero gives

$[\Sigma F_x = 0]$ $\quad A_x - 19.61 \cos 25° = 0 \quad A_x = 17.77 \text{ kN}$

$[\Sigma F_y = 0]$ $\quad A_y + 19.61 \sin 25° - 4.66 - 10 = 0 \quad A_y = 6.37 \text{ kN}$

③ $[A = \sqrt{A_x^2 + A_y^2}]$ $\quad A = \sqrt{(17.77)^2 + (6.37)^2} = 18.88 \text{ kN}$ *Ans.*

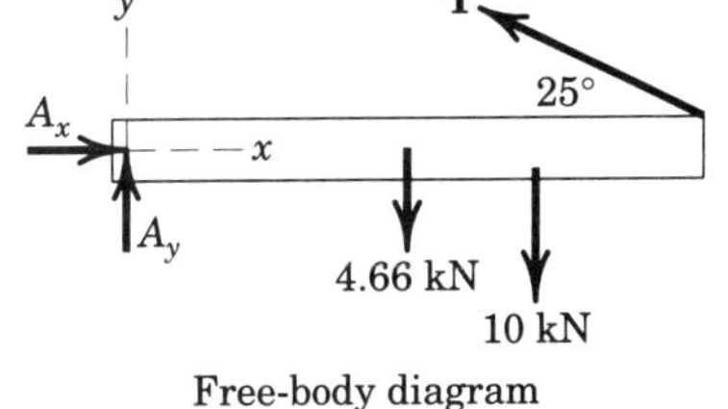

Free-body diagram

Helpful Hints

① The justification for this step is Varignon's theorem, explained in Art. 2/4. Be prepared to take full advantage of this principle frequently.

② The calculation of moments in two-dimensional problems is generally handled more simply by scalar algebra than by the vector cross product $\mathbf{r} \times \mathbf{F}$. In three dimensions, as we will see later, the reverse is often the case.

③ The direction of the force at A could be easily calculated if desired. However, in designing the pin A or in checking its strength, it is only the magnitude of the force that matters.

Graphical solution. The principle that three forces in equilibrium must be concurrent is utilized for a graphical solution by combining the two known vertical forces of 4.66 and 10 kN into a single 14.66-kN force, located as shown on the modified free-body diagram of the beam in the lower figure. The position of this resultant load may easily be determined graphically or algebraically. The intersection of the 14.66-kN force with the line of action of the unknown tension $\mathbf{T}$ defines the point of concurrency O through which the pin reaction $\mathbf{A}$ must pass. The unknown magnitudes of $\mathbf{T}$ and $\mathbf{A}$ may now be found by adding the forces head-to-tail to form the closed equilibrium polygon of forces, thus satisfying their zero vector sum. After the known vertical load is laid off to a convenient scale, as shown in the lower part of the figure, a line representing the given direction of the tension $\mathbf{T}$ is drawn through the tip of the 14.66-kN vector. Likewise a line representing the direction of the pin reaction $\mathbf{A}$, determined from the concurrency established with the free-body diagram, is drawn through the tail of the 14.66-kN vector. The intersection of the lines representing vectors $\mathbf{T}$ and $\mathbf{A}$ establishes the magnitudes T and A necessary to make the vector sum of the forces equal to zero. These magnitudes are scaled from the diagram. The x- and y-components of $\mathbf{A}$ may be constructed on the force polygon if desired.

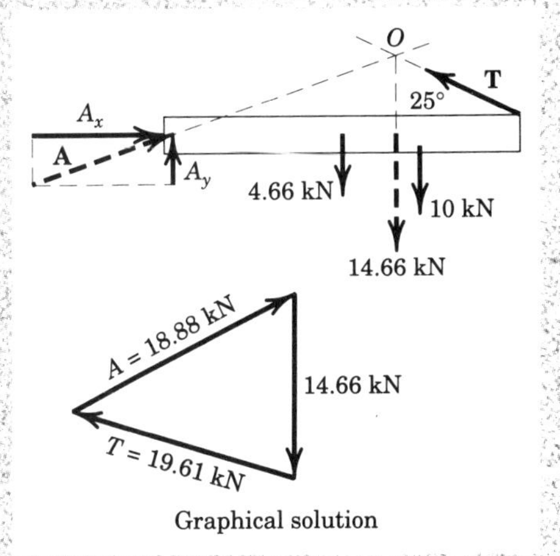

Graphical solution

PROBLEMS

Introductory Problems

3/1 The mass center G of the 1400-kg rear-engine car is located as shown in the figure. Determine the normal force under each tire when the car is in equilibrium. State any assumptions.

Ans. $N_f = 2820$ N, $N_r = 4050$ N

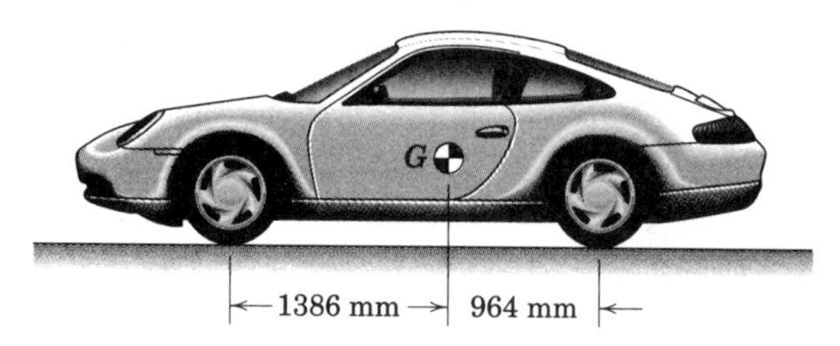

Problem 3/1

3/2 A carpenter carries a 6-kg uniform board as shown. What downward force does he feel on his shoulder at A?

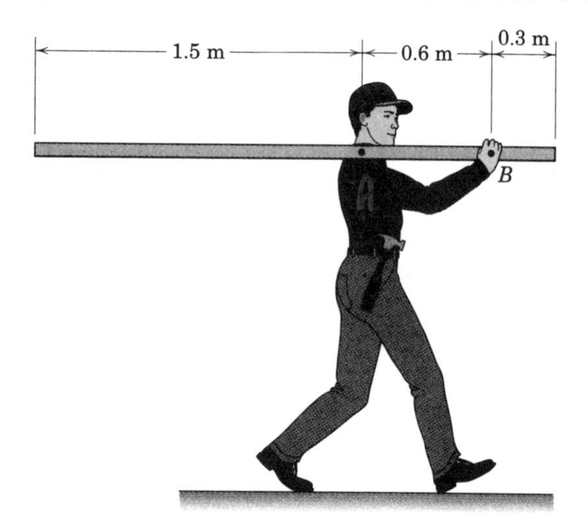

Problem 3/2

3/3 A carpenter holds a 6-kg uniform board as shown. If he exerts vertical forces on the board, determine the forces at A and B.

Ans. $N_A = 58.9$ N down, $N_B = 117.7$ N up

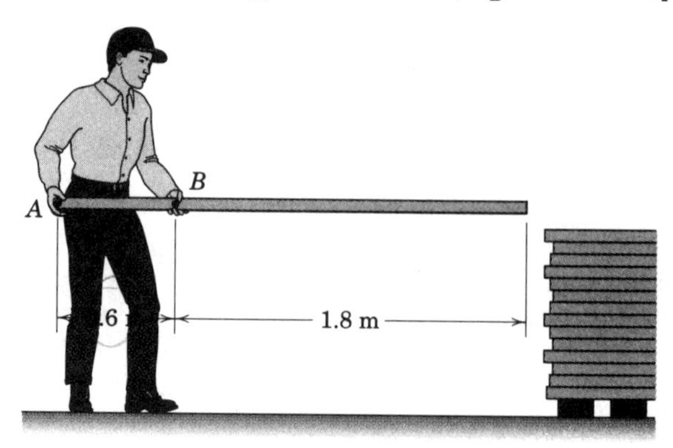

Problem 3/3

3/4 The 450-kg uniform I-beam supports the load shown. Determine the reactions at the supports.

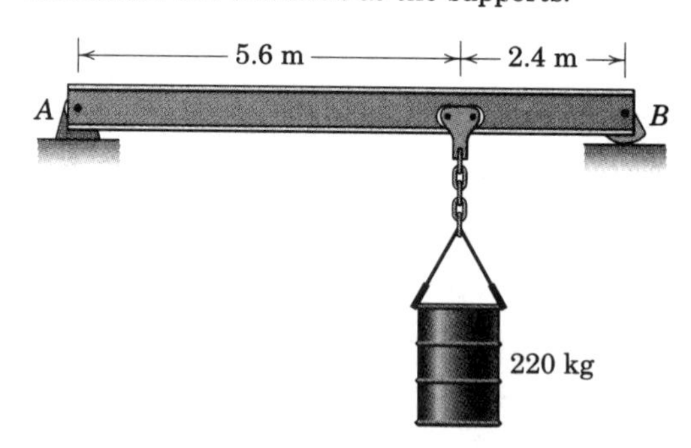

Problem 3/4

3/5 The 20-kg homogeneous smooth sphere rests on the two inclines as shown. Determine the contact forces at A and B.

Ans. $N_A = 101.6$ N, $N_B = 196.2$ N

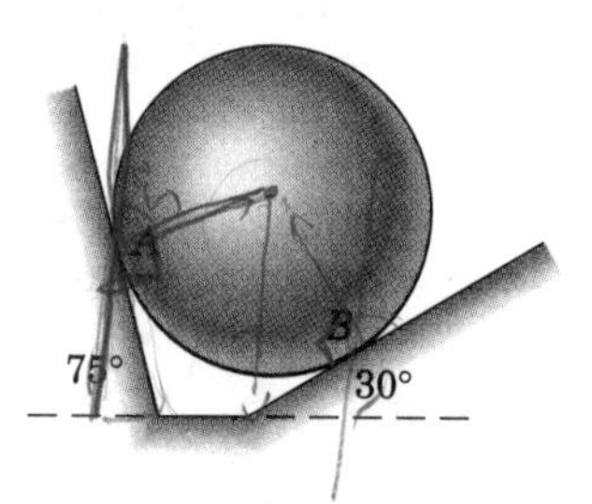

Problem 3/5

3/6 With what force magnitude T must the person pull on the cable in order to cause the scale A to read 2000 N? The weights of the pulleys and cables are negligible. State any assumptions.

Problem 3/6

3/7 What horizontal force P must a worker exert on the rope to position the 50-kg crate directly over the trailer?

Ans. $P = 126.6$ N

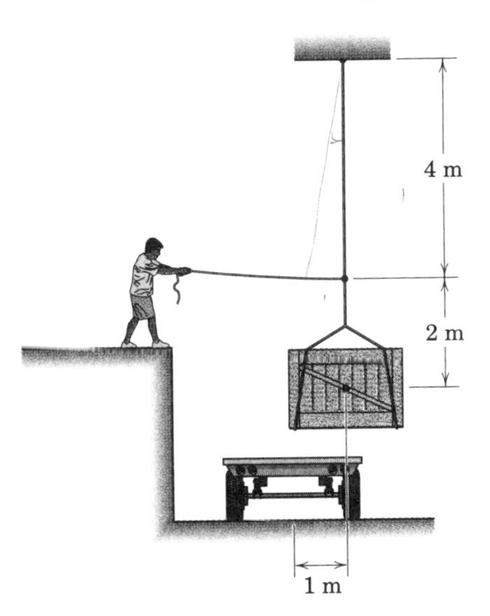

Problem 3/7

3/8 The 275-kg drum is being hoisted by the lifting device which hooks over the end lips of the drum. Determine the tension T in each of the equal-length rods which form the two U-shaped members of the device.

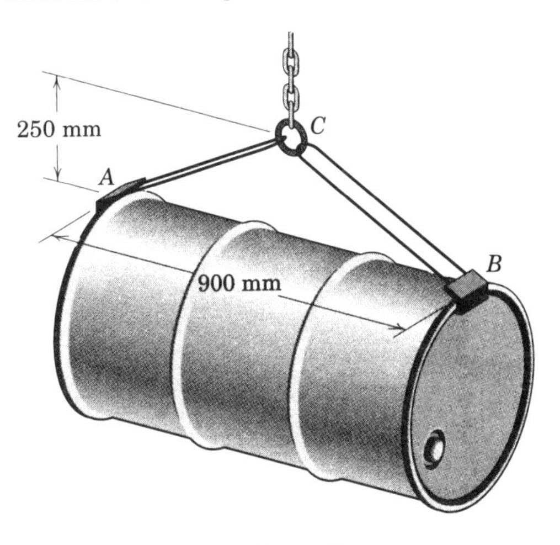

Problem 3/8

3/9 What fraction n of the weight W of a jet airplane is the net thrust (nozzle thrust T minus air resistance R) in order for the airplane to climb with a constant speed at an angle θ with the horizontal?

Ans. $n = \sin \theta$

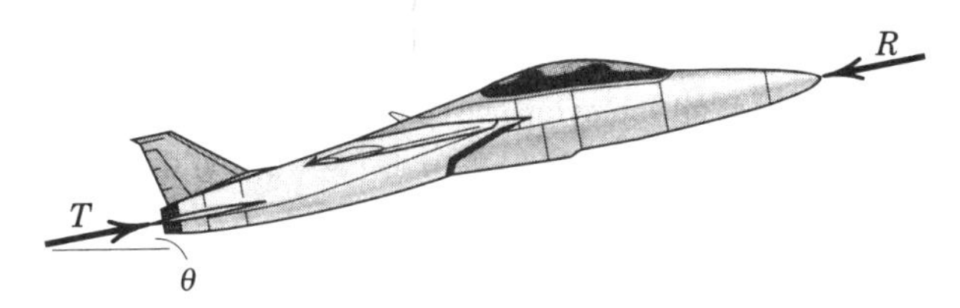

Problem 3/9

3/10 Determine the force magnitude P required to lift one end of the 250-kg crate with the lever dolly as shown. State any assumptions.

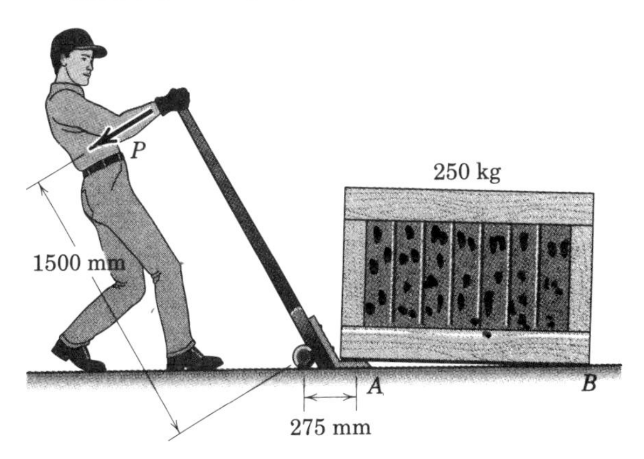

Problem 3/10

3/11 Find the angle of tilt θ with the horizontal so that the contact force at B will be one-half that at A for the smooth cylinder.

Ans. $\theta = 18.43°$

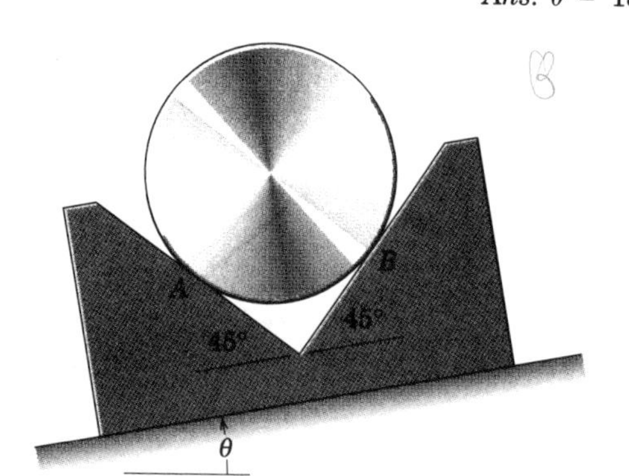

Problem 3/11

3/12 Determine the magnitude P of the vertical force required to lift the wheelbarrow free of the ground at point B. The combined mass of the wheelbarrow and its load is 110 kg with center of mass at G.

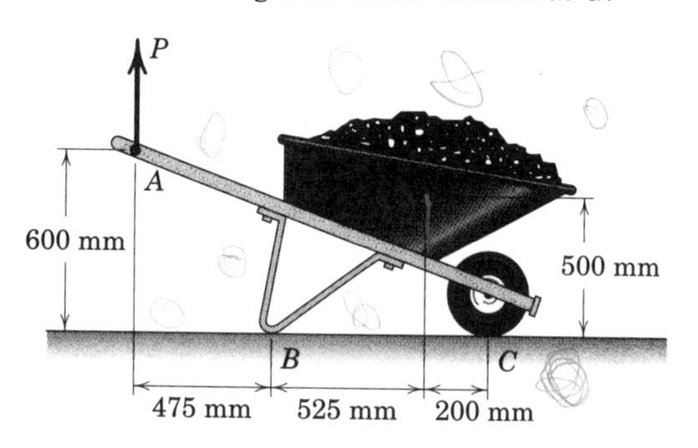

Problem 3/12

3/13 To facilitate shifting the position of a lifting hook when it is not under load, the sliding hanger shown is used. The projections at A and B engage the flanges of a box beam when a load is supported, and the hook projects through a horizontal slot in the beam. Compute the forces at A and B when the hook supports a 300-kg mass.

Ans. $A = 4.91$ kN, $B = 1.962$ kN

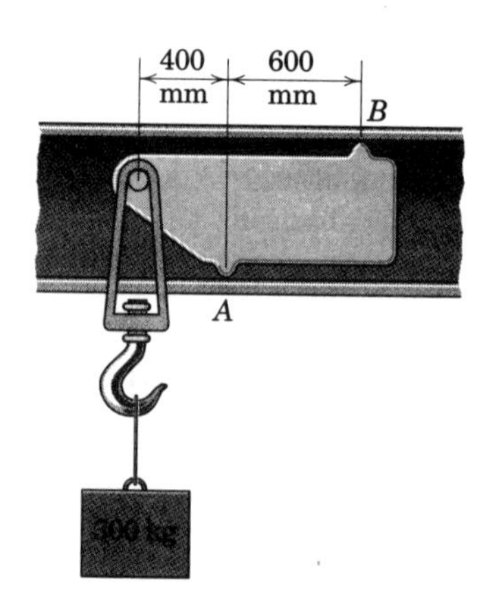

Problem 3/13

3/14 Three cables are joined at the junction ring C. Determine the tensions in cables AC and BC caused by the weight of the 30-kg cylinder.

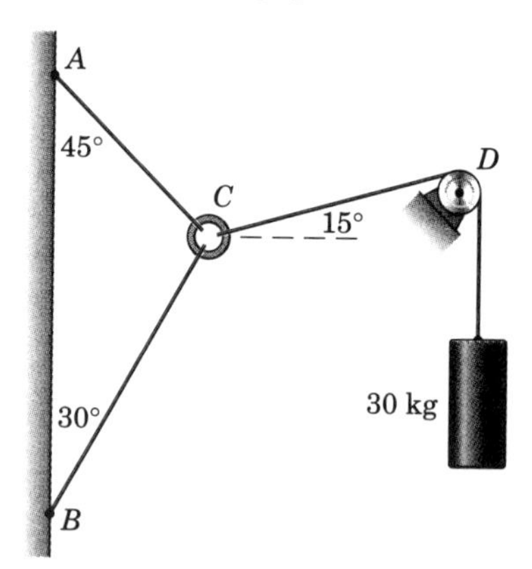

Problem 3/14

3/15 The 100-kg wheel rests on a rough surface and bears against the roller A when the couple M is applied. If $M = 60$ N·m and the wheel does not slip, compute the reaction on the roller A.

Ans. $F_A = 231$ N

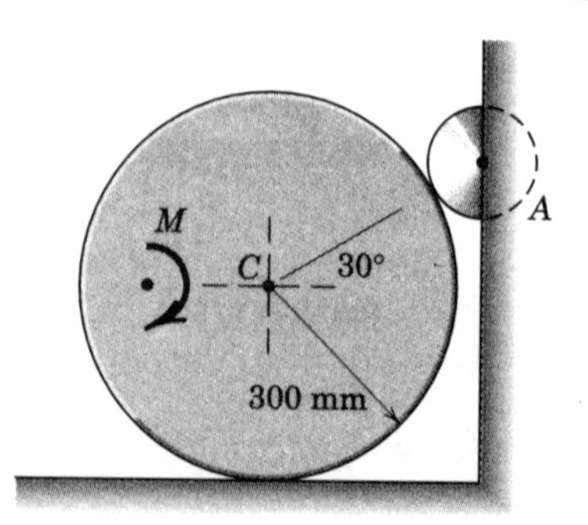

Problem 3/15

3/16 The uniform beam has a mass of 50 kg per meter of length. Compute the reactions at the support O. The force loads shown lie in a vertical plane.

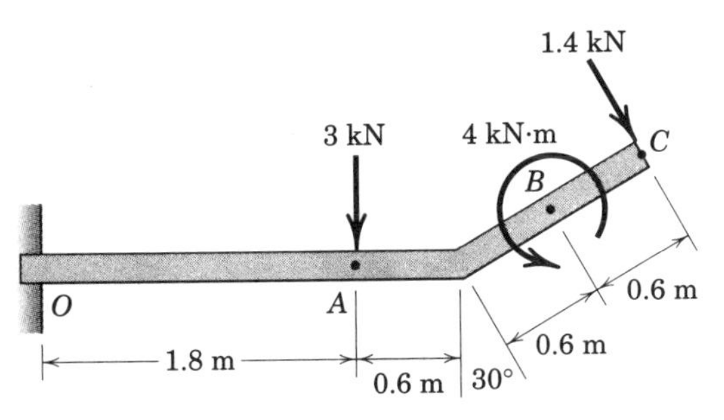

Problem 3/16

3/17 To accommodate the rise and fall of the tide, a walkway from a pier to a float is supported by two rollers as shown. If the mass center of the 300-kg walkway is at G, calculate the tension T in the horizontal cable which is attached to the cleat and find the force under the roller at A.

Ans. $T = 850$ N, $A = 1472$ N

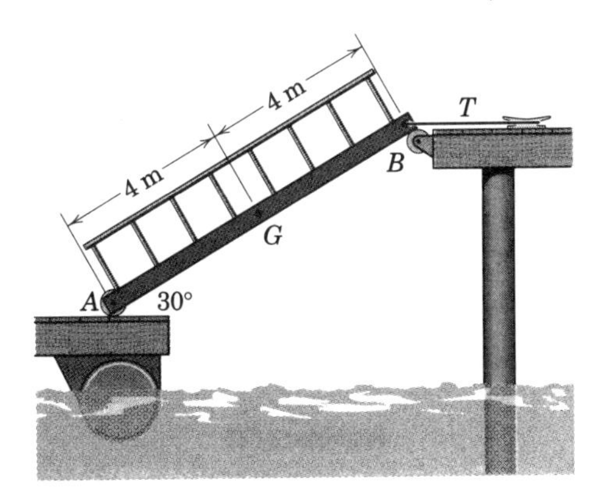

Problem 3/17

Representative Problems

3/18 Determine the magnitude P of the force which the man must exert perpendicular to the handle of the high-pressure washer in order to cause loss of contact at the front support B. Note that the operator prevents movement of the wheel with his left foot. The 60-kg machine has its mass center at point G. Treat the problem as two-dimensional.

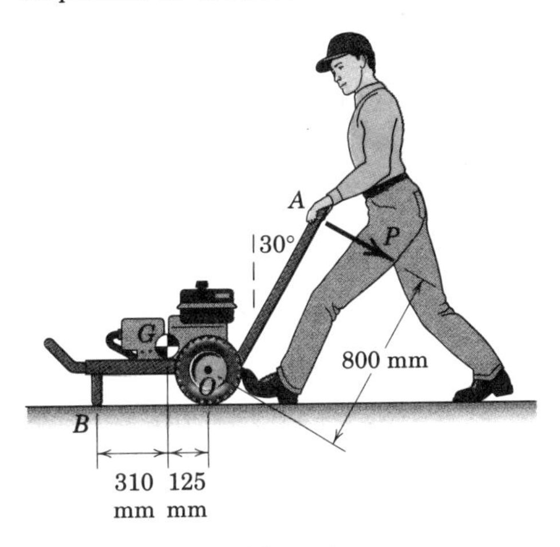

Problem 3/18

3/19 If the screw B of the wood clamp is tightened so that the two blocks are under a compression of 500 N, determine the force in screw A. (*Note:* The force supported by each screw may be taken in the direction of the screw.)

Ans. $A = 1250$ N

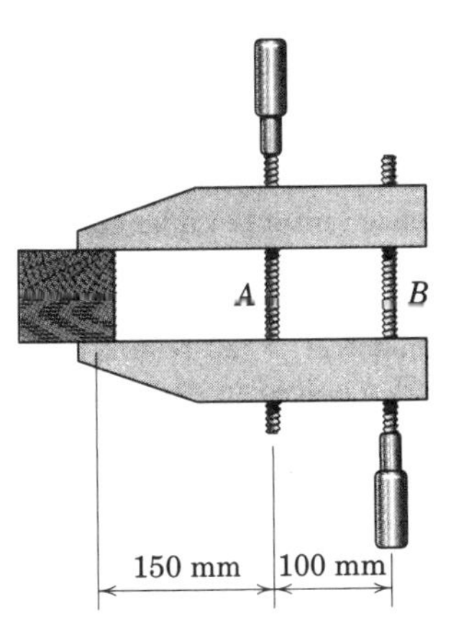

Problem 3/19

3/20 The uniform 15-m pole has a mass of 150 kg and is supported by its smooth ends against the vertical walls and by the tension T in the vertical cable. Compute the reactions at A and B.

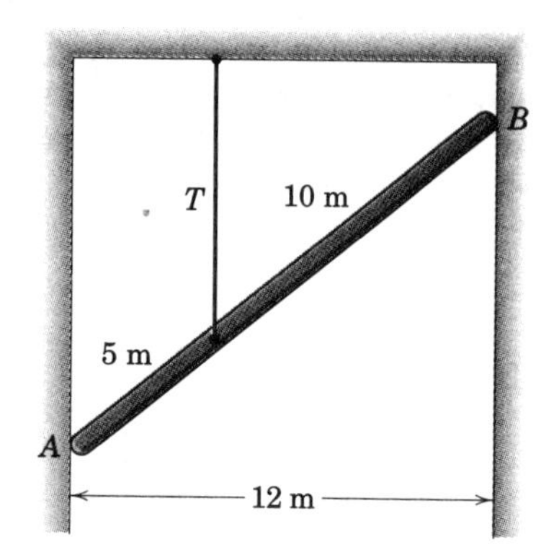

Problem 3/20

3/21 Determine the force P required to begin rolling the uniform cylinder of mass m over the obstruction of height h.

Ans. $P = \dfrac{mg\sqrt{2rh - h^2}}{r - h}$

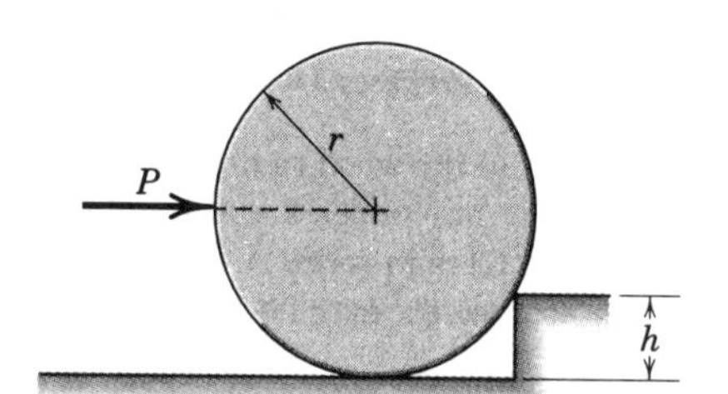

Problem 3/21

3/22 The elements of a heavy-duty fluid valve are shown in the figure. When the member OB rotates clockwise about the fixed pivot O under the action of the force P, the element S slides freely upward in its slot, releasing the flow. If an internal torsional spring exerts a moment $M = 20$ N·m as shown, determine the force P required to open the valve. Neglect all friction.

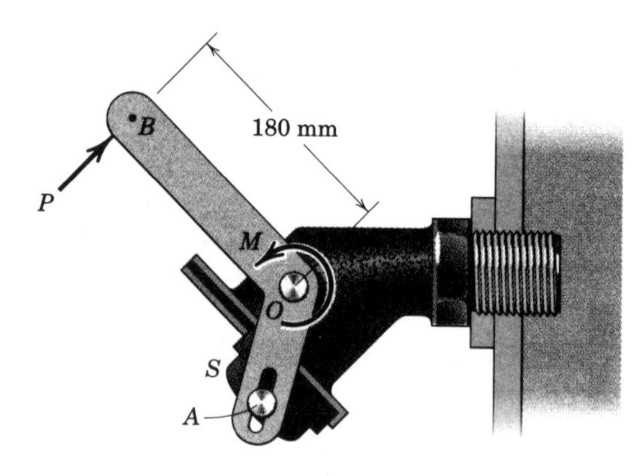

Problem 3/22

3/23 The spring of modulus $k = 3.5$ kN/m is stretched 10 mm when the disk center O is in the leftmost position $x = 0$. Determine the tension T required to position the disk center at $x = 150$ mm. At that position, what force N is exerted on the horizontal slotted guide? The mass of the disk is 3 kg.

Ans. $T = 328$ N, $N = 203$ N up

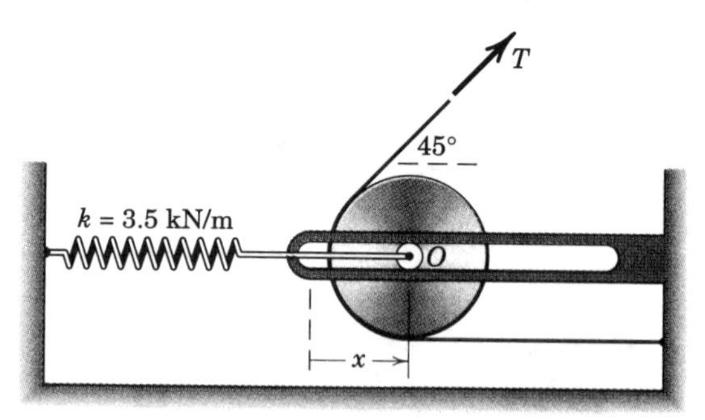

Problem 3/23

3/24 A block placed under the head of the claw hammer as shown greatly facilitates the extraction of the nail. If a 200-N pull on the handle is required to pull the nail, calculate the tension T in the nail and the magnitude A of the force exerted by the hammer head on the block. The contacting surfaces at A are sufficiently rough to prevent slipping.

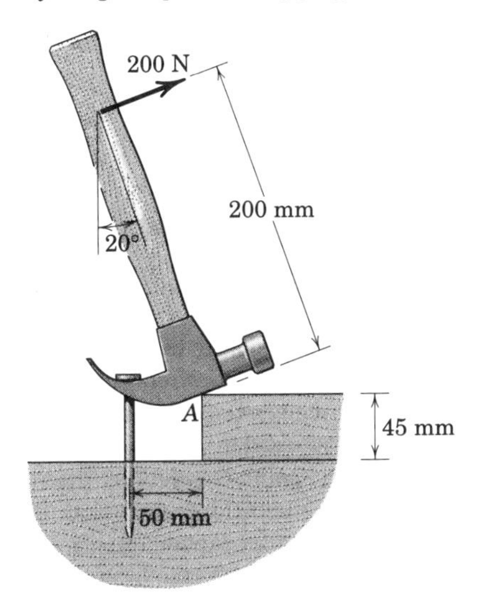

Problem 3/24

3/25 The indicated location of the center of mass of the 1600-kg pickup truck is for the unladen condition. If a load whose center of mass is $x = 400$ mm behind the rear axle is added to the truck, determine the mass m_L of the load for which the normal forces under the front and rear wheels are equal.

Ans. $m_L = 244$ kg

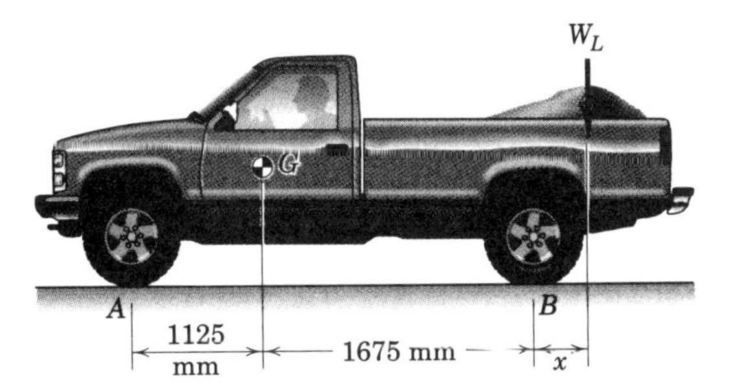

Problem 3/25

3/26 To test the validity of aerodynamic assumptions made in the design of the aircraft, its model is being tested in a wind tunnel. The support bracket is connected to a force and moment balance, which is zeroed when there is no airflow. Under test conditions, the lift L, drag D, and pitching moment M_G act as shown. The force balance records the lift, drag, and a moment M_P. Determine M_G in terms of L, D, and M_P.

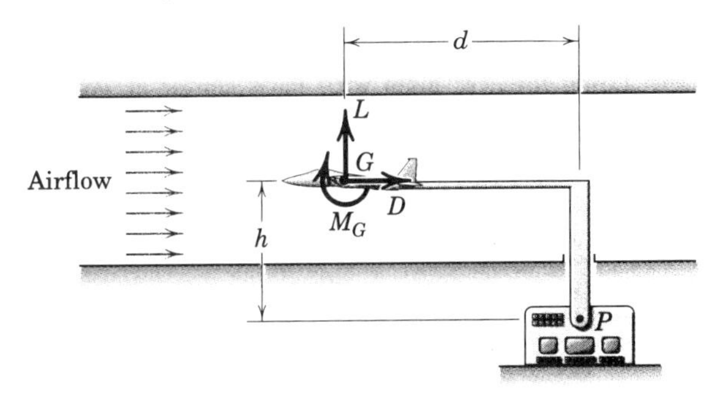

Problem 3/26

3/27 In a procedure to evaluate the strength of the triceps muscle, a person pushes down on a load cell with the palm of his hand as indicated in the figure. If the load-cell reading is 160 N, determine the vertical tensile force F generated by the triceps muscle. The mass of the lower arm is 1.5 kg with mass center at G. State any assumptions.

Ans. $F = 1832$ N

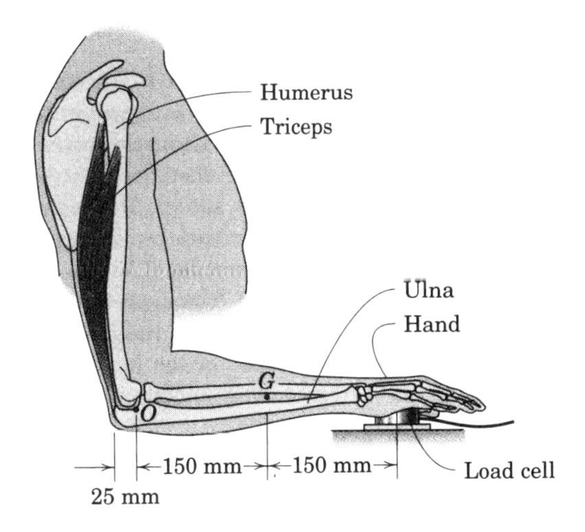

Problem 3/27

3/28 A person is performing slow arm curls with a 10-kg dumbbell as indicated in the figure. The brachialis muscle group (consisting of the biceps and brachialis muscles) is the major factor in this exercise. Determine the magnitude F of the brachialis-muscle-group force and the magnitude E of the elbow joint reaction at point E for the forearm position shown in the figure. Take the dimensions shown to locate the effective points of application of the two muscle groups; these points are 200 mm directly above E and 50 mm directly to the right of E. Include the effect of the 1.5-kg forearm mass with mass center at point G. State any assumptions.

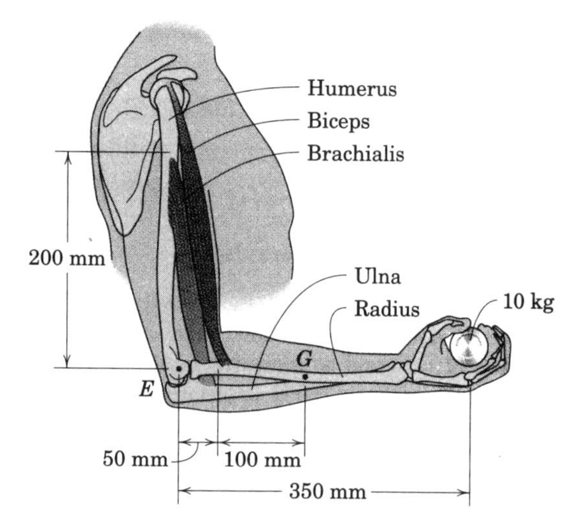

Problem 3/28

3/29 A woman is holding a 3.6-kg sphere in her hand with the entire arm held horizontally as shown in the figure. A tensile force in the deltoid muscle prevents the arm from rotating about the shoulder joint O; this force acts at the 21° angle shown. Determine the force exerted by the deltoid muscle on the upper arm at A and the x- and y-components of the force reaction at the shoulder joint O. The mass of the upper arm is m_U = 1.9 kg, the mass of the lower arm is m_L = 1.1 kg, and the mass of the hand is m_H = 0.4 kg; all the corresponding weights act at the locations shown in the figure.

Ans. F_D = 710 N, O_x = 662 N, O_y = −185.6 N

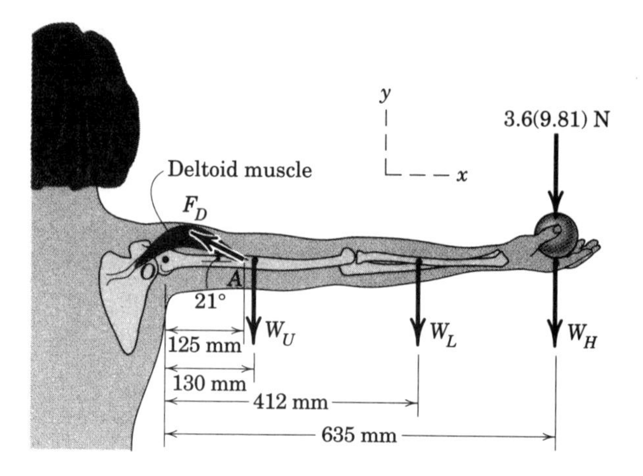

Problem 3/29

3/30 With his weight W equally distributed on both feet, a man begins to slowly rise from a squatting position as indicated in the figure. Determine the tensile force F in the patellar tendon and the magnitude of the force reaction at point O, which is the contact area between the tibia and the femur. Note that the line of action of the patellar tendon force is along its midline. Neglect the weight of the lower leg.

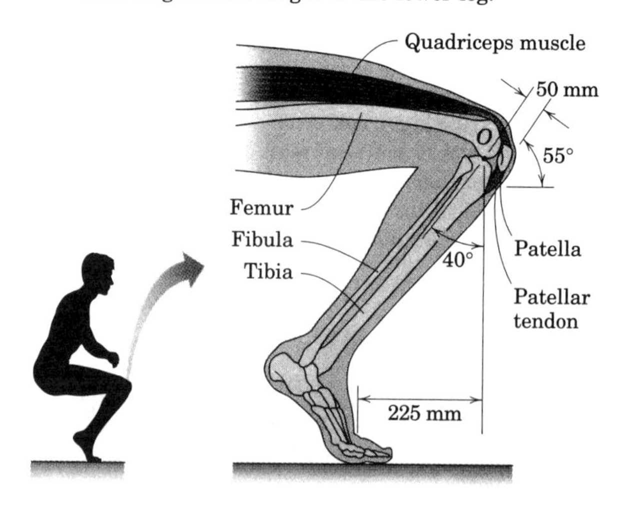

Problem 3/30

3/31 For the design of the belt-tensioning device, determine the dimension l if the mass m maintains a specified tension T in the belt for the position shown. Neglect the mass of the arm and central pulley compared with m. Also determine the magnitude R of the force supported by the pin at O.

Ans. $l = \dfrac{Tb\sqrt{3}}{mg}$, $R = \sqrt{3T^2 + m^2g^2}$

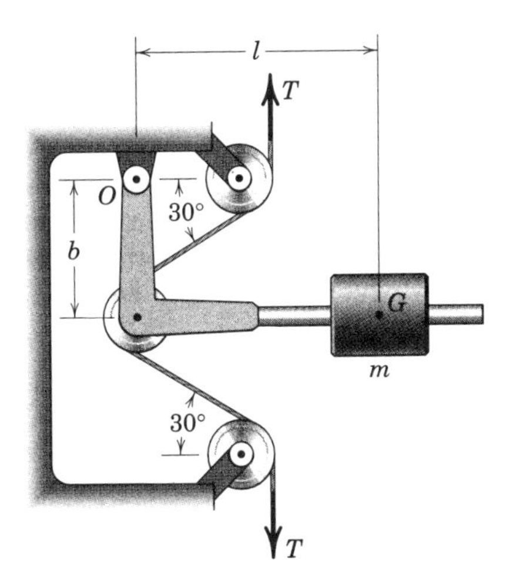

Problem 3/31

3/32 The uniform 18-kg bar OA is held in the position shown by the smooth pin at O and the cable AB. Determine the tension T in the cable and the magnitude and direction of the external pin reaction at O.

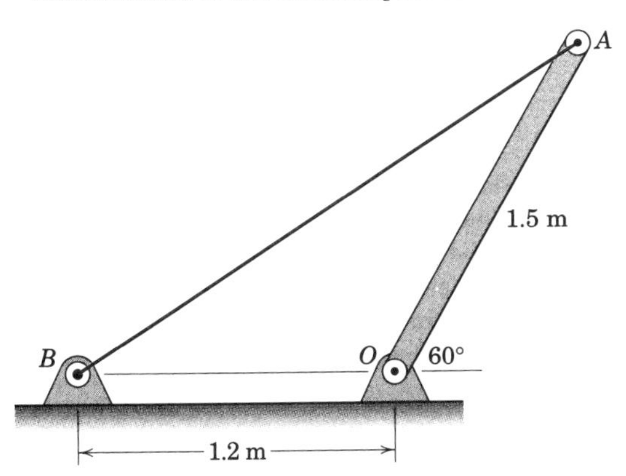

Problem 3/32

3/33 The exercise machine is designed with a lightweight cart which is mounted on small rollers so that it is free to move along the inclined ramp. Two cables are attached to the cart—one for each hand. If the hands are together so that the cables are parallel and if each cable lies essentially in a vertical plane, determine the force P which each hand must exert on its cable in order to maintain an equilibrium position. The mass of the person is 70 kg, the ramp angle θ is 15°, and the angle β is 18°. In addition, calculate the force R which the ramp exerts on the cart.

Ans. P = 45.5 N, R = 691 N

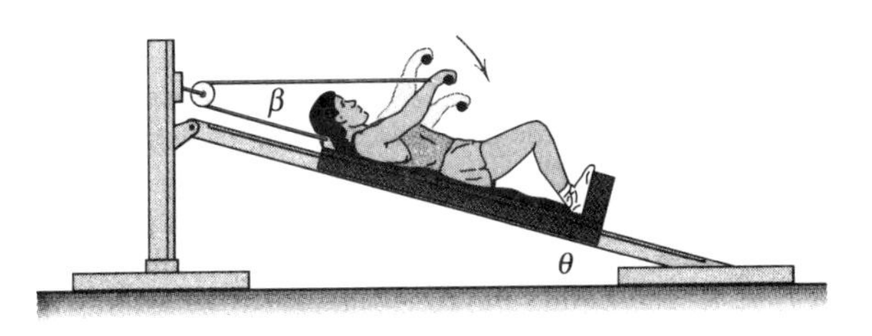

Problem 3/33

3/34 Calculate the magnitude of the force supported by the pin at C under the action of the 900-N load applied to the bracket. Neglect friction in the slot.

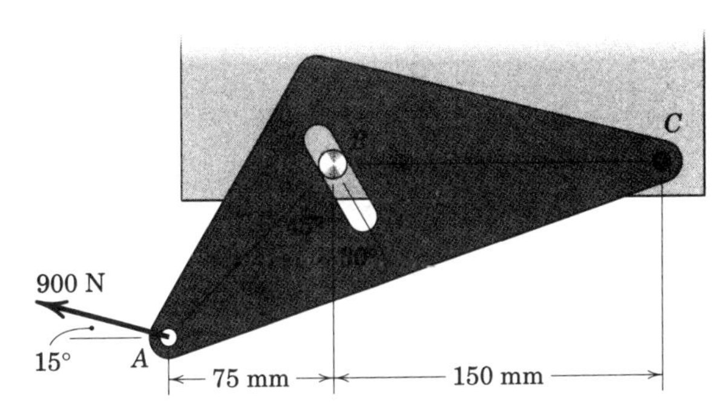

Problem 3/34

3/35 A uniform ring of mass m and radius r carries an eccentric mass m_0 at a radius b and is in an equilibrium position on the incline, which makes an angle α with the horizontal. If the contacting surfaces are rough enough to prevent slipping, write the expression for the angle θ which defines the equilibrium position.

Ans. $\theta = \sin^{-1}\left[\frac{r}{b}\left(1 + \frac{m}{m_0}\right)\sin\alpha\right]$

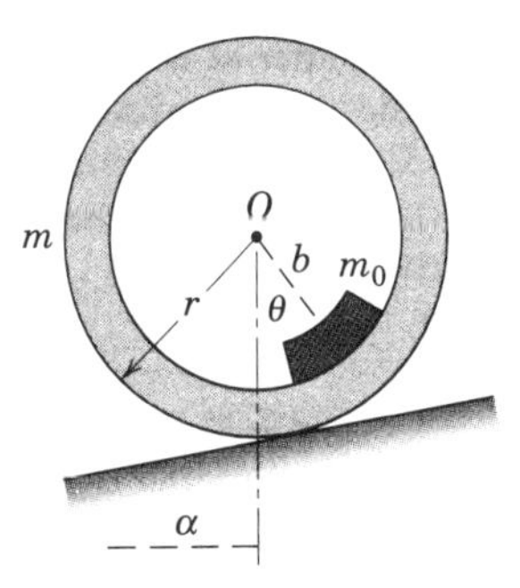

Problem 3/35

3/36 The concrete hopper and its load have a combined mass of 4 metric tons (1 metric ton equals 1000 kg) with mass center at G and is being elevated at constant velocity along its vertical guide by the cable tension T. The design calls for two sets of guide rollers at A, one on each side of the hopper, and two sets at B. Determine the force supported by each of the two pins at A and by each of the two pins at B.

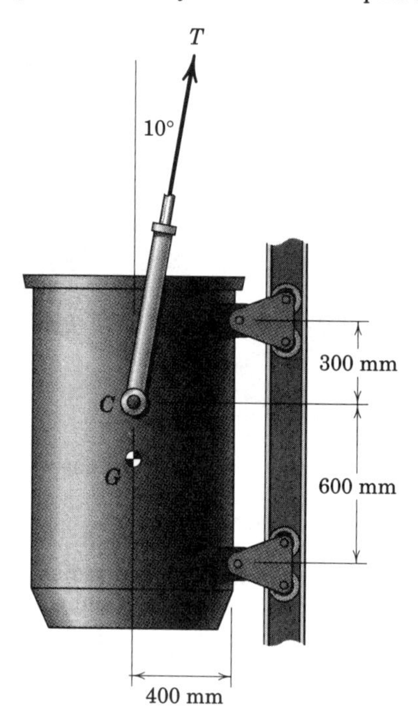

Problem 3/36

3/37 During an engine test on the ground, a propeller thrust $T = 3000$ N is generated on the 1800-kg airplane with mass center at G. The main wheels at B are locked and do not skid; the small tail wheel at A has no brake. Compute the percent change n in the normal forces at A and B as compared with their "engine-off" values.

Ans. $n_A = -32.6\%$, $n_B = 2.28\%$

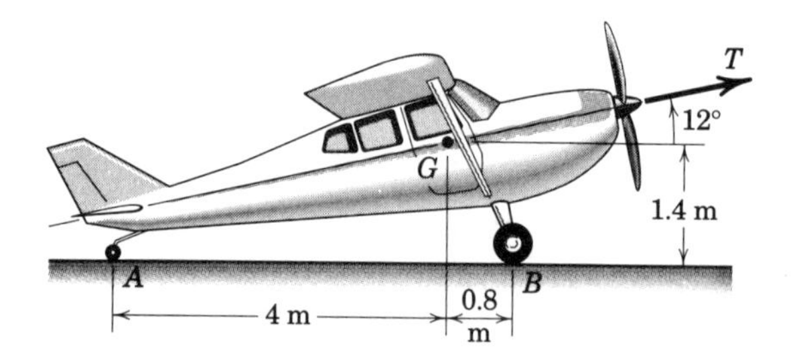

Problem 3/37

3/38 The elements of a wall-mounted swing-away stool are shown in the figure. The hinge pin P fits loosely through the frame tube, and the frame tube has a slight clearance between the supports A and B. Determine the reactions on the frame tube at A and B associated with the weight L of an 80-kg person. Also, calculate the changes in the horizontal reactions at C and D due to the same load L. State any assumptions.

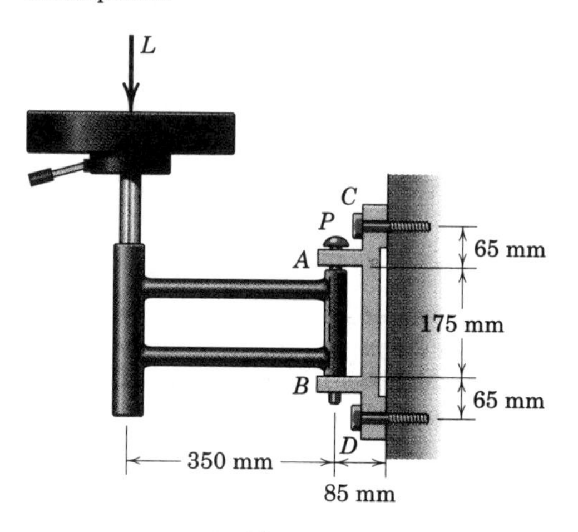

Problem 3/38

3/39 The hook wrench or pin spanner is used to turn shafts and collars. If a moment of 80 N·m is required to turn the 200-mm-diameter collar about its center O under the action of the applied force P, determine the contact force R on the smooth surface at A. Engagement of the pin at B may be considered to occur at the periphery of the collar.

Ans. $R = 1047$ N

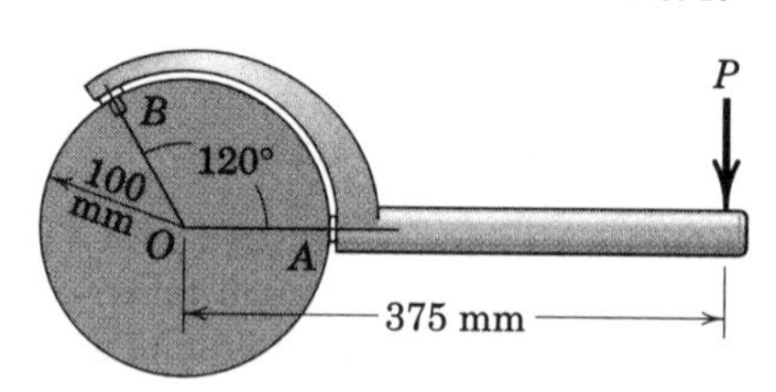

Problem 3/39

3/40 In sailing at a constant speed with the wind, the sailboat is driven by a 4-kN force against its mainsail and a 1.6-kN force against its staysail as shown. The total resistance due to fluid friction through the water is the force R. Determine the resultant of the lateral forces perpendicular to motion applied to the hull by the water.

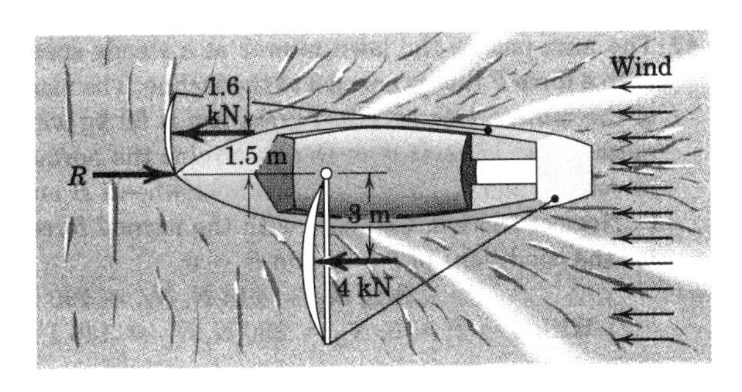

Problem 3/40

3/41 A portion of the shifter mechanism for a manual car transmission is shown in the figure. For the 8-N force exerted on the shift knob, determine the corresponding force P exerted by the shift link BC on the transmission (not shown). Neglect friction in the ball-and-socket joint at O, in the joint at B, and in the slip tube near support D. Note that a soft rubber bushing at D allows the slip tube to self-align with link BC.

Ans. $P = 26.3$ N

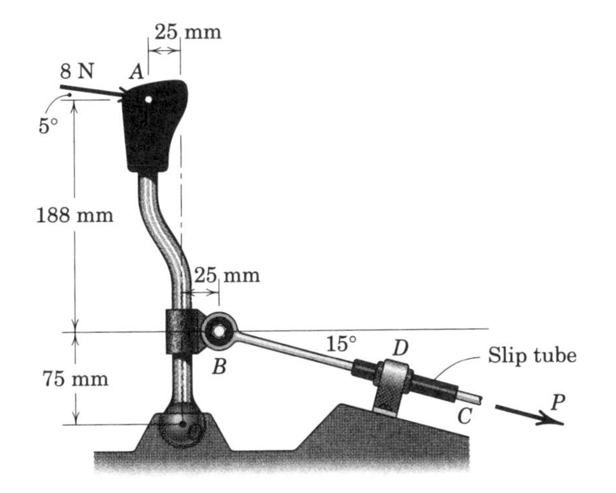

Problem 3/41

3/42 A torque (moment) of 24 N·m is required to turn the bolt about its axis. Determine P and the forces between the smooth hardened jaws of the wrench and the corners A and B of the hexagonal head. Assume that the wrench fits easily on the bolt so that contact is made at corners A and B only.

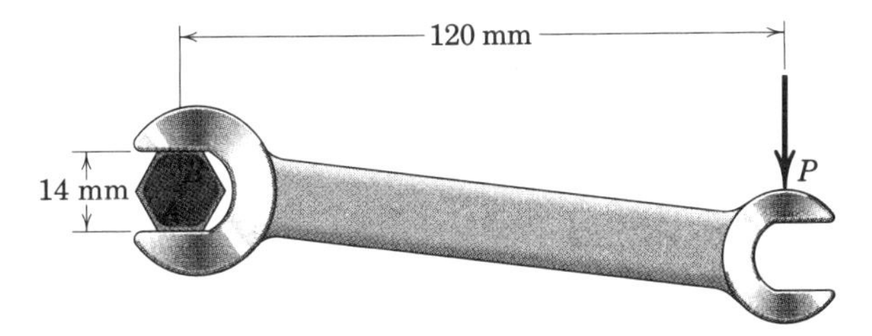

Problem 3/42

3/43 The car complete with driver has a mass of 815 kg and without the two airfoils has a 50%–50% front–rear weight distribution at a certain speed at which there is no lift on the car. It is estimated that at this speed each of the airfoils A_1 and A_2 will generate 2 kN of downward force L and 250 N of drag force D on the car. Specify the vertical reactions N_A and N_B under the two pairs of wheels at that speed when the airfoils are added. Assume that the addition of the airfoils does not affect the drag and zero-lift conditions of the car body itself and that the engine has sufficient power for equilibrium at that speed. The weight of the airfoils may be neglected.

Ans. $N_A = 5750$ N (48.0%), $N_B = 6240$ N (52.0%)

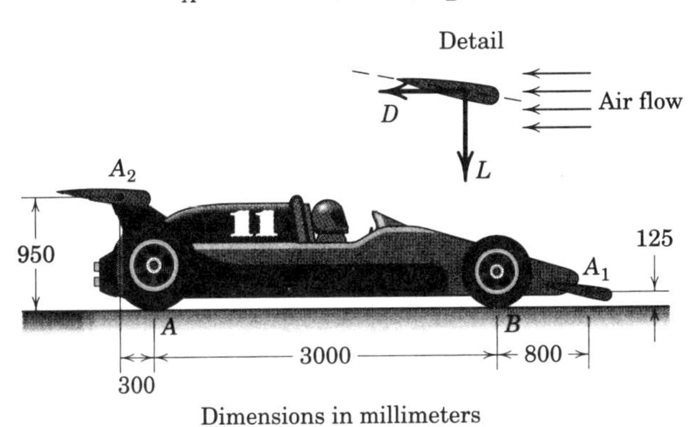

Problem 3/43

3/44 Determine the external reactions at A and F for the roof truss loaded as shown. The vertical loads represent the effect of the supported roofing materials, while the 400-N force represents a wind load.

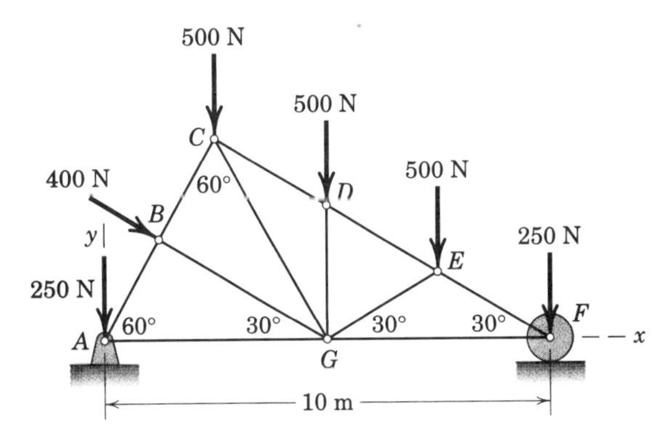

Problem 3/44

3/45 Calculate the normal forces associated with the front and rear wheel pairs of the 1600-kg front-wheel-drive van. Then repeat the calculations when the van (*a*) climbs a 10-percent grade and (*b*) descends a 10-percent grade, both at constant speed. Compute the percent changes n_A and n_B in the normal forces compared with the nominal values. Be sure to recognize that propulsive and braking forces are present for cases (*a*) and (*b*).

Ans. $N_A = 9420$ N, $N_B = 6280$ N
(*a*) $N_A = 9030$ N (-4.14%), $N_B = 6590$ N $(+4.98\%)$
(*b*) $N_A = 9710$ N $(+3.15\%)$, $N_B = 5900$ N (-5.97%)

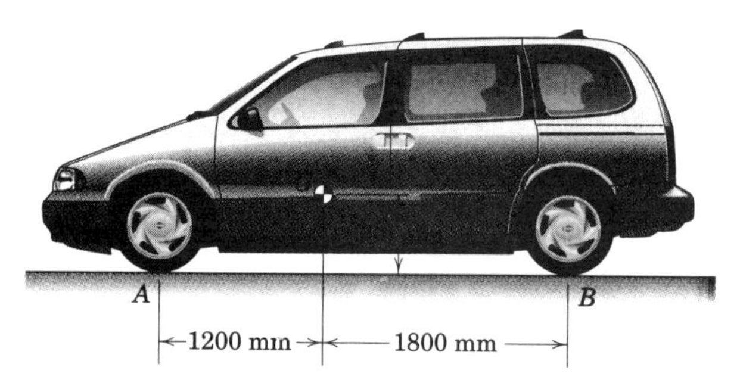

Problem 3/45

3/46 It is desired that a person be able to begin closing the van hatch from the open position shown with a 40-N vertical force *P*. As a design exercise, determine the necessary force in each of the two hydraulic struts *AB*. The mass center of the 40-kg door is 37.5 mm directly below point *A*. Treat the problem as two-dimensional.

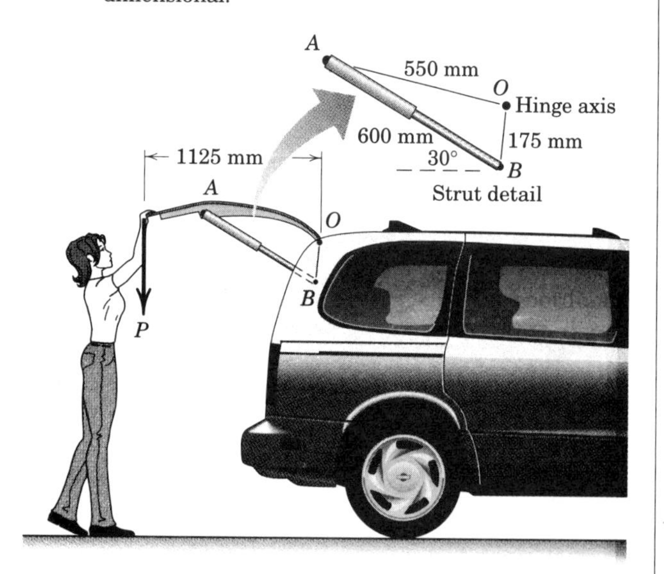

Problem 3/46

3/47 The man pushes the lawn mower at a steady speed with a force *P* that is parallel to the incline. The mass of the mower with attached grass bag is 50 kg with mass center at *G*. If $\theta = 15°$, determine the normal forces N_B and N_C under each pair of wheels *B* and *C*. Neglect friction. Compare with the normal forces for the conditions of $\theta = 0$ and $P = 0$.

Ans. $N_B = 214$ N, $N_C = 260$ N
With $\theta = P = 0$: $N_B = 350$ N, $N_C = 140.1$ N

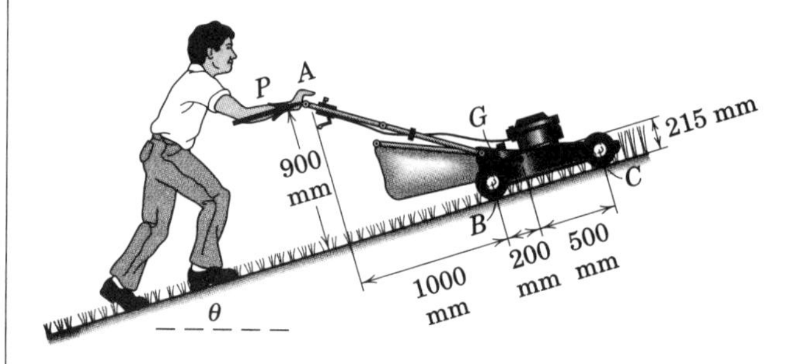

Problem 3/47

3/48 The small crane is mounted on one side of the bed of a pickup truck. For the position $\theta = 40°$, determine the magnitude of the force supported by the pin at *O* and the oil pressure *p* against the 50-mm-diameter piston of the hydraulic cylinder *BC*.

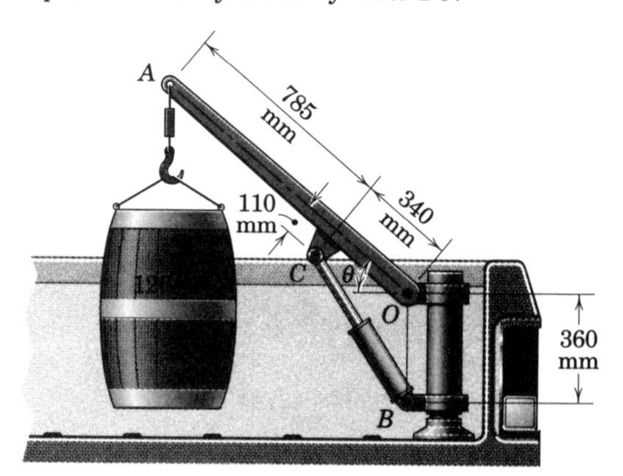

Problem 3/48

3/49 The pin A, which connects the 200-kg steel beam with center of gravity at G to the vertical column, is welded both to the beam and to the column. To test the weld, the 80-kg man loads the beam by exerting a 300-N force on the rope which passes through a hole in the beam as shown. Calculate the torque (couple) M supported by the pin.

Ans. $M = 4.94$ kN·m CCW

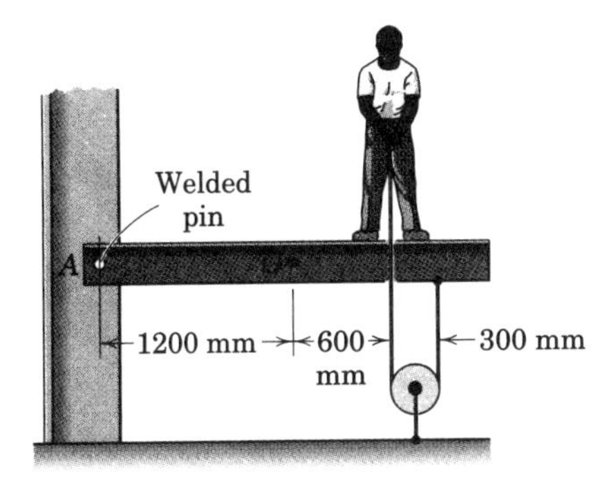

Problem 3/49

3/50 The cargo door for an airplane of circular fuselage section consists of the uniform semicircular cowling AB of mass m. Determine the compression C in the horizontal strut at B to hold the door open in the position shown. Also find an expression for the total force supported by the hinge at A. (Consult Table D/3 of Appendix D for the position of the centroid or mass center of the cowling.)

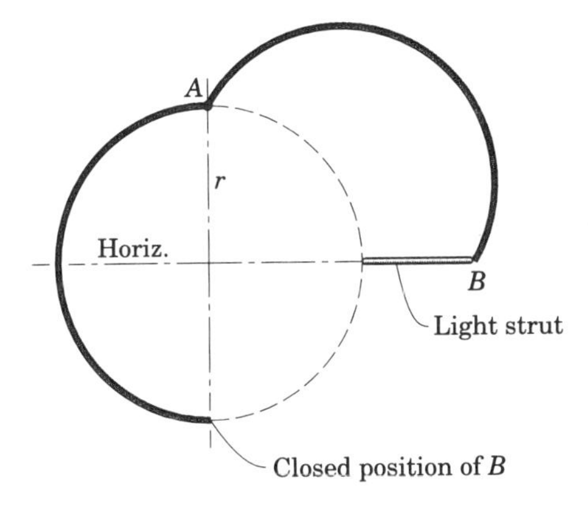

Problem 3/50

3/51 The cargo door for an airplane of circular fuselage section consists of the uniform quarter-circular segment AB of mass m. A detent in the hinge at A holds the door open in the position shown. Determine the moment exerted by the hinge on the door.

Ans. $M_A = 0.709mgr$ CCW

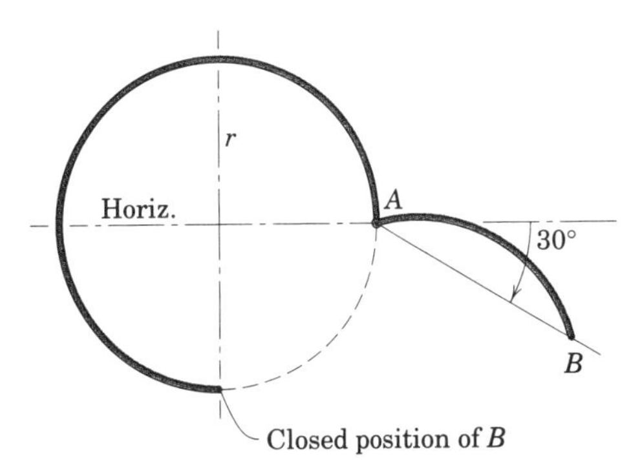

Problem 3/51

3/52 The rubber-tired tractor shown has a mass of 13.5 Mg with center of mass at G and is used for pushing or pulling heavy loads. Determine the load P which the tractor can pull at a constant speed of 5 km/h up the 15-percent grade if the driving force exerted by the ground on each of its four wheels is 80 percent of the normal force under that wheel. Also find the total normal reaction N_B under the rear pair of wheels at B.

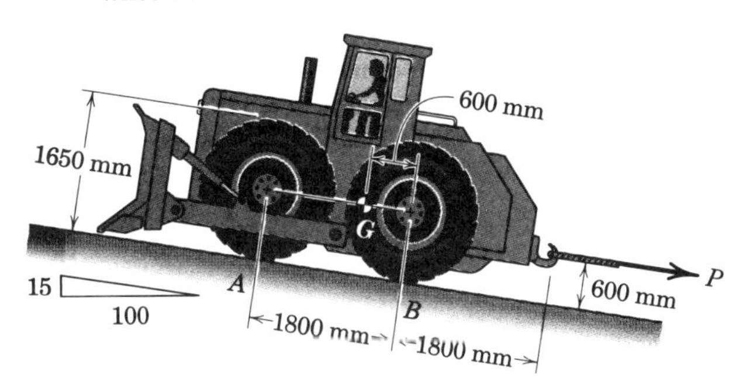

Problem 3/52

3/53 Pulley A delivers a steady torque (moment) of 100 N·m to a pump through its shaft at C. The tension in the lower side of the belt is 600 N. The driving motor B has a mass of 100 kg and rotates clockwise. As a design consideration, determine the magnitude R of the force on the supporting pin at O.

Ans. $R = 1.167$ kN

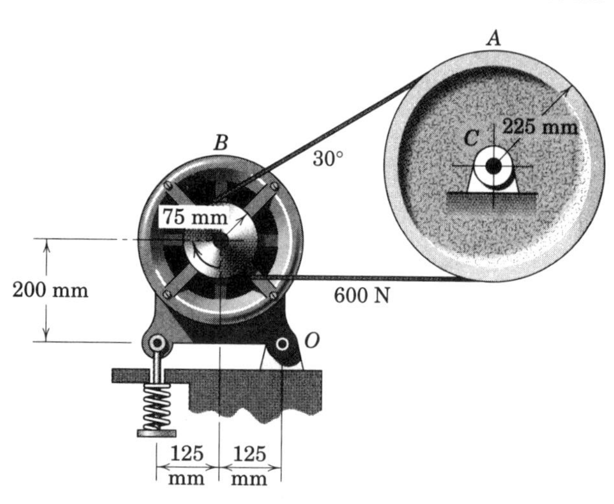

Problem 3/53

3/54 The receiving unit for a wireless microphone system, exclusive of the antenna, has a mass of 1100 grams with mass center at G. A single 375-g half-wave antenna with mass center at C is mounted to the receiver at point O as shown. Plot the reaction forces at A and B and their sum as functions of the antenna angle θ over the range $0 \leq \theta \leq 90°$. Physically interpret your plot. Treat the problem as two-dimensional.

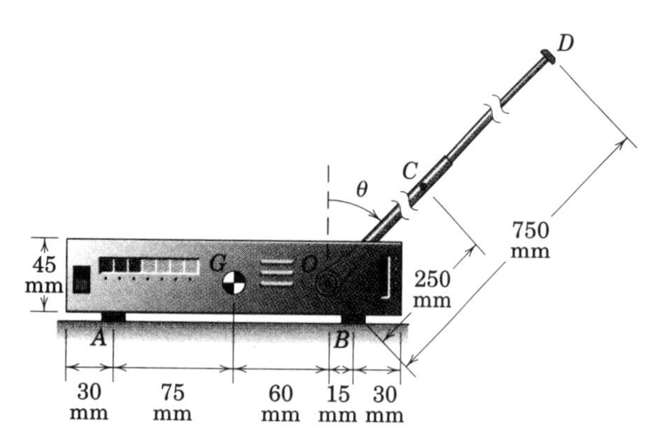

Problem 3/54

3/55 A slender rod of mass m_1 is welded to the horizontal edge of a uniform semicylindrical shell of mass m_2. Determine an expression for the angle θ with the horizontal made by the diameter of the shell through m_1. (Consult Table D/3 in Appendix D to locate the center of gravity of the semicircular section.)

Ans. $\theta = \tan^{-1} \dfrac{\pi m_1}{2m_2}$

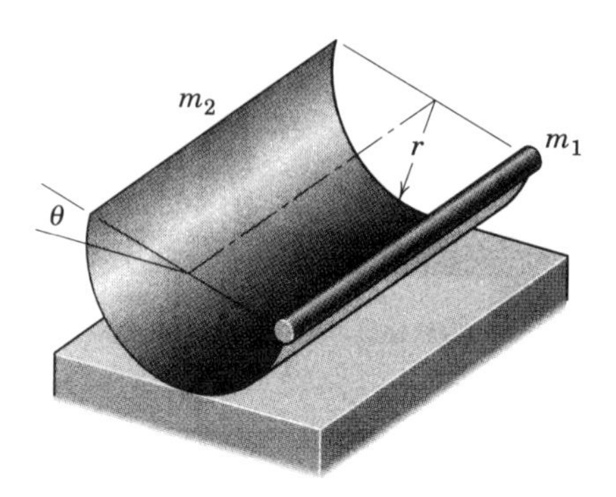

Problem 3/55

3/56 When setting the anchor so that it will dig into the sandy bottom, the engine of the 40-Mg cruiser with center of gravity at G is run in reverse to produce a horizontal thrust T of 2 kN. If the anchor chain makes an angle of 60° with the horizontal, determine the forward shift b of the center of buoyancy from its position when the boat is floating free. The center of buoyancy is the point through which the resultant of the buoyant forces passes.

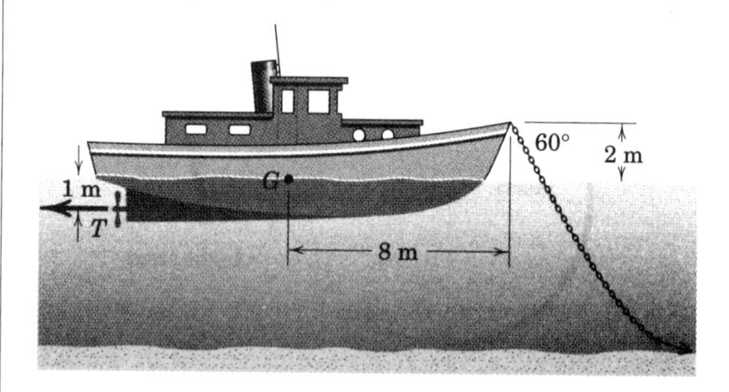

Problem 3/56

3/57 The uniform 400-kg drum is mounted on a line of rollers at A and a line of rollers at B. An 80-kg man moves slowly a distance of 700 mm from the vertical centerline before the drum begins to rotate. All rollers are perfectly free to rotate, except one of them at B which must overcome appreciable friction in its bearing. Calculate the friction force F exerted by that one roller tangent to the drum and find the magnitude R of the force exerted by all rollers at A on the drum for this condition.

Ans. $F = 305$ N, $R = 3770$ N

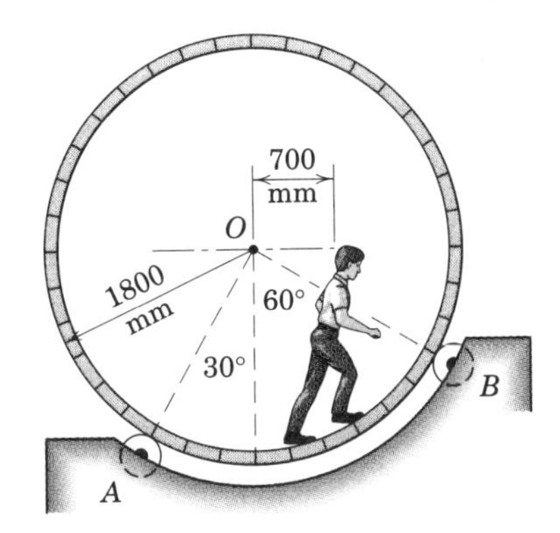

Problem 3/57

3/58 The pipe bender consists of two grooved pulleys mounted and free to turn on a fixed frame. The pipe is bent into the shape shown by a force $P = 300$ N. Calculate the forces supported by the bearings of the pulleys.

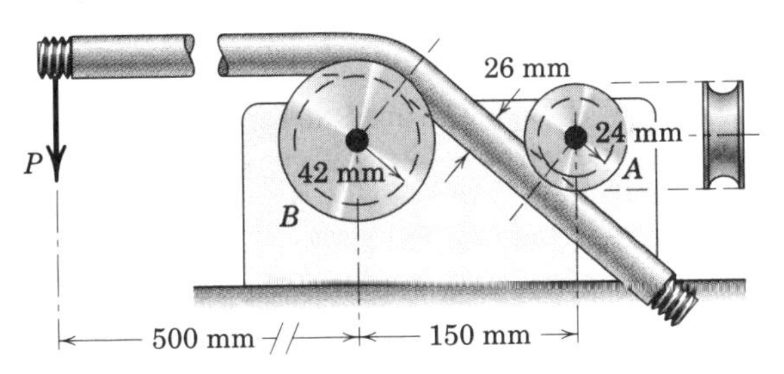

Problem 3/58

▶3/59 A special jig is designed to position large concrete pipe sections (shown in gray) and consists of an 80-Mg sector mounted on a line of rollers at A and a line of rollers at B. One of the rollers at B is a gear which meshes with a ring of gear teeth on the sector so as to turn the sector about its geometric center O. When $\alpha = 0$, a counterclockwise torque of 2460 N·m must be applied to the gear at B to keep the assembly from rotating. When $\alpha = 30°$, a clockwise torque of 4680 N·m is required to prevent rotation. Locate the mass center G of the jig by calculating $\bar{r}$ and θ. Note that the mass center of the pipe section is at O.

Ans. $\bar{r} = 367$ mm, $\theta = 79.8°$

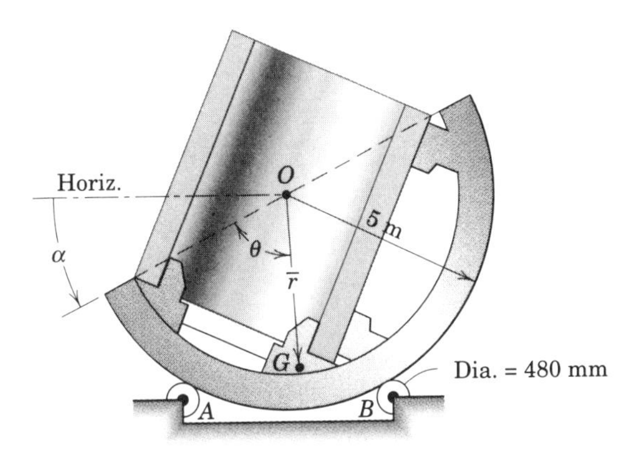

Problem 3/59

▶3/60 The lumbar portion of the human spine supports the entire weight of the upper torso and the force load imposed on it. We consider here the disk (shaded red) between the lowest vertebra of the lumbar region (L_5) and the uppermost vertebra of the sacrum region. (*a*) For the case $L = 0$, determine the compressive force C and the shear force S supported by this disk in terms of the body weight W. The weight W_u of the upper torso (above the disk in question) is 68% of the total body weight W and acts at G_1. The vertical force F which the rectus muscles of the back exert on the upper torso acts as shown in the figure. (*b*) Repeat for the case when the person holds a weight of magnitude $L = W/3$ as shown. State any assumptions.

Ans. (*a*) $C = 0.770W$, $S = 0.669W$
(*b*) $C = 2.53W$, $S = 2.20W$

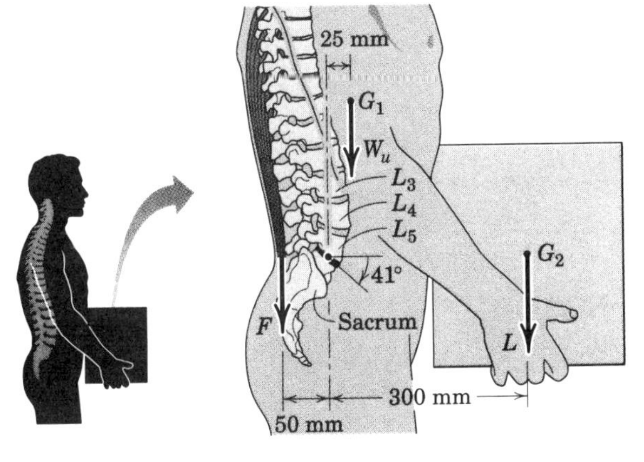

Problem 3/60

Chapter

4

STRUCTURES

CHAPTER OUTLINE

4/1 INTRODUCTION

In Chapter 3 we studied the equilibrium of a single rigid body or a system of connected members treated as a single rigid body. We first drew a free-body diagram of the body showing all forces external to the isolated body and then we applied the force and moment equations of equilibrium. In Chapter 4 we focus on the determination of the forces internal to a structure, that is, forces of action and reaction between the connected members. An engineering structure is any connected system of members built to support or transfer forces and to safely withstand the loads applied to it. To determine the forces internal to an engineering structure, we must dismember the structure and analyze separate free-body diagrams of individual members or combinations of members. This analysis requires careful application of Newton's third law, which states that each action is accompanied by an equal and opposite reaction.

In Chapter 4 we analyze the internal forces acting in several types of structures, namely, trusses, frames, and machines. In this treatment we consider only *statically determinate* structures, which do not have more supporting constraints than are necessary to maintain an equilibrium configuration. Thus, as we have already seen, the equations of equilibrium are adequate to determine all unknown reactions.

The analysis of trusses, frames and machines, and beams under concentrated loads constitutes a straightforward application of the material developed in the previous two chapters. The basic procedure developed in Chapter 3 for isolating a body by constructing a correct free-body diagram is essential for the analysis of statically determinate structures.

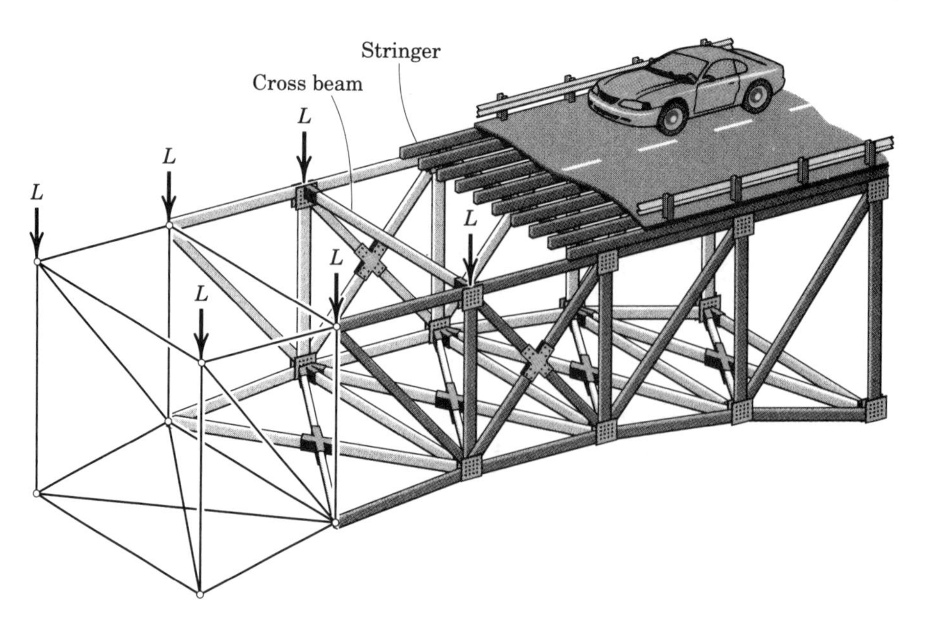

Figure 4/1

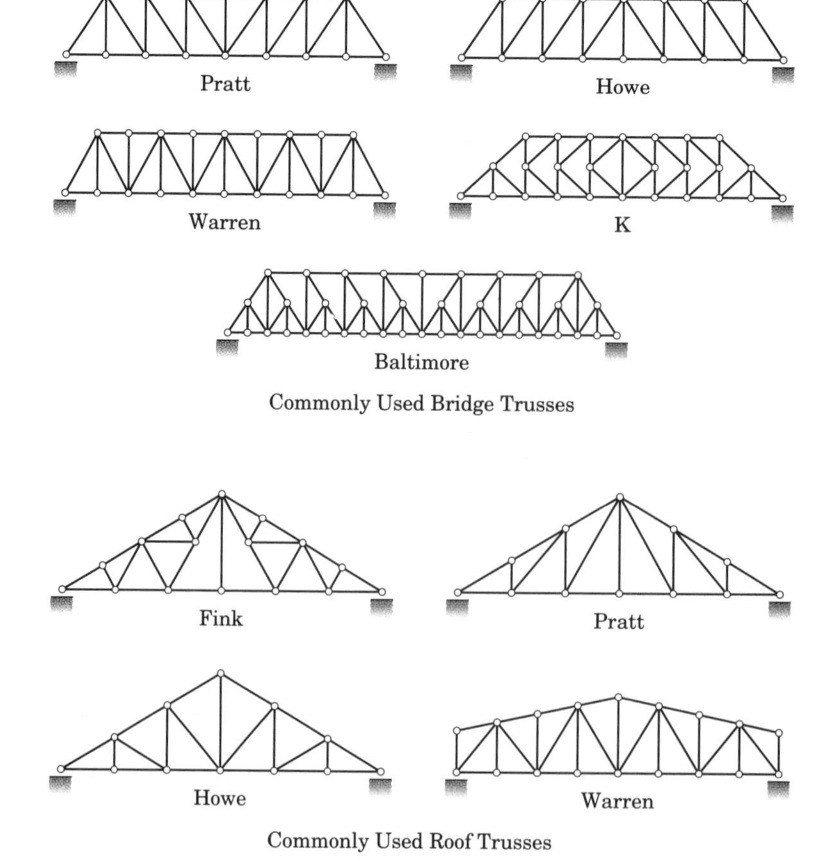

Figure 4/2

4/2 PLANE TRUSSES

A framework composed of members joined at their ends to form a rigid structure is called a *truss*. Bridges, roof supports, derricks, and other such structures are common examples of trusses. Structural members commonly used are I-beams, channels, angles, bars, and special shapes which are fastened together at their ends by welding, riveted connections, or large bolts or pins. When the members of the truss lie essentially in a single plane, the truss is called a *plane truss*.

For bridges and similar structures, plane trusses are commonly utilized in pairs with one truss assembly placed on each side of the structure. A section of a typical bridge structure is shown in Fig. 4/1. The combined weight of the roadway and vehicles is transferred to the longitudinal stringers, then to the cross beams, and finally, with the weights of the stringers and cross beams accounted for, to the upper joints of the two plane trusses which form the vertical sides of the structure. A simplified model of the truss structure is indicated at the left side of the illustration; the forces L represent the joint loadings.

Several examples of commonly used trusses which can be analyzed as plane trusses are shown in Fig. 4/2.

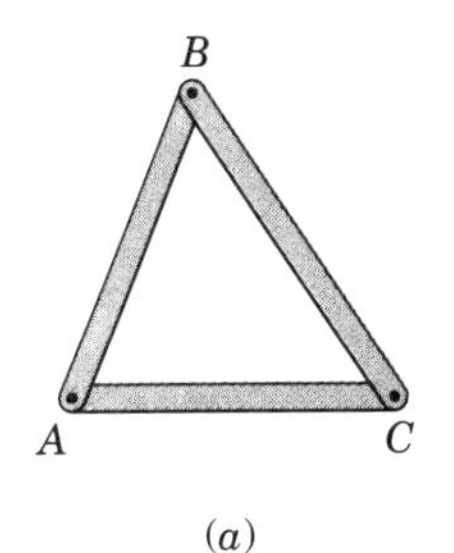

(*a*)

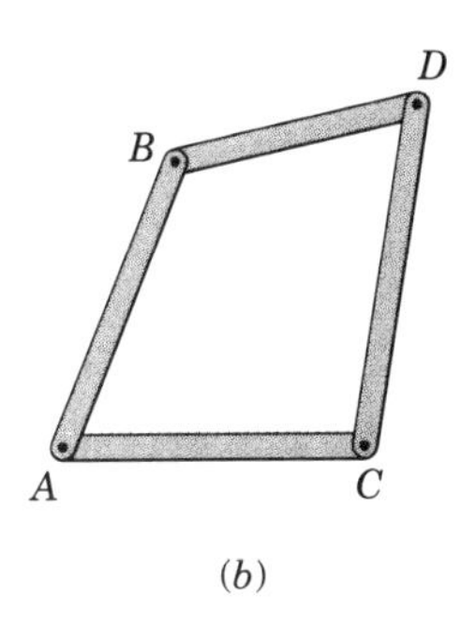

(*b*)

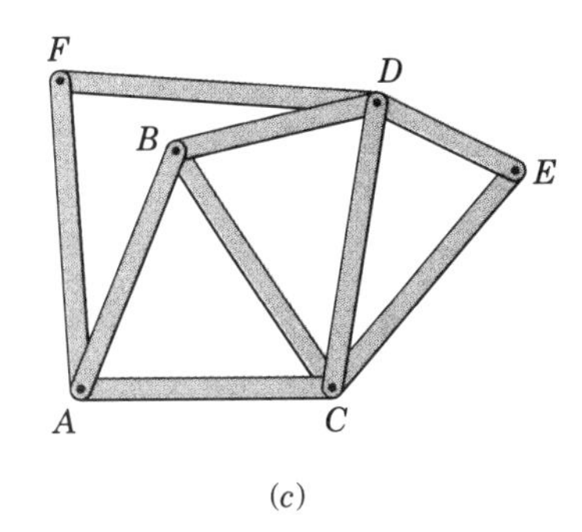

(*c*)

Figure 4/3

Simple Trusses

The basic element of a plane truss is the triangle. Three bars joined by pins at their ends, Fig. 4/3*a*, constitute a rigid frame. The term *rigid* is used to mean noncollapsible and also to mean that deformation of the members due to induced internal strains is negligible. On the other hand, four or more bars pin-jointed to form a polygon of as many sides constitute a nonrigid frame. We can make the nonrigid frame in Fig. 4/3*b* rigid, or stable, by adding a diagonal bar joining A and D or B and C and thereby forming two triangles. We can extend the structure by adding additional units of two end-connected bars, such as DE and CE or AF and DF, Fig. 4/3*c*, which are pinned to two fixed joints. In this way the entire structure will remain rigid.

Structures built from a basic triangle in the manner described are known as *simple trusses*. When more members are present than are needed to prevent collapse, the truss is statically indeterminate. A statically indeterminate truss cannot be analyzed by the equations of equilibrium alone. Additional members or supports which are not necessary for maintaining the equilibrium configuration are called *redundant*.

To design a truss we must first determine the forces in the various members and then select appropriate sizes and structural shapes to withstand the forces. Several assumptions are made in the force analysis of simple trusses. First, we assume all members to be *two-force members*. A two-force member is one in equilibrium under the action of two forces only, as defined in general terms with Fig. 3/4 in Art. 3/3. Each member of a truss is normally a straight link joining the two points of application of force. The two forces are applied at the ends of the member and are necessarily equal, opposite, and *collinear* for equilibrium.

The member may be in tension or compression, as shown in Fig. 4/4. When we represent the equilibrium of a portion of a two-force member, the tension T or compression C acting on the cut section is the same

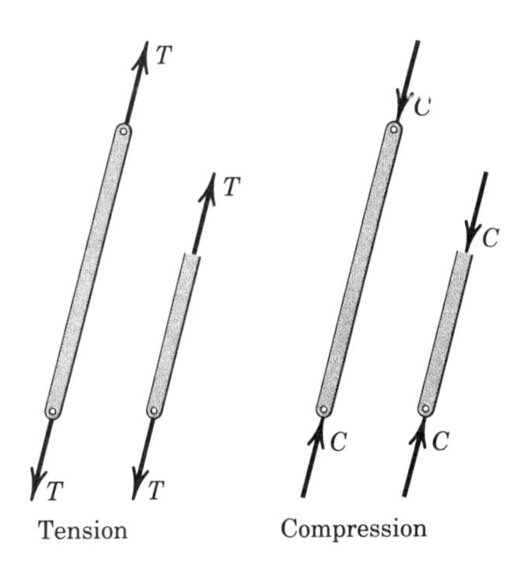

Figure 4/4

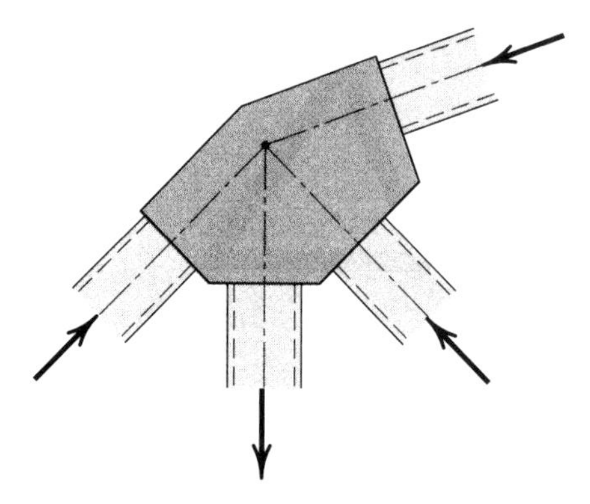

Figure 4/5

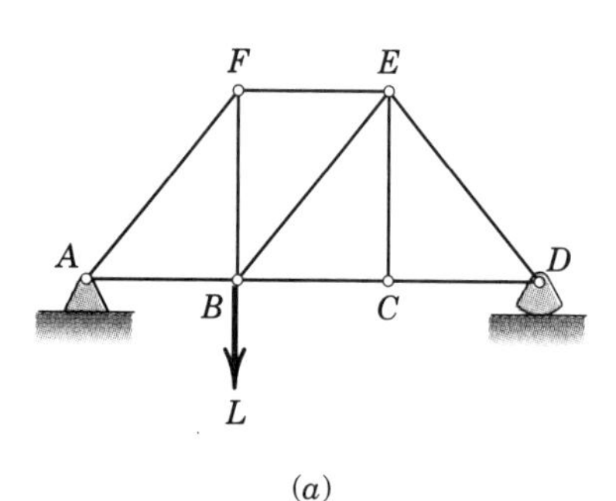

(a)

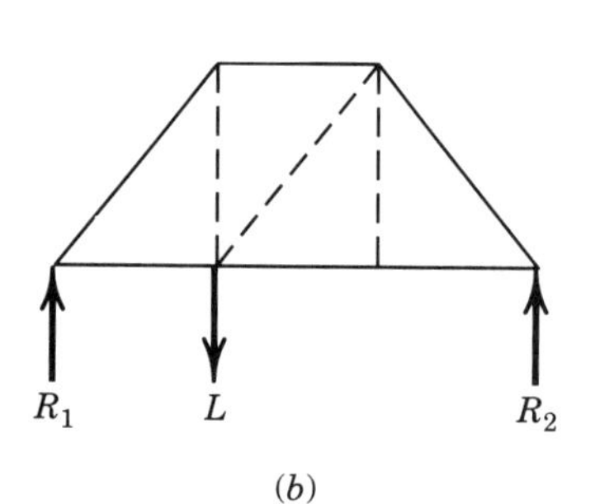

(b)

Figure 4/6

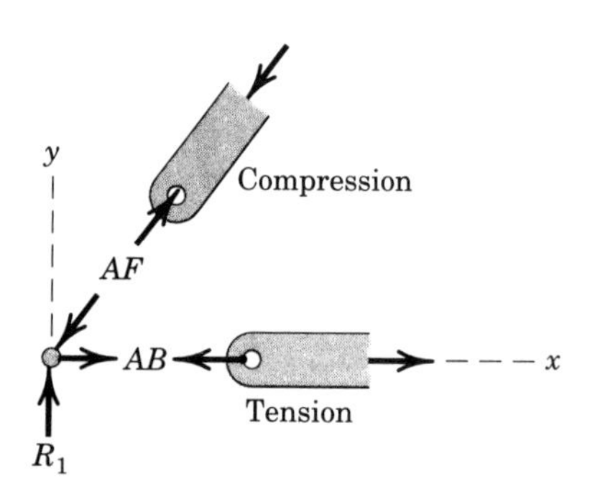

Figure 4/7

for all sections. We assume here that the weight of the member is small compared with the force it supports. If it is not, or if we must account for the small effect of the weight, we can replace the weight W of the member by two forces, each $W/2$ if the member is uniform, with one force acting at each end of the member. These forces, in effect, are treated as loads externally applied to the pin connections. Accounting for the weight of a member in this way gives the correct result for the average tension or compression along the member but will not account for the effect of bending of the member.

Truss Connections and Supports

When welded or riveted connections are used to join structural members, we may usually assume that the connection is a pin joint if the centerlines of the members are concurrent at the joint as in Fig. 4/5.

We also assume in the analysis of simple trusses that all external forces are applied at the pin connections. This condition is satisfied in most trusses. In bridge trusses the deck is usually laid on cross beams which are supported at the joints, as shown in Fig. 4/1.

For large trusses, a roller, rocker, or some kind of slip joint is used at one of the supports to provide for expansion and contraction due to temperature changes and for deformation from applied loads. Trusses and frames in which no such provision is made are statically indeterminate, as explained in Art. 3/3. Fig. 3/1 shows examples of such joints.

Two methods for the force analysis of simple trusses will be given. Each method will be explained for the simple truss shown in Fig. 4/6*a*. The free-body diagram of the truss as a whole is shown in Fig. 4/6*b*. The external reactions are usually determined first, by applying the equilibrium equations to the truss as a whole. Then the force analysis of the remainder of the truss is performed.

4/3 Method of Joints

This method for finding the forces in the members of a truss consists of satisfying the conditions of equilibrium for the forces acting on the connecting pin of each joint. The method therefore deals with the equilibrium of concurrent forces, and only two independent equilibrium equations are involved.

We begin the analysis with any joint where at least one known load exists and where not more than two unknown forces are present. The solution may be started with the pin at the left end. Its free-body diagram is shown in Fig. 4/7. With the joints indicated by letters, we usually designate the force in each member by the two letters defining the ends of the member. The proper directions of the forces should be evident by inspection for this simple case. The free-body diagrams of portions of members AF and AB are also shown to clearly indicate the mechanism of the action and reaction. The member AB actually makes contact on the left side of the pin, although the force AB is drawn from the right side and is shown acting away from the pin. Thus, if we consistently draw the force arrows on the *same* side of the pin as the member, then tension (such as AB) will always be indicated by an arrow *away* from the pin, and compression (such as AF) will always be indicated by an arrow *toward* the pin. The magnitude of AF is obtained from the equation $\Sigma F_y = 0$ and AB is then found from $\Sigma F_x = 0$.

Joint F may be analyzed next, since it now contains only two unknowns, EF and BF. Proceeding to the next joint having no more than two unknowns, we subsequently analyze joints B, C, E, and D in that order. Fig. 4/8 shows the free-body diagram of each joint and its corresponding force polygon, which represents graphically the two equilibrium conditions $\Sigma F_x = 0$ and $\Sigma F_y = 0$. The numbers indicate the order in which the joints are analyzed. We note that, when joint D is finally reached, the computed reaction R_2 must be in equilibrium with the forces in members CD and ED, which were determined previously from the two neighboring joints. This requirement provides a check on the correctness of our work. Note that isolation of joint C shows that the force in CE is zero when the equation $\Sigma F_y = 0$ is applied. The force in

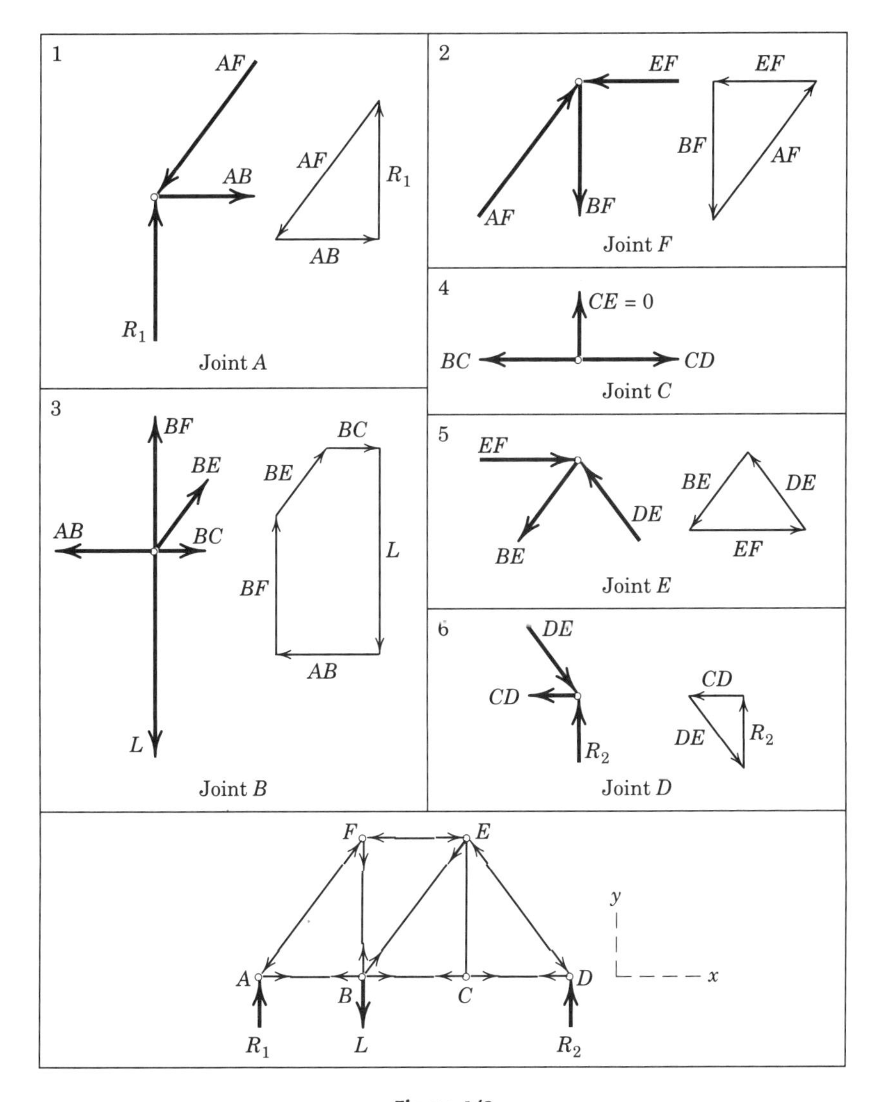

Figure 4/8

this member would not be zero, of course, if an external vertical load were applied at C.

It is often convenient to indicate the tension T and compression C of the various members directly on the original truss diagram by drawing arrows away from the pins for tension and toward the pins for compression. This designation is illustrated at the bottom of Fig. 4/8.

Sometimes we cannot initially assign the correct direction of one or both of the unknown forces acting on a given pin. If so, we may make an arbitrary assignment. A negative computed force value indicates that the initially assumed direction is incorrect.

Internal and External Redundancy

If a plane truss has more external supports than are necessary to ensure a stable equilibrium configuration, the truss as a whole is statically indeterminate, and the extra supports constitute *external* redundancy. If a truss has more internal members than are necessary to prevent collapse when the truss is removed from its supports, then the extra members constitute *internal* redundancy and the truss is again statically indeterminate.

For a truss which is statically determinate externally, there is a definite relation between the number of its members and the number of its joints necessary for internal stability without redundancy. Because we can specify the equilibrium of each joint by two scalar force equations, there are in all $2j$ such equations for a truss with j joints. For the entire truss composed of m two-force members and having the maximum of three unknown support reactions, there are in all $m + 3$ unknowns (m tension or compression forces and three reactions). Thus, for any plane truss, the equation $m + 3 = 2j$ will be satisfied if the truss is statically determinate internally.

A *simple* plane truss, formed by starting with a triangle and adding two new members to locate each new joint with respect to the existing structure, satisfies the relation automatically. The condition holds for the initial triangle, where $m = j = 3$, and m increases by 2 for each added joint while j increases by 1. Some other (nonsimple) statically determinate trusses, such as the K-truss in Fig. 4/2, are arranged differently, but can be seen to satisfy the same relation.

This equation is a necessary condition for stability but it is not a sufficient condition, since one or more of the m members can be arranged in such a way as not to contribute to a stable configuration of the entire truss. If $m + 3 > 2j$, there are more members than independent equations, and the truss is statically indeterminate internally with redundant members present. If $m + 3 < 2j$, there is a deficiency of internal members, and the truss is unstable and will collapse under load.

Special Conditions

We often encounter several special conditions in the analysis of trusses. When two collinear members are under compression, as indicated in Fig. 4/9a, it is necessary to add a third member to maintain

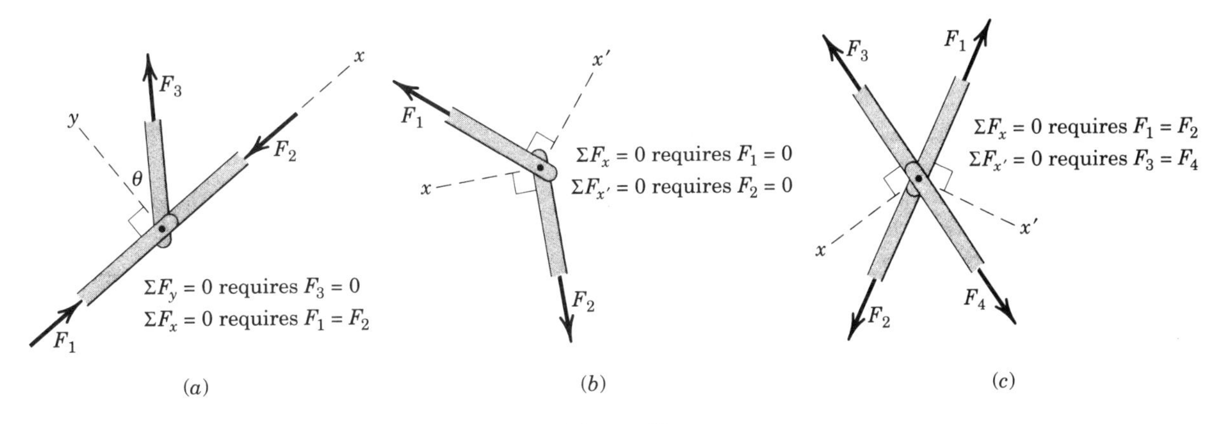

Figure 4/9

alignment of the two members and prevent buckling. We see from a force summation in the y-direction that the force F_3 in the third member must be zero and from the x-direction that $F_1 = F_2$. This conclusion holds regardless of the angle θ and holds also if the collinear members are in tension. If an external force with a component in the y-direction were applied to the joint, then F_3 would no longer be zero.

When two noncollinear members are joined as shown in Fig. 4/9*b*, then in the absence of an externally applied load at this joint, the forces in both members must be zero, as we can see from the two force summations.

When two pairs of collinear members are joined as shown in Fig. 4/9*c*, the forces in each pair must be equal and opposite. This conclusion follows from the force summations indicated in the figure.

Truss panels are frequently cross-braced as shown in Fig. 4/10*a*. Such a panel is statically indeterminate if each brace can support either tension or compression. However, when the braces are flexible members incapable of supporting compression, as are cables, then only the tension member acts and we can disregard the other member. It is usually evident from the asymmetry of the loading how the panel will deflect. If the deflection is as indicated in Fig. 4/10*b*, then member *AB* should be retained and *CD* disregarded. When this choice cannot be made by inspection, we may arbitrarily select the member to be retained. If the assumed tension turns out to be positive upon calculation, then the choice was correct. If the assumed tension force turns out to be negative, then the opposite member must be retained and the calculation redone.

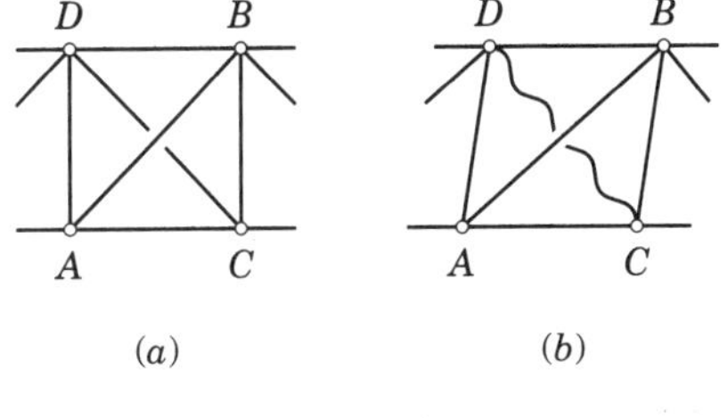

Figure 4/10

We can avoid simultaneous solution of the equilibrium equations for two unknown forces at a joint by a careful choice of reference axes. Thus, for the joint indicated schematically in Fig. 4/11 where L is known and F_1 and F_2 are unknown, a force summation in the x-direction eliminates reference to F_1 and a force summation in the x'-direction eliminates reference to F_2. When the angles involved are not easily found, then a simultaneous solution of the equations using one set of reference directions for both unknowns may be preferable.

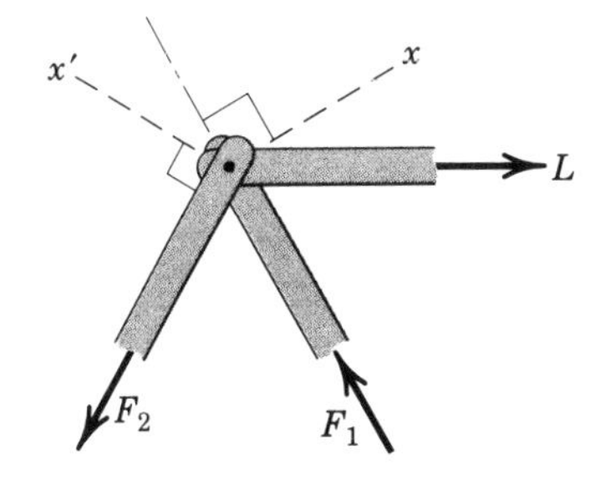

Figure 4/11

Sample Problem 4/1

Compute the force in each member of the loaded cantilever truss by the method of joints.

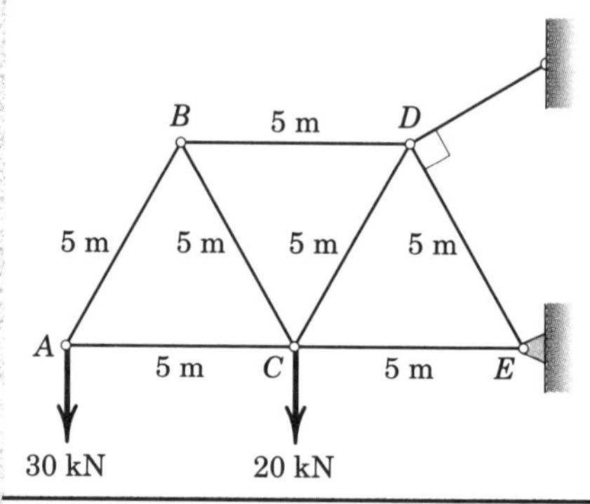

Solution. If it were not desired to calculate the external reactions at D and E, the analysis for a cantilever truss could begin with the joint at the loaded end. However, this truss will be analyzed completely, so the first step will be to compute the external forces at D and E from the free-body diagram of the truss as a whole. The equations of equilibrium give

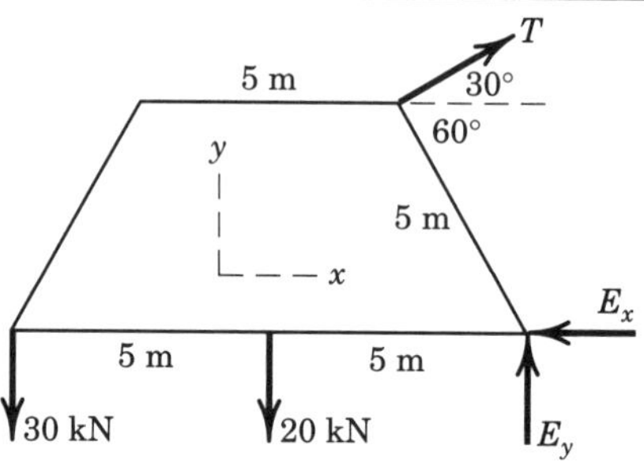

$[\Sigma M_E = 0] \qquad 5T - 20(5) - 30(10) = 0 \qquad T = 80 \text{ kN}$

$[\Sigma F_x = 0] \qquad 80 \cos 30° - E_x = 0 \qquad E_x = 69.3 \text{ kN}$

$[\Sigma F_y = 0] \qquad 80 \sin 30° + E_y - 20 - 30 = 0 \qquad E_y = 10 \text{ kN}$

Next we draw free-body diagrams showing the forces acting on each of the connecting pins. The correctness of the assigned directions of the forces is verified when each joint is considered in sequence. There should be no question about the correct direction of the forces on joint A. Equilibrium requires

$[\Sigma F_y = 0] \qquad 0.866AB - 30 = 0 \qquad AB = 34.6 \text{ kN } T$ *Ans.*

$[\Sigma F_x = 0] \qquad AC - 0.5(34.6) = 0 \qquad AC = 17.32 \text{ kN } C$ *Ans.*

① where T stands for tension and C stands for compression.

Joint B must be analyzed next, since there are more than two unknown forces on joint C. The force BC must provide an upward component, in which case BD must balance the force to the left. Again the forces are obtained from

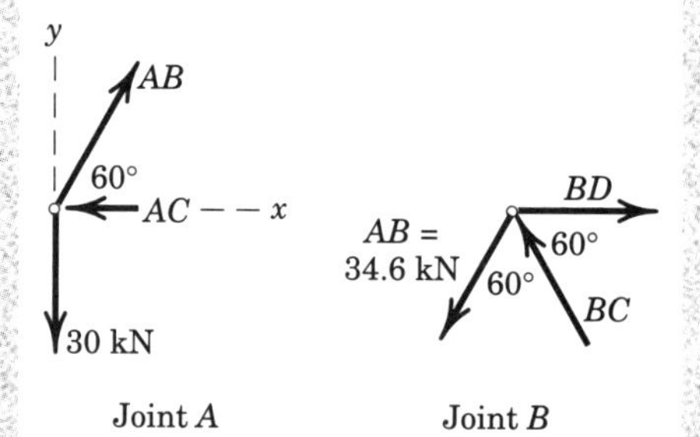

$[\Sigma F_y = 0] \qquad 0.866BC - 0.866(34.6) = 0 \qquad BC = 34.6 \text{ kN } C$ *Ans.*

$[\Sigma F_x = 0] \qquad BD - 2(0.5)(34.6) = 0 \qquad BD = 34.6 \text{ kN } T$ *Ans.*

Joint C now contains only two unknowns, and these are found in the same way as before:

$[\Sigma F_y = 0] \qquad 0.866CD - 0.866(34.6) - 20 = 0$

$CD = 57.7 \text{ kN } T$ *Ans.*

$[\Sigma F_x = 0] \qquad CE - 17.32 - 0.5(34.6) - 0.5(57.7) = 0$

$CE = 63.5 \text{ kN } C$ *Ans.*

Finally, from joint E there results

$[\Sigma F_y = 0] \qquad 0.866DE = 10 \qquad DE = 11.55 \text{ kN } C$ *Ans.*

and the equation $\Sigma F_x = 0$ checks.

Helpful Hint

① It should be stressed that the tension/compression designation refers to the *member*, not the joint. Note that we draw the force arrow on the same side of the joint as the member which exerts the force. In this way tension (arrow away from the joint) is distinguished from compression (arrow toward the joint).

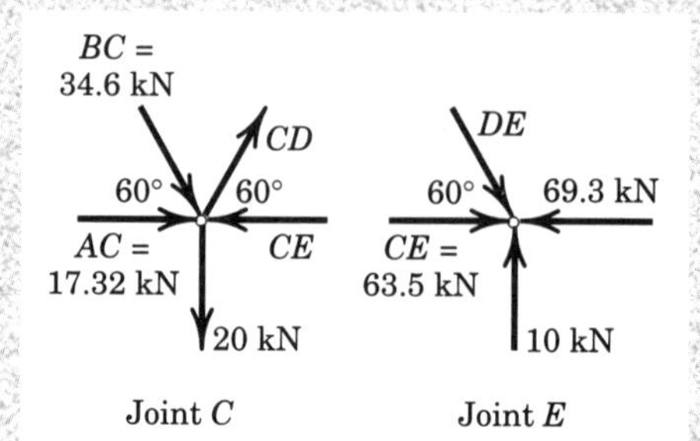

PROBLEMS

Introductory Problems

4/1 Determine the force in each member of the simple equilateral truss.
Ans. $AB = 736$ N T, $AC = 368$ N T, $BC = 736$ N C

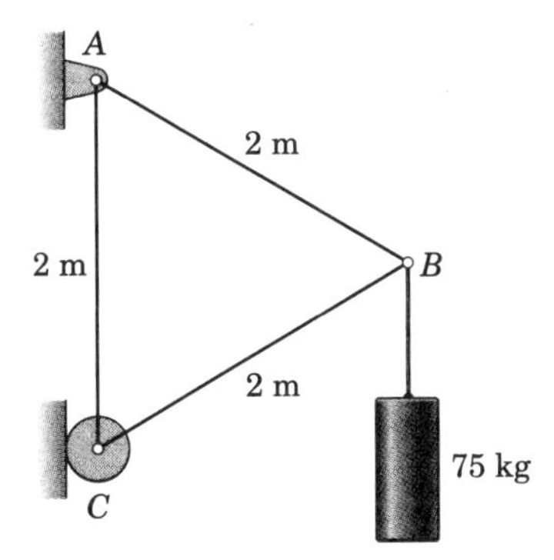

Problem 4/1

4/2 Determine the force in each member of the loaded truss. Discuss the effects of varying the angle of the 45° support surface at C.

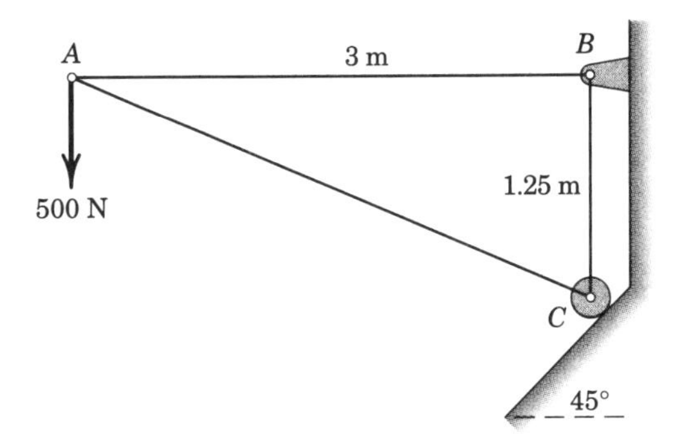

Problem 4/2

4/3 Determine the force in each member of the truss. Note the presence of any zero-force members.
Ans. $AB = 5$ kN T, $BC = 5\sqrt{2}$ kN C
$CD = 15$ kN C, $AC = 5\sqrt{5}$ kN T, $AD = 0$

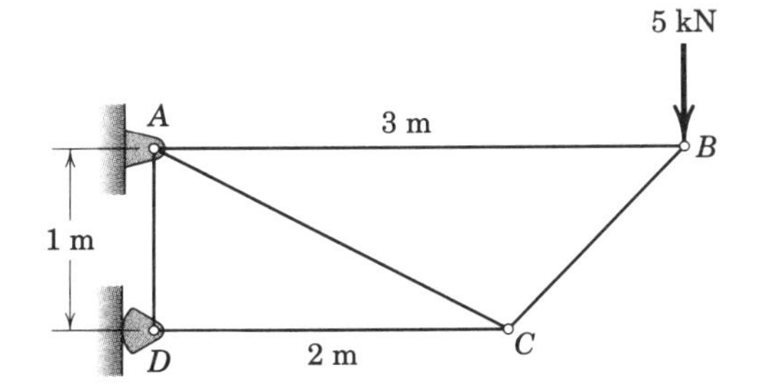

Problem 4/3

4/4 Calculate the forces in members BE and BD of the loaded truss.

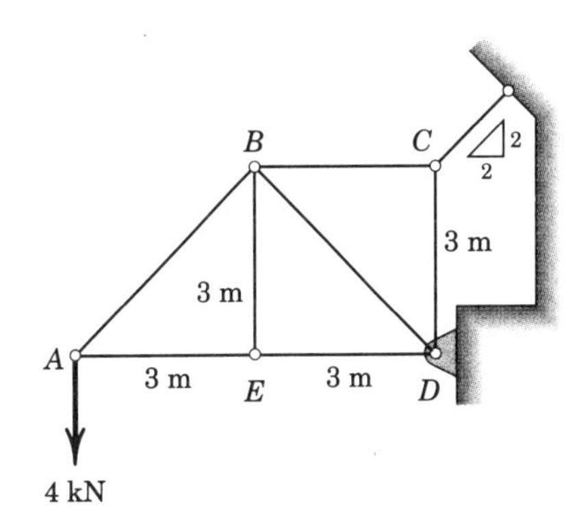

Problem 4/4

4/5 Determine the force in each member of the loaded truss.
Ans. $AB = 12$ kN T, $AE = 3$ kN C
$BC = 5.20$ kN T, $BD = 6$ kN T
$BE = 5.20$ kN C, $CD = DE = 6$ kN C

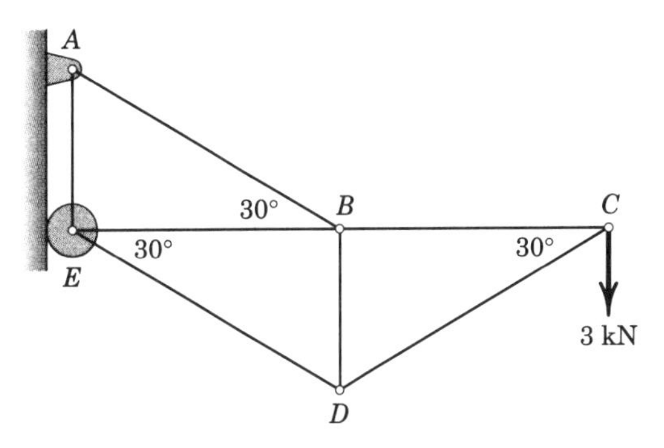

Problem 4/5

4/6 Calculate the force in each member of the loaded truss.

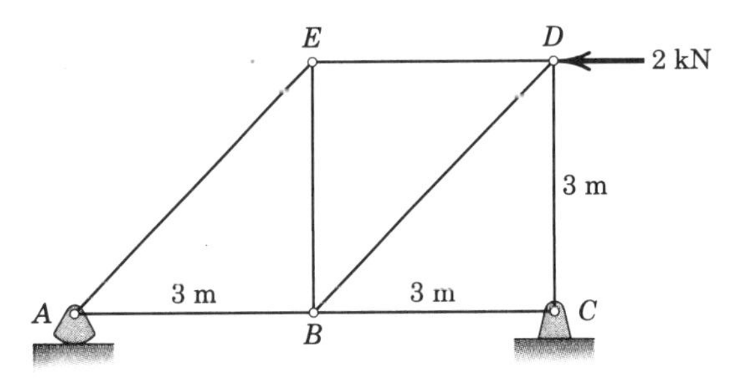

Problem 4/6

4/7 Determine the force in each member of the loaded truss. Make use of the symmetry of the truss and of the loading.

Ans. $AB = DE = 96.0$ kN C
$AH = EF = 75$ kN T, $BC = CD = 75$ kN C
$BH = CG = DF = 60$ kN T
$CH = CF = 48.0$ kN C, $GH = FG = 112.5$ kN T

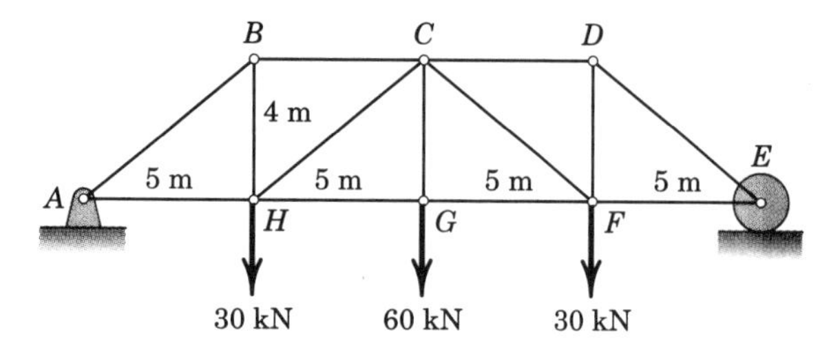

Problem 4/7

4/8 Determine the force in each member of the loaded truss. All triangles are isosceles.

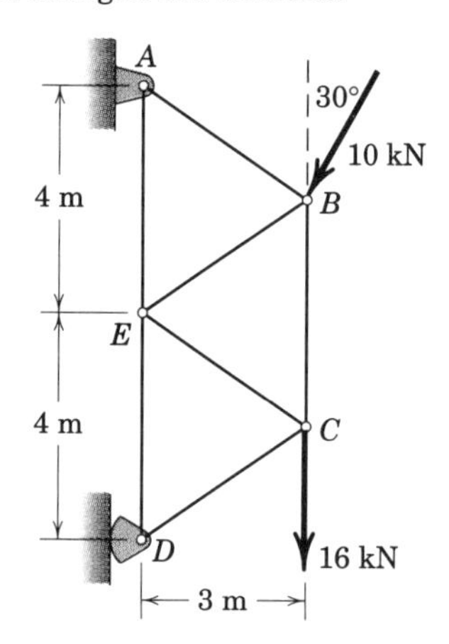

Problem 4/8

4/9 Determine the force in each member of the loaded truss. All triangles are equilateral.

Ans. $AB = 9\sqrt{3}$ kN C, $AE = 5\sqrt{3}$ kN T
$BC = \frac{26}{3}\sqrt{3}$ kN C, $BD = 3\sqrt{3}$ kN C, $BE = \frac{7}{3}\sqrt{3}$ kN C
$CD = \frac{13}{3}\sqrt{3}$ kN T, $DE = \frac{11}{3}\sqrt{3}$ kN T

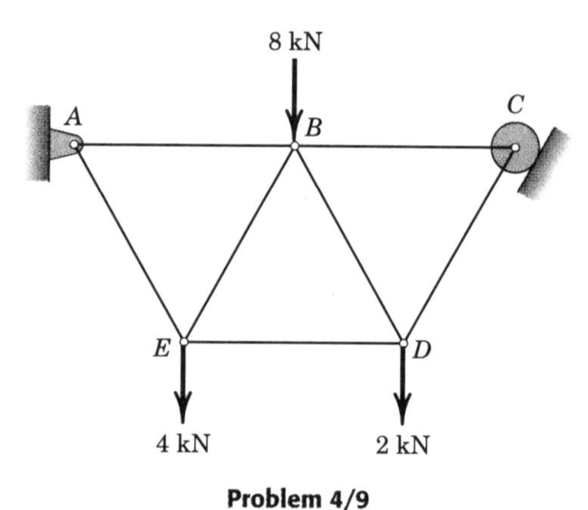

Problem 4/9

Representative Problems

4/10 Solve for the forces in members BE and BD of the truss which supports the load L. All interior angles are 60° or 120°.

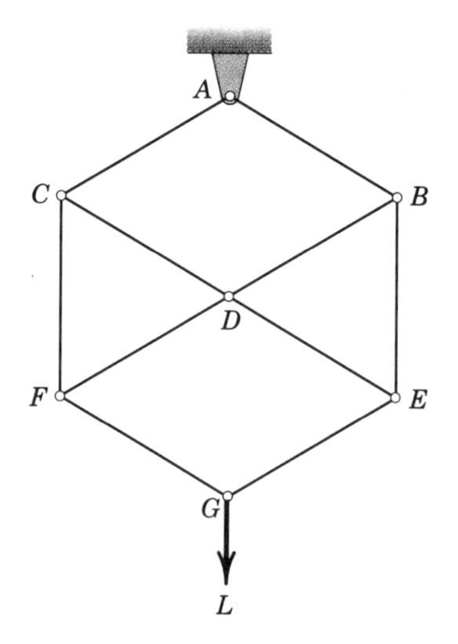

Problem 4/10

4/11 Determine the force in member AC of the loaded truss. The two quarter-circular members act as two-force members.

Ans. $AC = \dfrac{L}{2}\,T$

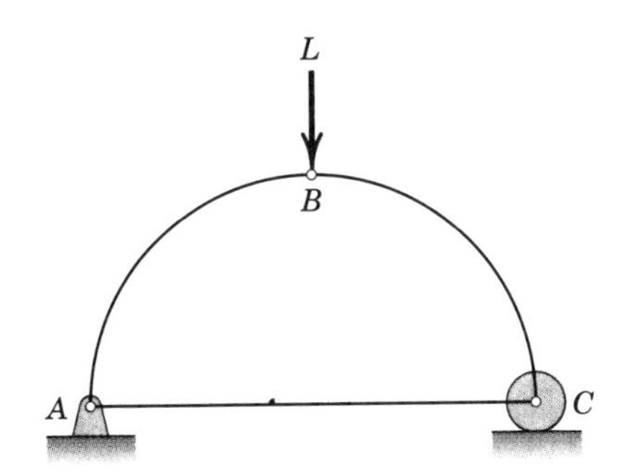

Problem 4/11

4/12 Calculate the forces in members CG and CF for the truss shown.

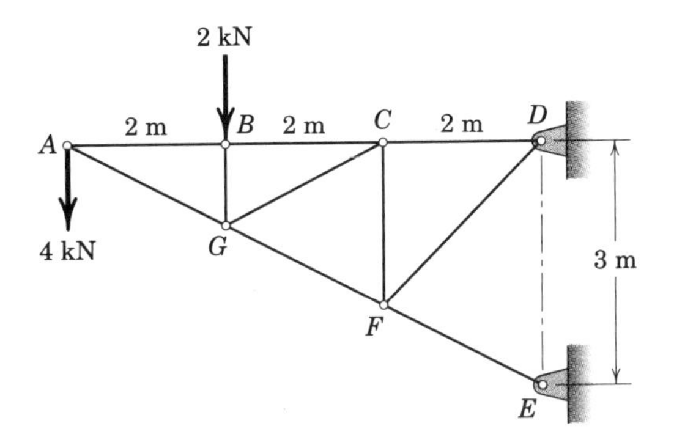

Problem 4/12

4/13 Each member of the truss is a uniform 8-m bar with a mass of 400 kg. Calculate the average tension or compression in each member due to the weights of the members.

Ans. $AB = BC = 5.66$ kN T
$AE = CD = 11.33$ kN C
$BD = BE = 4.53$ kN T
$ED = 7.93$ kN C

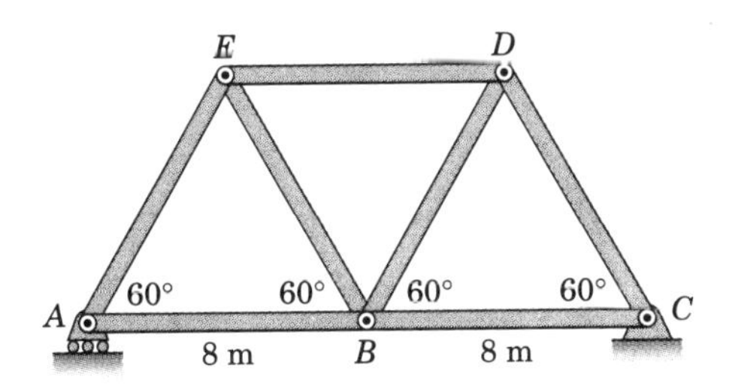

Problem 4/13

4/14 A drawbridge is being raised by a cable EI. The four joint loadings shown result from the weight of the roadway. Determine the forces in members EF, DE, DF, CD, and FG.

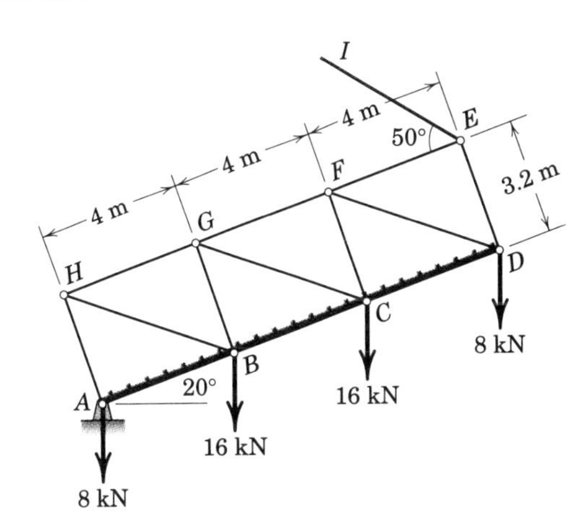

Problem 4/14

4/15 The equiangular truss is loaded and supported as shown. Determine the forces in all members in terms of the horizontal load L.

Ans. $AB = BC = L\ T$, $AF = EF = L\ C$
$DE = CD = L/2\ T$, $BF = DF = BD = 0$

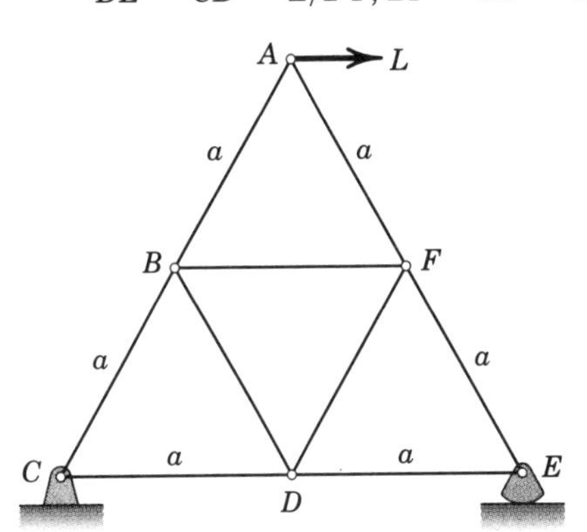

Problem 4/15

4/16 Determine the forces in members BI, CI, and HI for the loaded truss. All angles are 30°, 60°, or 90°.

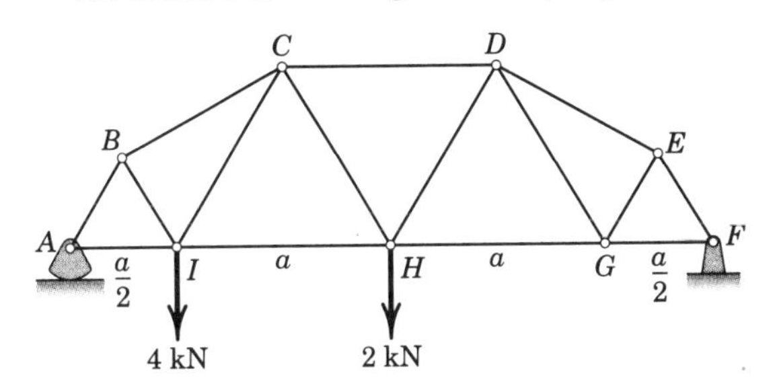

Problem 4/16

4/17 A snow load transfers the forces shown to the upper joints of a Pratt roof truss. Neglect any horizontal reactions at the supports and solve for the forces in all members.

Ans. $AB = DE = BC = CD = 3.35$ kN C
$AH = EF = 3$ kN T, $BH = DF = 1$ kN C
$CF = CH = 1.414$ kN T, $FG = GH = 2$ kN T

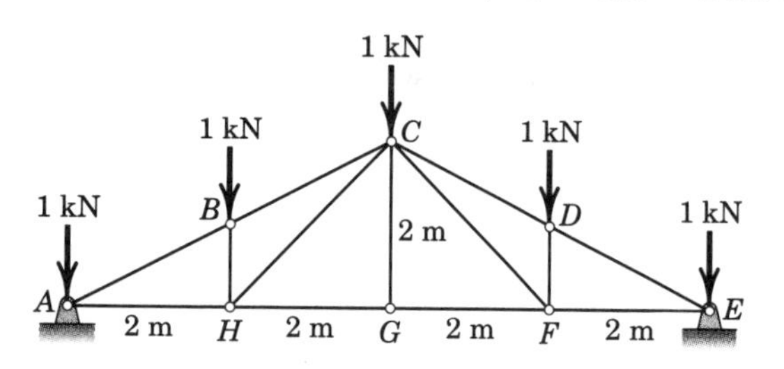

Problem 4/17

4/18 The loading of Prob. 4/17 is shown applied to a Howe roof truss. Neglect any horizontal reactions at the supports and solve for the forces in all members. Compare with the results of Prob. 4/17.

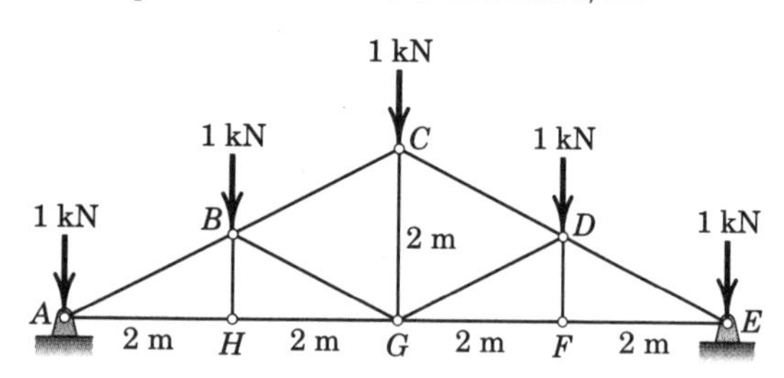

Problem 4/18

4/19 Calculate the forces in members CF, CG, and EF of the loaded truss.

Ans. $CF = 9.23$ kN C, $CG = 25.0$ kN T, $EF = 0$

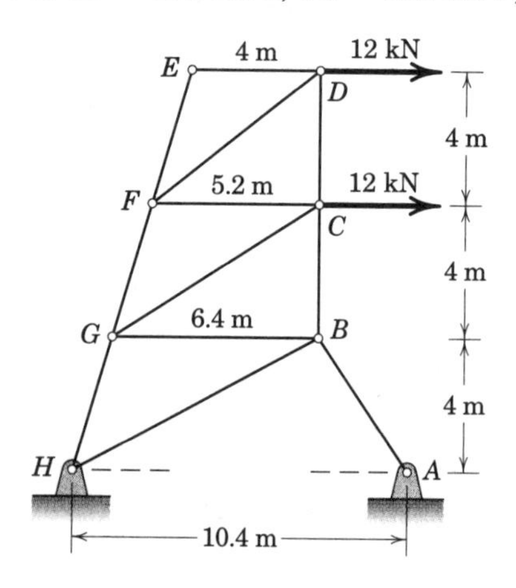

Problem 4/19

4/20 Determine the force in each member of the pair of trusses which support the 20-kN load at their common joint C.

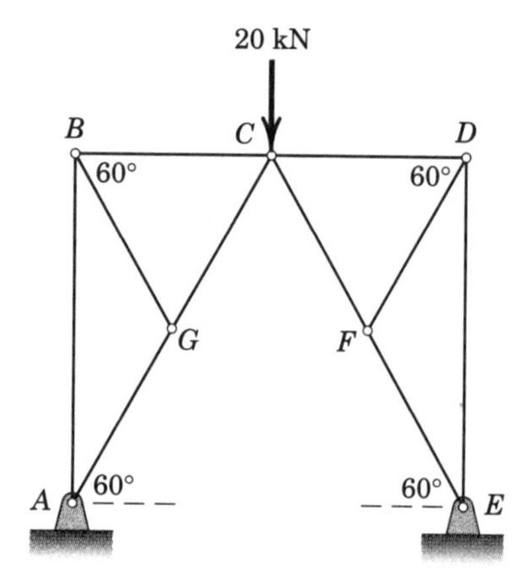

Problem 4/20

4/21 The rectangular frame is composed of four perimeter two-force members and two cables AC and BD which are incapable of supporting compression. Determine the forces in all members due to the load L in position (a) and then in position (b).

Ans. (a) $AB = AD = BD = 0$, $BC = L$ C

$AC = \frac{5L}{3}$ T, $CD = \frac{4L}{3}$ C

(b) $AB = AD = BC = BD = 0$

$AC = \frac{5L}{3}$ T, $CD = \frac{4L}{3}$ C

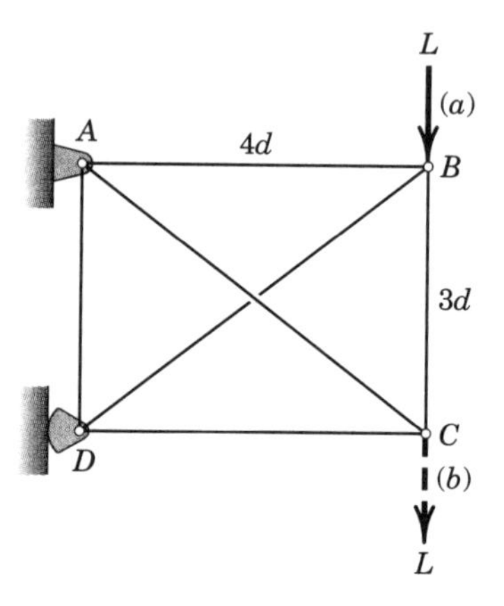

Problem 4/21

4/22 Determine the forces in members AB, CG, and DE of the loaded truss.

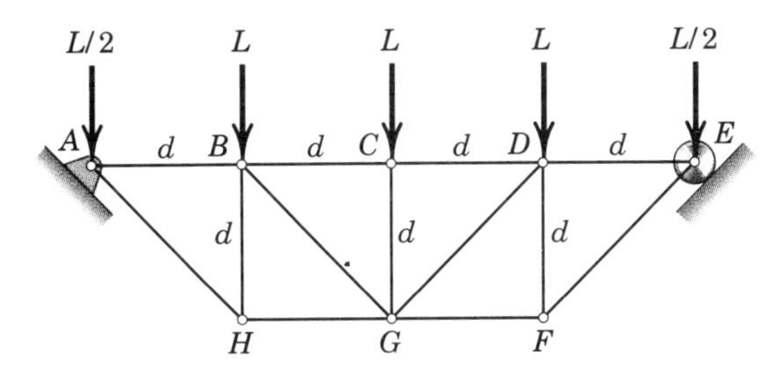

Problem 4/22

4/23 The movable gantry is used to erect and prepare a 500-Mg rocket for firing. The primary structure of the gantry is approximated by the symmetrical plane truss shown, which is statically indeterminate. As the gantry is positioning a 60-Mg section of the rocket suspended from A, strain-gage measurements indicate a compressive force of 50 kN in member AB and a tensile force of 120 kN in member CD due to the 60-Mg load. Calculate the corresponding forces in members BF and EF.

Ans. BF = 188.4 kN C, EF = 120 kN T

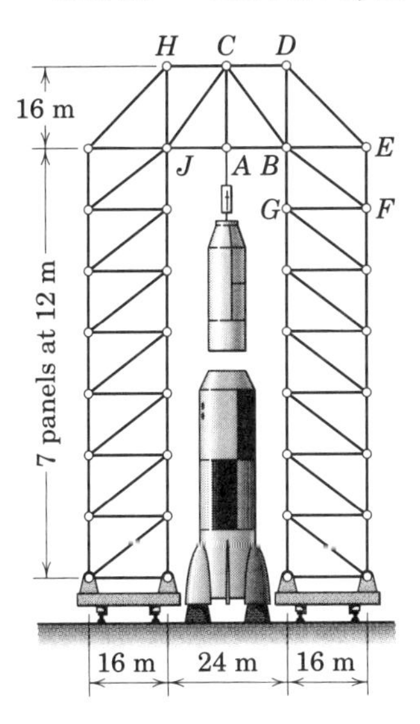

Problem 4/23

4/24 Verify the fact that each of the trusses contains one or more elements of redundancy and propose two separate changes, either one of which would remove the redundancy and produce complete statical determinacy. All members can support compression as well as tension.

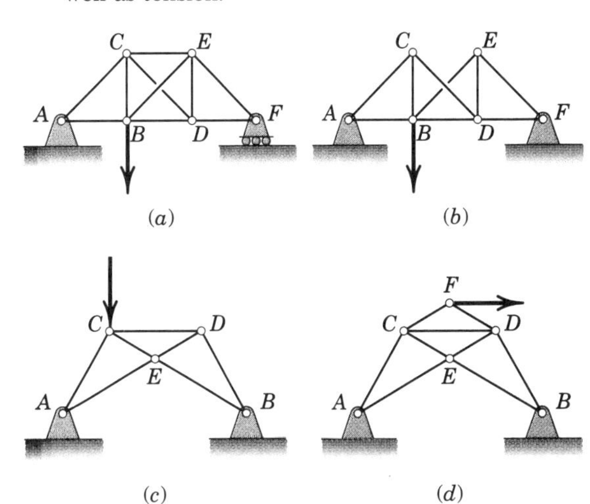

Problem 4/24

4/25 Analysis of the wind acting on a small Hawaiian church, which withstood the 280-km/h winds of Hurricane Iniki in 1992, showed the forces transmitted to each roof truss panel to be as shown. Treat the structure as a symmetrical simple truss and neglect any horizontal component of the support reaction at A. Identify the truss member which supports the largest force, tension or compression, and calculate this force.

Ans. FD = 97.8 kN T

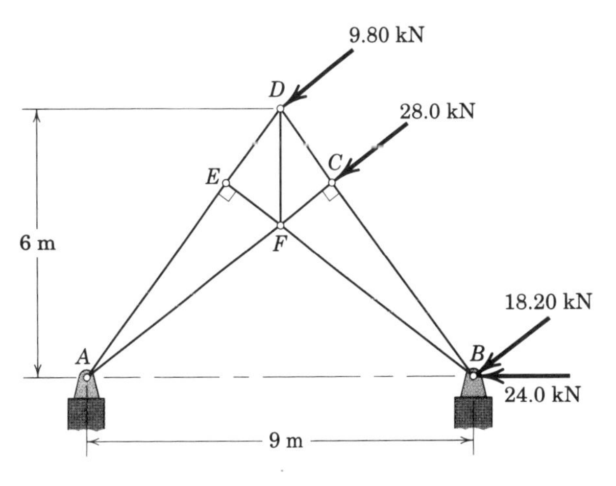

Problem 4/25

4/26 The 72-m structure is used to provide various support services to launch vehicles prior to liftoff. In a test, an 18-Mg mass is suspended from joints *F* and *G*, with its weight equally divided between the two joints. Determine the forces in members *GJ* and *GI*. What would be your path of joint analysis for members in the vertical tower, such as *AB* or *KL*?

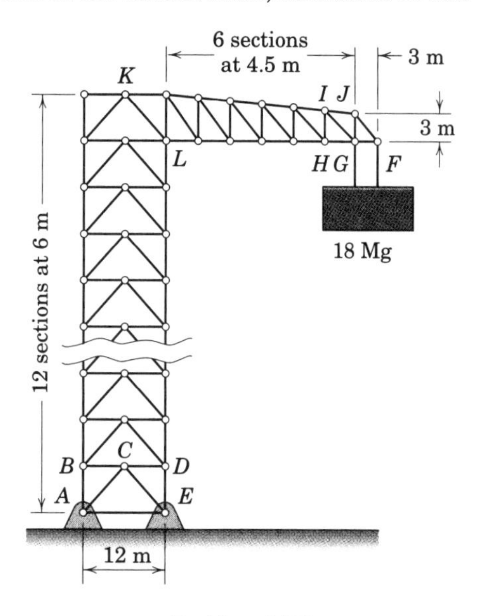

Problem 4/26

▶**4/27** The tower for a transmission line is modeled by the truss shown. The crossed members in the center sections of the truss may be assumed to be capable of supporting tension only. For the loads of 1.8 kN applied in the vertical plane, compute the forces induced in members *AB*, *DB*, and *CD*.

Ans. $AB = 3.89$ kN *C*, $DB = 0$, $CD = 0.932$ kN *C*

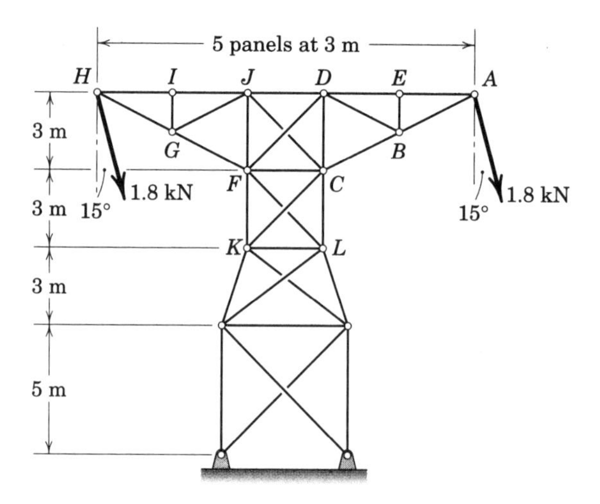

Problem 4/27

▶**4/28** Find the forces in members *EF*, *KL*, and *GL* for the Fink truss shown.

Ans. $EF = 75.1$ kN *C*, $KL = 40$ kN *T*
$GL = 20$ kN *T*

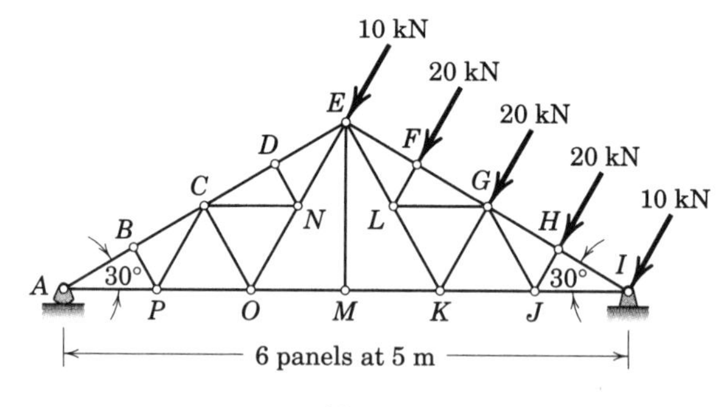

Problem 4/28

local variations. The values 9.81 m/s^2 in SI units and 32.2 ft/sec^2 in U.S. customary units are used for the sea-level value of g.

Apparent Weight

The gravitational attraction of the earth on a body of mass m may be calculated from the results of a simple gravitational experiment. The body is allowed to fall freely in a vacuum, and its absolute acceleration is measured. If the gravitational force of attraction or true weight of the body is W, then, because the body falls with an absolute acceleration g, Eq. 1/1 gives

$$\mathbf{W} = m\mathbf{g} \tag{1/3}$$

The *apparent weight* of a body as determined by a spring balance, calibrated to read the correct force and attached to the surface of the earth, will be slightly less than its true weight. The difference is due to the rotation of the earth. The ratio of the apparent weight to the apparent or relative acceleration due to gravity still gives the correct value of mass. The apparent weight and the relative acceleration due to gravity are, of course, the quantities which are measured in experiments conducted on the surface of the earth.

1/6 Dimensions

A given dimension such as length can be expressed in a number of different units such as meters, millimeters, or kilometers. Thus, a *dimension* is different from a *unit*. The *principle of dimensional homogeneity* states that all physical relations must be dimensionally homogeneous, that is, the dimensions of all terms in an equation must be the same. It is customary to use the symbols L, M, T, and F to stand for length, mass, time, and force, respectively. In SI units force is a derived quantity and from Eq. 1/1 has the dimensions of mass times acceleration or

$$F = ML/T^2$$

One important use of the dimensional homogeneity principle is to check the dimensional correctness of some derived physical relation. We can derive the following expression for the velocity v of a body of mass m which is moved from rest a horizontal distance x by a force F:

$$Fx = \tfrac{1}{2}mv^2$$

where the $\frac{1}{2}$ is a dimensionless coefficient resulting from integration. This equation is dimensionally correct because substitution of L, M, and T gives

$$[MLT^{-2}][L] = [M][LT^{-1}]^2$$

Dimensional homogeneity is a necessary condition for correctness of a physical relation, but it is not sufficient, since it is possible to construct an equation which is dimensionally correct but does not represent a correct relation. You should perform a dimensional check on the answer to every problem whose solution is carried out in symbolic form.

1/7 Solving Problems in Dynamics

The study of dynamics concerns the understanding and description of the motions of bodies. This description, which is largely mathematical, enables predictions of dynamical behavior to be made. A dual thought process is necessary in formulating this description. It is necessary to think in terms of both the physical situation and the corresponding mathematical description. This repeated transition of thought between the physical and the mathematical is required in the analysis of every problem.

One of the greatest difficulties encountered by students is the inability to make this transition freely. You should recognize that the mathematical formulation of a physical problem represents an ideal and limiting description, or model, which approximates but never quite matches the actual physical situation.

In Art. 1/8 of *Vol. 1 Statics* we extensively discussed the approach to solving problems in statics. We assume, therefore, that you are familiar with this approach, which we summarize here as applied to dynamics.

Approximations in Mathematical Models

Construction of an idealized mathematical model for a given engineering problem always requires approximations to be made. Some of these approximations may be mathematical, whereas others will be physical. For instance, it is often necessary to neglect small distances, angles, or forces compared with large distances, angles, or forces. If the change in velocity of a body with time is nearly uniform, then an assumption of constant acceleration may be justified. An interval of motion which cannot be easily described in its entirety is often divided into small increments, each of which can be approximated.

As another example, the retarding effect of bearing friction on the motion of a machine may often be neglected if the friction forces are small compared with the other applied forces. However, these same friction forces cannot be neglected if the purpose of the inquiry is to determine the decrease in efficiency of the machine due to the friction process. Thus, the type of assumptions you make depends on what information is desired and on the accuracy required.

You should be constantly alert to the various assumptions called for in the formulation of real problems. The ability to understand and make use of the appropriate assumptions when formulating and solving engineering problems is certainly one of the most important characteristics of a successful engineer.

Along with the development of the principles and analytical tools needed for modern dynamics, one of the major aims of this book is to provide many opportunities to develop the ability to formulate good mathematical models. Strong emphasis is placed on a wide range of practical problems which not only require you to apply theory but also force you to make relevant assumptions.

Method of Attack

An effective method of attack is essential in the solution of dynamics problems, as for all engineering problems. Development of good habits in formulating problems and in representing their solutions will be an invaluable asset. Each solution should proceed with a logical sequence of steps from hypothesis to conclusion. The following sequence of steps is useful in the construction of problem solutions.

1. Formulate the problem:
   (*a*) State the given data.
   (*b*) State the desired result.
   (*c*) State your assumptions and approximations.
2. Develop the solution:
   (*a*) Draw any diagrams which you need to understand the relationships.
   (*b*) State the governing principles to be applied to your solution.
   (*c*) Make your calculations.
   (*d*) Ensure that your calculations are consistent with the accuracy justified by the data.
   (*e*) Be sure that you have used consistent units throughout your calculations.
   (*f*) Ensure that your answers are reasonable in terms of magnitudes, directions, common sense, etc.
   (*g*) Draw conclusions.

The arrangement of your work should be neat and orderly. This will help your thought process and enable others to understand your work. The discipline of doing orderly work will help you develop skill in problem formulation and analysis. Problems which seem complicated at first often become clear when you approach them with logic and discipline.

Application of Basic Principles

The subject of dynamics is based on a surprisingly few fundamental concepts and principles which, however, can be extended and applied over a wide range of conditions. The study of dynamics is valuable partly because it provides experience in reasoning from fundamentals. This experience cannot be obtained merely by memorizing the kinematic and dynamic equations which describe various motions. It must be obtained through exposure to a wide variety of problem situations which require the choice, use, and extension of basic principles to meet the given conditions.

In describing the relations between forces and the motions they produce, it is essential to define clearly the system to which a principle is to be applied. At times a single particle or a rigid body is the system to be isolated, whereas at other times two or more bodies taken together constitute the system.

The definition of the system to be analyzed is made clear by constructing its ***free-body diagram***. This diagram consists of a closed outline of the external boundary of the system. All bodies which contact and exert forces on the system but are not a part of it are removed and replaced by vectors representing the forces they exert on the isolated system. In this way, we make a clear distinction between the action and reaction of each force, and all forces on and external to the system are accounted for. We assume that you are familiar with the technique of drawing free-body diagrams from your prior work in statics.

Numerical versus Symbolic Solutions

In applying the laws of dynamics, we may use numerical values of the involved quantities, or we may use algebraic symbols and leave the answer as a formula. When numerical values are used, the magnitudes of all quantities expressed in their particular units are evident at each stage of the calculation. This approach is useful when we need to know the magnitude of each term.

The symbolic solution, however, has several advantages over the numerical solution:

1. The use of symbols helps to focus attention on the connection between the physical situation and its related mathematical description.
2. A symbolic solution enables you to make a dimensional check at every step, whereas dimensional homogeneity cannot be checked when only numerical values are used.
3. We can use a symbolic solution repeatedly for obtaining answers to the same problem with different units or different numerical values.

Thus, facility with both forms of solution is essential, and you should practice each in the problem work.

In the case of numerical solutions, we repeat from *Vol. 1 Statics* our convention for the display of results. All given data is taken to be exact, and results are generally displayed to three significant figures, unless the leading digit is a one, in which case four significant figures are displayed.

Solution Methods

Solutions to the various equations of dynamics can be obtained in one of three ways.

1. Obtain a direct mathematical solution by hand calculation, using either algebraic symbols or numerical values. We can solve the large majority of the problems this way.
2. Obtain graphical solutions for certain problems, such as the determination of velocities and accelerations of rigid bodies in two-dimensional relative motion.
3. Solve the problem by computer. A number of problems in *Vol. 2 Dynamics* are designated as *Computer-Oriented Problems*. They ap-

pear at the end of the Review Problem sets and were selected to illustrate the type of problem for which solution by computer offers a distinct advantage.

The choice of the most expedient method of solution is an important aspect of the experience to be gained from the problem work. We emphasize, however, that the most important experience in learning mechanics lies in the formulation of problems, as distinct from their solution per se.

CHAPTER REVIEW

This chapter has introduced the concepts, definitions, and units used in dynamics, and has given an overview of the approach used to formulate and solve problems in dynamics. Now that you have finished this chapter, you should be able to do the following:

1. State Newton's laws of motion.
2. Perform calculations using SI and U.S. customary units.
3. Express the law of gravitation and calculate the weight of an object.
4. Discuss the effects of altitude and the rotation of the earth on the acceleration due to gravity.
5. Apply the principle of dimensional homogeneity to a given physical relation.
6. Describe the methodology used to formulate and solve dynamics problems.

Sample Problem 1/1

A space-shuttle payload module has a mass of 50 kg and rests on the surface of the earth at a latitude of 45° north.

(*a*) Determine the surface-level weight of the module in both newtons and pounds, and its mass in slugs.

(*b*) Now suppose the module is taken to an altitude of 300 kilometers above the surface of the earth and released there with no velocity relative to the center of the earth. Determine its weight under these conditions in both newtons and pounds.

(*c*) Finally, suppose the module is fixed inside the cargo bay of a space shuttle. The shuttle is in a circular orbit at an altitude of 300 kilometers above the surface of the earth. Determine the weight of the module in both newtons and pounds under these conditions.

For the surface-level value of the acceleration of gravity relative to a rotating earth, use $g = 9.80665 \text{ m/s}^2$ (32.1740 ft/sec^2). For the absolute value relative to a nonrotating earth, use $g = 9.825 \text{ m/s}^2$ (32.234 ft/sec^2). Round off all answers using the rules of this textbook.

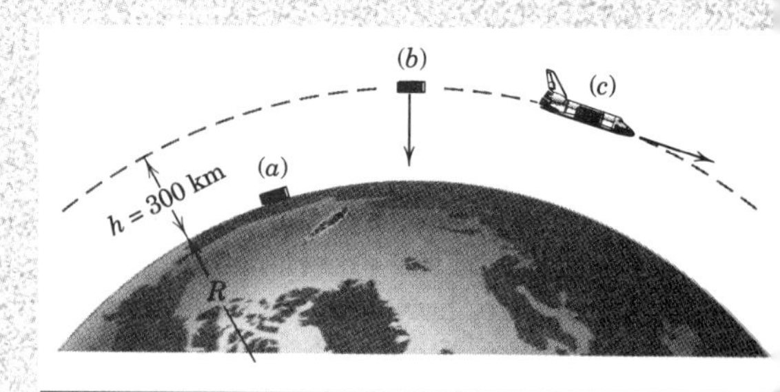

Solution. (*a*) From relationship 1/3, we have

① $[W = mg]$ $\qquad W = (50 \text{ kg})(9.80665 \text{ m/s}^2) = 490 \text{ N}$ *Ans.*

Here we have used the acceleration of gravity relative to the rotating earth, because that is the condition of the module in part (*a*). Note that we are using more significant figures in the acceleration of gravity than will normally be required in this textbook (9.81 m/s^2 and 32.2 ft/sec^2 will normally suffice).

From the table of conversion factors inside the front cover of the textbook, we see that 4.4482 newtons is equal to 1 pound. Thus, the weight of the module in pounds is

② $$W = 490 \text{ N}\left[\frac{1 \text{ lb}}{4.4482 \text{ N}}\right] = 110.2 \text{ lb}$$ *Ans.*

Finally, its mass in slugs is

③ $[W = mg]$ $$m = \frac{W}{g} = \frac{110.2 \text{ lb}}{32.1740 \text{ ft/sec}^2} = 3.43 \text{ slugs}$$ *Ans.*

As another route to the last result, we may convert from kilograms to slugs. Again using the table inside the front cover, we have

$$m = 50 \text{ kg}\left[\frac{1 \text{ slug}}{14.594 \text{ kg}}\right] = 3.43 \text{ slugs}$$

(Note on lb force, lb mass, and slug: We recall that 1 lbm is the amount of mass which under standard conditions has a weight of 1 lb of force. We rarely refer to the U.S. mass unit lbm in this textbook series, but rather use the slug for mass. The sole use of slug, rather than the unnecessary use of two units for mass, proves to be powerful and simple in U.S. units.)

Helpful Hints

① Our calculator indicates a result of 490.3325 · · · newtons. Using the rules of significant-figure display used in this textbook, we round the written result to three significant figures, or 490 newtons. Had the numerical result begun with the digit 1, we would have rounded the displayed answer to four significant figures.

② A good practice with unit conversion is to multiply by a factor such as $\left[\frac{1 \text{ lb}}{4.4482 \text{ N}}\right]$, which has a value of 1, because the numerator and the denominator are equivalent. Be sure that cancellation of the units leaves the units desired—here the units of N cancel, leaving the desired units of lb.

③ Note that we are using a previously calculated result (110.2 lb). We must be sure that when a calculated number is needed in subsequent calculations, it is retained in the calculator to its full accuracy (110.2316 · · ·). If necessary, numbers must be stored in a calculator storage register and then brought out of the register when needed. We must not merely punch 110.2 into our calculator and proceed to divide by 32.1740—this practice will result in loss of numerical accuracy. Some individuals like to place a small indication of the storage register used in the right margin of the work paper, directly beside the number stored.

Sample Problem 1/1 (Continued)

(*b*) We begin by calculating the absolute acceleration of gravity (relative to the nonrotating earth) at an altitude of 300 kilometers.

$$\left[g = g_0 \frac{R^2}{(R + h)^2}\right] \qquad g_h = 9.825\left[\frac{6371^2}{(6371 + 300)^2}\right] = 8.96 \text{ m/s}^2$$

The weight at an altitude of 300 kilometers is then

$$W_h = mg_h = 50(8.96) = 448 \text{ N} \qquad \textit{Ans.}$$

We now convert W_h to units of pounds.

$$W_h = 448 \text{ N}\left[\frac{1 \text{ lb}}{4.4482 \text{ N}}\right] = 100.7 \text{ lb} \qquad \textit{Ans.}$$

As an alternative solution to part (*b*), we may use Newton's law of universal gravitation. In SI units,

$$\left[F = \frac{Gm_1m_2}{r^2}\right] \qquad W_h = \frac{Gm_em}{(R + h)^2} = \frac{[6.673(10^{-11})][5.976(10^{24})][50]}{[(6371 + 300)(1000)]^2}$$
$$= 448 \text{ N}$$

which agrees with our earlier result. We note that the weight of the module when at an altitude of 300 km is about 90% of its surface-level weight—it is *not* weightless. We will study the effects of this weight on the motion of the module in Chapter 3.

(*c*) The weight of an object (the force of gravitational attraction) does not depend on the motion of the object. Thus the answers for part (*c*) are the same as those in part (*b*).

$$W_h = 448 \text{ N} \quad \text{or} \quad 100.7 \text{ lb} \qquad \textit{Ans.}$$

This Sample Problem has served to eliminate certain commonly held and persistent misconceptions. First, just because a body is raised to a typical shuttle altitude, it does not become weightless. This is true whether the body is released with no velocity relative to the center of the earth, is inside the orbiting shuttle, or is in its own arbitrary trajectory. And second, the acceleration of gravity is not zero at such altitudes. The only way to reduce both the acceleration of gravity and the corresponding weight of a body to zero is to take the body to an infinite distance from the earth.

PROBLEMS

(Refer to Table D/2 in Appendix D for relevant solar-system values.)

1/1 Determine the weight in newtons of a car which has a mass of 1500 kg. Convert the given mass of the car to slugs and calculate the corresponding weight in pounds.
Ans. $W = 14\,720$ N, $m = 102.8$ slugs, $W = 3310$ lb

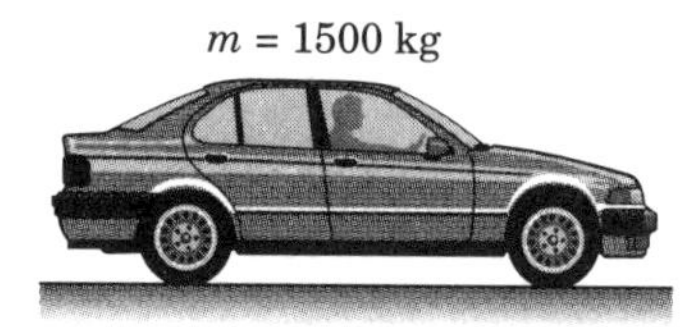

Problem 1/1

1/2 If a man's mass is 80 kg, determine his weight in newtons and calculate his corresponding mass in slugs.

1/3 The mass of one dozen apples is 2 kg. Determine the average weight of one apple in both SI and U.S. units. In the present case, how applicable is the "rule of thumb" that an average apple weighs 1 N?
Ans. $W = 1.635$ N, $W = 0.368$ lb

1/4 For the given vectors $\mathbf{V}_1$ and $\mathbf{V}_2$, determine $V_1 + V_2$, $\mathbf{V}_1 + \mathbf{V}_2$, $\mathbf{V}_1 - \mathbf{V}_2$, $\mathbf{V}_1 \times \mathbf{V}_2$, and $\mathbf{V}_1 \cdot \mathbf{V}_2$. Consider the vectors to be nondimensional.

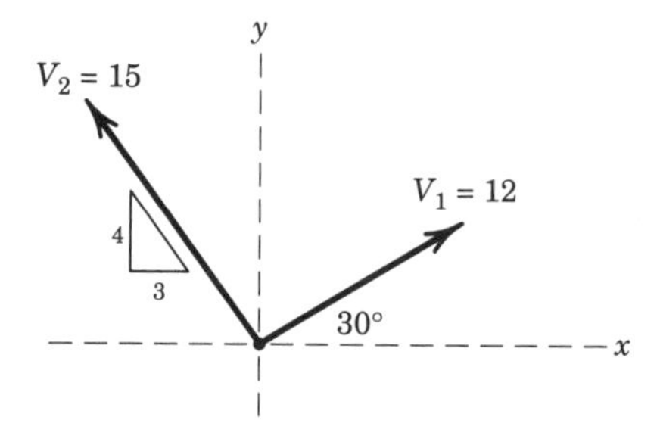

Problem 1/4

1/5 The two 100-mm-diameter spheres constructed of different metals are located in deep space. Determine the gravitational force **F** which the copper sphere exerts on the titanium sphere if (*a*) $d = 2$ m, and (*b*) $d = 4$ m.
Ans. (*a*) $\mathbf{F} = -1.255(10^{-10})\mathbf{i}$ N
(*b*) $\mathbf{F} = -3.14(10^{-11})\mathbf{i}$ N

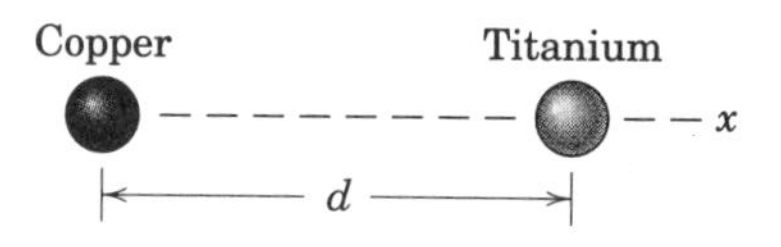

Problem 1/5

1/6 A space shuttle is in a circular orbit at an altitude of 250 km. Calculate the absolute value of g at this altitude and determine the corresponding weight of a shuttle passenger who weighs 880 N when standing on the surface of the earth at a latitude of 45°. Are the terms "zero-g" and "weightless," which are sometimes used to describe conditions aboard orbiting spacecraft, correct in the absolute sense?

1/7 At what altitude h above the north pole is the weight of an object reduced to one-half of its earth-surface value? Assume a spherical earth of radius R and express h in terms of R.
Ans. $h = 0.414R$

1/8 Determine the absolute weight and the weight relative to the rotating earth of a 90-kg man if he is standing on the surface of the earth at a latitude of 40°.

1/9 Calculate the force F_s exerted by the sun on a 90-kg man as he stands on the surface of the moon. Compare F_s with the force F_m exerted on him by the moon.
Ans. $F_s = 0.534$ N, $F_m = 146$ N

1/10 Calculate the distance d from the center of the sun at which a particle experiences equal attractions from the earth and the sun. The particle is restricted to the line which joins the centers of the earth and the sun. Justify the two solutions physically.

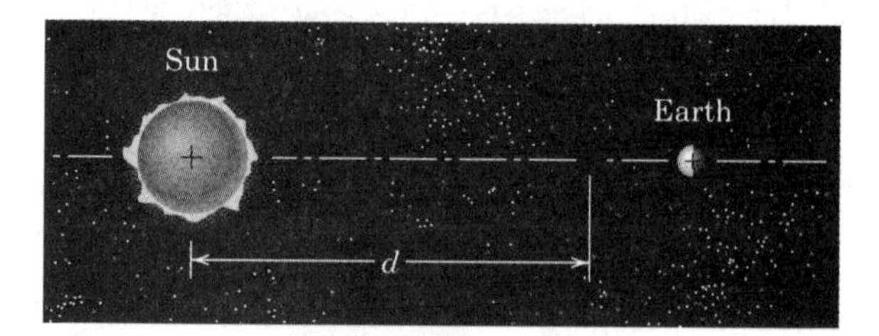

Problem 1/10

1/11 Determine the ratio R_A of the force exerted by the sun on the moon to that exerted by the earth on the moon for position A of the moon. Repeat for moon position B.

Ans. $R_A = 2.19$, $R_B = 2.21$

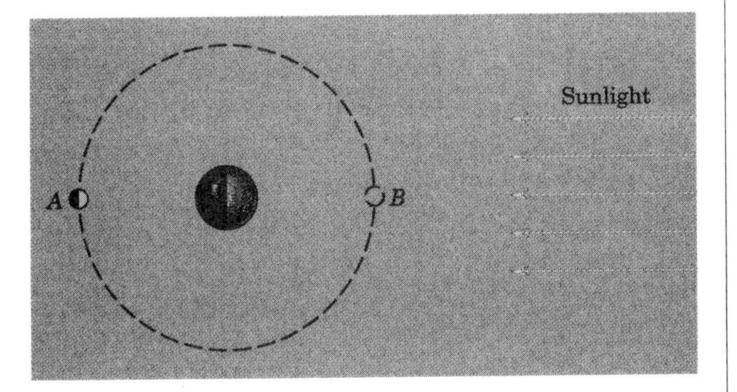

Problem 1/11

1/12 Check the following equation for dimensional homogeneity:

$$mv = \int_{t_1}^{t_2} (F \cos \theta)\, dt$$

where m is mass, v is velocity, F is force, θ is an angle, and t is time.

Chapter

2

KINEMATICS OF PARTICLES

CHAPTER OUTLINE

2/1 INTRODUCTION

Kinematics is the branch of dynamics which describes the motion of bodies without reference to the forces which either cause the motion or are generated as a result of the motion. Kinematics is often described as the "geometry of motion." Some engineering applications of kinematics include the design of cams, gears, linkages, and other machine elements to control or produce certain desired motions, and the calculation of flight trajectories for aircraft, rockets, and spacecraft. A thorough working knowledge of kinematics is a prerequisite to kinetics, which is the study of the relationships between motion and the corresponding forces which cause or accompany the motion.

Particle Motion

We begin our study of kinematics by first discussing in this chapter the motions of points or particles. A particle is a body whose physical dimensions are so small compared with the radius of curvature of its path that we may treat the motion of the particle as that of a point. For example, the wingspan of a jet transport flying between Los Angeles and New York is of no consequence compared with the radius of curvature

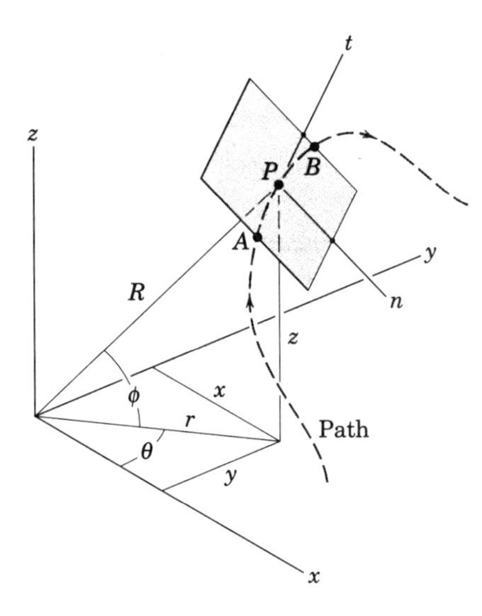

Figure 2/1

of its flight path, and thus the treatment of the airplane as a particle or point is an acceptable approximation.

We can describe the motion of a particle in a number of ways, and the choice of the most convenient or appropriate way depends a great deal on experience and on how the data are given. Let us obtain an overview of the several methods developed in this chapter by referring to Fig. 2/1, which shows a particle P moving along some general path in space. If the particle is confined to a specified path, as with a bead sliding along a fixed wire, its motion is said to be *constrained*. If there are no physical guides, the motion is said to be *unconstrained*. A small rock tied to the end of a string and whirled in a circle undergoes constrained motion until the string breaks, after which instant its motion is unconstrained.

Choice of Coordinates

The position of particle P at any time t can be described by specifying its rectangular coordinates* x, y, z, its cylindrical coordinates r, θ, z, or its spherical coordinates R, θ, ϕ. The motion of P can also be described by measurements along the tangent t and normal n to the curve. The direction of n lies in the local plane of the curve.† These last two measurements are called *path variables*.

The motion of particles (or rigid bodies) can be described by using coordinates measured from fixed reference axes (*absolute-motion* analysis) or by using coordinates measured from moving reference axes (*relative-motion* analysis). Both descriptions will be developed and applied in the articles which follow.

With this conceptual picture of the description of particle motion in mind, we restrict our attention in the first part of this chapter to the case of *plane motion* where all movement occurs in or can be represented as occurring in a single plane. A large proportion of the motions of machines and structures in engineering can be represented as plane motion. Later, in Chapter 7, an introduction to three-dimensional motion is presented. We begin our discussion of plane motion with *rectilinear motion,* which is motion along a straight line, and follow it with a description of motion along a plane curve.

2/2 Rectilinear Motion

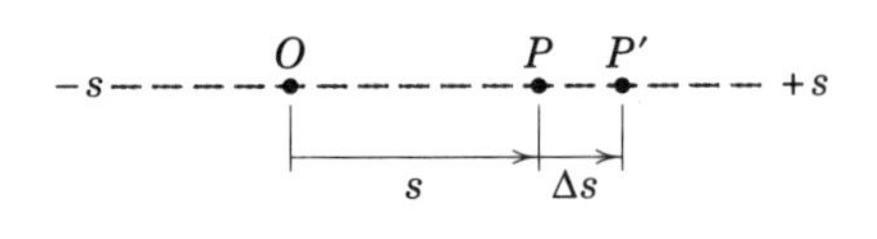

Figure 2/2

Consider a particle P moving along a straight line, Fig. 2/2. The position of P at any instant of time t can be specified by its distance s measured from some convenient reference point O fixed on the line. At time $t + \Delta t$ the particle has moved to P' and its coordinate becomes $s + \Delta s$. The change in the position coordinate during the interval Δt is called the *displacement* Δs of the particle. The displacement would be negative if the particle moved in the negative s-direction.

*Often called *Cartesian* coordinates, named after René Descartes (1596–1650), a French mathematician who was one of the inventors of analytic geometry.

†This plane is called the *osculating* plane, which comes from the Latin word *osculari* meaning "to kiss." The plane which contains P and the two points A and B, one on either side of P, becomes the osculating plane as the distances between the points approach zero.

Velocity and Acceleration

The average velocity of the particle during the interval Δt is the displacement divided by the time interval or $v_{av} = \Delta s/\Delta t$. As Δt becomes smaller and approaches zero in the limit, the average velocity approaches the *instantaneous velocity* of the particle, which is $v = \lim_{\Delta t \to 0} \frac{\Delta s}{\Delta t}$ or

$$v = \frac{ds}{dt} = \dot{s} \qquad (2/1)$$

Thus, the velocity is the time rate of change of the position coordinate s. The velocity is positive or negative depending on whether the corresponding displacement is positive or negative.

The average acceleration of the particle during the interval Δt is the change in its velocity divided by the time interval or $a_{av} = \Delta v/\Delta t$. As Δt becomes smaller and approaches zero in the limit, the average acceleration approaches the *instantaneous acceleration* of the particle, which is $a = \lim_{\Delta t \to 0} \frac{\Delta v}{\Delta t}$ or

$$a = \frac{dv}{dt} = \dot{v} \quad \text{or} \quad a = \frac{d^2s}{dt^2} = \ddot{s} \qquad (2/2)$$

The acceleration is positive or negative depending on whether the velocity is increasing or decreasing. Note that the acceleration would be positive if the particle had a negative velocity which was becoming less negative. If the particle is slowing down, the particle is said to be *decelerating*.

Velocity and acceleration are actually vector quantities, as we will see for curvilinear motion beginning with Art. 2/3. For rectilinear motion in the present article, where the direction of the motion is that of the given straight-line path, the sense of the vector along the path is described by a plus or minus sign. In our treatment of curvilinear motion, we will account for the changes in direction of the velocity and acceleration vectors as well as their changes in magnitude.

By eliminating the time dt between Eq. 2/1 and the first of Eqs. 2/2, we obtain a differential equation relating displacement, velocity, and acceleration.* This equation is

$$v\,dv = a\,ds \quad \text{or} \quad \dot{s}\,d\dot{s} = \ddot{s}\,ds \qquad (2/3)$$

Equations 2/1, 2/2, and 2/3 are the differential equations for the rectilinear motion of a particle. Problems in rectilinear motion involving finite changes in the motion variables are solved by integration of these basic differential relations. The position coordinate s, the velocity v, and the acceleration a are algebraic quantities, so that their signs, positive

*Differential quantities can be multiplied and divided in exactly the same way as other algebraic quantities.

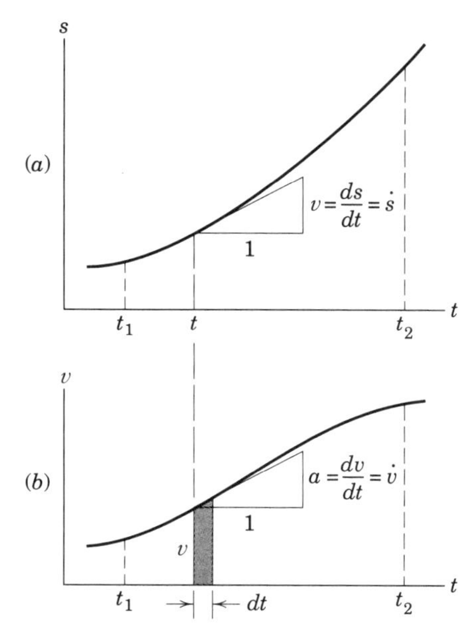

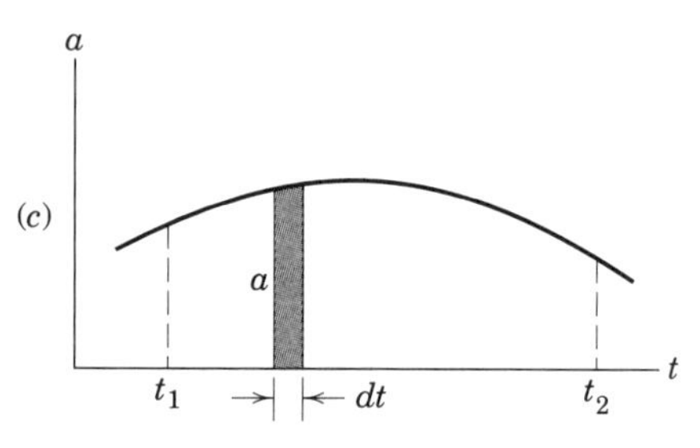

Figure 2/3

or negative, must be carefully observed. Note that the positive directions for v and a are the same as the positive direction for s.

Graphical Interpretations

Interpretation of the differential equations governing rectilinear motion is considerably clarified by representing the relationships among s, v, a, and t graphically. Figure 2/3*a* is a schematic plot of the variation of s with t from time t_1 to time t_2 for some given rectilinear motion. By constructing the tangent to the curve at any time t, we obtain the slope, which is the velocity $v = ds/dt$. Thus, the velocity can be determined at all points on the curve and plotted against the corresponding time as shown in Fig. 2/3*b*. Similarly, the slope dv/dt of the v-t curve at any instant gives the acceleration at that instant, and the a-t curve can therefore be plotted as in Fig. 2/3*c*.

We now see from Fig. 2/3*b* that the area under the v-t curve during time dt is $v\,dt$, which from Eq. 2/1 is the displacement ds. Consequently, the net displacement of the particle during the interval from t_1 to t_2 is the corresponding area under the curve, which is

$$\int_{s_1}^{s_2} ds = \int_{t_1}^{t_2} v\,dt \qquad \text{or} \qquad s_2 - s_1 = (\text{area under } v\text{-}t \text{ curve})$$

Similarly, from Fig. 2/3*c* we see that the area under the a-t curve during time dt is $a\,dt$, which, from the first of Eqs. 2/2, is dv. Thus, the net change in velocity between t_1 and t_2 is the corresponding area under the curve, which is

$$\int_{v_1}^{v_2} dv = \int_{t_1}^{t_2} a\,dt \qquad \text{or} \qquad v_2 - v_1 = (\text{area under } a\text{-}t \text{ curve})$$

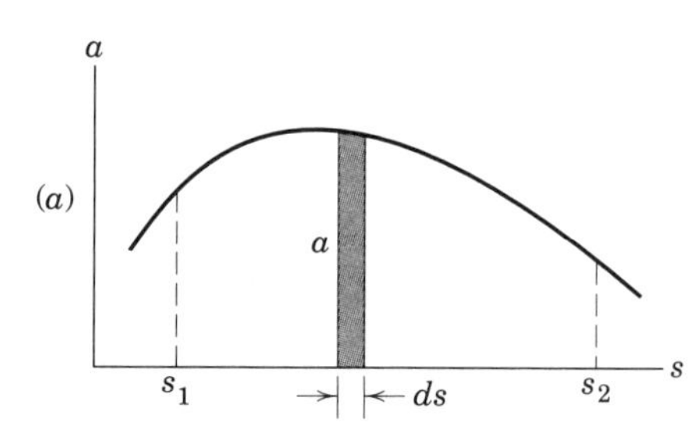

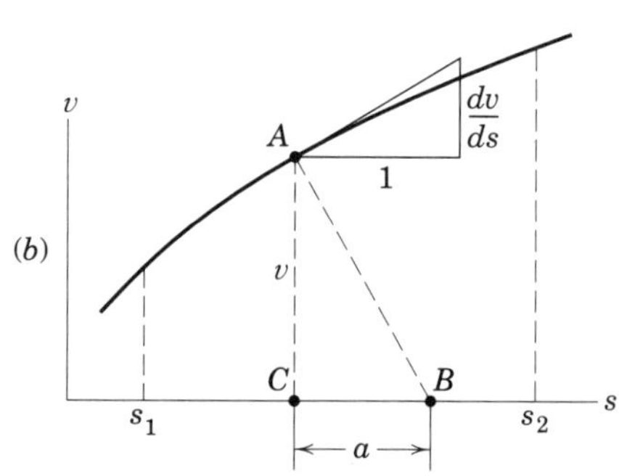

Figure 2/4

Note two additional graphical relations. When the acceleration a is plotted as a function of the position coordinate s, Fig. 2/4*a*, the area under the curve during a displacement ds is $a\,ds$, which, from Eq. 2/3, is $v\,dv = d(v^2/2)$. Thus, the net area under the curve between position coordinates s_1 and s_2 is

$$\int_{v_1}^{v_2} v\,dv = \int_{s_1}^{s_2} a\,ds \quad \text{or} \quad \tfrac{1}{2}(v_2{}^2 - v_1{}^2) = (\text{area under } a\text{-}s \text{ curve})$$

When the velocity v is plotted as a function of the position coordinate s, Fig. 2/4*b*, the slope of the curve at any point A is dv/ds. By constructing the normal AB to the curve at this point, we see from the similar triangles that $\overline{CB}/v = dv/ds$. Thus, from Eq. 2/3, $\overline{CB} = v(dv/ds) = a$, the acceleration. It is necessary that the velocity and position coordinate axes have the same numerical scales so that the acceleration read on the position coordinate scale in meters (or feet), say, will represent the actual acceleration in meters (or feet) per second squared.

The graphical representations described are useful not only in visualizing the relationships among the several motion quantities but also in obtaining approximate results by graphical integration or differentiation. The latter case occurs when a lack of knowledge of the mathematical relationship prevents its expression as an explicit mathematical function which can be integrated or differentiated. Experimental data and motions which involve discontinuous relationships between the variables are frequently analyzed graphically.

Analytical Integration

If the position coordinate s is known for all values of the time t, then successive mathematical or graphical differentiation with respect to t gives the velocity v and acceleration a. In many problems, however, the functional relationship between position coordinate and time is unknown, and we must determine it by successive integration from the acceleration. Acceleration is determined by the forces which act on moving bodies and is computed from the equations of kinetics discussed in subsequent chapters. Depending on the nature of the forces, the acceleration may be specified as a function of time, velocity, or position coordinate, or as a combined function of these quantities. The procedure for integrating the differential equation in each case is indicated as follows.

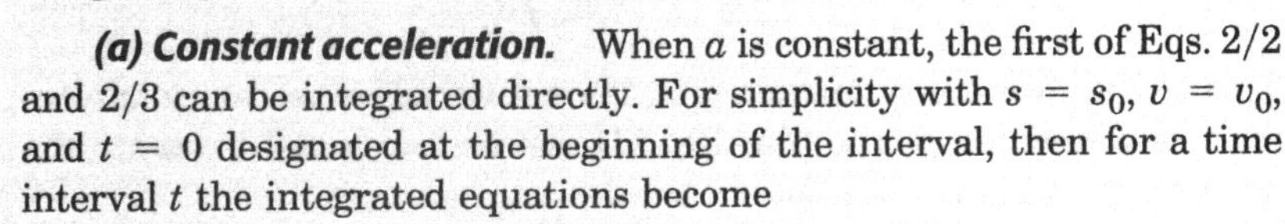

(a) Constant acceleration. When a is constant, the first of Eqs. 2/2 and 2/3 can be integrated directly. For simplicity with $s = s_0$, $v = v_0$, and $t = 0$ designated at the beginning of the interval, then for a time interval t the integrated equations become

$$\int_{v_0}^{v} dv = a \int_0^t dt \qquad \text{or} \qquad v = v_0 + at$$

$$\int_{v_0}^{v} v\,dv = a \int_{s_0}^{s} ds \qquad \text{or} \qquad v^2 = v_0{}^2 + 2a(s - s_0)$$

Substitution of the integrated expression for v into Eq. 2/1 and integration with respect to t give

$$\int_{s_0}^{s} ds = \int_0^t (v_0 + at)\,dt \qquad \text{or} \qquad s = s_0 + v_0 t + \tfrac{1}{2}at^2$$

These relations are necessarily restricted to the special case where the acceleration is constant. The integration limits depend on the initial and final conditions, which for a given problem may be different from those used here. It may be more convenient, for instance, to begin the integration at some specified time t_1 rather than at time $t = 0$.

Caution: The foregoing equations have been integrated for constant acceleration only. A common mistake is to use these equations for problems involving variable acceleration, where they do not apply.

(b) Acceleration given as a function of time, $a = f(t)$. Substitution of the function into the first of Eqs. 2/2 gives $f(t) = dv/dt$. Multiplying by dt separates the variables and permits integration. Thus,

$$\int_{v_0}^{v} dv = \int_0^t f(t)\,dt \qquad \text{or} \qquad v = v_0 + \int_0^t f(t)\,dt$$

From this integrated expression for v as a function of t, the position coordinate s is obtained by integrating Eq. 2/1, which, in form, would be

$$\int_{s_0}^{s} ds = \int_0^t v\,dt \quad \text{or} \quad s = s_0 + \int_0^t v\,dt$$

If the indefinite integral is employed, the end conditions are used to establish the constants of integration. The results are identical with those obtained by using the definite integral.

If desired, the displacement s can be obtained by a direct solution of the second-order differential equation $\ddot{s} = f(t)$ obtained by substitution of $f(t)$ into the second of Eqs. 2/2.

(c) Acceleration given as a function of velocity, $a = f(v)$. Substitution of the function into the first of Eqs. 2/2 gives $f(v) = dv/dt$, which permits separating the variables and integrating. Thus,

$$t = \int_0^t dt = \int_{v_0}^{v} \frac{dv}{f(v)}$$

This result gives t as a function of v. Then it would be necessary to solve for v as a function of t so that Eq. 2/1 can be integrated to obtain the position coordinate s as a function of t.

Another approach is to substitute the function $a = f(v)$ into the first of Eqs. 2/3, giving $v\,dv = f(v)\,ds$. The variables can now be separated and the equation integrated in the form

$$\int_{v_0}^{v} \frac{v\,dv}{f(v)} = \int_{s_0}^{s} ds \quad \text{or} \quad s = s_0 + \int_{v_0}^{v} \frac{v\,dv}{f(v)}$$

Note that this equation gives s in terms of v without explicit reference to t.

(d) Acceleration given as a function of displacement, $a = f(s)$. Substituting the function into Eq. 2/3 and integrating give the form

$$\int_{v_0}^{v} v\,dv = \int_{s_0}^{s} f(s)\,ds \quad \text{or} \quad v^2 = v_0{}^2 + 2\int_{s_0}^{s} f(s)\,ds$$

Next we solve for v to give $v = g(s)$, a function of s. Now we can substitute ds/dt for v, separate variables, and integrate in the form

$$\int_{s_0}^{s} \frac{ds}{g(s)} = \int_0^t dt \quad \text{or} \quad t = \int_{s_0}^{s} \frac{ds}{g(s)}$$

which gives t as a function of s. Finally, we can rearrange to obtain s as a function of t.

In each of the foregoing cases when the acceleration varies according to some functional relationship, the possibility of solving the equations by direct mathematical integration will depend on the form of the function. In cases where the integration is excessively awkward or difficult, integration by graphical, numerical, or computer methods can be utilized.

Sample Problem 2/1

The position coordinate of a particle which is confined to move along a straight line is given by $s = 2t^3 - 24t + 6$, where s is measured in meters from a convenient origin and t is in seconds. Determine (a) the time required for the particle to reach a velocity of 72 m/s from its initial condition at $t = 0$, (b) the acceleration of the particle when $v = 30$ m/s, and (c) the net displacement of the particle during the interval from $t = 1$ s to $t = 4$ s.

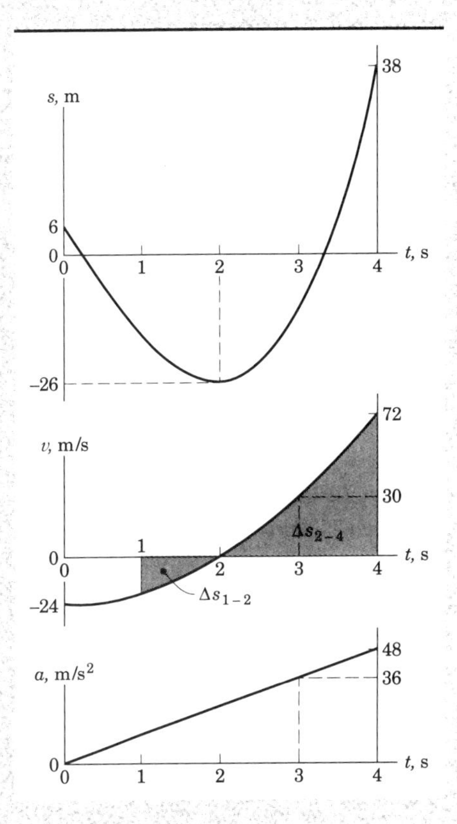

Solution. The velocity and acceleration are obtained by successive differentiation of s with respect to the time. Thus,

$[v = \dot{s}]$ $\qquad v = 6t^2 - 24$ m/s

$[a = \dot{v}]$ $\qquad a = 12t$ m/s^2

(a) Substituting $v = 72$ m/s into the expression for v gives us $72 = 6t^2 - 24$, from which $t = \pm 4$ s. The negative root describes a mathematical ① solution for t before the initiation of motion, so this root is of no physical interest. Thus, the desired result is

$$t = 4 \text{ s} \qquad \textit{Ans.}$$

(b) Substituting $v = 30$ m/s into the expression for v gives $30 = 6t^2 - 24$, from which the positive root is $t = 3$ s, and the corresponding acceleration is

$$a = 12(3) = 36 \text{ m/s}^2 \qquad \textit{Ans.}$$

(c) The net displacement during the specified interval is

$$\Delta s = s_4 - s_1 \quad \text{or}$$

$$\Delta s = [2(4^3) - 24(4) + 6] - [2(1^3) - 24(1) + 6]$$

$$= 54 \text{ m} \qquad \textit{Ans.}$$

② which represents the net advancement of the particle along the s-axis from the position it occupied at $t = 1$ s to its position at $t = 4$ s.

To help visualize the motion, the values of s, v, and a are plotted against the time t as shown. Because the area under the v-t curve represents displace- ③ ment, we see that the net displacement from $t = 1$ s to $t = 4$ s is the positive area Δs_{2-4} less the negative area Δs_{1-2}.

Helpful Hints

① Be alert to the proper choice of sign when taking a square root. When the situation calls for only one answer, the positive root is not always the one you may need.

② Note carefully the distinction between italic s for the position coordinate and the vertical s for seconds.

③ Note from the graphs that the values for v are the slopes ($\dot{s}$) of the s-t curve and that the values for a are the slopes ($\dot{v}$) of the v-t curve. *Suggestion:* Integrate $v\,dt$ for each of the two intervals and check the answer for Δs. Show that the total distance traveled during the interval $t = 1$ s to $t = 4$ s is 74 m.

Sample Problem 2/2

A particle moves along the x-axis with an initial velocity $v_x = 50$ m/s at the origin when $t = 0$. For the first 4 seconds it has no acceleration, and thereafter it is acted on by a retarding force which gives it a constant acceleration $a_x = -10$ m/s^2. Calculate the velocity and the x-coordinate of the particle for ① the conditions of $t = 8$ s and $t = 12$ s and find the maximum positive x-coordinate reached by the particle.

Helpful Hints

① Learn to be flexible with symbols. The position coordinate x is just as valid as s.

Solution. The velocity of the particle after $t = 4$ s is computed from

② $$\left[\int dv = \int a\,dt\right] \qquad \int_{50}^{v_x} dv_x = -10\int_4^t dt \qquad v_x = 90 - 10t \text{ m/s}$$

② Note that we integrate to a general time t and then substitute specific values.

and is plotted as shown. At the specified times, the velocities are

$$t = 8 \text{ s}, \qquad v_x = 90 - 10(8) = 10 \text{ m/s}$$

$$t = 12 \text{ s}, \qquad v_x = 90 - 10(12) = -30 \text{ m/s} \qquad \textit{Ans.}$$

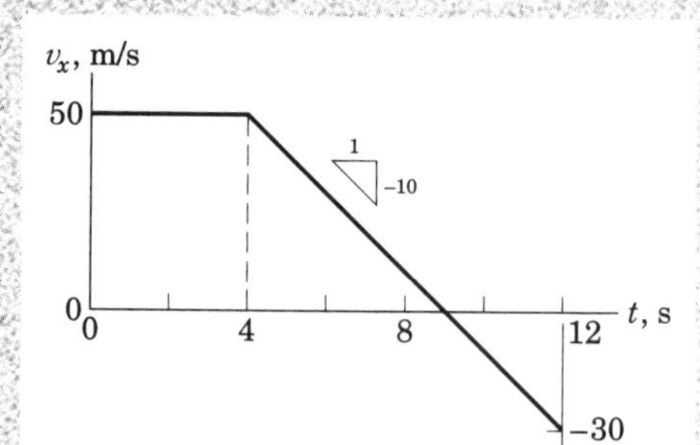

The x-coordinate of the particle at any time greater than 4 seconds is the distance traveled during the first 4 seconds plus the distance traveled after the discontinuity in acceleration occurred. Thus,

$$\left[\int ds = \int v\,dt\right] \qquad x = 50(4) + \int_4^t (90 - 10t)\,dt = -5t^2 + 90t - 80 \text{ m}$$

For the two specified times,

$$t = 8 \text{ s}, \qquad x = -5(8^2) + 90(8) - 80 = 320 \text{ m}$$

$$t = 12 \text{ s}, \qquad x = -5(12^2) + 90(12) - 80 = 280 \text{ m} \qquad \textit{Ans.}$$

The x-coordinate for $t = 12$ s is less than that for $t = 8$ s since the motion is in the negative x-direction after $t = 9$ s. The maximum positive x-coordinate is, then, the value of x for $t = 9$ s which is

$$x_{max} = -5(9^2) + 90(9) - 80 = 325 \text{ m} \qquad \textit{Ans.}$$

③ These displacements are seen to be the net positive areas under the v-t graph up to the values of t in question.

③ Show that the total distance traveled by the particle in the 12 s is 370 m.

Sample Problem 2/3

The spring-mounted slider moves in the horizontal guide with negligible friction and has a velocity v_0 in the s-direction as it crosses the mid-position where $s = 0$ and $t = 0$. The two springs together exert a retarding force to the motion of the slider, which gives it an acceleration proportional to the displacement but oppositely directed and equal to $a = -k^2 s$, where k is constant. (The constant is arbitrarily squared for later convenience in the form of the expressions.) Determine the expressions for the displacement s and velocity v as functions of the time t.

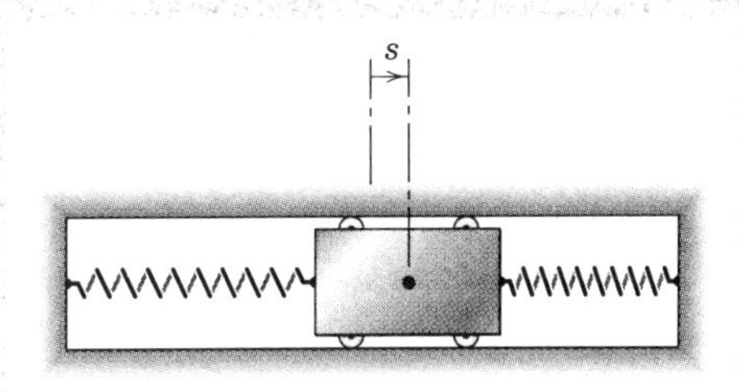

Solution I. Since the acceleration is specified in terms of the displacement, the differential relation $v\,dv = a\,ds$ may be integrated. Thus,

① $$\int v\,dv = \int -k^2 s\,ds + C_1 \text{ a constant,} \quad \text{or} \quad \frac{v^2}{2} = -\frac{k^2 s^2}{2} + C_1$$

When $s = 0$, $v = v_0$, so that $C_1 = v_0^2/2$, and the velocity becomes

$$v = +\sqrt{v_0^2 - k^2 s^2}$$

The plus sign of the radical is taken when v is positive (in the plus s-direction). This last expression may be integrated by substituting $v = ds/dt$. Thus,

② $$\int \frac{ds}{\sqrt{v_0^2 - k^2 s^2}} = \int dt + C_2 \text{ a constant,} \quad \text{or} \quad \frac{1}{k}\sin^{-1}\frac{ks}{v_0} = t + C_2$$

With the requirement of $t = 0$ when $s = 0$, the constant of integration becomes $C_2 = 0$, and we may solve the equation for s so that

$$s = \frac{v_0}{k}\sin kt \qquad \textit{Ans.}$$

The velocity is $v = \dot{s}$, which gives

$$v = v_0 \cos kt \qquad \textit{Ans.}$$

Solution II. Since $a = \ddot{s}$, the given relation may be written at once as

$$\ddot{s} + k^2 s = 0$$

This is an ordinary linear differential equation of second order for which the solution is well known and is

$$s = A \sin Kt + B \cos Kt$$

where A, B, and K are constants. Substitution of this expression into the differential equation shows that it satisfies the equation, provided that $K = k$. The velocity is $v = \dot{s}$, which becomes

$$v = Ak \cos kt - Bk \sin kt$$

The initial condition $v = v_0$ when $t = 0$ requires that $A = v_0/k$, and the condition $s = 0$ when $t = 0$ gives $B = 0$. Thus, the solution is

③ $$s = \frac{v_0}{k}\sin kt \quad \text{and} \quad v = v_0 \cos kt \qquad \textit{Ans.}$$

Helpful Hints

① We have used an indefinite integral here and evaluated the constant of integration. For practice, obtain the same results by using the definite integral with the appropriate limits.

② Again try the definite integral here as above.

③ This motion is called *simple harmonic motion* and is characteristic of all oscillations where the restoring force, and hence the acceleration, is proportional to the displacement but opposite in sign.

Sample Problem 2/4

A freighter is moving at a speed of 8 knots when its engines are suddenly ① stopped. If it takes 10 minutes for the freighter to reduce its speed to 4 knots, determine and plot the distance s in nautical miles moved by the ship and its speed v in knots as functions of the time t during this interval. The deceleration of the ship is proportional to the square of its speed, so that $a = -kv^2$.

Solution. The speeds and the time are given, so we may substitute the expression for acceleration directly into the basic definition $a = dv/dt$ and integrate. Thus,

$$-kv^2 = \frac{dv}{dt} \qquad \frac{dv}{v^2} = -k\,dt \qquad \int_8^v \frac{dv}{v^2} = -k\int_0^t dt$$

②
$$-\frac{1}{v} + \frac{1}{8} = -kt \qquad v = \frac{8}{1 + 8kt}$$

Now we substitute the end limits of $v = 4$ knots and $t = \frac{10}{60} = \frac{1}{6}$ hour and get

$$4 = \frac{8}{1 + 8k(1/6)} \qquad k = \frac{3}{4}\,\text{mi}^{-1} \qquad v = \frac{8}{1 + 6t} \qquad \textit{Ans.}$$

The speed is plotted against the time as shown.

The distance is obtained by substituting the expression for v into the definition $v = ds/dt$ and integrating. Thus,

$$\frac{8}{1 + 6t} = \frac{ds}{dt} \qquad \int_0^t \frac{8\,dt}{1 + 6t} = \int_0^s ds \qquad s = \frac{4}{3}\ln(1 + 6t) \qquad \textit{Ans.}$$

The distance s is also plotted against the time as shown, and we see that the ship has moved through a distance $s = \frac{4}{3}\ln(1 + \frac{6}{6}) = \frac{4}{3}\ln 2 = 0.924$ mi (nautical) during the 10 minutes.

Helpful Hints

① Recall that one knot is the speed of one nautical mile (1852 m) per hour. Work directly in the units of nautical miles and hours.

② We choose to integrate to a general value of v and its corresponding time t so that we may obtain the variation of v with t.

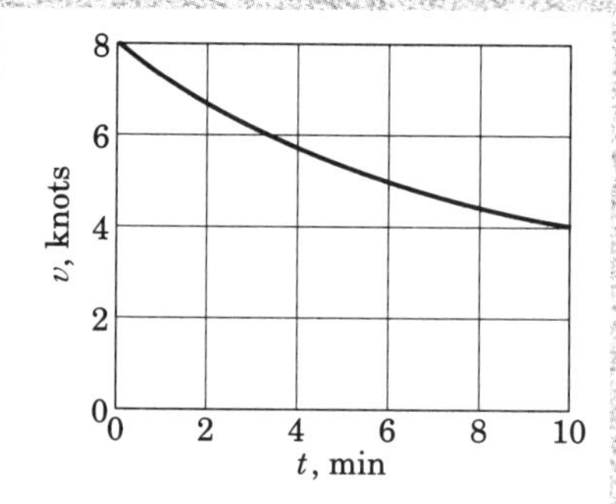

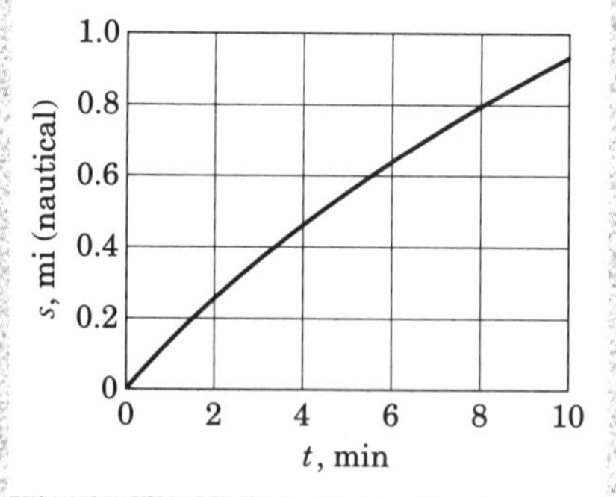

PROBLEMS

Problems 2/1 through 2/7 treat the motion of a particle which moves along the s-axis shown in the figure.

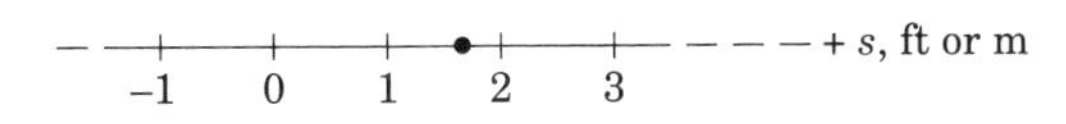

Problems 2/1–2/7

Introductory Problems

2/1 The velocity of a particle is given by $v = 20t^2 - 100t + 50$, where v is in meters per second and t is in seconds. Plot the velocity v and acceleration a versus time for the first 6 seconds of motion and evaluate the velocity when a is zero.

Ans. $v = -75$ m/s

2/2 The displacement of a particle is given by $s = 2t^3 - 30t^2 + 100t - 50$, where s is in meters and t is in seconds. Plot the displacement, velocity, and acceleration as functions of time for the first 12 seconds of motion. Determine the time at which the velocity is zero.

2/3 The velocity of a particle which moves along the s-axis is given by $v = 2 + 5t^{3/2}$, where t is in seconds and v is in meters per second. Evaluate the displacement s, velocity v, and acceleration a when $t = 4$ s. The particle is at the origin $s = 0$ when $t = 0$.

Ans. $s = 72$ m, $v = 42$ m/s, $a = 15$ m/s^2

2/4 The velocity of a particle along the s-axis is given by $v = 5s^{3/2}$, where s is in millimeters and v is in millimeters per second. Determine the acceleration when s is 2 millimeters.

2/5 The position of a particle in millimeters is given by $s = 27 - 12t + t^2$, where t is in seconds. Plot the s-t and v-t relationships for the first 9 seconds. Determine the net displacement Δs during that interval and the total distance D traveled. By inspection of the s-t relationship, what conclusion can you reach regarding the acceleration?

Ans. $\Delta s = -27$ mm, $D = 45$ mm, $a =$ constant

2/6 A particle moves along a straight line with a velocity in millimeters per second given by $v = 400 - 16t^2$ where t is in seconds. Calculate the net displacement Δs and total distance D traveled during the first 6 seconds of motion.

2/7 The acceleration of a particle is given by $a = 4t - 30$, where a is in meters per second squared and t is in seconds. Determine the velocity and displacement as functions of time. The initial displacement at $t = 0$ is $s_0 = -5$ m, and the initial velocity is $v_0 = 3$ m/s.

Ans. $v = 3 - 30t + 2t^2$ m/s
$s = -5 + 3t - 15t^2 + \frac{2}{3}t^3$ m

2/8 A rocket is fired vertically up from rest. If it is designed to maintain a constant upward acceleration of $1.5g$, calculate the time t required for it to reach an altitude of 30 km and its velocity at that position.

2/9 A car comes to a complete stop from an initial speed of 80 km/h in a distance of 30 m. With the same constant acceleration, what would be the stopping distance s from an initial speed of 110 km/h?

Ans. $s = 56.7$ m

2/10 Calculate the constant acceleration a in g's which the catapult of an aircraft carrier must provide to produce a launch velocity of 300 km/h in a distance of 100 m. Assume that the carrier is at anchor.

2/11 The pilot of a jet transport brings the engines to full takeoff power before releasing the brakes as the aircraft is standing on the runway. The jet thrust remains constant, and the aircraft has a near-constant acceleration of $0.4g$. If the takeoff speed is 200 km/h, calculate the distance s and time t from rest to takeoff.

Ans. $s = 393$ m, $t = 14.16$ s

2/12 A jet aircraft with a landing speed of 200 km/h has a maximum of 600 m of available runway after touchdown in which to reduce its ground speed to 30 km/h. Compute the average acceleration a required of the aircraft during braking.

2/13 In the final stages of a moon landing, the lunar module descends under retrothrust of its descent engine to within $h = 5$ m of the lunar surface where it has a downward velocity of 2 m/s. If the descent engine is cut off abruptly at this point, compute the impact velocity of the landing gear with the moon. Lunar gravity is $\frac{1}{6}$ of the earth's gravity.

Ans. $v = 4.51$ m/s

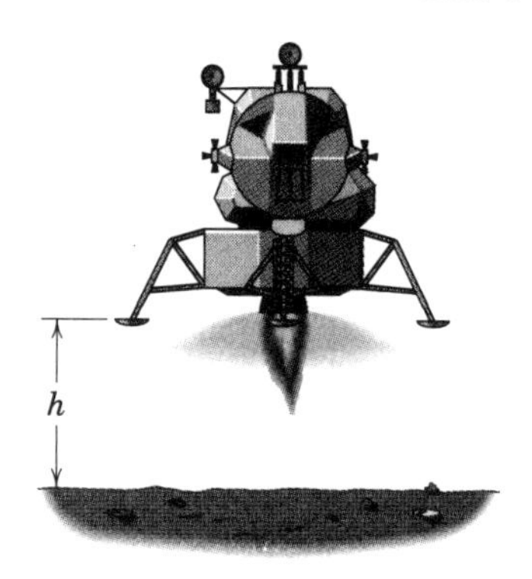

Problem 2/13

2/14 A projectile is fired vertically with an initial velocity of 200 m/s. Calculate the maximum altitude h reached by the projectile and the time t after firing for it to return to the ground. Neglect air resistance and take the gravitational acceleration to be constant at 9.81 m/s^2.

Representative Problems

2/15 A ball is thrown vertically upward with an initial speed of 25 m/s from the base A of a 15-m cliff. Determine the distance h by which the ball clears the top of the cliff and the time t after release for the ball to land at B. Also, calculate the impact velocity v_B. Neglect air resistance and the small horizontal motion of the ball.

Ans. $h = 16.86$ m, $t = 4.40$ s
$v_B = 18.19$ m/s downward

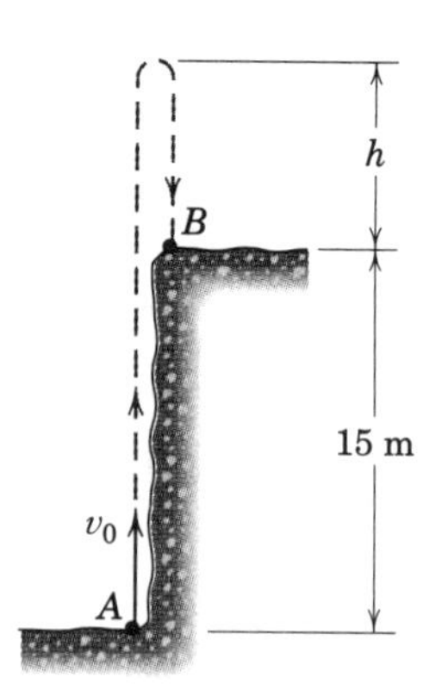

Problem 2/15

2/16 The main elevator A of the CN Tower in Toronto rises about 350 m and for most of its run has a constant speed of 22 km/h. Assume that both the acceleration and deceleration have a constant magnitude of $\frac{1}{4}g$ and determine the time duration t of the elevator run.

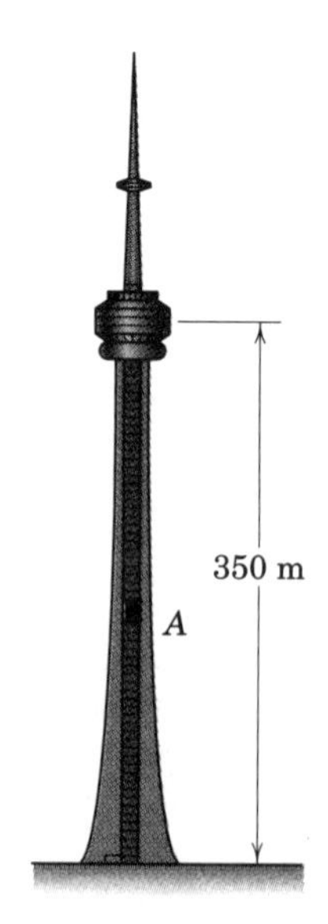

Problem 2/16

2/17 Experimental data for the motion of a particle along a straight line yield measured values of the velocity v for various position coordinates s. A smooth curve is drawn through the points as shown in the graph. Determine the acceleration of the particle when $s = 20$ m.

Ans. $a = 1.2\text{ m/s}^2$

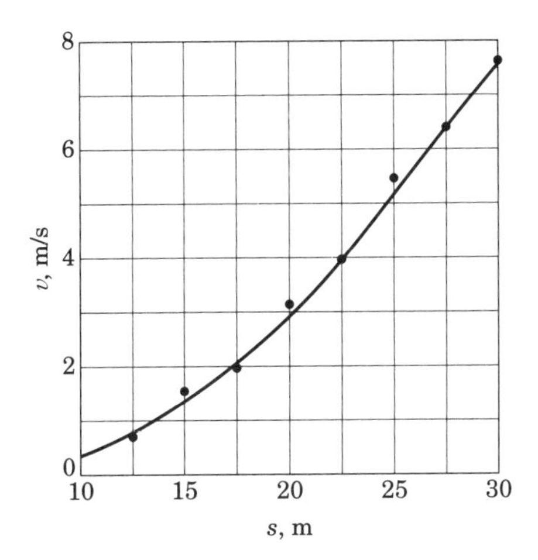

Problem 2/17

2/18 The graph shows the displacement-time history for the rectilinear motion of a particle during an 8-second interval. Determine the average velocity v_{av} during the interval and, to within reasonable limits of accuracy, find the instantaneous velocity v when $t = 4$ s.

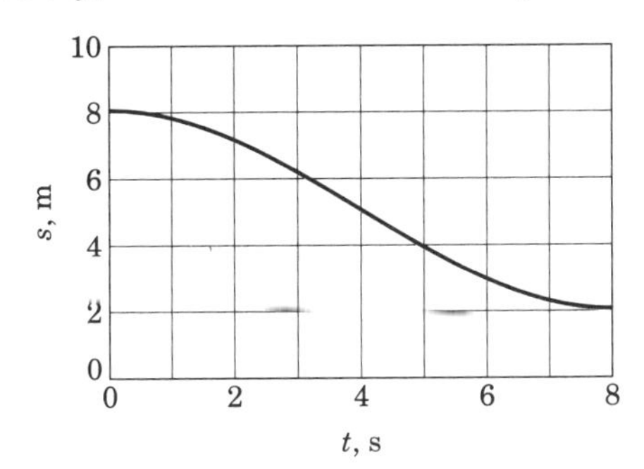

Problem 2/18

2/19 Small steel balls fall from rest through the opening at A at the steady rate of two per second. Find the vertical separation h of two consecutive balls when the lower one has dropped 3 meters. Neglect air resistance.

Ans. $h = 2.61$ m

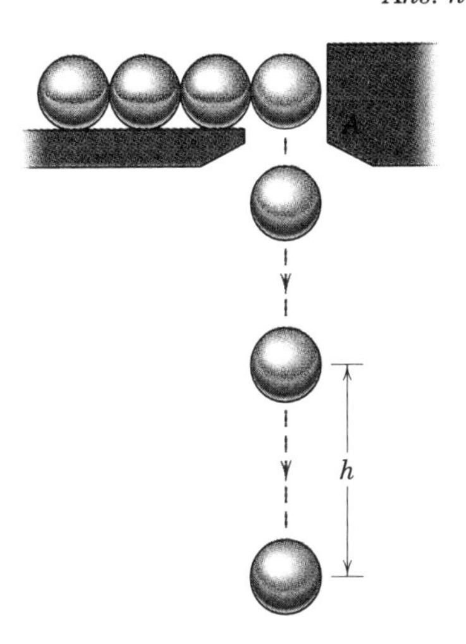

Problem 2/19

2/20 In traveling a distance of 3 km between points A and D, a car is driven at 100 km/h from A to B for t seconds and 60 km/h from C to D also for t seconds. If the brakes are applied for 4 seconds between B and C to give the car a uniform deceleration, calculate t and the distance s between A and B.

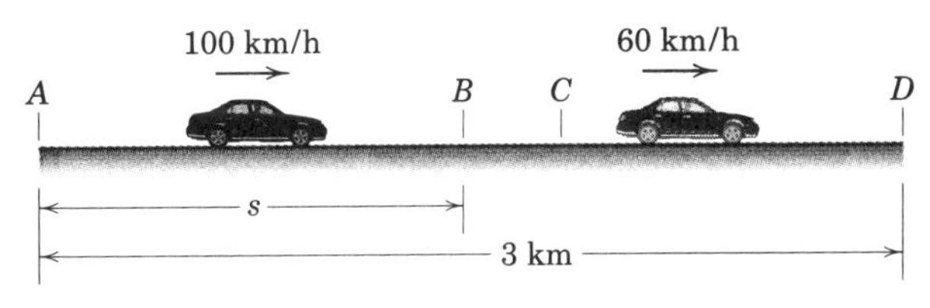

Problem 2/20

2/21 During an 8-second interval the velocity of a particle moving in a straight line varies with time as shown. Within reasonable limits of accuracy, determine the amount Δa by which the acceleration at $t = 4$ s exceeds the average acceleration during the interval. What is the displacement Δs during the interval?

Ans. $\Delta a = 0.50 \text{ m/s}^2$, $\Delta s = 64$ m

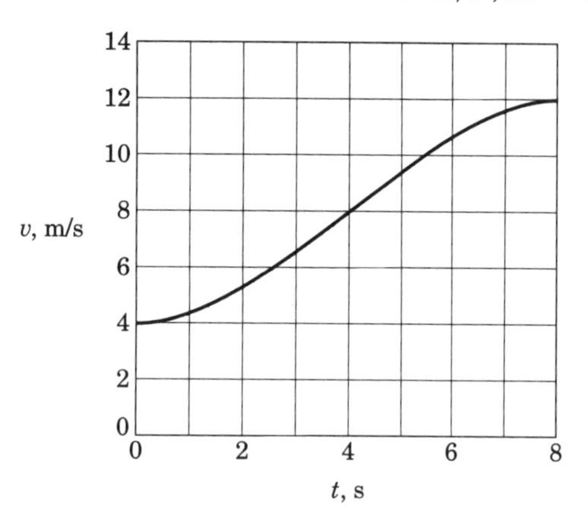

Problem 2/21

2/22 A particle moves along the positive x-axis with an acceleration a_x in meters per second squared which increases linearly with x expressed in millimeters, as shown on the graph for an interval of its motion. If the velocity of the particle at $x = 40$ mm is 0.4 m/s, determine the velocity at $x = 120$ mm.

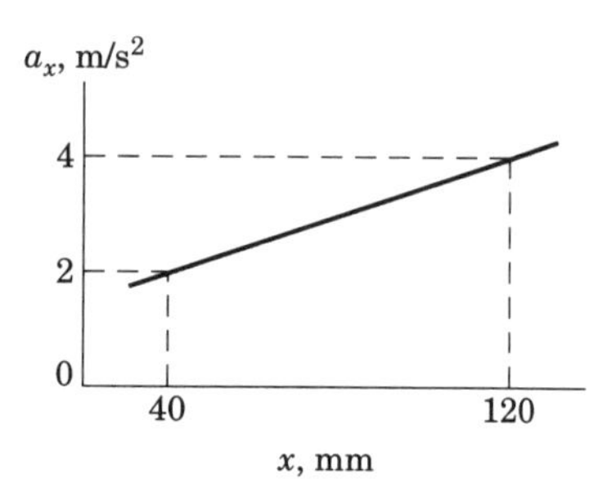

Problem 2/22

2/23 A girl rolls a ball up an incline and allows it to return to her. For the angle θ and ball involved, the acceleration of the ball along the incline is constant at $0.25g$, directed down the incline. If the ball is released with a speed of 4 m/s, determine the distance s it moves up the incline before reversing its direction and the total time t required for the ball to return to the child's hand.

Ans. $s = 3.26$ m, $t = 3.26$ s

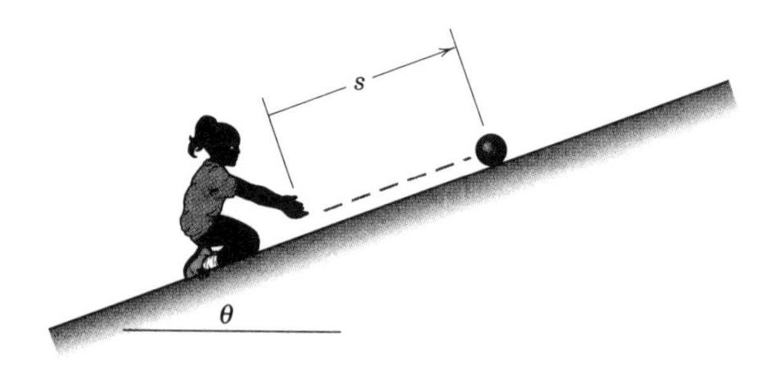

Problem 2/23

2/24 The 350-mm spring is compressed to a 200-mm length, where it is released from rest and accelerates the sliding block A. The acceleration has an initial value of 130 m/s^2 and then decreases linearly with the x-movement of the block, reaching zero when the spring regains its original 350-mm length. Calculate the time t for the block to go (a) 75 mm and (b) 150 mm.

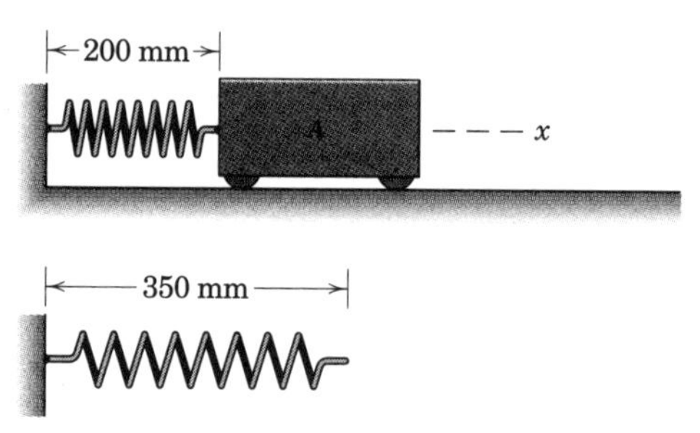

Problem 2/24

2/25 The car is traveling at a constant speed $v_0 = 100$ km/h on the level portion of the road. When the 6-percent ($\tan \theta = 6/100$) incline is encountered, the driver does not change the throttle setting and consequently the car decelerates at the constant rate $g \sin \theta$. Determine the speed of the car (a) 10 seconds after passing point A and (b) when $s = 100$ m.

Ans. (a) $v = 21.9$ m/s, (b) $v = 25.6$ m/s

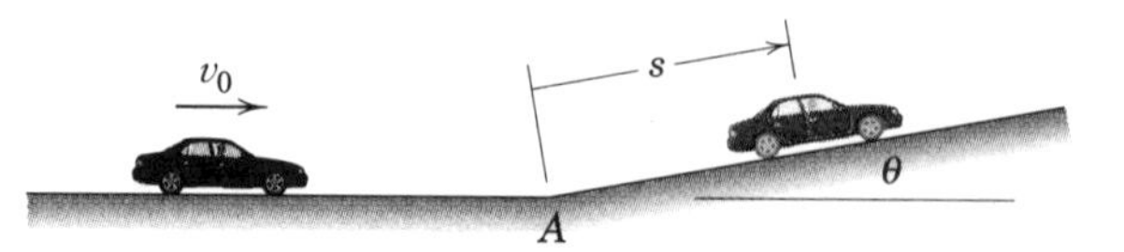

Problem 2/25

2/26 A train which is traveling at 130 km/h applies its brakes as it reaches point A and slows down with a constant deceleration. Its decreased velocity is observed to be 96 km/h as it passes a point 0.8 km beyond A. A car moving at 80 km/h passes point B at the same instant that the train reaches point A. In an unwise effort to beat the train to the crossing, the driver "steps on the gas." Calculate the constant acceleration a that the car must have in order to beat the train to the crossing by 4 s and find the velocity v of the car as it reaches the crossing.

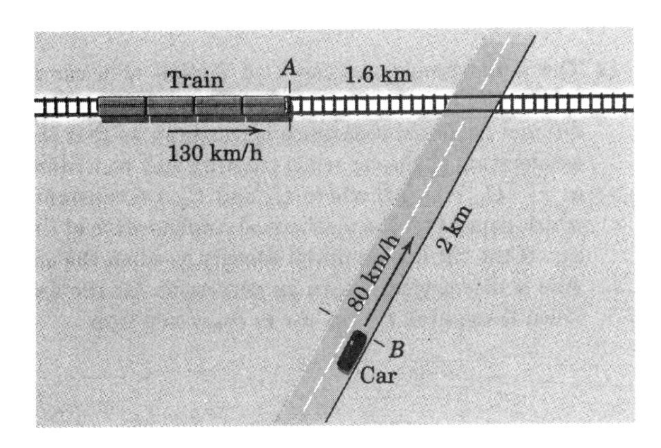

Problem 2/26

2/27 A particle moves along the x-axis with a constant acceleration. When $t = 0$, $x = 4$ m and $\dot{x} = 3$ m/s. Also, when $t = 4$ s, a maximum value of x is observed. Determine x_{max} and the value of x when $t = 12$ s. Plot x versus t.

Ans. $x_{max} = 10$ m, $x_{12} = -14$ m

2/28 A single-stage rocket is launched vertically from rest, and its thrust is programmed to give the rocket a constant upward acceleration of 6 m/s^2. If the fuel is exhausted 20 s after launch, calculate the maximum velocity v_m and the subsequent maximum altitude h reached by the rocket.

2/29 A motorcycle patrolman starts from rest at A two seconds after a car, speeding at the constant rate of 120 km/h, passes point A. If the patrolman accelerates at the rate of 6 m/s^2 until he reaches his maximum permissible speed of 150 km/h, which he maintains, calculate the distance s from point A to the point at which he overtakes the car.

Ans. $s = 912$ m

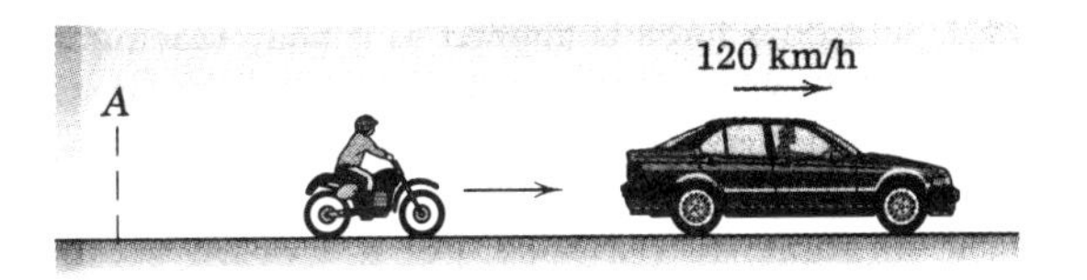

Problem 2/29

2/30 A sprinter reaches his maximum speed v_{max} in 2.5 s from rest with constant acceleration. He then maintains that speed and finishes the 100 meters in the overall time of 10.40 s. Determine his maximum speed v_{max}.

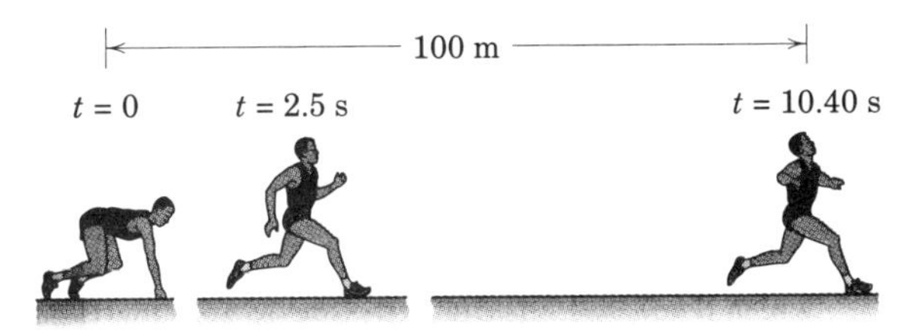

Problem 2/30

2/31 A particle starts from rest at $x = -2$ m and moves along the x-axis with the velocity history shown. Plot the corresponding acceleration and the displacement histories for the 2 seconds. Find the time t when the particle crosses the origin.

Ans. $t = 0.917$ s

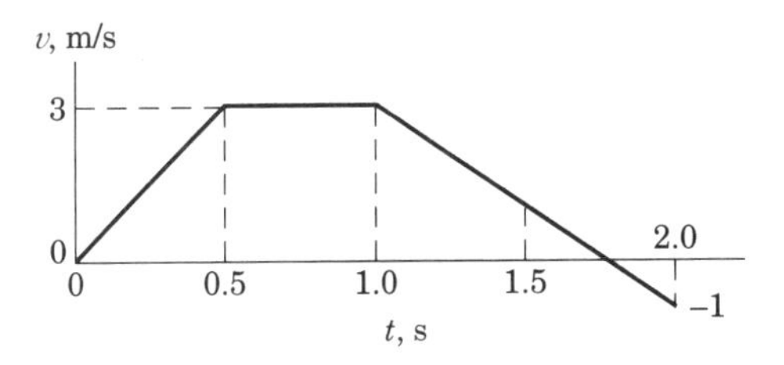

Problem 2/31

2/32 A vacuum-propelled capsule for a high-speed tube transportation system of the future is being designed for operation between two stations A and B, which are 10 km apart. If the acceleration and deceleration are to have a limiting magnitude of 0.6g and if velocities are to be limited to 400 km/h, determine the minimum time t for the capsule to make the 10-km trip.

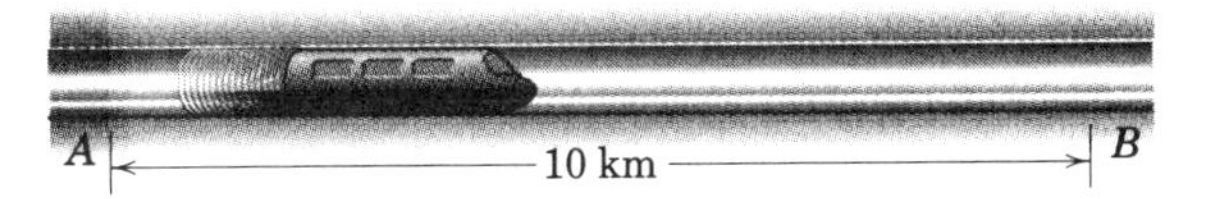

Problem 2/32

2/33 A retarding force is applied to a body moving in a straight line so that, during an interval of its motion, its speed v decreases with increased position coordinate s according to the relation $v^2 = k/s$, where k is a constant. If the body has a forward speed of 50 mm/s and its position coordinate is 225 mm at time $t = 0$, determine the speed v at $t = 3$ s.

Ans. $v = 39.7$ mm/s

2/34 Particle 1 is subjected to an acceleration $a = -kv$, particle 2 is subjected to $a = -kt$, and particle 3 is subjected to $a = -ks$. All three particles start at the origin $s = 0$ with an initial velocity $v_0 = 10$ m/s at time $t = 0$, and the magnitude of k is 0.1 for all three particles (note that the units of k vary from case to case). Plot the position, velocity, and acceleration versus time for each particle over the range $0 \leq t \leq 10$ s.

2/35 A car starts from rest with an acceleration of 6 m/s^2 which decreases linearly with time to zero in 10 seconds, after which the car continues at a constant speed. Determine the time t required for the car to travel 400 m from the start.

Ans. $t = 16.67$ s

2/36 The body falling with speed v_0 strikes and maintains contact with the platform supported by a nest of springs. The acceleration of the body after impact is $a = g - cy$, where c is a positive constant and y is measured from the original platform position. If the maximum compression of the springs is observed to be y_m, determine the constant c.

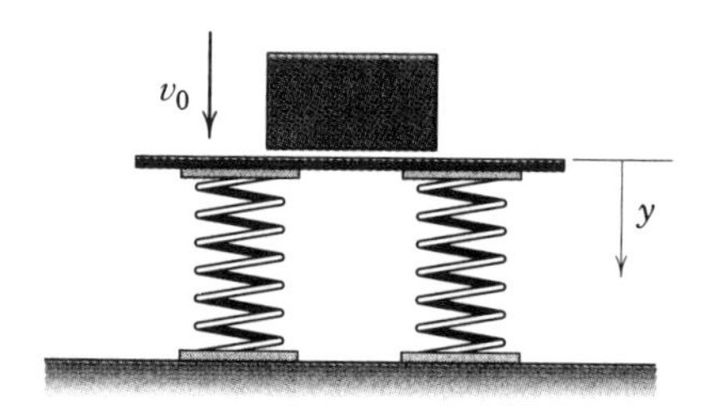

Problem 2/36

2/37 A certain lake is proposed as a landing area for large jet aircraft. The touchdown speed of 160 km/h upon contact with the water is to be reduced to 30 km/h in a distance of 400 m. If the deceleration is proportional to the square of the velocity of the aircraft through the water, $a = -Kv^2$, find the value of the design parameter K, which would be a measure of the size and shape of the landing gear vanes that plow through the water. Also find the time t elapsed during the specified interval.

Ans. $K = 4.18(10^{-3})$ m^{-1}, $t = 23.3$ s

2/38 The aerodynamic resistance to motion of a car is nearly proportional to the square of its velocity. Additional frictional resistance is constant, so that the acceleration of the car when coasting may be written $a = -C_1 - C_2v^2$, where C_1 and C_2 are constants which depend on the mechanical configuration of the car. If the car has an initial velocity v_0 when the engine is disengaged, derive an expression for the distance D required for the car to coast to a stop.

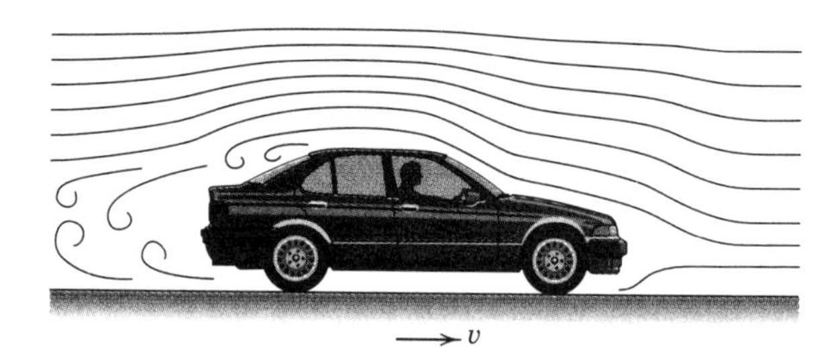

Problem 2/38

2/39 Compute the impact speed of a body released from rest at an altitude $h = 800$ km. (*a*) Assume a constant gravitational acceleration $g_0 = 9.81$ m/s^2 and (*b*) account for the variation of g with altitude (refer to Art. 1/5). Neglect the effects of atmospheric drag.

Ans. (*a*) $v = 3960$ m/s, (*b*) $v = 3730$ m/s

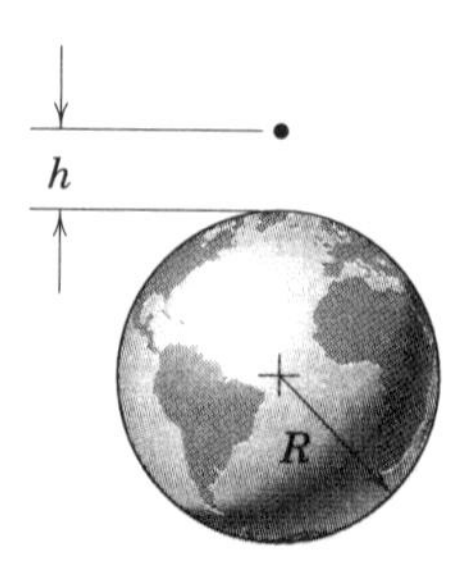

Problem 2/39

2/40 The horizontal motion of the plunger and shaft is arrested by the resistance of the attached disk which moves through the oil bath. If the velocity of the plunger is v_0 in the position A where $x = 0$ and $t = 0$, and if the deceleration is proportional to v so that $a = -kv$, derive expressions for the velocity v and position coordinate x in terms of the time t. Also express v in terms of x.

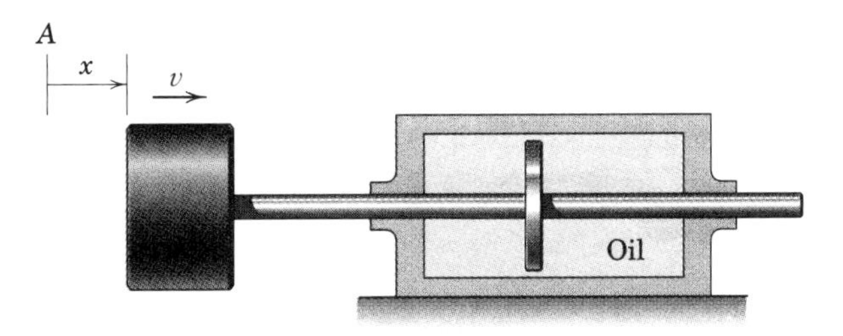

Problem 2/40

2/41 On its takeoff roll, the airplane starts from rest and accelerates according to $a = a_0 - kv^2$, where a_0 is the constant acceleration resulting from the engine thrust and $-kv^2$ is the acceleration due to aerodynamic drag. If $a_0 = 2$ m/s^2, $k = 0.00004$ m^{-1}, and v is in meters per second, determine the design length of runway required for the airplane to reach the takeoff speed of 250 km/h if the drag term is (*a*) excluded and (*b*) included.

Ans. (*a*) $s = 1206$ m, (*b*) $s = 1268$ m

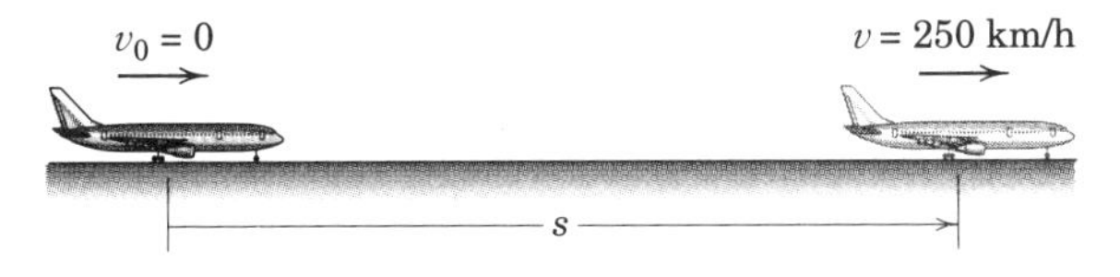

Problem 2/41

2/42 Packages enter the 3-m chute at A with a speed of 1.2 m/s and have a $0.3g$ acceleration from A to B. If the packages come to rest at C, calculate the constant acceleration a of the packages from B to C. Also find the time required for the packages to go from A to C.

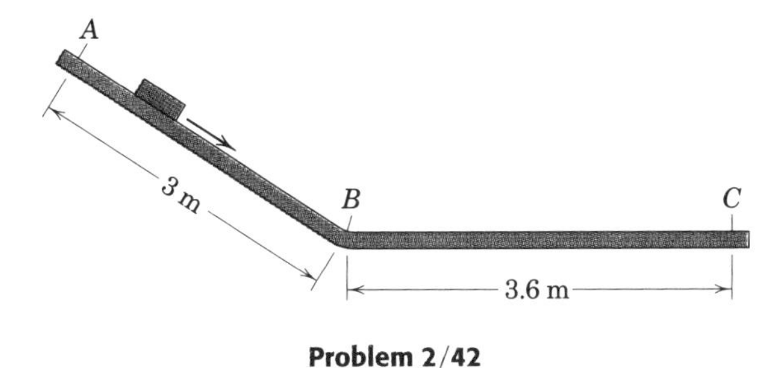

Problem 2/42

2/43 The steel ball A of diameter D slides freely on the horizontal rod which leads to the pole face of the electromagnet. The force of attraction obeys an inverse-square law, and the resulting acceleration of the ball is $a = K/(L - x)^2$, where K is a measure of the strength of the magnetic field. If the ball is released from rest at $x = 0$, determine the velocity v with which it strikes the pole face.

Ans. $v = 2\sqrt{\dfrac{K(L - D/2)}{LD}}$

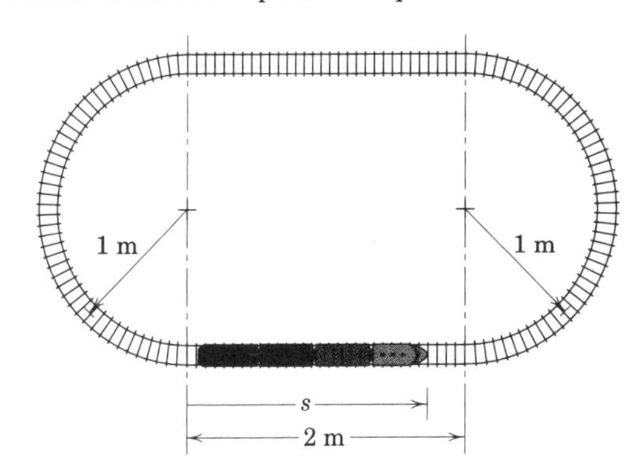

Problem 2/43

2/44 The electronic throttle control of a model train is programmed so that the train speed varies with position as shown in the plot. Determine the time t required for the train to complete one lap.

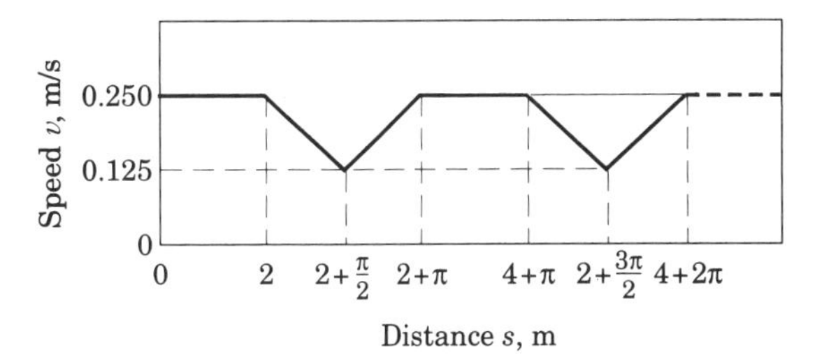

Problem 2/44

2/45 A subway train travels between two of its station stops with the acceleration schedule shown. Determine the time interval Δt during which the train brakes to a stop with a deceleration of 2 m/s^2 and find the distance s between stations.

Ans. $\Delta t = 10$ s, $s = 416$ m

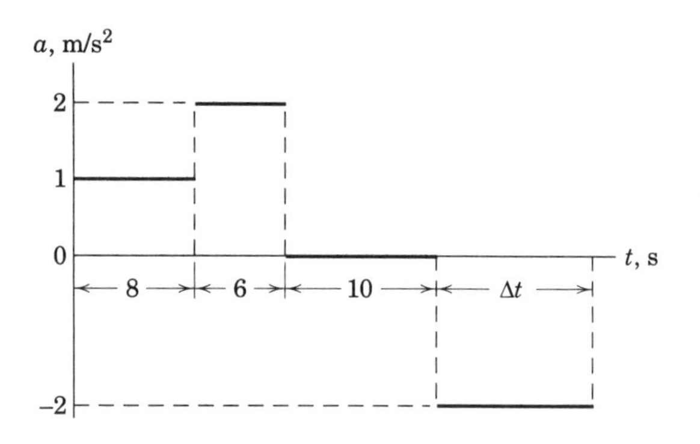

Problem 2/45

2/46 A test projectile is fired horizontally into a viscous liquid with a velocity v_0. The retarding force is proportional to the square of the velocity, so that the acceleration becomes $a = -kv^2$. Derive expressions for the distance D traveled in the liquid and the corresponding time t required to reduce the velocity to $v_0/2$. Neglect any vertical motion.

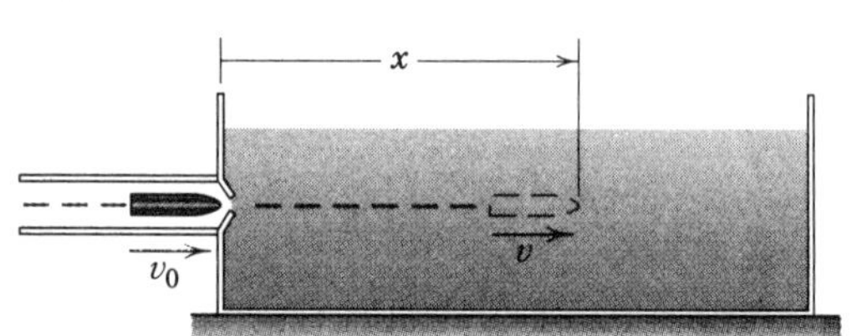

Problem 2/46

2/47 A projectile is fired horizontally into a resisting medium with a velocity v_0, and the resulting deceleration is equal to cv^n, where c and n are constants and v is the velocity within the medium. Find the expression for the velocity v of the projectile in terms of the time t of penetration.

Ans. $v = [v_0^{\,1-n} + c(n - 1)t]^{1/(1-n)}$

2/48 To a close approximation the pressure behind a rifle bullet varies inversely with the position x of the bullet along the barrel. Thus the acceleration of the bullet may be written as $a = k/x$ where k is a constant. If the bullet starts from rest at $x = 7.5$ mm and if the muzzle velocity of the bullet is 600 m/s at the end of the 750-mm barrel, compute the acceleration of the bullet as it passes the midpoint of the barrel at $x = 375$ mm.

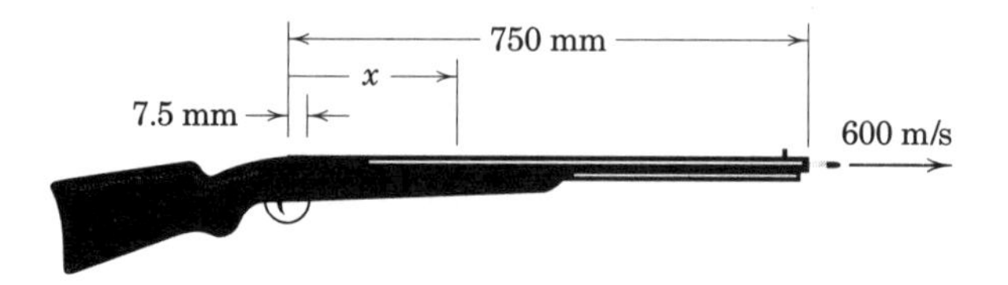

Problem 2/48

2/49 The driver of a car, which is initially at rest at the top A of the grade, releases the brakes and coasts down the grade with an acceleration in feet per second squared given by $a = 0.981 - 0.013v^2$, where v is the velocity in meters per second. Determine the velocity v_B at the bottom B of the grade.

Ans. $v_B = 8.66$ m/s

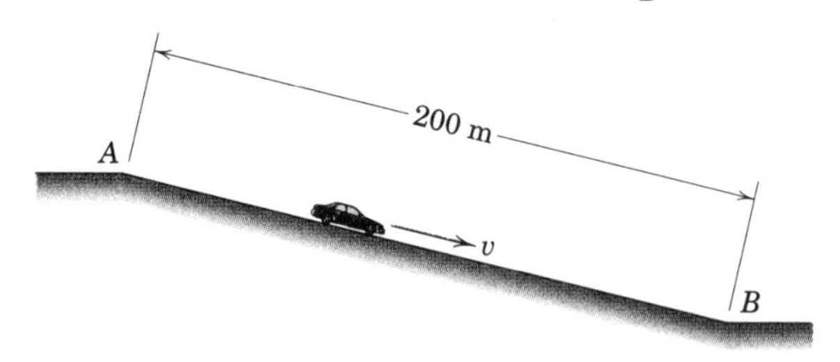

Problem 2/49

2/50 A bumper, consisting of a nest of three springs, is used to arrest the horizontal motion of a large mass which is traveling at 40 m/s as it contacts the bumper. The two outer springs cause a deceleration proportional to the spring deformation. The center spring increases the deceleration rate when the compression exceeds 0.5 m as shown on the graph. Determine the maximum compression x of the outer springs.

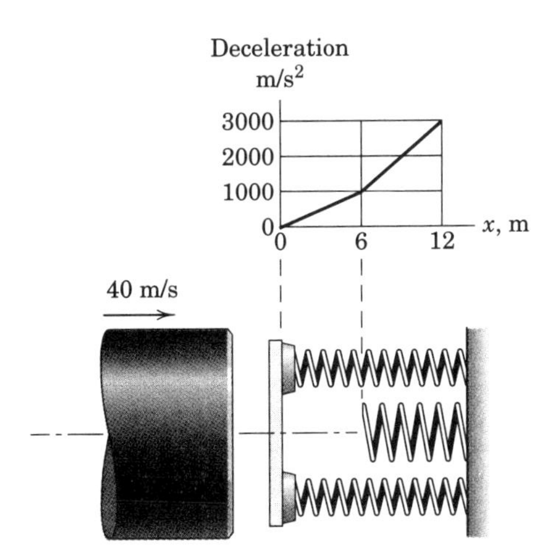

Problem 2/50

2/51 When the effect of aerodynamic drag is included, the y-acceleration of a baseball moving vertically upward is $a_u = -g - kv^2$, while the acceleration when the ball is moving downward is $a_d = -g + kv^2$, where k is a positive constant and v is the speed in meters per second. If the ball is thrown upward at 30 m/s from essentially ground level, compute its maximum height h and its speed v_f upon impact with the ground. Take k to be 0.006 m^{-1} and assume that g is constant.

Ans. $h = 36.5$ m, $v_f = 24.1$ m/s

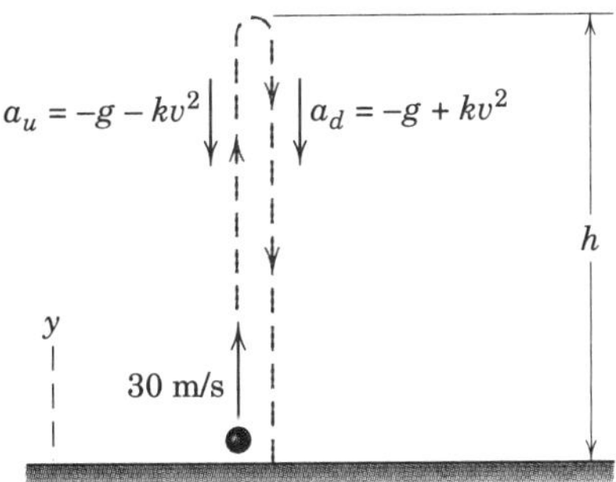

Problem 2/51

2/52 For the baseball of Prob. 2/51 thrown upward with an initial speed of 30 m/s, determine the time t_u from ground to apex and the time t_d from apex to ground.

2/53 The fuel of a model rocket is burned so quickly that one may assume that the rocket acquires its burnout velocity of 120 m/s while essentially still at ground level. The rocket then coasts vertically upward to the trajectory apex. With the inclusion of aerodynamic drag, the y-acceleration is $a_y = -g - 0.0005v^2$ during this motion, where the units are meters and seconds. At apex a parachute pops out of the nose cone, and the rocket quickly acquires a constant downward speed of 4 m/s. Estimate the flight time t.

Ans. $t = 147.7$ s

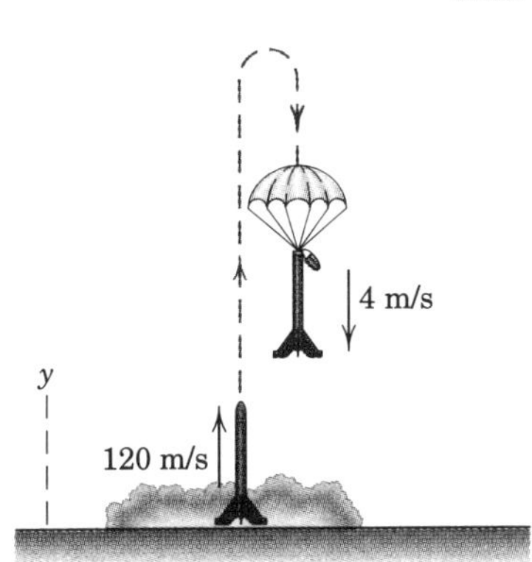

Problem 2/53

2/54 The stories of a tall building are uniformly 3 meters in height. A ball A is dropped from the rooftop position shown. Determine the times required for it to pass the 3 meters of the first, tenth, and one-hundredth stories (counted from the top). Neglect aerodynamic drag.

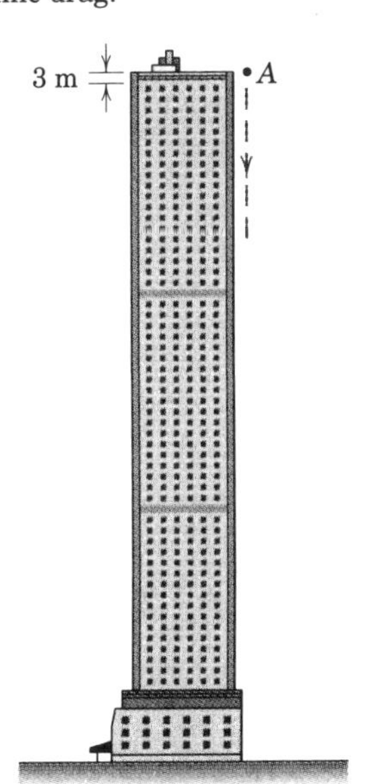

Problem 2/54

2/55 Repeat Prob. 2/54, except now include the effects of aerodynamic drag. The drag force causes an acceleration component in m/s^2 of $0.016v^2$ in the direction opposite the velocity vector, where v is in m/s.

Ans. $t_1 = 0.788$ s, $t_{10} = 0.1567$ s
$t_{100} = 0.1212$ s

►2/56 The vertical acceleration of a certain solid-fuel rocket is given by $a = ke^{-bt} - cv - g$, where k, b, and c are constants, v is the vertical velocity acquired, and g is the gravitational acceleration, essentially constant for atmospheric flight. The exponential term represents the effect of a decaying thrust as fuel is burned, and the term $-cv$ approximates the retardation due to atmospheric resistance. Determine the expression for the vertical velocity of the rocket t seconds after firing.

Ans. $v = \frac{g}{c}(e^{-ct} - 1) + \frac{k}{c - b}(e^{-bt} - e^{-ct})$

►2/57 A particle which is constrained to move in a straight line is subjected to an accelerating force which increases with time and a retarding force which increases directly with the position coordinate x. The resulting acceleration is $a = Kt - k^2x$, where K and k are positive constants and where both x and $\dot{x}$ are zero when $t = 0$. Determine x as a function of t.

Ans. $x = \frac{K}{k^3}(kt - \sin kt)$

►2/58 The preliminary design for a rapid-transit system calls for the train velocity to vary with time as shown in the plot as the train runs the 3.2 km between stations A and B. The slopes of the cubic transition curves (which are of form $a + bt + ct^2 + dt^3$) are zero at the end points. Determine the total run time t between the stations and the maximum acceleration.

Ans. $t = 103.6$ s, $a_{max} = 3.61$ m/s^2

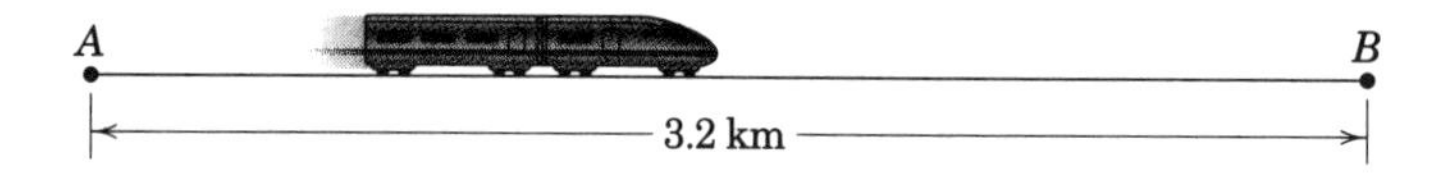

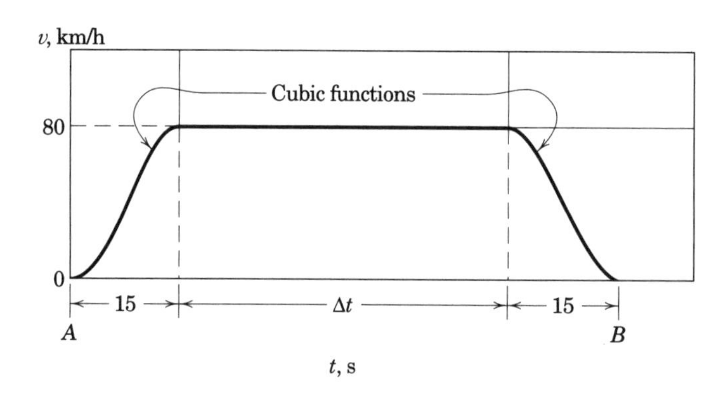

Problem 2/58

2/3 Plane Curvilinear Motion

We now treat the motion of a particle along a curved path which lies in a single plane. This motion is a special case of the more general three-dimensional motion introduced in Art. 2/1 and illustrated in Fig. 2/1. If we let the plane of motion be the x-y plane, for instance, then the coordinates z and ϕ of Fig. 2/1 are both zero, and R becomes the same as r. As mentioned previously, the vast majority of the motions of points or particles encountered in engineering practice can be represented as plane motion.

Before pursuing the description of plane curvilinear motion in any specific set of coordinates, we will first use vector analysis to describe the motion, since the results will be independent of any particular coordinate system. What follows in this article constitutes one of the most basic concepts in dynamics, namely, the *time derivative of a vector*. Much analysis in dynamics utilizes the time rates of change of vector quantities. You are therefore well advised to master this topic at the outset because you will have frequent occasion to use it.

Consider now the continuous motion of a particle along a plane curve as represented in Fig. 2/5. At time t the particle is at position A, which is located by the *position vector* $\mathbf{r}$ measured from some convenient fixed origin O. If both the magnitude and direction of $\mathbf{r}$ are known at time t, then the position of the particle is completely specified. At time $t + \Delta t$, the particle is at A', located by the position vector $\mathbf{r} + \Delta\mathbf{r}$. We note, of course, that this combination is vector addition and not scalar addition. The *displacement* of the particle during time Δt is the vector $\Delta\mathbf{r}$ which represents the vector change of position and is clearly independent of the choice of origin. If an origin were chosen at some different location, the position vector $\mathbf{r}$ would be changed, but $\Delta\mathbf{r}$ would be unchanged. The *distance* actually traveled by the particle as it moves along the path from A to A' is the scalar length Δs measured along the path. Thus, we distinguish between the vector displacement $\Delta\mathbf{r}$ and the scalar distance Δs.

Velocity

The *average velocity* of the particle between A and A' is defined as $\mathbf{v}_{av} = \Delta\mathbf{r}/\Delta t$, which is a vector whose direction is that of $\Delta\mathbf{r}$ and whose

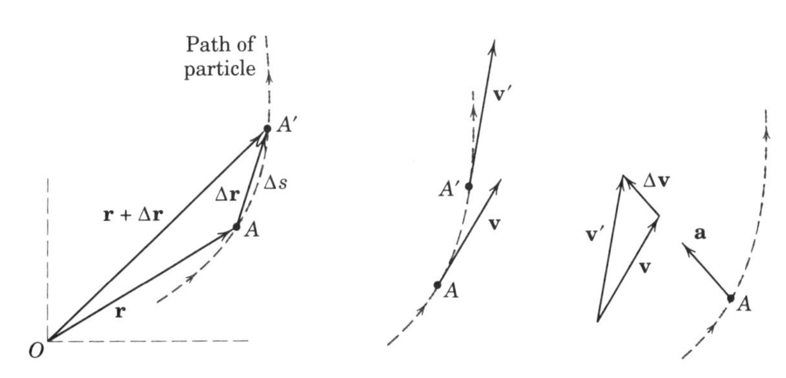

Figure 2/5

magnitude is the magnitude of $\Delta \mathbf{r}$ divided by Δt. The average speed of the particle between A and A' is the scalar quotient $\Delta s/\Delta t$. Clearly, the magnitude of the average velocity and the speed approach one another as the interval Δt decreases and A and A' become closer together.

The *instantaneous velocity* $\mathbf{v}$ of the particle is defined as the limiting value of the average velocity as the time interval approaches zero. Thus,

$$\mathbf{v} = \lim_{\Delta t \to 0} \frac{\Delta \mathbf{r}}{\Delta t}$$

We observe that the direction of $\Delta \mathbf{r}$ approaches that of the tangent to the path as Δt approaches zero and, thus, the velocity $\mathbf{v}$ is always a vector tangent to the path.

We now extend the basic definition of the derivative of a scalar quantity to include a vector quantity and write

$$\mathbf{v} = \frac{d\mathbf{r}}{dt} = \dot{\mathbf{r}} \qquad \textbf{(2/4)}$$

The derivative of a vector is itself a vector having both a magnitude and a direction. The magnitude of $\mathbf{v}$ is called the *speed* and is the scalar

$$v = |\mathbf{v}| = \frac{ds}{dt} = \dot{s}$$

At this point we make a careful distinction between the *magnitude of the derivative* and the *derivative of the magnitude.* The magnitude of the derivative can be written in any one of the several ways $|d\mathbf{r}/dt| = |\dot{\mathbf{r}}| = \dot{s} = |\mathbf{v}| = v$ and represents the magnitude of the velocity, or the speed, of the particle. On the other hand, the derivative of the magnitude is written $d|\mathbf{r}|/dt = dr/dt = \dot{r}$, and represents the rate at which the length of the position vector $\mathbf{r}$ is changing. Thus, these two derivatives have two entirely different meanings, and we must be extremely careful to distinguish between them in our thinking and in our notation. For this and other reasons, you are urged to adopt a consistent notation for handwritten work for all vector quantities to distinguish them from scalar quantities. For simplicity the underline $\underline{v}$ is recommended. Other handwritten symbols such as $\vec{v}$, $\underset{\sim}{v}$, and $\hat{v}$ are sometimes used.

With the concept of velocity as a vector established, we return to Fig. 2/5 and denote the velocity of the particle at A by the tangent vector $\mathbf{v}$ and the velocity at A' by the tangent $\mathbf{v}'$. Clearly, there is a vector change in the velocity during the time Δt. The velocity $\mathbf{v}$ at A plus (vectorially) the change $\Delta \mathbf{v}$ must equal the velocity at A', so we can write $\mathbf{v}' - \mathbf{v} = \Delta \mathbf{v}$. Inspection of the vector diagram shows that $\Delta \mathbf{v}$ depends both on the change in magnitude (length) of $\mathbf{v}$ and on the change in direction of $\mathbf{v}$. These two changes are fundamental characteristics of the derivative of a vector.

Acceleration

The *average acceleration* of the particle between A and A' is defined as $\Delta\mathbf{v}/\Delta t$, which is a vector whose direction is that of $\Delta\mathbf{v}$. The magnitude of this average acceleration is the magnitude of $\Delta\mathbf{v}$ divided by Δt.

The *instantaneous acceleration* **a** of the particle is defined as the limiting value of the average acceleration as the time interval approaches zero. Thus,

$$\mathbf{a} = \lim_{\Delta t \to 0} \frac{\Delta\mathbf{v}}{\Delta t}$$

By definition of the derivative, then, we write

$$\mathbf{a} = \frac{d\mathbf{v}}{dt} = \dot{\mathbf{v}} \qquad \textbf{(2/5)}$$

As the interval Δt becomes smaller and approaches zero, the direction of the change $\Delta\mathbf{v}$ approaches that of the differential change $d\mathbf{v}$ and, thus, of **a**. The acceleration **a**, then, includes the effects of both the change in magnitude of **v** and the change of direction of **v**. It is apparent, in general, that the direction of the acceleration of a particle in curvilinear motion is neither tangent to the path nor normal to the path. We do observe, however, that the acceleration component which is normal to the path points toward the center of curvature of the path.

Visualization of Motion

A further approach to the visualization of acceleration is shown in Fig. 2/6, where the position vectors to three arbitrary positions on the path of the particle are shown for illustrative purpose. There is a velocity vector tangent to the path corresponding to each position vector, and the relation is $\mathbf{v} = \dot{\mathbf{r}}$. If these velocity vectors are now plotted from some arbitrary point C, a curve, called the *hodograph,* is formed. The derivatives of these velocity vectors will be the acceleration vectors $\mathbf{a} = \dot{\mathbf{v}}$ which are tangent to the hodograph. We see that the acceleration has the same relation to the velocity as the velocity has to the position vector.

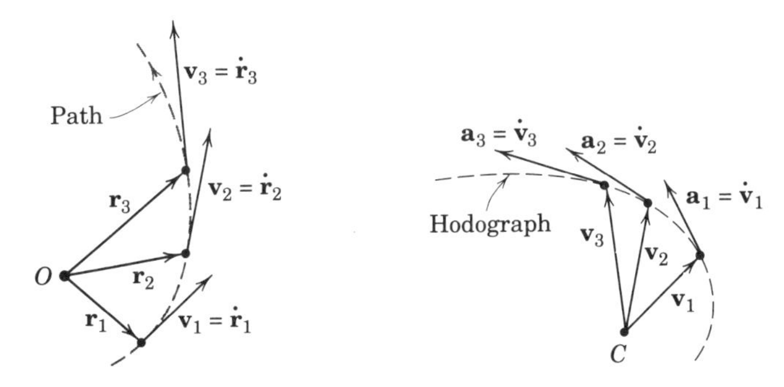

Figure 2/6

The geometric portrayal of the derivatives of the position vector $\mathbf{r}$ and velocity vector $\mathbf{v}$ in Fig. 2/5 can be used to describe the derivative of any vector quantity with respect to t or with respect to any other scalar variable. Now that we have used the definitions of velocity and acceleration to introduce the concept of the derivative of a vector, it is important to establish the rules for differentiating vector quantities. These rules are the same as for the differentiation of scalar quantities, except for the case of the cross product where the order of the terms must be preserved. These rules are covered in Art. C/7 of Appendix C and should be reviewed at this point.

Three different coordinate systems are commonly used for describing the vector relationships for curvilinear motion of a particle in a plane: rectangular coordinates, normal and tangential coordinates, and polar coordinates. An important lesson to be learned from the study of these coordinate systems is the proper choice of a reference system for a given problem. This choice is usually revealed by the manner in which the motion is generated or by the form in which the data are specified. Each of the three coordinate systems will now be developed and illustrated.

2/4 Rectangular Coordinates (x-y)

This system of coordinates is particularly useful for describing motions where the x- and y-components of acceleration are independently generated or determined. The resulting curvilinear motion is then obtained by a vector combination of the x- and y-components of the position vector, the velocity, and the acceleration.

Vector Representation

The particle path of Fig. 2/5 is shown again in Fig. 2/7 along with x- and y-axes. The position vector $\mathbf{r}$, the velocity $\mathbf{v}$, and the acceleration $\mathbf{a}$ of the particle as developed in Art. 2/3 are represented in Fig. 2/7 together with their x- and y-components. With the aid of the unit vectors $\mathbf{i}$ and $\mathbf{j}$, we can write the vectors $\mathbf{r}$, $\mathbf{v}$, and $\mathbf{a}$ in terms of their x- and y-components. Thus,

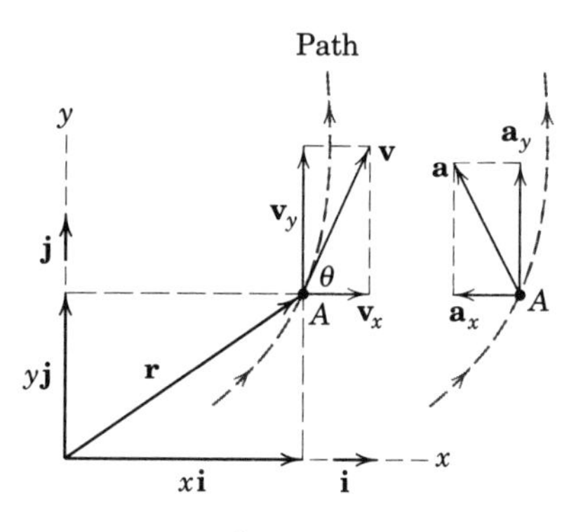

Figure 2/7

$$
\begin{aligned}
\mathbf{r} &= x\mathbf{i} + y\mathbf{j} \\
\mathbf{v} = \dot{\mathbf{r}} &= \dot{x}\mathbf{i} + \dot{y}\mathbf{j} \\
\mathbf{a} = \dot{\mathbf{v}} = \ddot{\mathbf{r}} &= \ddot{x}\mathbf{i} + \ddot{y}\mathbf{j}
\end{aligned}
\qquad (2/6)
$$

As we differentiate with respect to time, we observe that the time derivatives of the unit vectors are zero because their magnitudes and directions remain constant. The scalar values of the components of $\mathbf{v}$ and $\mathbf{a}$ are merely $v_x = \dot{x}$, $v_y = \dot{y}$ and $a_x = \dot{v}_x = \ddot{x}$, $a_y = \dot{v}_y = \ddot{y}$. (As drawn in Fig. 2/7, a_x is in the negative x-direction, so that $\ddot{x}$ would be a negative number.)

As observed previously, the direction of the velocity is always tangent to the path, and from the figure it is clear that

$$v^2 = v_x{}^2 + v_y{}^2 \qquad v = \sqrt{v_x{}^2 + v_y{}^2} \qquad \tan\theta = \frac{v_y}{v_x}$$

$$a^2 = a_x{}^2 + a_y{}^2 \qquad a = \sqrt{a_x{}^2 + a_y{}^2}$$

If the angle θ is measured counterclockwise from the x-axis to $\mathbf{v}$ for the configuration of axes shown, then we can also observe that $dy/dx = \tan\theta = v_y/v_x$.

If the coordinates x and y are known independently as functions of time, $x = f_1(t)$ and $y = f_2(t)$, then for any value of the time we can combine them to obtain $\mathbf{r}$. Similarly, we combine their first derivatives $\dot{x}$ and $\dot{y}$ to obtain $\mathbf{v}$ and their second derivatives $\ddot{x}$ and $\ddot{y}$ to obtain $\mathbf{a}$. On the other hand, if the acceleration components a_x and a_y are given as functions of the time, we can integrate each one separately with respect to time, once to obtain v_x and v_y and again to obtain $x = f_1(t)$ and $y = f_2(t)$. Elimination of the time t between these last two parametric equations gives the equation of the curved path $y = f(x)$.

From the foregoing discussion we can see that the rectangular-coordinate representation of curvilinear motion is merely the superposition of the components of two simultaneous rectilinear motions in the x- and y-directions. Therefore, everything covered in Art. 2/2 on rectilinear motion can be applied separately to the x-motion and to the y-motion.

Projectile Motion

An important application of two-dimensional kinematic theory is the problem of projectile motion. For a first treatment of the subject, we neglect aerodynamic drag and the curvature and rotation of the earth, and we assume that the altitude change is small enough so that the acceleration due to gravity can be considered constant. With these assumptions, rectangular coordinates are useful for the trajectory analysis.

For the axes shown in Fig. 2/8, the acceleration components are

$$a_x = 0 \qquad a_y = -g$$

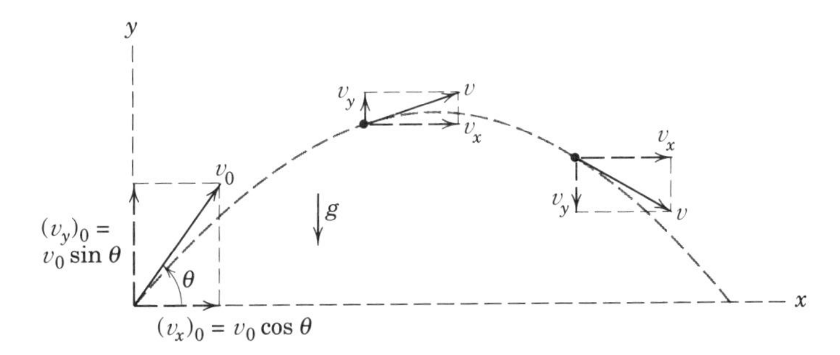

Figure 2/8

Integration of these accelerations follows the results obtained previously in Art. 2/2*a* for constant acceleration and yields

$$v_x = (v_x)_0 \qquad v_y = (v_y)_0 - gt$$

$$x = x_0 + (v_x)_0 t \qquad y = y_0 + (v_y)_0 t - \tfrac{1}{2}gt^2$$

$$v_y^2 = (v_y)_0^2 - 2g(y - y_0)$$

In all these expressions, the subscript zero denotes initial conditions, frequently taken as those at launch where, for the case illustrated, $x_0 = y_0 = 0$. Note that the quantity g is taken to be positive throughout this text.

We can see that the x- and y-motions are independent for the simple projectile conditions under consideration. Elimination of the time t between the x- and y-displacement equations shows the path to be parabolic (see Sample Problem 2/6). If we were to introduce a drag force which depends on the speed squared (for example), then the x- and y-motions would be coupled (interdependent), and the trajectory would be nonparabolic.

When the projectile motion involves large velocities and high altitudes, to obtain accurate results we must account for the shape of the projectile, the variation of g with altitude, the variation of the air density with altitude, and the rotation of the earth. These factors introduce considerable complexity into the motion equations, and numerical integration of the acceleration equations is usually necessary.

Sample Problem 2/5

The curvilinear motion of a particle is defined by $v_x = 50 - 16t$ and $y = 100 - 4t^2$, where v_x is in meters per second, y is in meters, and t is in seconds. It is also known that $x = 0$ when $t = 0$. Plot the path of the particle and determine its velocity and acceleration when the position $y = 0$ is reached.

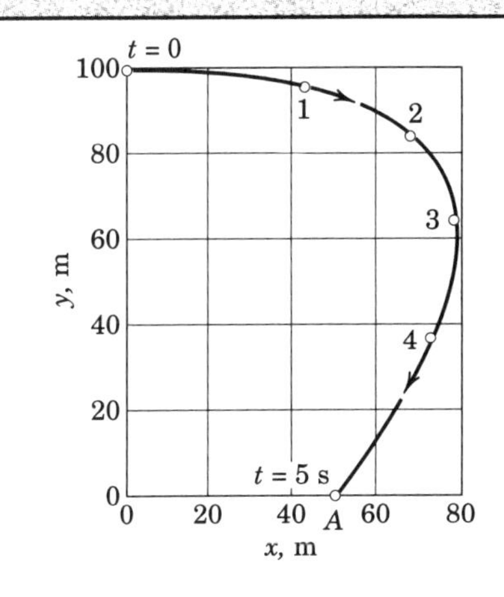

Solution. The x-coordinate is obtained by integrating the expression for v_x, and the x-component of the acceleration is obtained by differentiating v_x. Thus,

$$\left[\int dx = \int v_x\,dt\right] \qquad \int_0^x dx = \int_0^t (50 - 16t)\,dt \qquad x = 50t - 8t^2 \text{ m}$$

$$[a_x = \dot{v}_x] \qquad a_x = \frac{d}{dt}(50 - 16t) \qquad a_x = -16 \text{ m/s}^2$$

The y-components of velocity and acceleration are

$$[v_y = \dot{y}] \qquad v_y = \frac{d}{dt}(100 - 4t^2) \qquad v_y = -8t \text{ m/s}$$

$$[a_y = \dot{v}_y] \qquad a_y = \frac{d}{dt}(-8t) \qquad a_y = -8 \text{ m/s}^2$$

We now calculate corresponding values of x and y for various values of t and plot x against y to obtain the path as shown.

When $y = 0$, $0 = 100 - 4t^2$, so $t = 5$ s. For this value of the time, we have

$$v_x = 50 - 16(5) = -30 \text{ m/s}$$
$$v_y = -8(5) = -40 \text{ m/s}$$
$$v = \sqrt{(-30)^2 + (-40)^2} = 50 \text{ m/s}$$
$$a = \sqrt{(-16)^2 + (-8)^2} = 17.89 \text{ m/s}^2$$

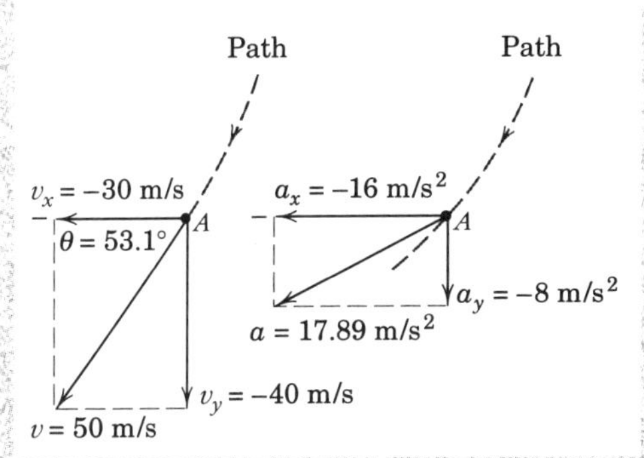

The velocity and acceleration components and their resultants are shown on the separate diagrams for point A, where $y = 0$. Thus, for this condition we may write

$$\mathbf{v} = -30\mathbf{i} - 40\mathbf{j} \text{ m/s} \qquad \textit{Ans.}$$

$$\mathbf{a} = -16\mathbf{i} - 8\mathbf{j} \text{ m/s}^2 \qquad \textit{Ans.}$$

Helpful Hint

We observe that the velocity vector lies along the tangent to the path as it should, but that the acceleration vector is not tangent to the path. Note especially that the acceleration vector has a component that points toward the inside of the curved path. We concluded from our diagram in Fig. 2/5 that it is impossible for the acceleration to have a component that points toward the outside of the curve.

Sample Problem 2/6

A rocket has expended all its fuel when it reaches position A, where it has a velocity $\mathbf{u}$ at an angle θ with respect to the horizontal. It then begins unpowered flight and attains a maximum added height h at position B after traveling a horizontal distance s from A. Determine the expressions for h and s, the time t of flight from A to B, and the equation of the path. For the interval concerned, assume a flat earth with a constant gravitational acceleration g and neglect any atmospheric resistance.

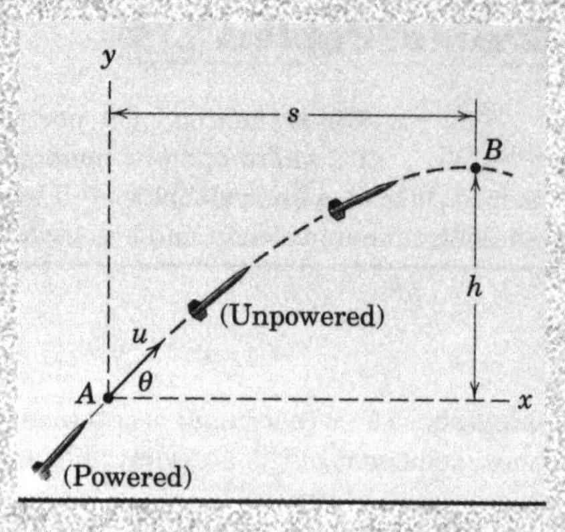

Solution. Since all motion components are directly expressible in terms of horizontal and vertical coordinates, a rectangular set of axes x-y will be employed. ① With the neglect of atmospheric resistance, $a_x = 0$ and $a_y = -g$, and the resulting motion is a direct superposition of two rectilinear motions with constant acceleration. Thus,

$$[dx = v_x\,dt] \qquad x = \int_0^t u\cos\theta\,dt \qquad x = ut\cos\theta$$

$$[dv_y = a_y\,dt] \qquad \int_{u\sin\theta}^{v_y} dv_y = \int_0^t (-g)\,dt \qquad v_y = u\sin\theta - gt$$

$$[dy = v_y\,dt] \qquad y = \int_0^t (u\sin\theta - gt)\,dt \qquad y = ut\sin\theta - \tfrac{1}{2}gt^2$$

Position B is reached when $v_y = 0$, which occurs for $0 = u\sin\theta - gt$ or

$$t = (u\sin\theta)/g \qquad \textit{Ans.}$$

Substitution of this value for the time into the expression for y gives the maximum added altitude

$$h = u\left(\frac{u\sin\theta}{g}\right)\sin\theta - \frac{1}{2}g\left(\frac{u\sin\theta}{g}\right)^2 \qquad h = \frac{u^2\sin^2\theta}{2g} \qquad \textit{Ans.}$$

The horizontal distance is seen to be

② $$s = u\left(\frac{u\sin\theta}{g}\right)\cos\theta \qquad s = \frac{u^2\sin 2\theta}{2g} \qquad \textit{Ans.}$$

which is clearly a maximum when $\theta = 45°$. The equation of the path is obtained by eliminating t from the expressions for x and y, which gives

$$y = x\tan\theta - \frac{gx^2}{2u^2}\sec^2\theta \qquad \textit{Ans.}$$

③ This equation describes a vertical parabola as indicated in the figure.

Helpful Hints

① Note that this problem is simply the description of projectile motion neglecting atmospheric resistance.

② We see that the total range and time of flight for a projectile fired above a horizontal plane would be twice the respective values of s and t given here.

③ If atmospheric resistance were to be accounted for, the dependency of the acceleration components on the velocity would have to be established before an integration of the equations could be carried out. This becomes a much more difficult problem.

PROBLEMS

(In the following problems where motion as a projectile in air is involved, neglect air resistance unless otherwise instructed and use $g = 9.81 \text{ m/s}^2$.)

Introductory Problems

2/59 The position vector of a particle moving in the x-y plane at time $t = 3.60$ s is $2.76\mathbf{i} - 3.28\mathbf{j}$ m. At $t = 3.62$ s its position vector has become $2.79\mathbf{i} - 3.33\mathbf{j}$ m. Determine the magnitude v of its average velocity during this interval and the angle θ made by the average velocity with the x-axis.

Ans. $v = 2.92$ m/s, $\theta = -59.0°$

2/60 A particle moving in the x-y plane has a velocity at time $t = 6$ s given by $4\mathbf{i} + 5\mathbf{j}$ m/s, and at $t = 6.1$ s its velocity has become $4.3\mathbf{i} + 5.4\mathbf{j}$ m/s. Calculate the magnitude a of its average acceleration during the 0.1-s interval and the angle θ it makes with the x-axis.

2/61 The velocity of a particle moving in the x-y plane is given by $6.12\mathbf{i} + 3.24\mathbf{j}$ m/s at time $t = 3.65$ s. Its average acceleration during the next 0.02 s is $4\mathbf{i} + 6\mathbf{j}$ m/s^2. Determine the velocity $\mathbf{v}$ of the particle at $t = 3.67$ s and the angle θ between the average-acceleration vector and the velocity vector at $t = 3.67$ s.

Ans. $\mathbf{v} = 6.20\mathbf{i} + 3.36\mathbf{j}$ m/s, $\theta = 27.9°$

2/62 A particle which moves with curvilinear motion has coordinates in millimeters which vary with the time t in seconds according to $x = 2t^2 - 4t$ and $y = 3t^2 - \frac{1}{3}t^3$. Determine the magnitudes of the velocity $\mathbf{v}$ and acceleration $\mathbf{a}$ and the angles which these vectors make with the x-axis when $t = 2$ s.

2/63 A roofer tosses a small tool toward a coworker on the ground. What is the minimum horizontal velocity v_0 necessary so that the tool clears point B? Locate the point of impact by specifying the distance s shown in the figure.

Ans. $v_0 = 6.64$ m/s, $s = 2.49$ m

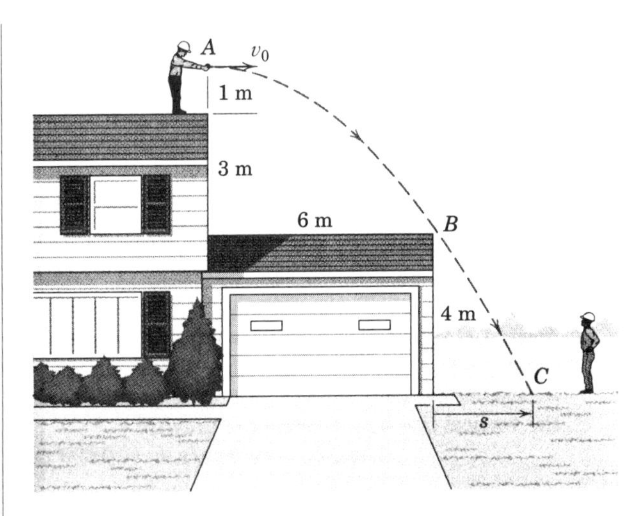

Problem 2/63

2/64 The particle P moves along the curved slot, a portion of which is shown. Its distance in meters measured along the slot is given by $s = t^2/4$, where t is in seconds. The particle is at A when $t = 2.00$ s and at B when $t = 2.20$ s. Determine the magnitude a_{av} of the average acceleration of P between A and B. Also express the acceleration as a vector $\mathbf{a}_{av}$ using unit vectors $\mathbf{i}$ and $\mathbf{j}$.

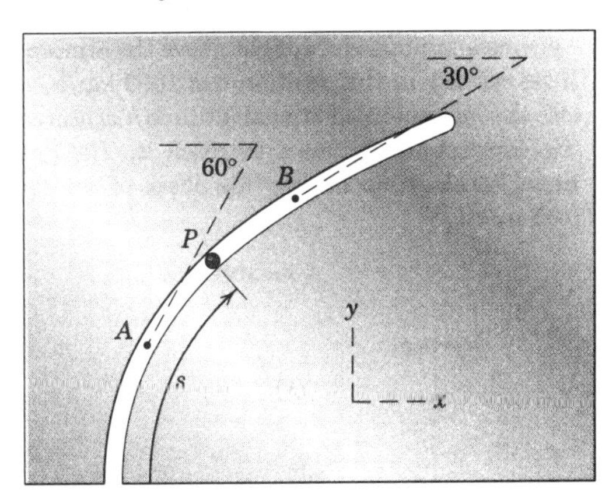

Problem 2/64

2/65 The y-coordinate of a particle in curvilinear motion is given by $y = 4t^3 - 3t$, where y is in meters and t is in seconds. Also, the particle has an acceleration in the x-direction given by $a_x = 12t$ m/s^2. If the velocity of the particle in the x-direction is 4 m/s when $t = 0$, calculate the magnitudes of the velocity $\mathbf{v}$ and acceleration $\mathbf{a}$ of the particle when $t = 1$ s. Construct $\mathbf{v}$ and $\mathbf{a}$ in your solution.

Ans. $v = 13.45$ m/s, $a = 26.8$ m/s^2

2/66 The position vector of a point which moves in the x-y plane is given by

$$\mathbf{r} = \left(\frac{2}{3}t^3 - \frac{3}{2}t^2\right)\mathbf{i} + \frac{t^4}{12}\mathbf{j}$$

where $\mathbf{r}$ is in meters and t is in seconds. Determine the angle between the velocity $\mathbf{v}$ and the acceleration $\mathbf{a}$ when (a) $t = 2$ s and (b) $t = 3$ s.

2/67 A longjumper approaches his takeoff board A with a horizontal velocity of 10 m/s. Determine the vertical component v_y of the velocity of his center of gravity at takeoff for him to make the jump shown. What is the vertical rise h of his center of gravity?

Ans. $v_y = 3.68$ m/s, $h = 0.690$ m

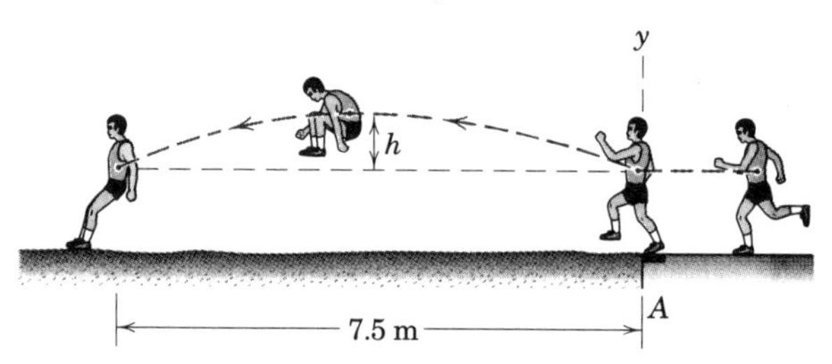

Problem 2/67

2/68 A rocket runs out of fuel in the position shown and continues in unpowered flight above the atmosphere. If its velocity in this position was 1000 km/h, calculate the maximum additional altitude h acquired and the corresponding time t to reach it. The gravitational acceleration during this phase of its flight is 9.39 m/s^2.

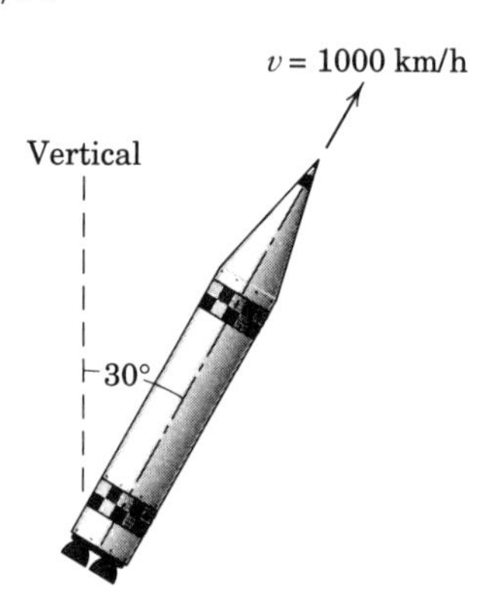

Problem 2/68

2/69 The center of mass G of a high jumper follows the trajectory shown. Determine the component v_0, measured in the vertical plane of the figure, of his takeoff velocity and angle θ if the apex of the trajectory just clears the bar at A. (In general, must the mass center G of the jumper clear the bar during a successful jump?)

Ans. $v_0 = 5.04$ m/s, $\theta = 64.7°$

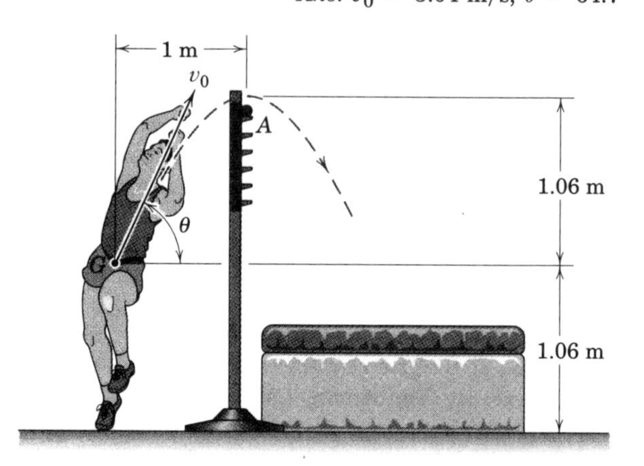

Problem 2/69

2/70 A particle moves in the x-y plane with a y-component of velocity in meters per second given by $v_y = 8t$ with t in seconds. The acceleration of the particle in the x-direction in meters per second squared is given by $a_x = 4t$ with t in seconds. When $t = 0$, $y = 2$ m, $x = 0$, and $v_x = 0$. Find the equation of the path of the particle and calculate the magnitude of the velocity $\mathbf{v}$ of the particle for the instant when its x-coordinate reaches 18 m.

2/71 With what minimum horizontal velocity u can a boy throw a rock at A and have it just clear the obstruction at B?

Ans. $u = 28.0$ m/s

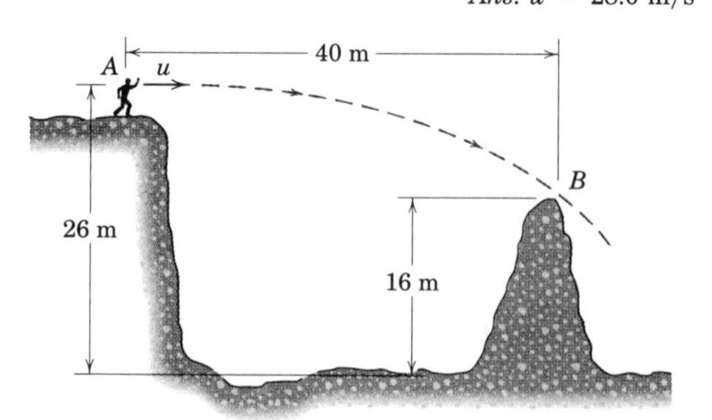

Problem 2/71

Representative Problems

2/72 The rifle is aimed at point A and fired. Calculate the distance b below A to the point B where the bullet strikes. The muzzle velocity is 800 m/s.

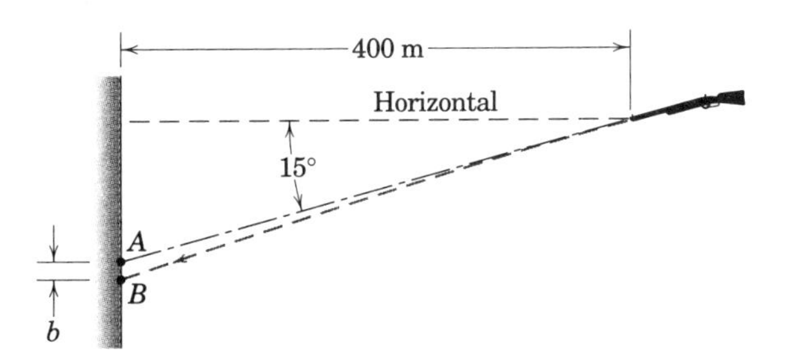

Problem 2/72

2/73 A particle is ejected from the tube at A with a velocity v at an angle θ with the vertical y-axis. A strong horizontal wind gives the particle a constant horizontal acceleration a in the x-direction. If the particle strikes the ground at a point directly under its released position, determine the height h of point A. The downward y-acceleration may be taken as the constant g.

Ans. $h = \dfrac{2v^2}{a} \sin\theta \left(\cos\theta + \dfrac{g}{a}\sin\theta\right)$

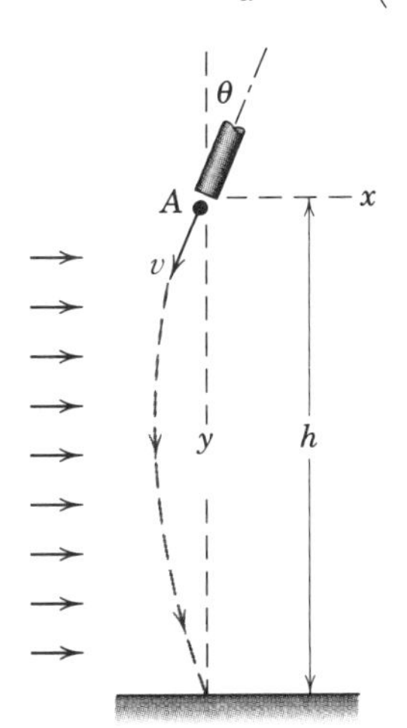

Problem 2/73

2/74 Prove the well-known result that, for a given launch speed v_0, the launch angle $\theta = 45°$ yields the maximum horizontal range R. Determine the maximum range. (Note that this result does not hold when aerodynamic drag is included in the analysis.)

2/75 Calculate the minimum possible magnitude u of the muzzle velocity which a projectile must have when fired from point A to reach a target B on the same horizontal plane 12 km away.

Ans. $u = 343$ m/s

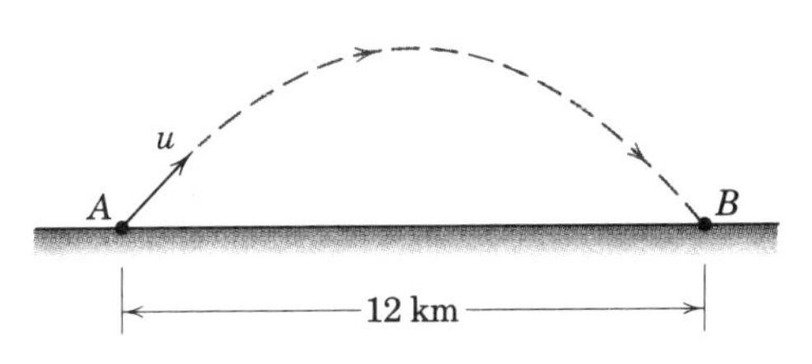

Problem 2/75

2/76 Electrons are emitted at A with a velocity v at the angle θ into the space between two charged plates. The electric field between the plates is in the direction E and repels the electrons approaching the upper plate. The field produces an acceleration of the electrons in the E-direction of eE/m, where e is the electron charge and m is its mass. Determine the field strength E which will permit the electrons to cross one-half of the gap between the plates. Also find the distance s.

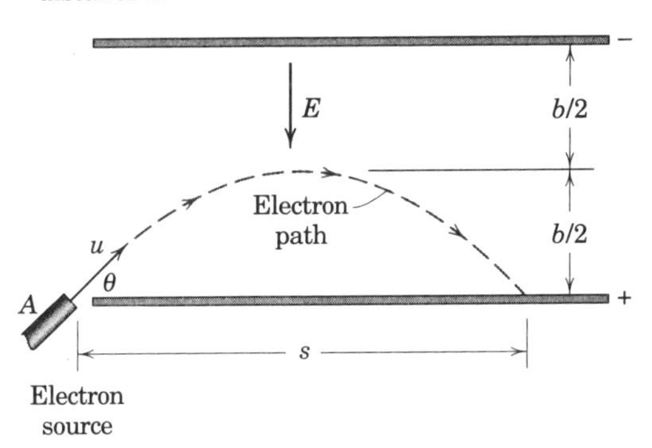

Problem 2/76

2/77 A rocket is released at point A from a jet aircraft flying horizontally at 1000 km/h at an altitude of 800 m. If the rocket thrust remains horizontal and gives the rocket a horizontal acceleration of $0.5g$, determine the angle θ from the horizontal to the line of sight to the target.

Ans. $\theta = 11.46°$

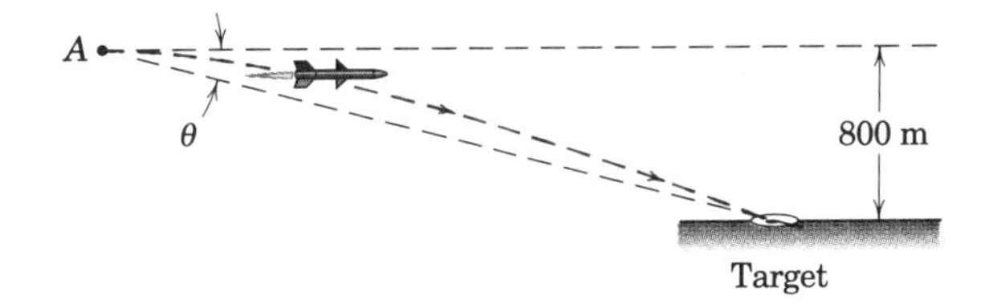

Problem 2/77

2/78 The water nozzle ejects water at a speed $v_0 = 14$ m/s at the angle $\theta = 40°$. Determine where, relative to the wall base point B, the water lands. Neglect the effects of the thickness of the wall.

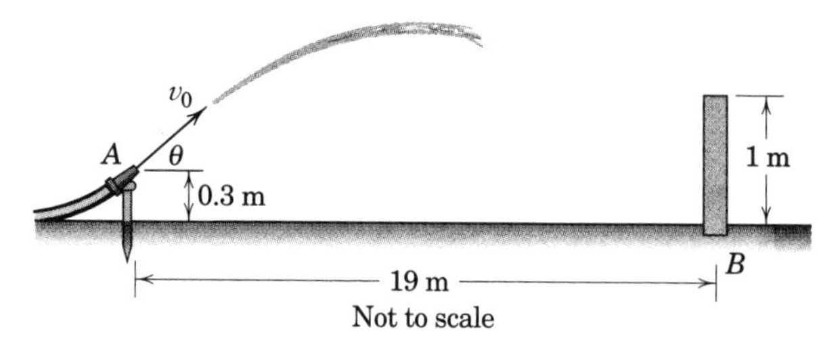

Problem 2/78

2/79 Water is ejected from the water nozzle of Prob. 2/78 with a speed $v_0 = 14$ m/s. For what value of the angle θ will the water land closest to the wall after clearing the top? Neglect the effects of wall thickness and air resistance. Where does the water land?

Ans. $\theta = 50.7°$, 0.835 m to the right of B

2/80 The pilot of an airplane carrying a package of mail to a remote outpost wishes to release the package at the right moment to hit the recovery location A. What angle θ with the horizontal should the pilot's line of sight to the target make at the instant of release? The airplane is flying horizontally at an altitude of 100 m with a velocity of 200 km/h.

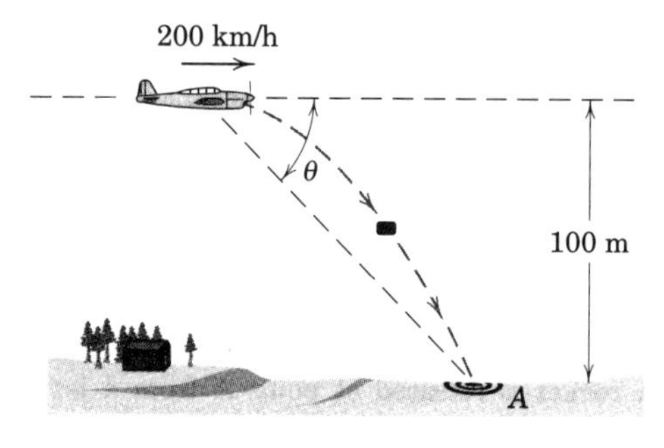

Problem 2/80

2/81 A rock is thrown horizontally from a tower at A and hits the ground 3.5 s later at B. The line of sight from A to B makes an angle of 50° with the horizontal. Compute the magnitude of the initial velocity **u** of the rock.

Ans. $u = 14.41$ m/s

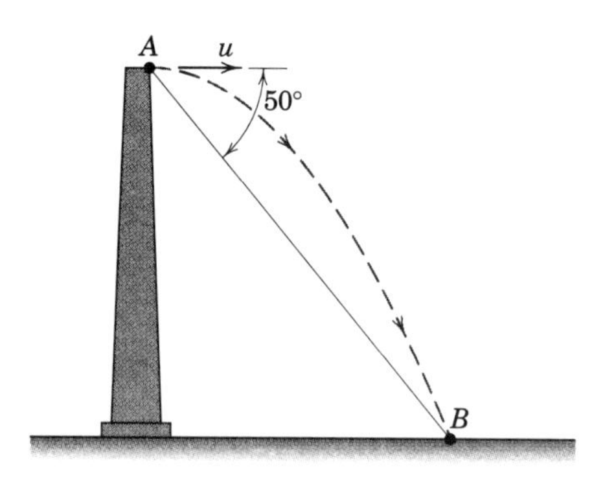

Problem 2/81

2/82 An outfielder experiments with two different trajectories for throwing to home plate from the position shown: (*a*) $v_0 = 42$ m/s with $\theta = 8°$ and (*b*) $v_0 = 36$ m/s with $\theta = 12°$. For each set of initial conditions, determine the time t required for the baseball to reach home plate and the altitude h as the ball crosses the plate.

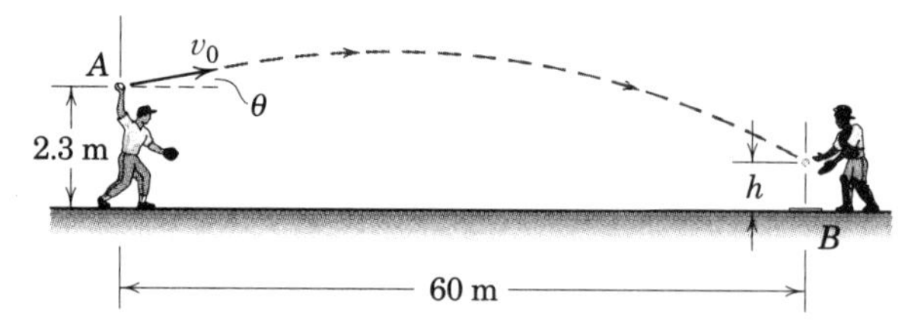

Problem 2/82

2/83 A football player attempts a 30-m field goal. If he is able to impart a velocity u of 30 m/s to the ball, compute the minimum angle θ for which the ball will clear the crossbar of the goal. (*Hint:* Let $m = \tan\theta$.)

Ans. $\theta = 15.43°$

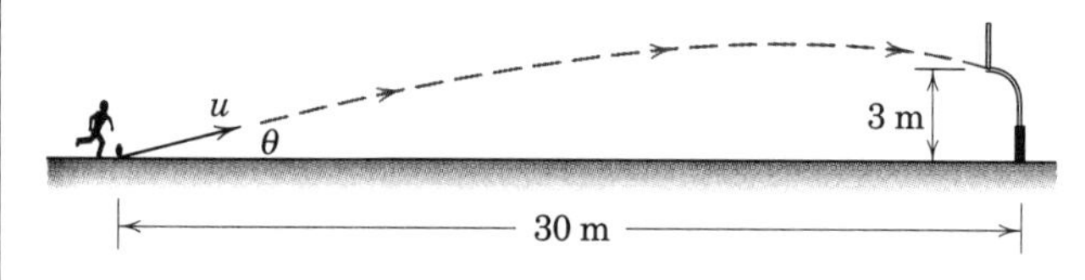

Problem 2/83

2/84 A projectile is launched with a speed $v_0 = 25$ m/s from the floor of a 5-m-high tunnel as shown. Determine the maximum horizontal range R of the projectile and the corresponding launch angle θ.

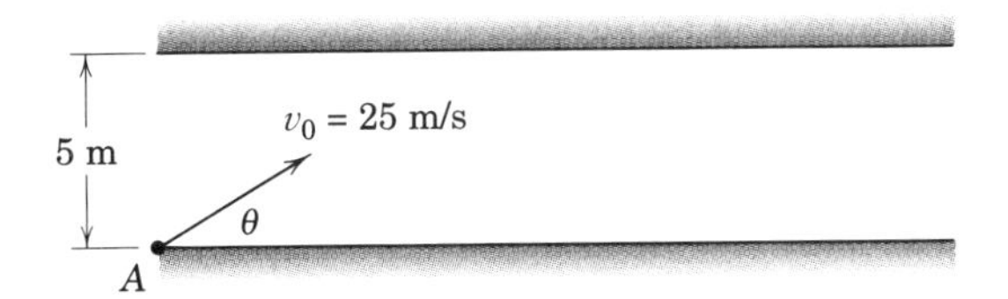

Problem 2/84

2/85 If the tennis player serves the ball horizontally ($\theta = 0$), calculate its velocity v if the center of the ball clears the 0.9-m net by 150 mm. Also find the distance s from the net to the point where the ball hits the court surface. Neglect air resistance and the effect of ball spin.

Ans. $v = 21.2$ m/s, $s = 3.55$ m

Problem 2/85

2/86 If the tennis player shown in Prob. 2/85 serves the ball with a velocity v of 130 km/h at the angle $\theta = 5°$, calculate the vertical clearance h of the center of the ball above the net and the distance s from the net where the ball hits the court surface. Neglect air resistance and the effect of ball spin.

2/87 The muzzle velocity of a long-range rifle at A is $u = 400$ m/s. Determine the two angles of elevation θ which will permit the projectile to hit the mountain target B.

Ans. $\theta_1 = 26.1°$, $\theta_2 = 80.6°$

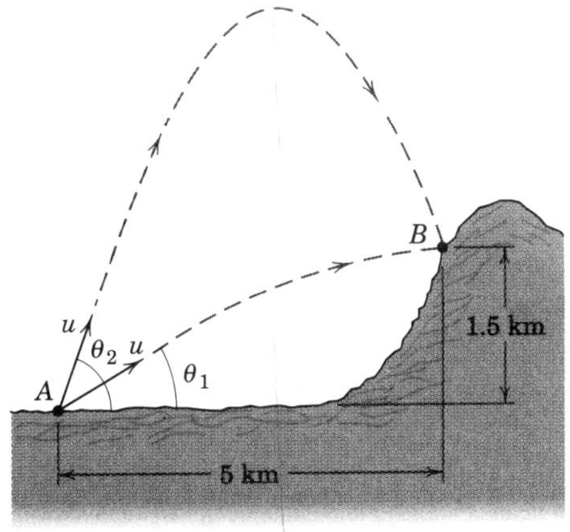

Problem 2/87

2/88 A team of engineering students is designing a catapult to launch a small ball at A so that it lands in the box. If it is known that the initial velocity vector makes a 30° angle with the horizontal, determine the range of launch speeds v_0 for which the ball will land inside the box.

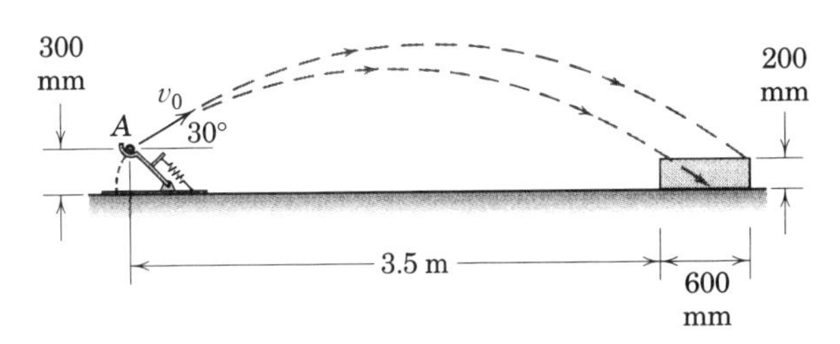

Problem 2/88

2/89 Ball bearings leave the horizontal trough with a velocity of magnitude u and fall through the 70-mm-diameter hole as shown. Calculate the permissible range of u which will enable the balls to enter the hole. Take the dotted positions to represent the limiting conditions.

Ans. $u_{max} = 1.135$ m/s, $u_{min} = 0.744$ m/s

Problem 2/89

2/90 A horseshoe player releases the horseshoe at A with an initial speed $v_0 = 11$ m/s. Determine the range for the launch angle θ for which the shoe will strike the 350-mm vertical stake.

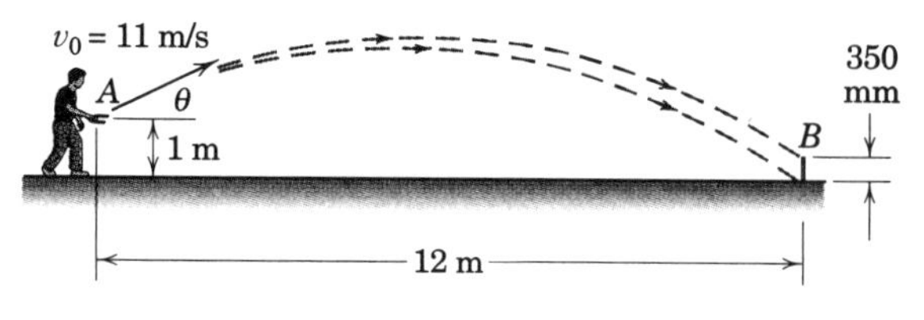

Problem 2/90

2/91 Determine the location h of the spot toward which the pitcher must throw if the ball is to hit the catcher's mitt. The ball is released with a speed of 40 m/s.

Ans. $h = 1.227$ m

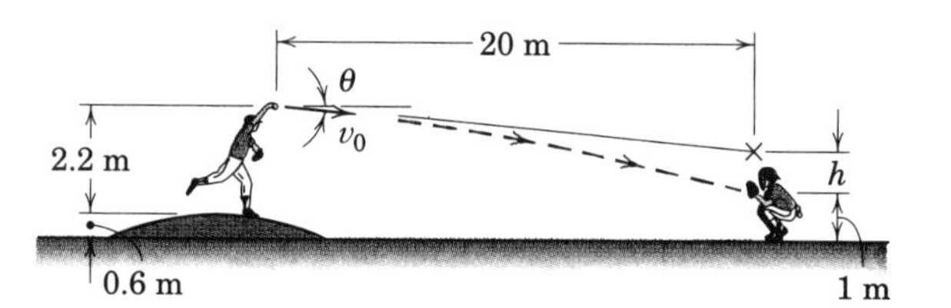

Problem 2/91

2/92 The basketball player likes to release his foul shots at an angle $\theta = 50°$ to the horizontal as shown. What initial speed v_0 will cause the ball to pass through the center of the rim?

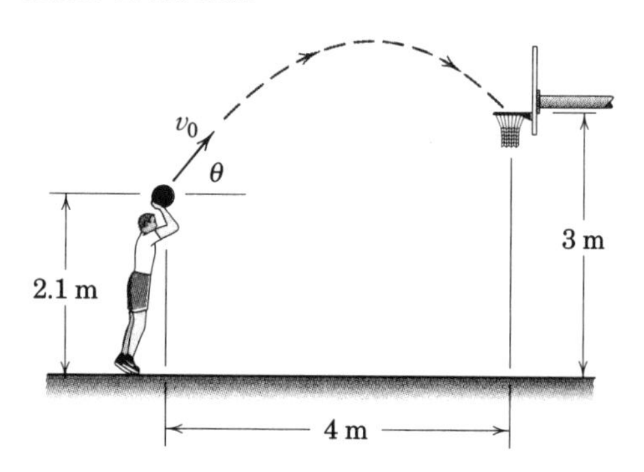

Problem 2/92

2/93 The pilot of an airplane pulls into a steep 45° climb at 300 km/h and releases a package at position A. Calculate the horizontal distance s and the time t from the point of release to the point at which the package strikes the ground.

Ans. $s = 1.046$ km, $t = 17.75$ s

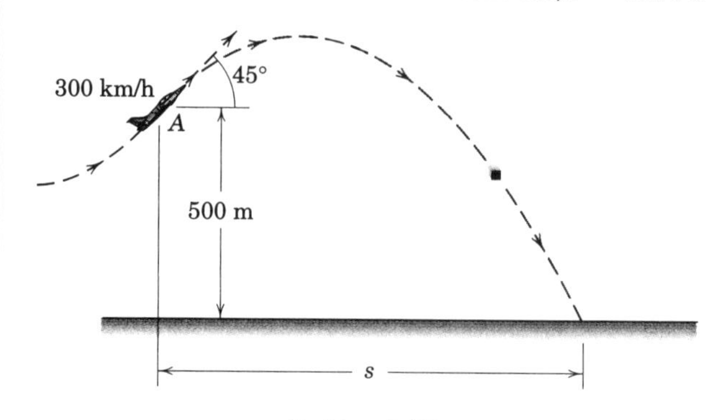

Problem 2/93

►**2/94** A projectile is ejected into an experimental fluid at time $t = 0$. The initial speed is v_0 and the angle to the horizontal is θ. The drag on the projectile results in an acceleration term $\mathbf{a}_D = -k\mathbf{v}$, where k is a constant and $\mathbf{v}$ is the velocity of the projectile. Determine the x- and y-components of both the velocity and displacement as functions of time. What is the terminal velocity? Include the effects of gravitational acceleration.

Ans. $v_x = (v_0 \cos\theta)e^{-kt}$, $x = \dfrac{v_0 \cos\theta}{k}(1 - e^{-kt})$

$$v_y = \left(v_0 \sin\theta + \frac{g}{k}\right)e^{-kt} - \frac{g}{k}$$

$$y = \frac{1}{k}\left(v_0 \sin\theta + \frac{g}{k}\right)(1 - e^{-kt}) - \frac{g}{k}t$$

$$v_x \rightarrow 0,\ v_y \rightarrow -\frac{g}{k}$$

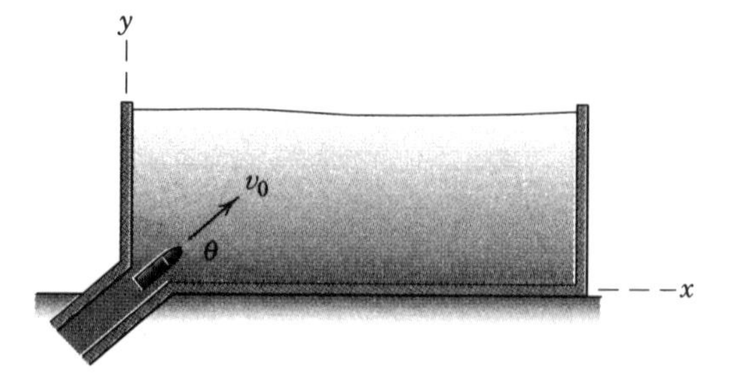

Problem 2/94

▶**2/95** A projectile is launched with speed v_0 from point A. Determine the launch angle θ which results in the maximum range R up the incline of angle α (where $0 \le \alpha \le 90°$). Evaluate your results for $\alpha = 0, 30°$, and $45°$.

Ans. $\theta = \dfrac{90° + \alpha}{2}$, $\theta = 45°, 60°, 67.5°$

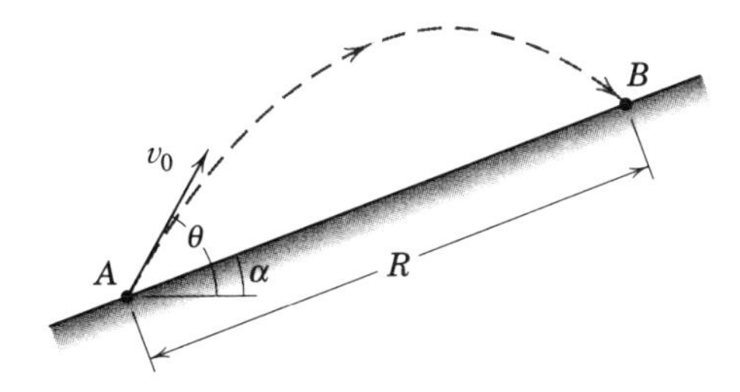

Problem 2/95

▶**2/96** Determine the equation for the envelope a of the parabolic trajectories of a projectile fired at any angle but with a fixed muzzle velocity u. (*Hint:* Substitute $m = \tan\theta$, where θ is the firing angle with the horizontal, into the equation of the trajectory. The two roots m_1 and m_2 of the equation written as a quadratic in m give the two firing angles for the two trajectories shown such that the shells pass through the same point A. Point A will approach the envelope a as the two roots approach equality.) Neglect air resistance and assume g is constant.

Ans. $y = \dfrac{u^2}{2g} - \dfrac{gx^2}{2u^2}$

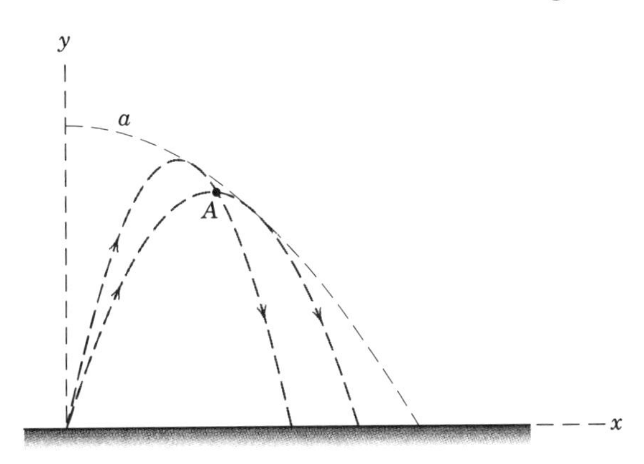

Problem 2/96

2/5 Normal and Tangential Coordinates (*n*-*t*)

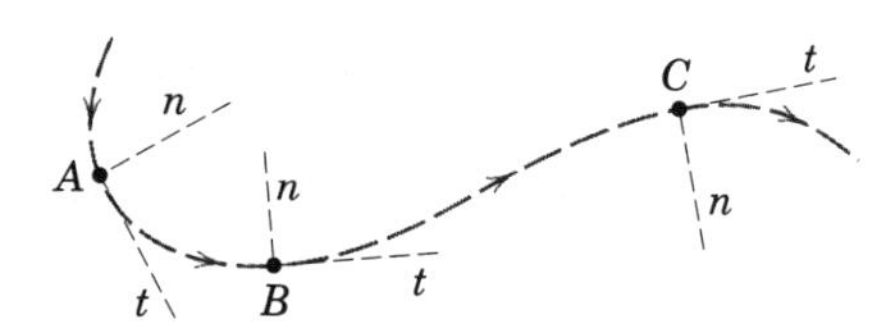

Figure 2/9

As we mentioned in Art. 2/1, one of the common descriptions of curvilinear motion uses path variables, which are measurements made along the tangent t and normal n to the path of the particle. These coordinates provide a very natural description for curvilinear motion and are frequently the most direct and convenient coordinates to use. The n- and t-coordinates are considered to move along the path with the particle, as seen in Fig. 2/9 where the particle advances from A to B to C. The positive direction for n at any position is always taken toward the center of curvature of the path. As seen from Fig. 2/9, the positive n-direction will shift from one side of the curve to the other side if the curvature changes direction.

Velocity and Acceleration

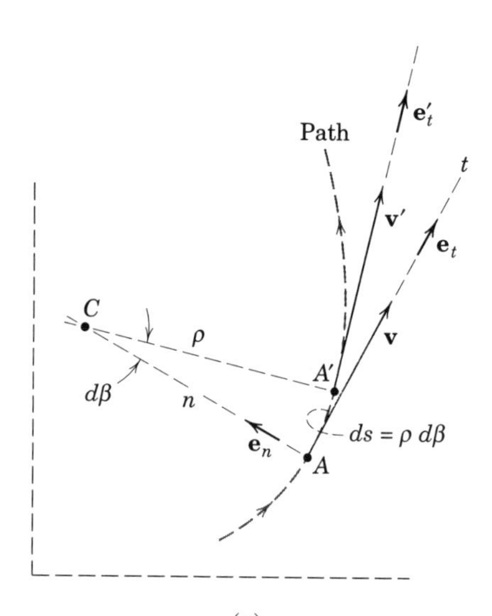

(a)

We now use the coordinates n and t to describe the velocity $\mathbf{v}$ and acceleration $\mathbf{a}$ which were introduced in Art. 2/3 for the curvilinear motion of a particle. For this purpose, we introduce unit vectors $\mathbf{e}_n$ in the n-direction and $\mathbf{e}_t$ in the t-direction, as shown in Fig. 2/10a for the position of the particle at point A on its path. During a differential increment of time dt, the particle moves a differential distance ds along the curve from A to A'. With the radius of curvature of the path at this position designated by ρ, we see that $ds = \rho\, d\beta$, where β is in radians. It is unnecessary to consider the differential change in ρ between A and A' because a higher-order term would be introduced which disappears in the limit. Thus, the magnitude of the velocity can be written $v = ds/dt = \rho\, d\beta/dt$, and we can write the velocity as the vector

$$\mathbf{v} = v\mathbf{e}_t = \rho\dot{\beta}\mathbf{e}_t \tag{2/7}$$

The acceleration $\mathbf{a}$ of the particle was defined in Art. 2/3 as $\mathbf{a} = d\mathbf{v}/dt$, and we observed from Fig. 2/5 that the acceleration is a vector which reflects both the change in magnitude and the change in direction of $\mathbf{v}$. We now differentiate $\mathbf{v}$ in Eq. 2/7 by applying the ordinary rule for the differentiation of the product of a scalar and a vector* and get

$$\mathbf{a} = \frac{d\mathbf{v}}{dt} = \frac{d(v\mathbf{e}_t)}{dt} = v\dot{\mathbf{e}}_t + \dot{v}\mathbf{e}_t \tag{2/8}$$

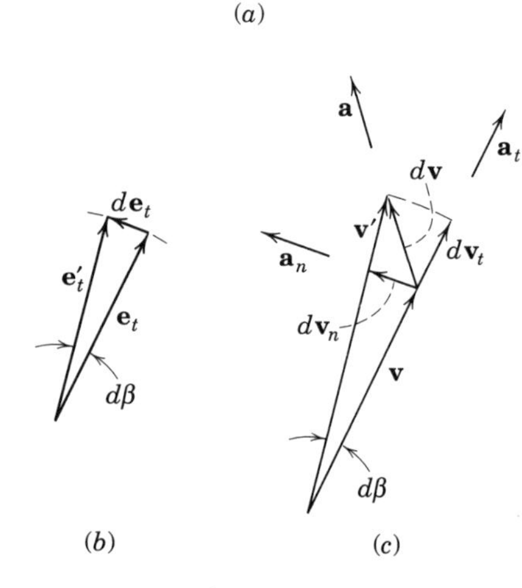

(b) (c)

Figure 2/10

where the unit vector $\mathbf{e}_t$ now has a nonzero derivative because its direction changes.

To find $\dot{\mathbf{e}}_t$ we analyze the change in $\mathbf{e}_t$ during a differential increment of motion as the particle moves from A to A' in Fig. 2/10a. The unit vector $\mathbf{e}_t$ correspondingly changes to $\mathbf{e}_t'$, and the vector difference $d\mathbf{e}_t$ is shown in part b of the figure. The vector $d\mathbf{e}_t$ in the limit has a magnitude equal to the length of the arc $|\mathbf{e}_t|\, d\beta = d\beta$ obtained by swing-

*See Art. C/7 of Appendix C.

ing the unit vector $\mathbf{e}_t$ through the angle $d\beta$ expressed in radians. The direction of $d\mathbf{e}_t$ is given by $\mathbf{e}_n$. Thus, we can write $d\mathbf{e}_t = \mathbf{e}_n\, d\beta$. Dividing by $d\beta$ gives

$$\frac{d\mathbf{e}_t}{d\beta} = \mathbf{e}_n$$

Dividing by dt gives $d\mathbf{e}_t/dt = (d\beta/dt)\mathbf{e}_n$, which can be written

$$\dot{\mathbf{e}}_t = \dot{\beta}\mathbf{e}_n \qquad \textbf{(2/9)}$$

With the substitution of Eq. 2/9 and $\dot{\beta}$ from the relation $v = \rho\dot{\beta}$, Eq. 2/8 for the acceleration becomes

$$\mathbf{a} = \frac{v^2}{\rho}\mathbf{e}_n + \dot{v}\mathbf{e}_t \qquad \textbf{(2/10)}$$

where

$$a_n = \frac{v^2}{\rho} = \rho\dot{\beta}^2 = v\dot{\beta}$$

$$a_t = \dot{v} = \ddot{s}$$

$$a = \sqrt{{a_n}^2 + {a_t}^2}$$

We may also note that $a_t = \dot{v} = d(\rho\dot{\beta})/dt = \rho\ddot{\beta} + \dot{\rho}\dot{\beta}$. This relation, however, finds little use because we seldom have reason to compute $\dot{\rho}$.

Geometric Interpretation

Full understanding of Eq. 2/10 comes only when we clearly see the geometry of the physical changes it describes. Figure 2/10*c* shows the velocity vector $\mathbf{v}$ when the particle is at A and $\mathbf{v}'$ when it is at A'. The vector change in the velocity is $d\mathbf{v}$, which establishes the direction of the acceleration $\mathbf{a}$. The n-component of $d\mathbf{v}$ is labeled $d\mathbf{v}_n$, and in the limit its magnitude equals the length of the arc generated by swinging the vector $\mathbf{v}$ as a radius through the angle $d\beta$. Thus, $|d\mathbf{v}_n| = v\, d\beta$ and the n-component of acceleration is $a_n = |d\mathbf{v}_n|/dt = v(d\beta/dt) = v\dot{\beta}$ as before. The t-component of $d\mathbf{v}$ is labeled $d\mathbf{v}_t$, and its magnitude is simply the change dv in the magnitude or length of the velocity vector. Therefore, the t-component of acceleration is $a_t = dv/dt = \dot{v} = \ddot{s}$ as before. The acceleration vectors resulting from the corresponding vector changes in velocity are shown in Fig. 2/10*c*.

It is especially important to observe that the normal component of acceleration a_n is *always directed toward the center of curvature C*. The tangential component of acceleration, on the other hand, will be in the positive t-direction of motion if the speed v is increasing and in the neg-

B B

A A

Speed increasing (*a*)

Speed decreasing (*b*)

Acceleration vectors for particle moving from A to B

Figure 2/11

ative t-direction if the speed is decreasing. In Fig. 2/11 are shown schematic representations of the variation in the acceleration vector for a particle moving from A to B with (a) increasing speed and (b) decreasing speed. At an inflection point on the curve, the normal acceleration v^2/ρ goes to zero because ρ becomes infinite.

Circular Motion

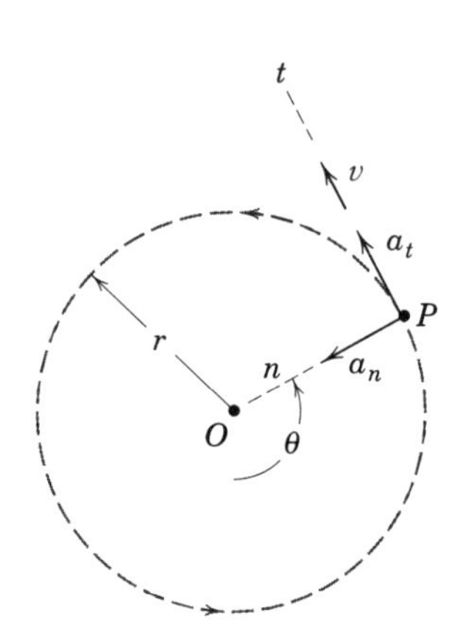

Figure 2/12

Circular motion is an important special case of plane curvilinear motion where the radius of curvature ρ becomes the constant radius r of the circle and the angle β is replaced by the angle θ measured from any convenient radial reference to OP, Fig. 2/12. The velocity and the acceleration components for the circular motion of the particle P become

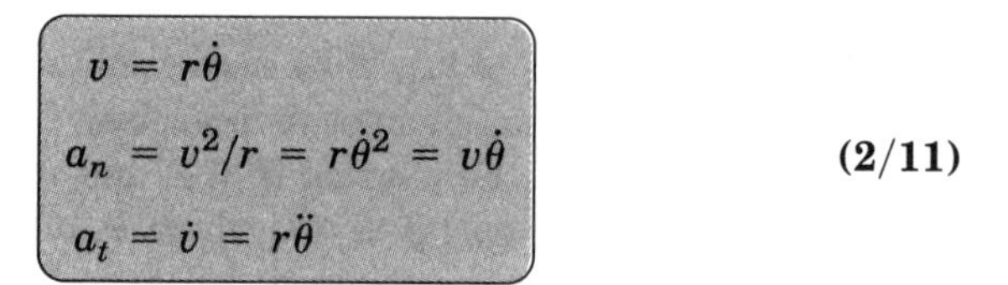

$$v = r\dot{\theta}$$

$$a_n = v^2/r = r\dot{\theta}^2 = v\dot{\theta} \qquad \textbf{(2/11)}$$

$$a_t = \dot{v} = r\ddot{\theta}$$

We find repeated use for Eqs. 2/10 and 2/11 in dynamics, so these relations and the principles behind them should be mastered.

Sample Problem 2/7

To anticipate the dip and hump in the road, the driver of a car applies her brakes to produce a uniform deceleration. Her speed is 100 km/h at the bottom A of the dip and 50 km/h at the top C of the hump, which is 120 m along the road from A. If the passengers experience a total acceleration of 3 m/s^2 at A and if the radius of curvature of the hump at C is 150 m, calculate (*a*) the radius of curvature ρ at A, (*b*) the acceleration at the inflection point B, and (*c*) the total acceleration at C.

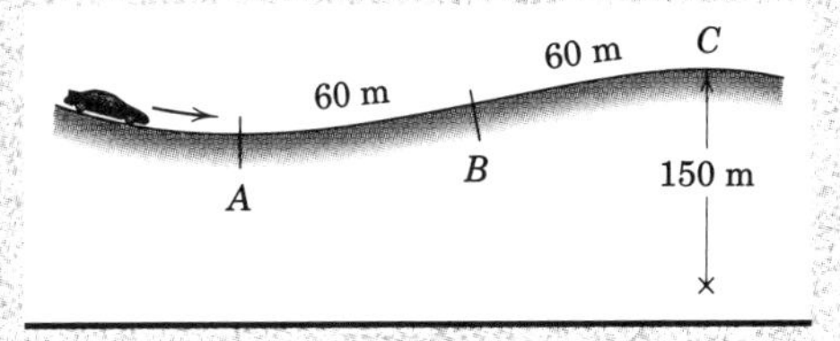

Solution. The dimensions of the car are small compared with those of the path, ① so we will treat the car as a particle. The velocities are

$$v_A = \left(100\,\frac{\text{km}}{\text{h}}\right)\left(\frac{1\text{ h}}{3600\text{ s}}\right)\left(1000\,\frac{\text{m}}{\text{km}}\right) = 27.8\text{ m/s}$$

$$v_C = 50\,\frac{1000}{3600} = 13.89\text{ m/s}$$

We find the constant deceleration along the path from

$$\left[\int v\,dv = \int a_t\,ds\right] \qquad \int_{v_A}^{v_C} v\,dv = a_t\int_0^s ds$$

$$a_t = \frac{1}{2s}(v_C{}^2 - v_A{}^2) = \frac{(13.89)^2 - (27.8)^2}{2(120)} = -2.41\text{ m/s}^2$$

Helpful Hint

① Actually, the radius of curvature to the road differs by about 1 m from that to the path followed by the center of mass of the passengers, but we have neglected this relatively small difference.

(a) Condition at A. With the total acceleration given and a_t determined, we can easily compute a_n and hence ρ from

$[a^2 = a_n{}^2 + a_t{}^2]$ $\quad a_n{}^2 = 3^2 - (2.41)^2 = 3.19 \quad a_n = 1.785\text{ m/s}^2$

$[a_n = v^2/\rho]$ $\quad \rho = v^2/a_n = (27.8)^2/1.785 = 432\text{ m}$ *Ans.*

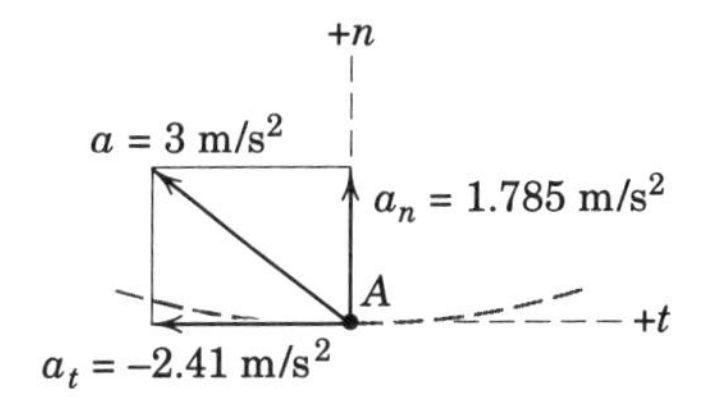

(b) Condition at B. Since the radius of curvature is infinite at the inflection point, $a_n = 0$ and

$$a = a_t = -2.41\text{ m/s}^2$$ *Ans.*

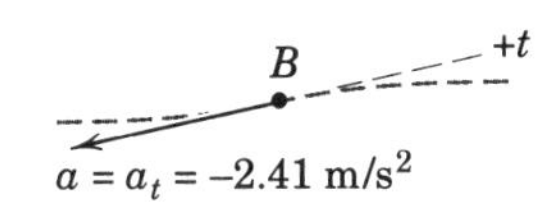

(c) Condition at C. The normal acceleration becomes

$[a_n = v^2/\rho]$ $\quad a_n = (13.89)^2/150 = 1.286\text{ m/s}^2$

With unit vectors $\mathbf{e}_n$ and $\mathbf{e}_t$ in the n- and t-directions, the acceleration may be written

$$\mathbf{a} = 1.286\mathbf{e}_n - 2.41\mathbf{e}_t\text{ m/s}^2$$

where the magnitude of $\mathbf{a}$ is

$[a = \sqrt{a_n{}^2 + a_t{}^2}]$ $\quad a = \sqrt{(1.286)^2 + (-2.41)^2} = 2.73\text{ m/s}^2$ *Ans.*

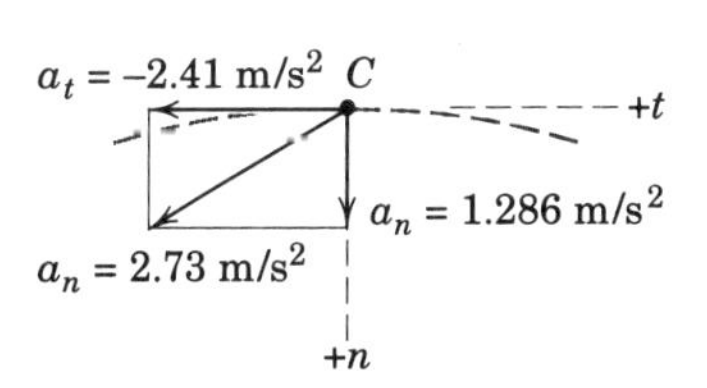

The acceleration vectors representing the conditions at each of the three points are shown for clarification.

Sample Problem 2/8

A certain rocket maintains a horizontal attitude of its axis during the powered phase of its flight at high altitude. The thrust imparts a horizontal component of acceleration of 6 m/s^2, and the downward acceleration component is the acceleration due to gravity at that altitude, which is $g = 9$ m/s^2. At the instant represented, the velocity of the mass center G of the rocket along the 15° direction of its trajectory is $20(10^3)$ km/h. For this position determine (*a*) the radius of curvature of the flight trajectory, (*b*) the rate at which the speed v is increasing, (*c*) the angular rate $\dot{\beta}$ of the radial line from G to the center of curvature C, and (*d*) the vector expression for the total acceleration **a** of the rocket.

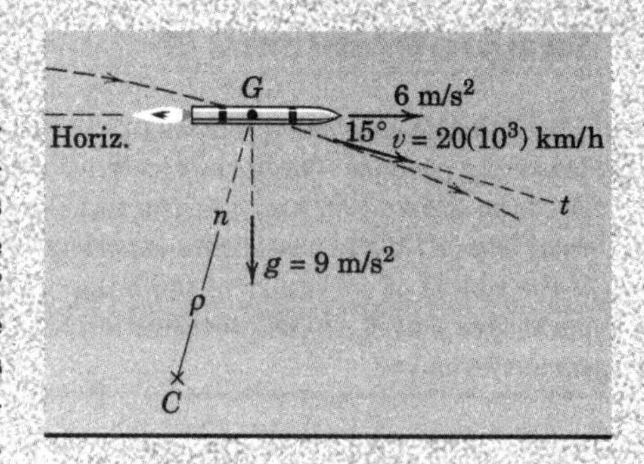

Solution. We observe that the radius of curvature appears in the expression for the normal component of acceleration, so we use n- and t-coordinates to describe ① the motion of G. The n- and t-components of the total acceleration are obtained by resolving the given horizontal and vertical accelerations into their n- and t-components and then combining. From the figure we get

$$a_n = 9 \cos 15° - 6 \sin 15° = 7.14 \text{ m/s}^2$$
$$a_t = 9 \sin 15° + 6 \cos 15° = 8.12 \text{ m/s}^2$$

(a) We may now compute the radius of curvature from

② $[a_n = v^2/\rho]$ $\qquad \rho = \dfrac{v^2}{a_n} = \dfrac{[20(10^3)/3.6]^2}{7.14} = 4.32(10^6) \text{ m}$ *Ans.*

(b) The rate at which v is increasing is simply the t-component of acceleration.

$[\dot{v} = a_t]$ $\qquad \dot{v} = 8.12 \text{ m/s}^2$ *Ans.*

(c) The angular rate $\dot{\beta}$ of line GC depends on v and ρ and is given by

$[v = \rho\dot{\beta}]$ $\qquad \dot{\beta} = v/\rho = \dfrac{20(10^3)/3.6}{4.32(10^6)} = 12.85(10^{-4}) \text{ rad/s}$ *Ans.*

(d) With unit vectors $\mathbf{e}_n$ and $\mathbf{e}_t$ for the n- and t-directions, respectively, the total acceleration becomes

$$\mathbf{a} = 7.14\mathbf{e}_n + 8.12\mathbf{e}_t \text{ m/s}^2$$ *Ans.*

Helpful Hints

① Alternatively, we could find the resultant acceleration and then resolve it into n- and t-components.

② To convert from km/h to m/s, multiply by $\dfrac{1000 \text{ m/km}}{3600 \text{ s/h}}$ or divide by 3.6, which is easily remembered.

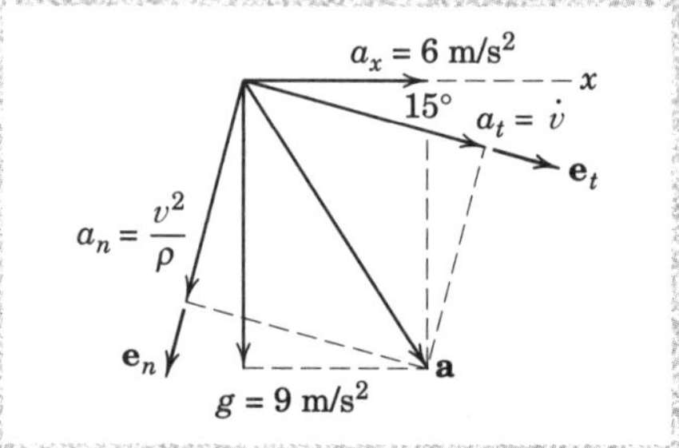

PROBLEMS

Introductory Problems

2/97 A particle moves in a circular path of 0.4-m radius. Calculate the magnitude a of the acceleration of the particle (a) if its speed is constant at 0.6 m/s and (b) if its speed is 0.6 m/s but is increasing at the rate of 1.2 m/s each second.

Ans. (a) $a = 0.9\ \text{m/s}^2$, (b) $a = 1.5\ \text{m/s}^2$

2/98 Six acceleration vectors are shown for the car whose velocity vector is directed forward. For each acceleration vector describe in words the instantaneous motion of the car.

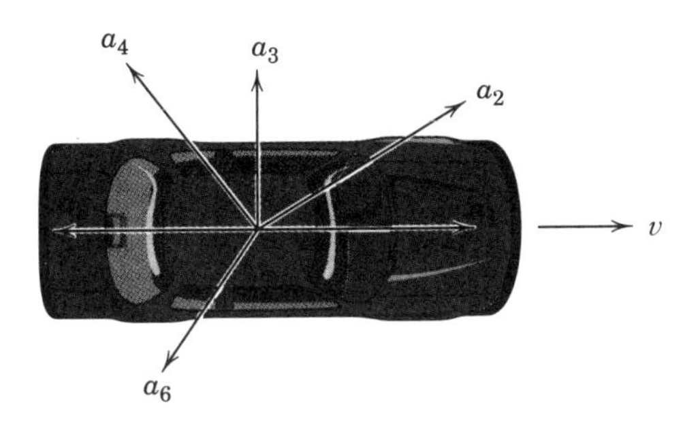

Problem 2/98

2/99 A car is traveling around a circular track of 240-m radius. If the magnitude of its total acceleration is 3 m/s^2 at the instant when its speed is 75 km/h, determine the rate at which the car is changing its speed.

Ans. $a_t = \pm 2.39\ \text{m/s}^2$

2/100 The car passes through a dip in the road at A with a constant speed which gives its mass center G an acceleration equal to $0.5g$. If the radius of curvature of the road at A is 100 m, and if the distance from the road to the mass center G of the car is 0.6 m, determine the speed v of the car.

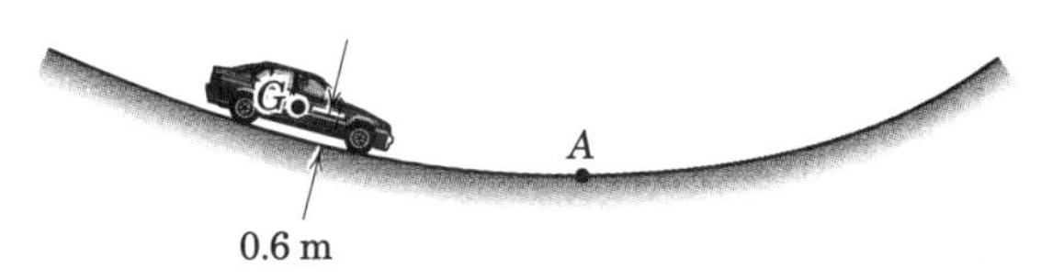

Problem 2/100

2/101 The driver of the truck has an acceleration of $0.4g$ as the truck passes over the top A of the hump in the road at constant speed. The radius of curvature of the road at the top of the hump is 98 m, and the center of mass G of the driver (considered a particle) is 2 m above the road. Calculate the speed v of the truck.

Ans. $v = 71.3$ km/h

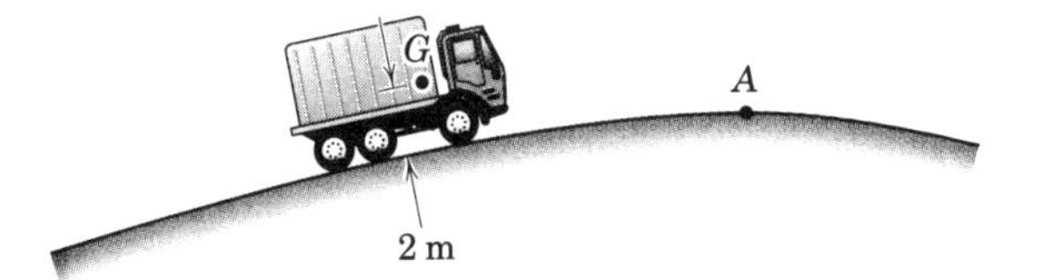

Problem 2/101

2/102 A ship which moves at a steady 20-knot speed (1 knot = 1.852 km/h) executes a turn to port by changing its compass heading at a constant counterclockwise rate. If it requires 60 s to alter course 90°, calculate the magnitude of the acceleration **a** of the ship during the turn.

2/103 A train enters a curved horizontal section of track at a speed of 100 km/h and slows down with constant deceleration to 50 km/h in 12 seconds. An accelerometer mounted inside the train records a horizontal acceleration of 2 m/s^2 when the train is 6 seconds into the curve. Calculate the radius of curvature ρ of the track for this instant.

Ans. $\rho = 266$ m

2/104 A particle moves along the curved path shown. If the particle has a speed of 12 m/s at A at time t_A and a speed of 13.5 m/s at B at time t_B, determine the average values of the acceleration of the particle between A and B, both normal and tangent to the path.

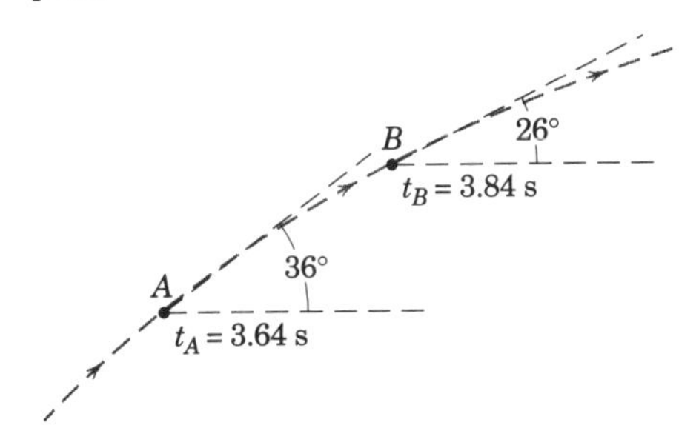

Problem 2/104

2/105 The speed of a car increases uniformly with time from 50 km/h at A to 100 km/h at B during 10 seconds. The radius of curvature of the hump at A is 40 m. If the magnitude of the total acceleration of the mass center of the car is the same at B as at A, compute the radius of curvature ρ_B of the dip in the road at B. The mass center of the car is 0.6 m from the road.

Ans. $\rho_B = 163.0$ m

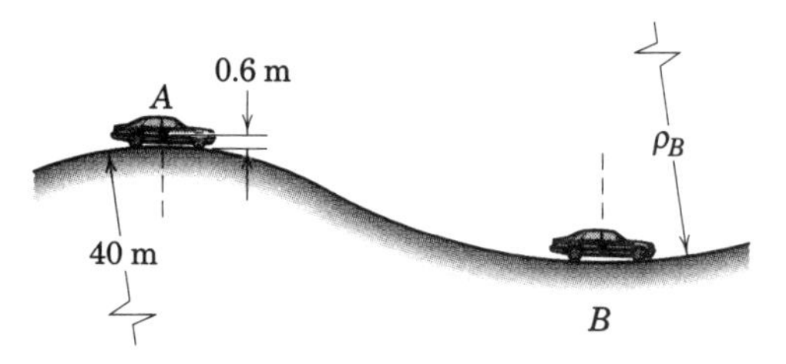

Problem 2/105

2/106 A particle moves on a circular path of radius $r = 0.8$ m with a constant speed of 2 m/s. The velocity undergoes a vector change $\Delta\mathbf{v}$ from A to B. Express the magnitude of $\Delta\mathbf{v}$ in terms of v and $\Delta\theta$ and divide it by the time interval Δt between A and B to obtain the magnitude of the average acceleration of the particle for (*a*) $\Delta\theta = 30°$, (*b*) $\Delta\theta = 15°$, and (*c*) $\Delta\theta = 5°$. In each case, determine the percentage difference from the instantaneous value of acceleration.

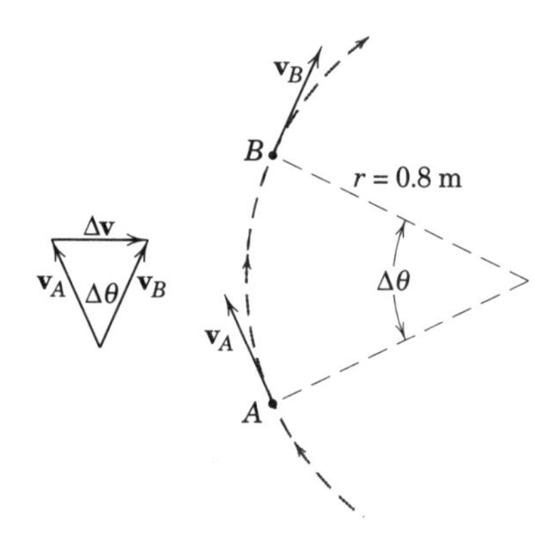

Problem 2/106

Representative Problems

2/107 A satellite travels with constant speed v in a circular orbit 320 km above the earth's surface. Calculate v knowing that the acceleration of the satellite is the gravitational acceleration at its altitude. (*Note:* Review Art. 1/5 as necessary and use the mean value of g and the mean value of the earth's radius. Also recognize that v is the magnitude of the velocity of the satellite with respect to the center of the earth.)

Ans. $v = 27.8(10^3)$ km/h

2/108 The figure shows two possible paths for negotiating an unbanked turn on a horizontal portion of a race course. Path AA follows the centerline of the road and has a radius of curvature $\rho_A = 85$ m, while path BB uses the width of the road to good advantage in increasing the radius of curvature to $\rho_B = 200$ m. If the drivers limit their speeds in their curves so that the lateral acceleration does not exceed $0.8g$, determine the maximum speed for each path.

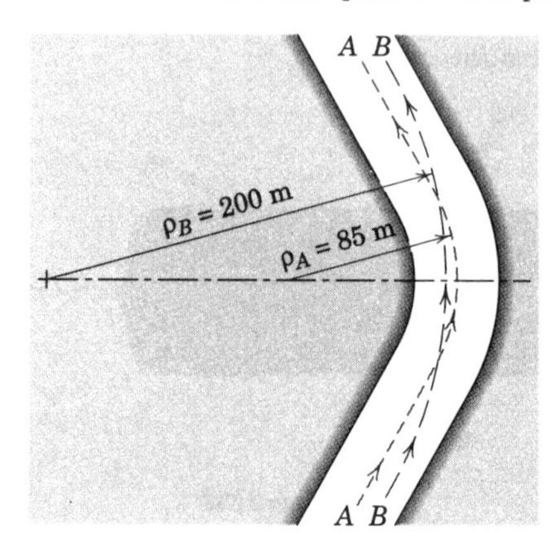

Problem 2/108

2/109 Consider the polar axis of the earth to be fixed in space and compute the magnitude of the acceleration $\mathbf{a}$ of a point P on the earth's surface at latitude 40° north. The mean diameter of the earth is 12 742 km and its angular velocity is $0.729(10^{-4})$ rad/s.

Ans. $a = 0.0259$ m/s^2

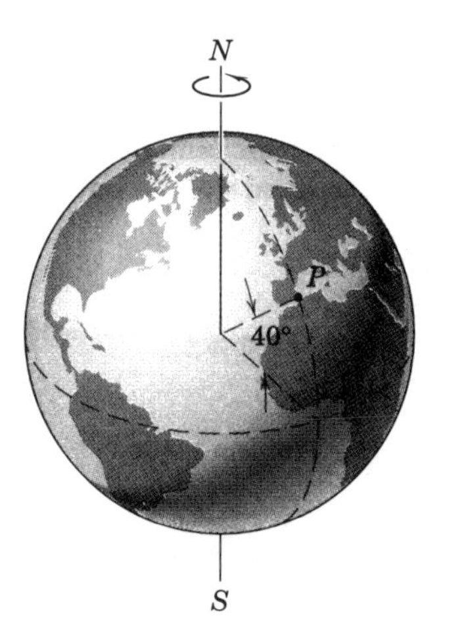

Problem 2/109

2/110 Write the vector expression for the acceleration **a** of the mass center G of the simple pendulum in both n-t and x-y coordinates for the instant when $\theta = 60°$ if $\dot{\theta} = 2.00$ rad/s and $\ddot{\theta} = 2.45$ rad/s^2.

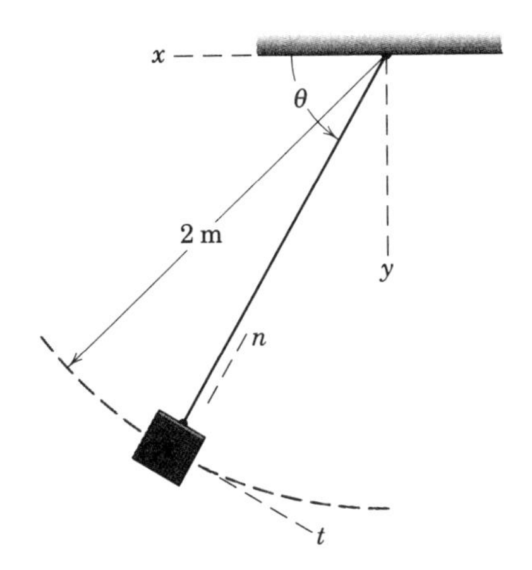

Problem 2/110

2/111 A space shuttle which moves in a circular orbit around the earth at a height $h = 240$ km above its surface must have a speed of 27 995 km/h. Calculate the gravitational acceleration g for this altitude. The mean radius of the earth is 6371 km. (Check your answer by computing g from the gravitational law $g = g_0 \left(\dfrac{R}{R + h}\right)^2$, where $g_0 = 9.821$ m/s^2 from Table D/2 in Appendix D.)

Ans. $a_n = g = 9.12$ m/s^2

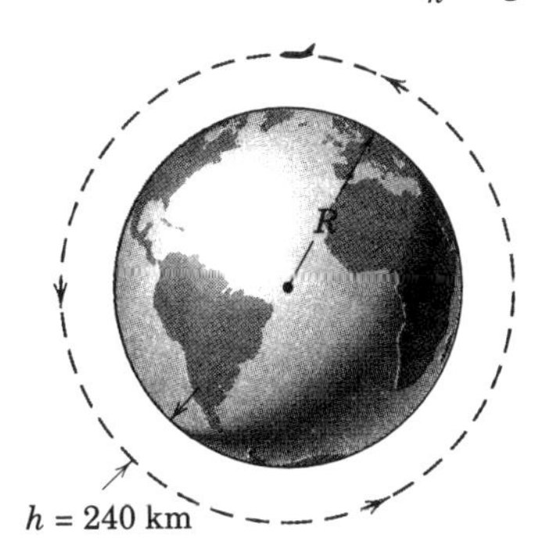

Problem 2/111

2/112 At the bottom A of the vertical inside loop, the magnitude of the total acceleration of the airplane is $3g$. If the airspeed is 800 km/h and is increasing at the rate of 20 km/h per second, calculate the radius of curvature ρ of the path at A.

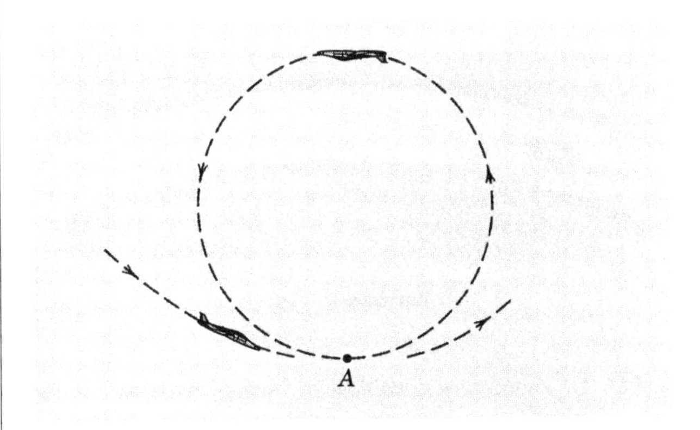

Problem 2/112

2/113 Magnetic tape is being transferred from reel A to reel B and passes around idler pulleys C and D. At a certain instant, point P_1 on the tape is in contact with pulley C and point P_2 is in contact with pulley D. If the normal component of acceleration of P_1 is 40 m/s^2 and the tangential component of acceleration of P_2 is 30 m/s^2 at this instant, compute the corresponding speed v of the tape, the magnitude of the total acceleration of P_1, and the magnitude of the total acceleration of P_2.

Ans. $v = 2$ m/s, $a_1 = 50$ m/s^2
$a_2 = 85.4$ m/s^2

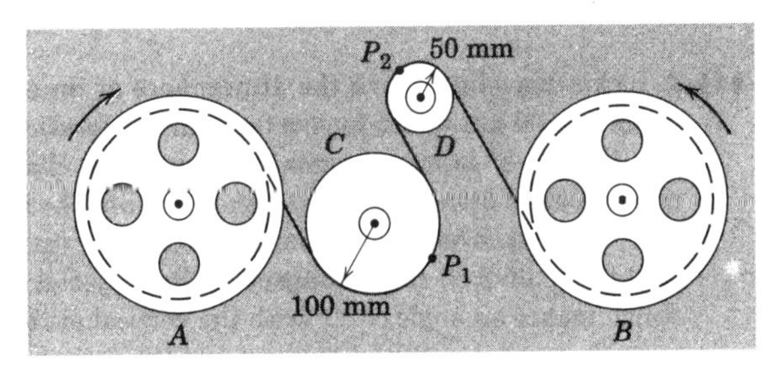

Problem 2/113

2/114 The car C increases its speed at the constant rate of 1.5 m/s^2 as it rounds the curve shown. If the magnitude of the total acceleration of the car is 2.5 m/s^2 at the point A where the radius of curvature is 200 m, compute the speed v of the car at this point.

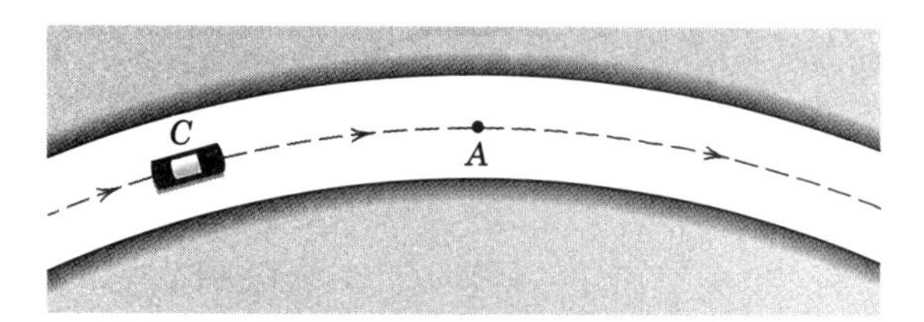

Problem 2/114

2/115 To simulate a condition of "weightlessness" in its cabin, a jet transport traveling at 800 km/h moves on a sustained vertical curve as shown. At what rate $\dot{\beta}$ in degrees per second should the pilot drop his longitudinal line of sight to effect the desired condition? The maneuver takes place at a mean altitude of 8 km, and the gravitational acceleration may be taken as 9.79 m/s^2.

Ans. $\dot{\beta} = 2.52$ deg/s

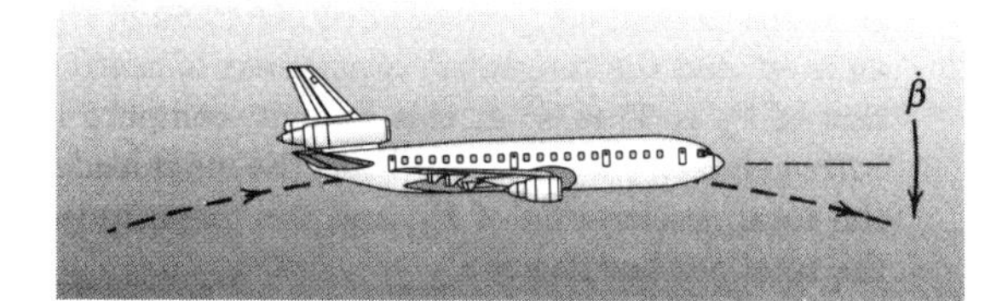

Problem 2/115

2/116 A rocket traveling above the atmosphere at an altitude of 500 km would have a free-fall acceleration $g = 8.43$ m/s^2 in the absence of forces other than gravitational attraction. Because of thrust, however, the rocket has an additional acceleration component a_1 of 8.80 m/s^2 tangent to its trajectory, which makes an angle of 30° with the vertical at the instant considered. If the velocity v of the rocket is 30 000 km/h at this position, compute the radius of curvature ρ of the trajectory and the rate at which v is changing with time.

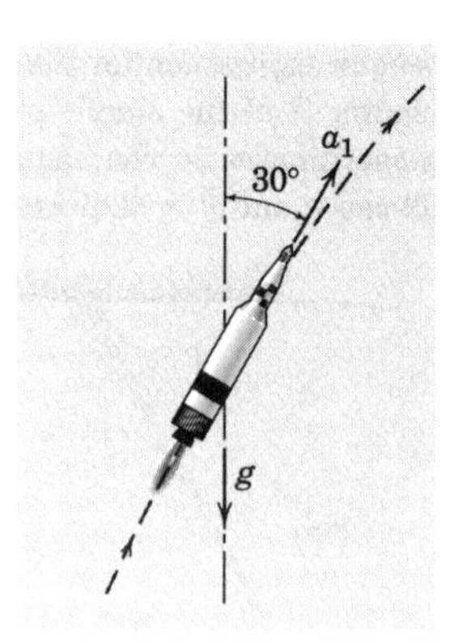

Problem 2/116

2/117 The preliminary design for a "small" space station to orbit the earth in a circular path consists of a ring (torus) with a circular cross section as shown. The living space within the torus is shown in section A, where the "ground level" is 6 m from the center of the section. Calculate the angular speed N in revolutions per minute required to simulate standard gravity at the surface of the earth (9.81 m/s^2). Recall that you would be unaware of a gravitational field if you were in a nonrotating spacecraft in a circular orbit around the earth.

Ans. $N = 3.32$ rev/min

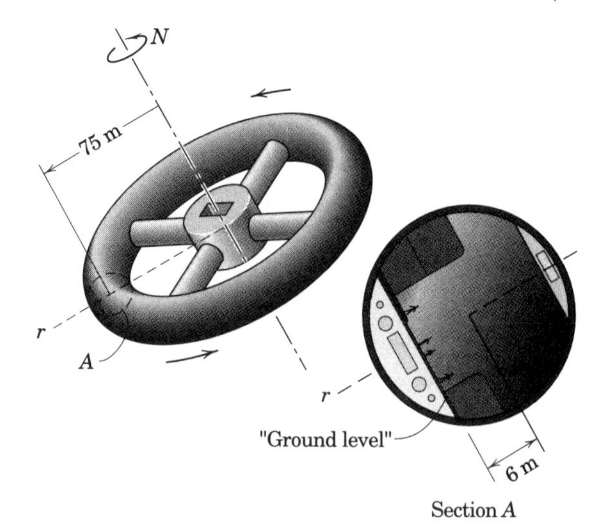

Problem 2/117

2/118 The design of a camshaft-drive system of a four-cylinder automobile engine is shown. As the engine is revved up, the belt speed v changes uniformly from 3 m/s to 6 m/s over a two-second interval. Calculate the magnitudes of the accelerations of points P_1 and P_2 halfway through this time interval.

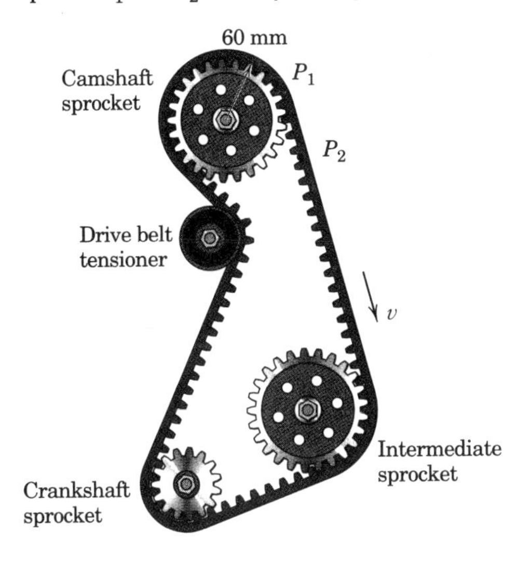

Problem 2/118

2/119 A particle moving in the x-y plane has a position vector given by $\mathbf{r} = \frac{3}{2}t^2\mathbf{i} + \frac{2}{3}t^3\mathbf{j}$, where $\mathbf{r}$ is in meters and t is in seconds. Calculate the radius of curvature ρ of the path for the position of the particle when $t = 2$ s. Sketch the velocity $\mathbf{v}$ and the curvature of the path for this particular instant.

Ans. $\rho = 41.7$ m

2/120 A baseball player releases a ball with the initial conditions shown in the figure. Determine the radius of curvature of the trajectory (*a*) just after release and (*b*) at the apex. For each case, compute the time rate of change of the speed.

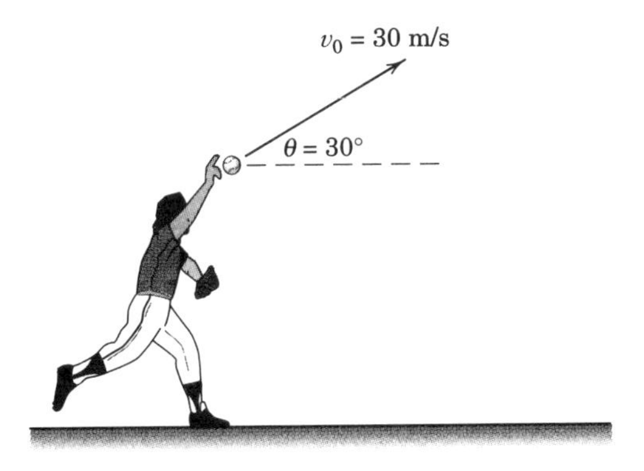

Problem 2/120

2/121 For the baseball of Prob. 2/120, determine the radius of curvature ρ of the path and the time rate of change $\dot{v}$ of the speed at times $t = 1$ s and $t = 2.5$ s, where $t = 0$ is the time of release from the player's hand.

Ans. (*a*) $\rho = 73.0$ m, $\dot{v} = -1.922$ m/s^2
(*b*) $\rho = 83.1$ m, $\dot{v} = 3.38$ m/s^2

2/122 A projectile is fired at an angle of 30° above the horizontal with a muzzle velocity of 460 m/s. Determine the radius of curvature ρ of the trajectory at the position of the particle 10 seconds after firing. Neglect atmospheric resistance so that the gravitational attraction is the only force to be considered. The acceleration of the projectile, consequently, is g down.

2/123 The command module of a lunar mission is orbiting the moon in a circular path at an altitude of 200 km above the moon's surface. Consult Table D/2, Appendix D, for any needed information about the moon and compute the magnitude of the orbital velocity $\mathbf{v}$ of the module with respect to the moon.

Ans. $v = 5720$ km/h

2/124 At a certain point in the reentry of the space shuttle into the earth's atmosphere, the total acceleration of the shuttle may be represented by two components. One component is the gravitational acceleration $g = 9.66$ m/s^2 at this altitude. The second component equals 12.90 m/s^2 due to atmospheric resistance and is directed opposite to the velocity. The shuttle is at an altitude of 48.2 km and has reduced its orbital velocity of 28 300 km/h to 15 450 km/h in the direction $\theta = 1.50°$. For this instant, calculate the radius of curvature ρ of the path and the rate $\dot{v}$ at which the speed is changing.

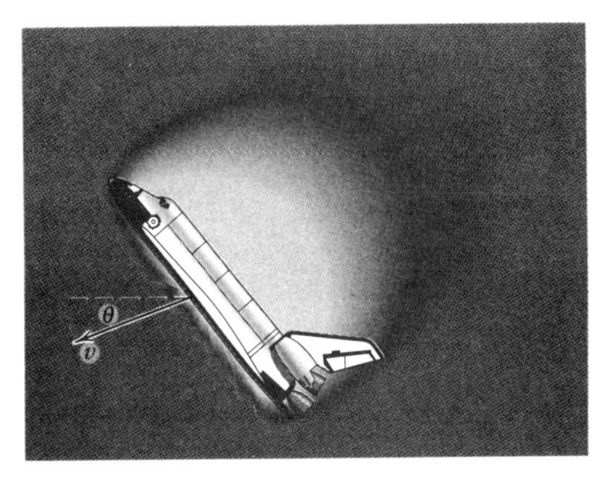

Problem 2/124

2/125 Pin P in the crank PO engages the horizontal slot in the guide C and controls its motion on the fixed vertical rod. Determine the velocity $\dot{y}$ and the acceleration $\ddot{y}$ of guide C for a given value of the angle θ if (a) $\dot{\theta} = \omega$ and $\ddot{\theta} = 0$ and (b) if $\dot{\theta} = 0$ and $\ddot{\theta} = \alpha$.

Ans. (a) $\dot{y} = r\omega \sin \theta, \ddot{y} = r\omega^2 \cos \theta$
(b) $\dot{y} = 0, \ddot{y} = r\alpha \sin \theta$

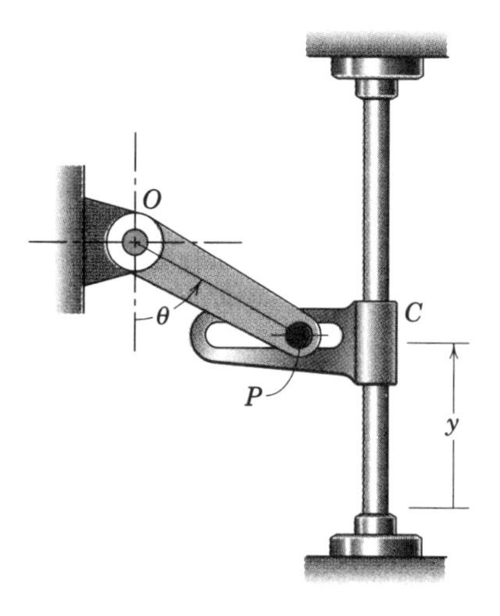

Problem 2/125

2/126 Race car A follows path a-a while race car B follows path b-b on the unbanked track. If each car has a constant speed limited to that corresponding to a lateral (normal) acceleration of $0.8g$, determine the times t_A and t_B for both cars to negotiate the turn as delimited by the line C-C.

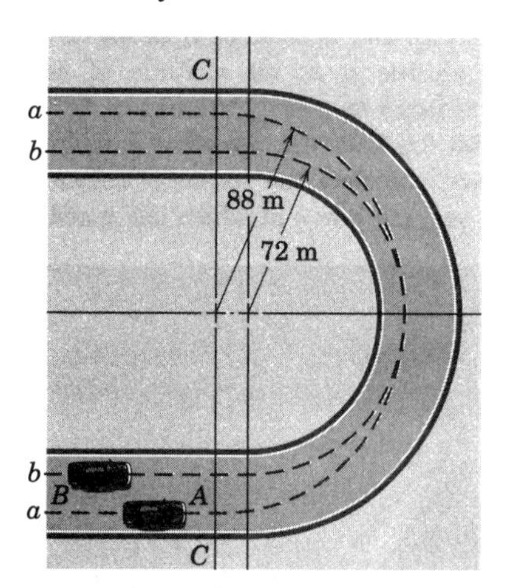

Problem 2/126

2/127 An earth satellite which moves in the elliptical equatorial orbit shown has a velocity v in space of 17 970 km/h when it passes the end of the semi-minor axis at A. The earth has an absolute surface value of g of 9.821 m/s^2 and has a radius of 6371 km. Determine the radius of curvature ρ of the orbit at A.

Ans. $\rho = 18\,480$ km

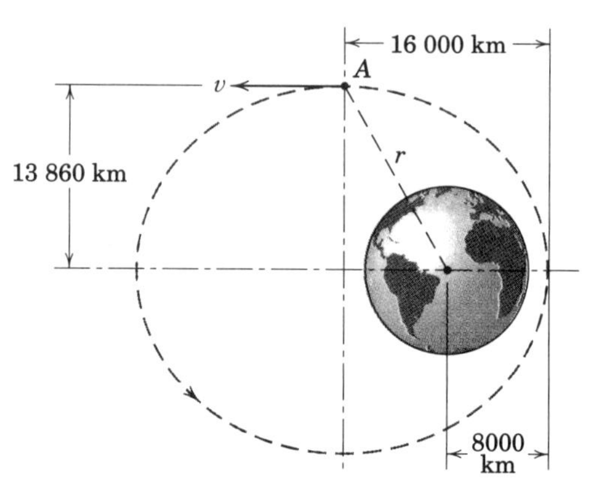

Problem 2/127

►2/128 The pin P is constrained to move in the slotted guides which move at right angles to one another. At the instant represented, A has a velocity to the right of 0.2 m/s which is decreasing at the rate of 0.75 m/s each second. At the same time, B is moving down with a velocity of 0.15 m/s which is decreasing at the rate of 0.5 m/s each second. For this instant determine the radius of curvature ρ of the path followed by P. Is it possible to determine also the time rate of change of ρ?

Ans. $\rho = 1.25$ m

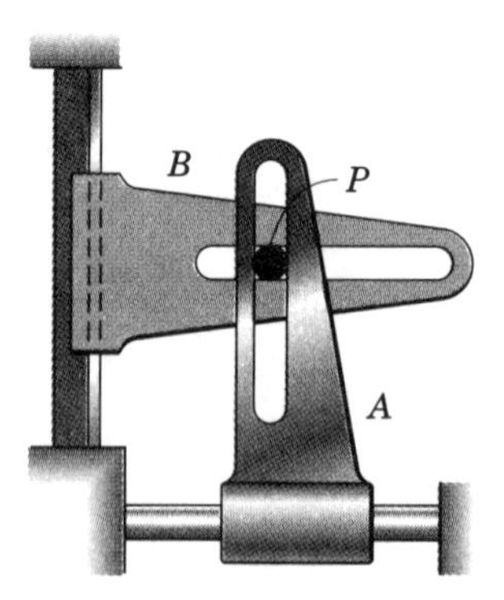

Problem 2/128

▶**2/129** In a handling test, a car is driven through the slalom course shown. It is assumed that the car path is sinusoidal and that the maximum lateral acceleration is $0.7g$. If the testers wish to design a slalom through which the maximum speed is 80 km/h, what cone spacing L should be used?

Ans. $L = 46.1$ m

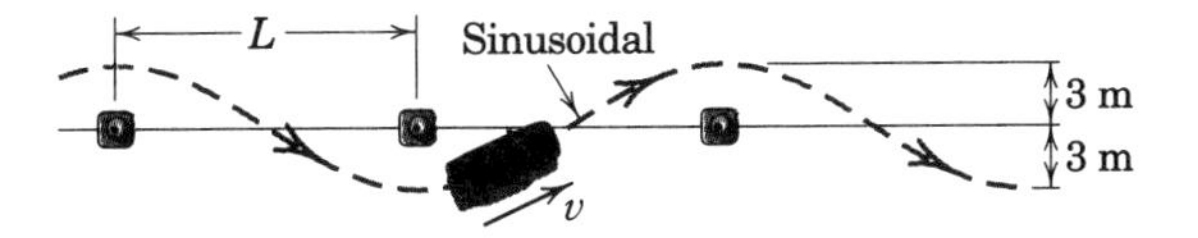

Problem 2/129

▶**2/130** A particle starts from rest at the origin and moves along the positive branch of the curve $y = 2x^{3/2}$ so that the distance s measured from the origin along the curve varies with the time t according to $s = 2t^3$, where x, y, and s are in millimeters and t is in seconds. Find the magnitude of the total acceleration $\mathbf{a}$ of the particle when $t = 1$ s. (Refer to Art. C/10 of Appendix C for an expression for ρ.)

Ans. $a = 12.17$ mm/s^2

Chapter

3

KINETICS OF PARTICLES

CHAPTER OUTLINE

3/1 INTRODUCTION

According to Newton's second law, a particle will accelerate when it is subjected to unbalanced forces. Kinetics is the study of the relations between unbalanced forces and the resulting changes in motion. In Chapter 3 we will study the kinetics of particles. This topic requires that we combine our knowledge of the properties of forces, which we developed in statics, and the kinematics of particle motion just covered in Chapter 2. With the aid of Newton's second law, we can combine these two topics and solve engineering problems involving force, mass, and motion.

The three general approaches to the solution of kinetics problems are: (A) direct application of Newton's second law (called the force-mass-acceleration method), (B) use of work and energy principles, and (C) solution by impulse and momentum methods. Each approach has its special characteristics and advantages, and Chapter 3 is subdivided into Sections A, B, and C, according to these three methods of solution. In addition, a fourth section, Section D, treats special applications and combinations of the three basic approaches. Before proceeding, you should review carefully the definitions and concepts of Chapter 1, because they are fundamental to the developments which follow.

SECTION A. FORCE, MASS, AND ACCELERATION

3/2 Newton's Second Law

The basic relation between force and acceleration is found in Newton's second law, Eq. 1/1, the verification of which is entirely experimental. We now describe the fundamental meaning of this law by considering an ideal experiment in which force and acceleration are assumed to be measured without error. We subject a mass particle to the action of a single force $\mathbf{F}_1$, and we measure the acceleration $\mathbf{a}_1$ of the particle in the primary inertial system.* The ratio F_1/a_1 of the magnitudes of the force and the acceleration will be some number C_1 whose value depends on the units used for measurement of force and acceleration. We then repeat the experiment by subjecting the same particle to a different force $\mathbf{F}_2$ and measuring the corresponding acceleration $\mathbf{a}_2$. The ratio F_2/a_2 of the magnitudes will again produce a number C_2. The experiment is repeated as many times as desired.

We draw two important conclusions from the results of these experiments. First, the ratios of applied force to corresponding acceleration all equal the *same* number, provided the units used for measurement are not changed in the experiments. Thus,

$$\frac{F_1}{a_1} = \frac{F_2}{a_2} = \cdots = \frac{F}{a} = C, \qquad \text{a constant}$$

We conclude that the constant C is a measure of some invariable property of the particle. This property is the *inertia* of the particle, which is its *resistance to rate of change of velocity*. For a particle of high inertia (large C), the acceleration will be small for a given force F. On the other hand, if the inertia is small, the acceleration will be large. The mass m is used as a quantitative measure of inertia, and therefore, we may write the expression $C = km$, where k is a constant introduced to account for the units used. Thus, we may express the relation obtained from the experiments as

$$F = kma \tag{3/1}$$

*The primary inertial system or astronomical frame of reference is an imaginary set of reference axes which are assumed to have no translation or rotation in space. See Art. 1/2, Chapter 1.

where F is the magnitude of the resultant force acting on the particle of mass m, and a is the magnitude of the resulting acceleration of the particle.

The second conclusion we draw from this ideal experiment is that the acceleration is always in the direction of the applied force. Thus, Eq. 3/1 becomes a *vector* relation and may be written

$$\mathbf{F} = km\mathbf{a} \tag{3/2}$$

Although an actual experiment cannot be performed in the ideal manner described, the same conclusions have been drawn from countless accurately performed experiments. One of the most accurate checks is given by the precise prediction of the motions of planets based on Eq. 3/2.

Inertial System

Although the results of the ideal experiment are obtained for measurements made relative to the "fixed" primary inertial system, they are equally valid for measurements made with respect to any nonrotating reference system which translates with a constant velocity with respect to the primary system. From our study of relative motion in Art. 2/8, we know that the acceleration measured in a system translating with no acceleration is the same as that measured in the primary system. Thus, Newton's second law holds equally well in a nonaccelerating system, so that we may define an *inertial system* as any system in which Eq. 3/2 is valid.

If the ideal experiment described were performed on the surface of the earth and all measurements were made relative to a reference system attached to the earth, the measured results would show a slight discrepancy from those predicted by Eq. 3/2, because the measured acceleration would not be the correct absolute acceleration. The discrepancy would disappear when we introduced the corrections due to the acceleration components of the earth. These corrections are negligible for most engineering problems which involve the motions of structures and machines on the surface of the earth. In such cases, the accelerations measured with respect to reference axes attached to the surface of the earth may be treated as "absolute," and Eq. 3/2 may be applied with negligible error to experiments made on the surface of the earth.*

An increasing number of problems occur, particularly in the fields of rocket and spacecraft design, where the acceleration components of the earth are of primary concern. For this work it is essential that the

*As an example of the magnitude of the error introduced by neglect of the motion of the earth, consider a particle which is allowed to fall from rest (relative to earth) at a height h above the ground. We can show that the rotation of the earth gives rise to an eastward acceleration (Coriolis acceleration) relative to the earth and, neglecting air resistance, that the particle falls to the ground a distance

$$x = \frac{2}{3}\omega\sqrt{\frac{2h^3}{g}}\cos\gamma$$

east of the point on the ground directly under that from which it was dropped. The angular velocity of the earth is $\omega = 0.729(10^{-4})$ rad/s, and the latitude, north or south, is γ. At a latitude of 45° and from a height of 200 m, this eastward deflection would be $x = 43.9$ mm.

fundamental basis of Newton's second law be thoroughly understood and that the appropriate absolute acceleration components be employed.

Before 1905 the laws of Newtonian mechanics had been verified by innumerable physical experiments and were considered the final description of the motion of bodies. The concept of *time*, considered an absolute quantity in the Newtonian theory, received a basically different interpretation in the theory of relativity announced by Einstein in 1905. The new concept called for a complete reformulation of the accepted laws of mechanics. The theory of relativity was subjected to early ridicule, but has been verified by experiment and is now universally accepted by scientists. Although the difference between the mechanics of Newton and that of Einstein is basic, there is a practical difference in the results given by the two theories only when velocities of the order of the speed of light (300×10^6 m/s) are encountered.* Important problems dealing with atomic and nuclear particles, for example, require calculations based on the theory of relativity.

Systems of Units

It is customary to take k equal to unity in Eq. 3/2, thus putting the relation in the usual form of Newton's second law

$$\mathbf{F} = m\mathbf{a} \qquad [1/1]$$

A system of units for which k is unity is known as a *kinetic* system. Thus, for a kinetic system the units of force, mass, and acceleration are not independent. In SI units, as explained in Art. 1/4, the units of force (newtons, N) are derived by Newton's second law from the base units of mass (kilograms, kg) times acceleration (meters per second squared, m/s^2). Thus, N = kg·m/s^2. This system is known as an *absolute* system since the unit for force is dependent on the absolute value of mass.

In U.S. customary units, on the other hand, the units of mass (slugs) are derived from the units of force (pounds force, lb) divided by acceleration (feet per second squared, ft/sec^2). Thus, the mass units are slugs = lb-sec^2/ft. This system is known as a *gravitational* system since mass is derived from force as determined from gravitational attraction.

For measurements made relative to the rotating earth, the relative value of g should be used. The internationally accepted value of g relative to the earth at sea level and at a latitude of 45° is 9.806 65 m/s^2. Except where greater precision is required, the value of 9.81 m/s^2 will be used for g. For measurements relative to a nonrotating earth, the absolute value of g should be used. At a latitude of 45° and at sea level, the absolute value is 9.8236 m/s^2. The sea-level variation in both the

*The theory of relativity demonstrates that there is no such thing as a preferred primary inertial system and that measurements of time made in two coordinate systems which have a velocity relative to one another are different. On this basis, for example, the principles of relativity show that a clock carried by the pilot of a spacecraft traveling around the earth in a circular polar orbit of 644 km altitude at a velocity of 27 080 km/h would be slow compared with a clock at the pole by 0.000 001 85 s for each orbit.

absolute and relative values of g with latitude is shown in Fig. 1/1 of Art. 1/5.

In the U.S. customary system, the standard value of g relative to the rotating earth at sea level and at a latitude of 45° is 32.1740 ft/sec^2. The corresponding value relative to a nonrotating earth is 32.2230 ft/sec^2.

Force and Mass Units

We need to use both SI units and U.S. customary units, so we must have a clear understanding of the correct force and mass units in each system. These units were explained in Art. 1/4, but it will be helpful to illustrate them here using simple numbers before applying Newton's second law. Consider, first, the free-fall experiment as depicted in Fig. 3/1*a* where we release an object from rest near the surface of the earth. We allow it to fall freely under the influence of the force of gravitational attraction W on the body. We call this force the *weight* of the body. In SI units for a mass $m = 1$ kg, the weight is $W = 9.81$ N, and the corresponding downward acceleration a is $g = 9.81$ m/s^2. In U.S. customary units for a mass $m = 1$ lbm (1/32.2 slug), the weight is $W = 1$ lbf and the resulting gravitational acceleration is $g = 32.2$ ft/sec^2. For a mass $m = 1$ slug (32.2 lbm), the weight is $W = 32.2$ lbf and the acceleration, of course, is also $g = 32.2$ ft/sec^2.

In Fig. 3/1*b* we illustrate the proper units with the simplest example where we accelerate an object of mass m along the horizontal with a

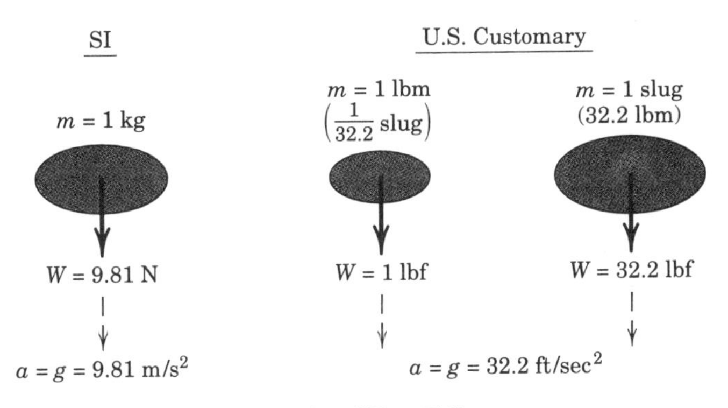

(*a*) Gravitational Free-Fall

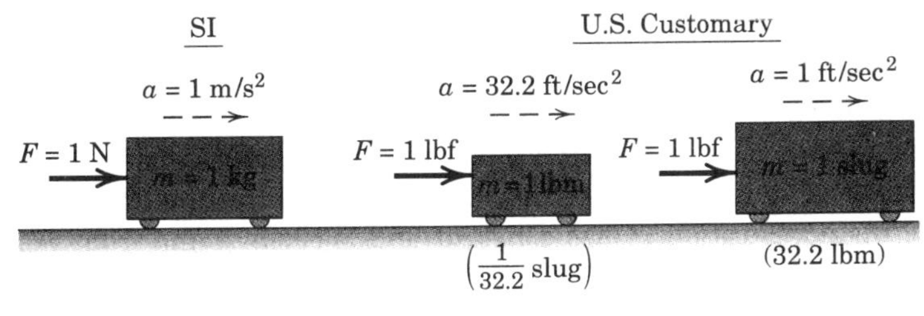

(*b*) Newton's Second Law

Figure 3/1

force F. In SI units (an absolute system), a force $F = 1$ N causes a mass $m = 1$ kg to accelerate at the rate $a = 1$ m/s^2. Thus, 1 N = 1 kg·m/s^2. In the U.S. customary system (a gravitational system), a force $F = 1$ lbf causes a mass $m = 1$ lbm (1/32.2 slug) to accelerate at the rate $a = 32.2$ ft/sec^2, whereas a force $F = 1$ lbf causes a mass $m = 1$ slug (32.2 lbm) to accelerate at the rate $a = 1$ ft/sec^2.

We note that in SI units where the mass is expressed in kilograms (kg), the weight W of the body in newtons (N) is given by $W = mg$, where $g = 9.81$ m/s^2. In U.S. customary units, the weight W of a body is expressed in pounds force (lbf), and the mass in slugs (lbf-sec^2/ft) is given by $m = W/g$, where $g = 32.2$ ft/sec^2.

In U.S. customary units, we frequently speak of the weight of a body when we really mean mass. It is entirely proper to specify the mass of a body in pounds (lbm) which must be converted to mass in slugs before substituting into Newton's second law. Unless otherwise stated, the pound (lb) is normally used as the unit of force (lbf).

3/3 Equation of Motion and Solution of Problems

When a particle of mass m is subjected to the action of concurrent forces $\mathbf{F}_1$, $\mathbf{F}_2$, $\mathbf{F}_3$, . . . whose vector sum is $\Sigma\mathbf{F}$, Eq. 1/1 becomes

$$\Sigma\mathbf{F} = m\mathbf{a} \tag{3/3}$$

When applying Eq. 3/3 to solve problems, we usually express it in scalar component form with the use of one of the coordinate systems developed in Chapter 2. The choice of an appropriate coordinate system depends on the type of motion involved and is a vital step in the formulation of any problem. Equation 3/3, or any one of the component forms of the force-mass-acceleration equation, is usually called the *equation of motion*. The equation of motion gives the instantaneous value of the acceleration corresponding to the instantaneous values of the forces which are acting.

Two Types of Dynamics Problems

We encounter two types of problems when applying Eq. 3/3. In the first type, the acceleration of the particle is either specified or can be determined directly from known kinematic conditions. We then determine the corresponding forces which act on the particle by direct substitution into Eq. 3/3. This problem is generally quite straightforward.

In the second type of problem, the forces acting on the particle are specified and we must determine the resulting motion. If the forces are constant, the acceleration is also constant and is easily found from Eq. 3/3. When the forces are functions of time, position, or velocity, Eq. 3/3 becomes a differential equation which must be integrated to determine the velocity and displacement.

Problems of this second type are often more formidable, as the integration may be difficult to carry out, particularly when the force is a mixed function of two or more motion variables. In practice, it is frequently necessary to resort to approximate integration techniques, either numerical or graphical, particularly when experimental data are involved. The procedures for a mathematical integration of the acceleration when it is a function of the motion variables were developed in Art. 2/2, and these same procedures apply when the force is a specified function of these same parameters, since force and acceleration differ only by the constant factor of the mass.

Constrained and Unconstrained Motion

There are two physically distinct types of motion, both described by Eq. 3/3. The first type is *unconstrained* motion where the particle is free of mechanical guides and follows a path determined by its initial motion and by the forces which are applied to it from external sources. An airplane or rocket in flight and an electron moving in a charged field are examples of unconstrained motion.

The second type is *constrained* motion where the path of the particle is partially or totally determined by restraining guides. An ice-hockey puck is partially constrained to move in the horizontal plane by the surface of the ice. A train moving along its track and a collar sliding along a fixed shaft are examples of more fully constrained motion. Some of the forces acting on a particle during constrained motion may be applied from outside sources, and others may be the reactions on the particle from the constraining guides. *All forces*, both applied and reactive, which act *on* the particle must be accounted for in applying Eq. 3/3.

The choice of an appropriate coordinate system is frequently indicated by the number and geometry of the constraints. Thus, if a particle is free to move in space, as is the center of mass of the airplane or rocket in free flight, the particle is said to have *three degrees of freedom* since three independent coordinates are required to specify the position of the particle at any instant. All three of the scalar components of the equation of motion would have to be integrated to obtain the space coordinates as a function of time.

If a particle is constrained to move along a surface, as is the hockey puck or a marble sliding on the curved surface of a bowl, only two coordinates are needed to specify its position, and in this case it is said to have *two degrees of freedom*. If a particle is constrained to move along a fixed linear path, as is the collar sliding along a fixed shaft, its position may be specified by the coordinate measured along the shaft. In this case, the particle would have only *one degree of freedom*.

Free-Body Diagram

When applying any of the force-mass-acceleration equations of motion, you must account correctly for *all* forces acting on the particle. The only forces which we may neglect are those whose magnitudes are negligible compared with other forces acting, such as the forces of mutual attraction between two particles compared with their attraction to a celestial body such as the earth. The vector sum $\Sigma\mathbf{F}$ of Eq. 3/3 means the vector sum of *all* forces acting *on* the particle in question. Likewise, the corresponding scalar force summation in any one of the component directions means the sum of the components of *all* forces acting *on* the particle in that particular direction.

The only reliable way to account accurately and consistently for every force is to *isolate* the particle under consideration from *all* contacting and influencing bodies and replace the bodies removed by the forces they exert on the particle isolated. The resulting *free-body diagram* is the means by which every force, known and unknown, which acts on the particle is represented and thus accounted for. Only after this vital step has been completed should you write the appropriate equation or equations of motion.

The free-body diagram serves the same key purpose in dynamics as it does in statics. This purpose is simply to establish a *thoroughly reliable method* for the correct evaluation of the resultant of all actual forces acting on the particle or body in question. In statics this resultant equals zero, whereas in dynamics it is equated to the product of mass and acceleration. When you use the vector form of the equation of motion, remember that it represents several scalar equations and that every equation must be satisfied.

Careful and consistent use of the *free-body method* is the *most important single lesson* to be learned in the study of engineering mechanics. When drawing a free-body diagram, clearly indicate the coordinate axes and their positive directions. When you write the equations of motion, make sure all force summations are consistent with the choice of these positive directions. As an aid to the identification of external forces which act on the body in question, these forces are shown as heavy red vectors in the illustrations in this book. Sample Problems 3/1 through 3/5 in the next article contain five examples of free-body diagrams. You should study these to see how the diagrams are constructed.

In solving problems, you may wonder how to get started and what sequence of steps to follow in arriving at the solution. This difficulty may be minimized by forming the habit of first recognizing some relationship between the desired unknown quantity in the problem and other quantities, known and unknown. Then determine additional relationships between these unknowns and other quantities, known and unknown. Finally, establish the dependence on the original data and develop the procedure for the analysis and computation. A few minutes spent organizing the plan of attack through recognition of the dependence of one quantity on another will be time well spent and will usually prevent groping for the answer with irrelevant calculations.

3/4 Rectilinear Motion

We now apply the concepts discussed in Arts. 3/2 and 3/3 to problems in particle motion, starting with rectilinear motion in this article and treating curvilinear motion in Art. 3/5. In both articles, we will analyze the motions of bodies which can be treated as particles. This simplification is possible as long as we are interested only in the motion of the mass center of the body. In this case we may treat the forces as concurrent through the mass center. We will account for the action of nonconcurrent forces on the motions of bodies when we discuss the kinetics of rigid bodies in Chapter 6.

If we choose the x-direction, for example, as the direction of the rectilinear motion of a particle of mass m, the acceleration in the y- and z-directions will be zero and the scalar components of Eq. 3/3 become

$$\begin{aligned} \Sigma F_x &= ma_x \\ \Sigma F_y &= 0 \\ \Sigma F_z &= 0 \end{aligned} \tag{3/4}$$

For cases where we are not free to choose a coordinate direction along the motion, we would have in the general case all three component equations

$$\begin{aligned} \Sigma F_x &= ma_x \\ \Sigma F_y &= ma_y \\ \Sigma F_z &= ma_z \end{aligned} \tag{3/5}$$

where the acceleration and resultant force are given by

$$\begin{aligned} \mathbf{a} &= a_x\mathbf{i} + a_y\mathbf{j} + a_z\mathbf{k} \\ a &= \sqrt{a_x^2 + a_y^2 + a_z^2} \\ \Sigma\mathbf{F} &= \Sigma F_x\mathbf{i} + \Sigma F_y\mathbf{j} + \Sigma F_z\mathbf{k} \\ |\Sigma\mathbf{F}| &= \sqrt{(\Sigma F_x)^2 + (\Sigma F_y)^2 + (\Sigma F_z)^2} \end{aligned}$$

Sample Problem 3/1

A 75-kg man stands on a spring scale in an elevator. During the first 3 seconds of motion from rest, the tension T in the hoisting cable is 8300 N. Find the reading R of the scale in newtons during this interval and the upward velocity v of the elevator at the end of the 3 seconds. The total mass of the elevator, man, and scale is 750 kg.

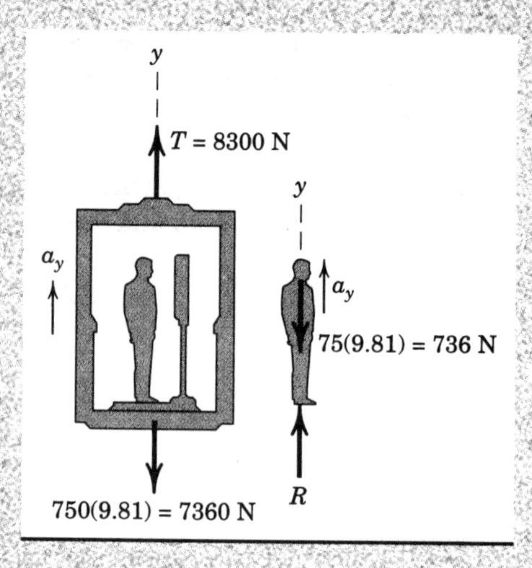

Solution. The force registered by the scale and the velocity both depend on the acceleration of the elevator, which is constant during the interval for which the forces are constant. From the free-body diagram of the elevator, scale, and man taken together, the acceleration is found to be

$$[\Sigma F_y = ma_y] \qquad 8300 - 7360 = 750a_y \qquad a_y = 1.257 \text{ m/s}^2$$

The scale reads the downward force exerted on it by the man's feet. The equal and opposite reaction R to this action is shown on the free-body diagram of the man alone together with his weight, and the equation of motion for him gives

① $$[\Sigma F_y = ma_y] \qquad R - 736 = 75(1.257) \qquad R = 830 \text{ N} \qquad \textit{Ans.}$$

The velocity reached at the end of the 3 seconds is

$$\left[\Delta v = \int a\, dt\right] \qquad v - 0 = \int_0^3 1.257\, dt \qquad v = 3.77 \text{ m/s} \qquad \textit{Ans.}$$

Helpful Hint

① If the scale is calibrated in kilograms it would read $830/9.81 = 84.6$ kg which, of course, is not his true mass since the measurement was made in a noninertial (accelerating) system. *Suggestion:* Rework this problem in U.S. customary units.

Sample Problem 3/2

A small inspection car with a mass of 200 kg runs along the fixed overhead cable and is controlled by the attached cable at A. Determine the acceleration of the car when the control cable is horizontal and under a tension $T = 2.4$ kN. Also find the total force P exerted by the supporting cable on the wheels.

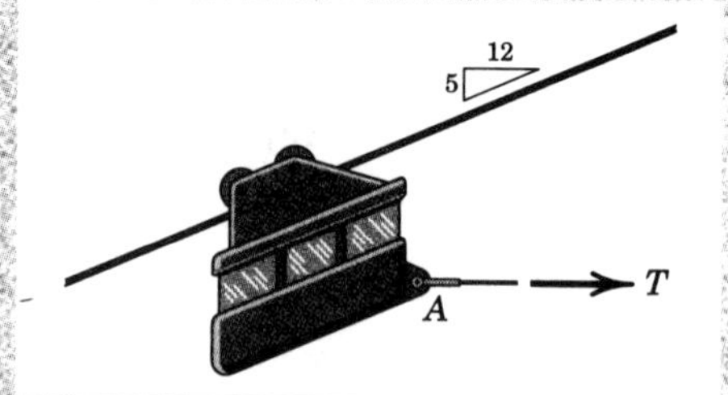

Solution. The free-body diagram of the car and wheels taken together and treated as a particle discloses the 2.4-kN tension T, the weight $W = mg = 200(9.81) = 1962$ N, and the force P exerted on the wheel assembly by the cable.

The car is in equilibrium in the y-direction since there is no acceleration in this direction. Thus,

$$[\Sigma F_y = 0] \qquad P - 2.4(\tfrac{5}{13}) - 1.962(\tfrac{12}{13}) = 0 \qquad P = 2.73 \text{ kN} \qquad \textit{Ans.}$$

① In the x-direction the equation of motion gives

$$[\Sigma F_x = ma_x] \quad 2400(\tfrac{12}{13}) - 1962(\tfrac{5}{13}) = 200a \qquad a = 7.30 \text{ m/s}^2 \qquad \textit{Ans.}$$

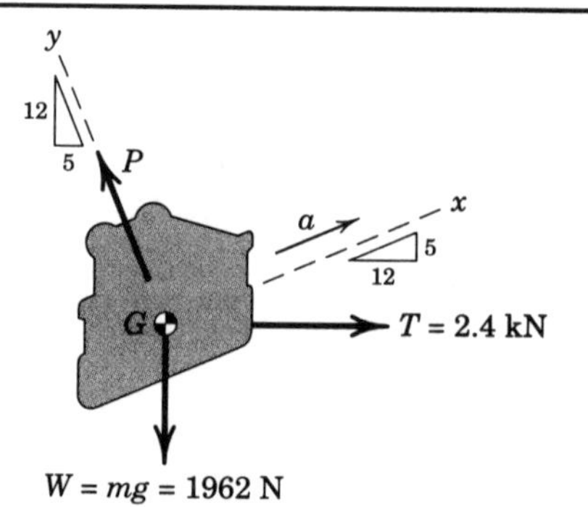

Helpful Hint

① By choosing our coordinate axes along and normal to the direction of the acceleration, we are able to solve the two equations independently. Would this be so if x and y were chosen as horizontal and vertical?

Sample Problem 3/3

The 125-kg concrete block A is released from rest in the position shown and pulls the 200-kg log up the 30° ramp. If the coefficient of kinetic friction between the log and the ramp is 0.5, determine the velocity of the block as it hits the ground at B.

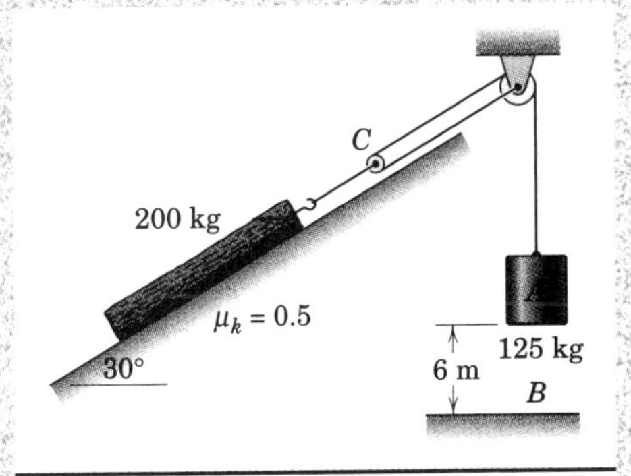

Solution. The motions of the log and the block A are clearly dependent. Although by now it should be evident that the acceleration of the log up the incline is half the downward acceleration of A, we may prove it formally. The constant total length of the cable is $L = 2s_C + s_A$ + constant, where the constant accounts for the cable portions wrapped around the pulleys. Differentiating twice with respect to time gives $0 = 2\ddot{s}_C + \ddot{s}_A$, or ①

$$0 = 2a_C + a_A$$

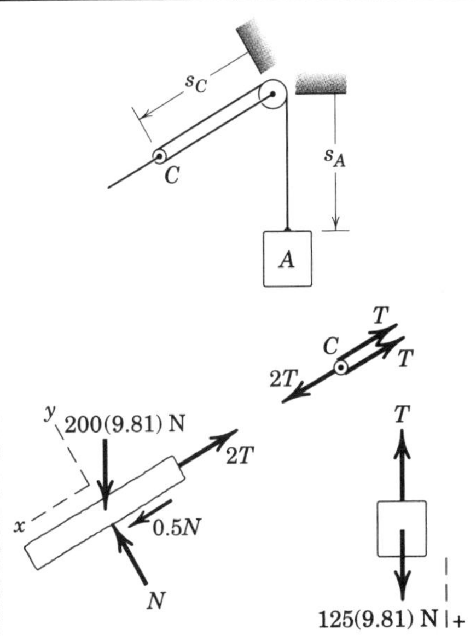

We assume here that the masses of the pulleys are negligible and that they turn with negligible friction. With these assumptions the free-body diagram of the pulley C discloses force and moment equilibrium. Thus, the tension in the cable attached to the log is twice that applied to the block. Note that the accelerations of the log and the center of pulley C are identical.

The free-body diagram of the log shows the friction force $\mu_k N$ for motion up the plane. Equilibrium of the log in the y-direction gives

② $[\Sigma F_y = 0]$ $\quad N - 200(9.81)\cos 30° = 0 \qquad N = 1699 \text{ N}$

and its equation of motion in the x-direction gives

$[\Sigma F_x = ma_x]$ $\quad 0.5(1699) - 2T + 200(9.81)\sin 30° = 200a_C$

For the block in the positive downward direction, we have

③ $[+\downarrow \Sigma F = ma]$ $\quad 125(9.81) - T = 125a_A$

Solving the three equations in a_C, a_A, and T gives us

$$a_A = 1.777 \text{ m/s}^2 \qquad a_C = -0.888 \text{ m/s}^2 \qquad T = 1004 \text{ N}$$

④ For the 6-m drop with constant acceleration, the block acquires a velocity

$[v^2 = 2ax]$ $\quad v_A = \sqrt{2(1.777)(6)} = 4.62 \text{ m/s}$ *Ans.*

Helpful Hints

① The coordinates used in expressing the final kinematic constraint relationship must be consistent with those used for the kinetic equations of motion.

② We can verify that the log will indeed move up the ramp by calculating the force in the cable necessary to initiate motion from the equilibrium condition. This force is $2T = 0.5N + 200(9.81)\sin 30° = 1831$ N or $T = 915$ N, which is less than the 1226-N weight of block A. Hence, the log will move up.

③ Note the serious error in assuming that $T = 125(9.81)$ N, in which case, block A would not accelerate.

④ Because the forces on this system remain constant, the resulting accelerations also remain constant.

Sample Problem 3/4

The design model for a new ship has a mass of 10 kg and is tested in an experimental towing tank to determine its resistance to motion through the water at various speeds. The test results are plotted on the accompanying graph, and the resistance R may be closely approximated by the dashed parabolic curve shown. If the model is released when it has a speed of 2 m/s, determine the time t required for it to reduce its speed to 1 m/s and the corresponding travel distance x.

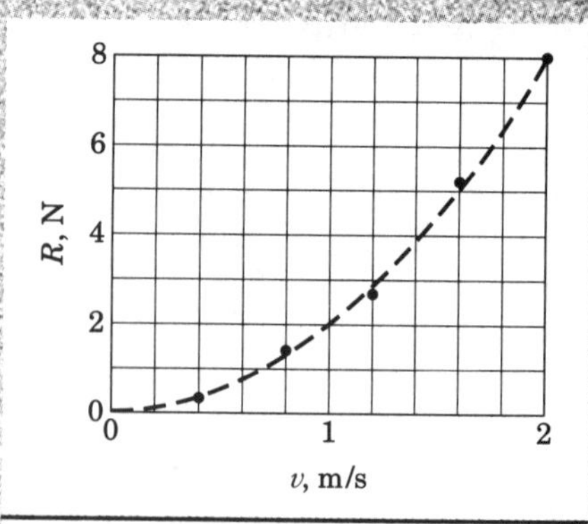

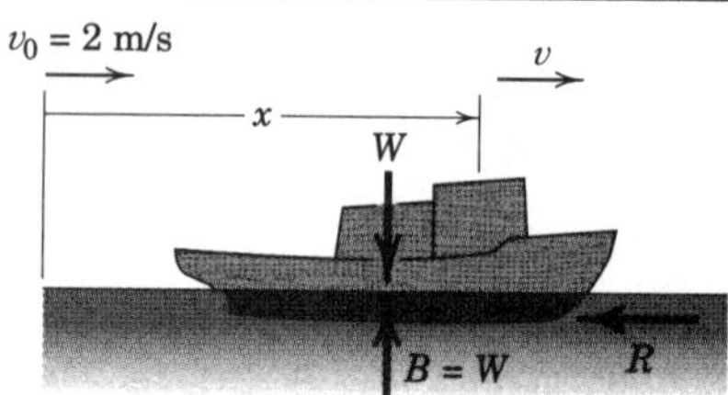

Solution. We approximate the resistance-velocity relation by $R = kv^2$ and find k by substituting $R = 8$ N and $v = 2$ m/s into the equation, which gives $k = 8/2^2 = 2\ \text{N}\cdot\text{s}^2/\text{m}^2$. Thus, $R = 2v^2$.

The only horizontal force on the model is R, so that

① $[\Sigma F_x = ma_x]$ $$-R = ma_x \quad \text{or} \quad -2v^2 = 10\frac{dv}{dt}$$

We separate the variables and integrate to obtain

$$\int_0^t dt = -5\int_2^v \frac{dv}{v^2} \qquad t = 5\left(\frac{1}{v} - \frac{1}{2}\right)\ \text{sec}$$

Thus, when $v = v_0/2 = 1$ m/s, the time is $t = 5(\frac{1}{1} - \frac{1}{2}) = 2.5$ s *Ans.*

The distance traveled during the 2.5 seconds is obtained by integrating $v = dx/dt$. Thus, $v = 10/(5 + 2t)$ so that

② $$\int_0^x dx = \int_0^{2.5} \frac{10}{5 + 2t}\,dt \qquad x = \frac{10}{2}\ln(5 + 2t)\Big|_0^{2.5} = 3.47\ \text{m}$$ *Ans.*

Helpful Hints

① Be careful to observe the minus sign for R.

② *Suggestion:* Express the distance x after release in terms of the velocity v and see if you agree with the resulting relation $x = 5\ln(v_0/v)$.

Sample Problem 3/5

The collar of mass m slides up the vertical shaft under the action of a force F of constant magnitude but variable direction. If $\theta = kt$ where k is a constant and if the collar starts from rest with $\theta = 0$, determine the magnitude F of the force which will result in the collar coming to rest as θ reaches $\pi/2$. The coefficient of kinetic friction between the collar and shaft is μ_k.

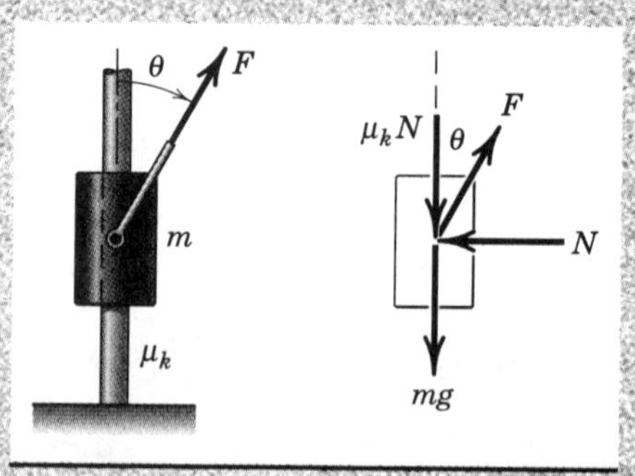

Solution. After drawing the free-body diagram, we apply the equation of motion in the y-direction to get

① $[\Sigma F_y = ma_y]$ $$F\cos\theta - \mu_k N - mg = m\frac{dv}{dt}$$

where equilibrium in the horizontal direction requires $N = F\sin\theta$. Substituting $\theta = kt$ and integrating first between general limits give

$$\int_0^t (F\cos kt - \mu_k F\sin kt - mg)\,dt = m\int_0^v dv$$

which becomes

$$\frac{F}{k}[\sin kt + \mu_k(\cos kt - 1)] - mgt = mv$$

For $\theta = \pi/2$ the time becomes $t = \pi/2k$, and $v = 0$ so that

② $$\frac{F}{k}[1 + \mu_k(0 - 1)] - \frac{mg\pi}{2k} = 0 \quad \text{and} \quad F = \frac{mg\pi}{2(1 - \mu_k)}$$ *Ans.*

Helpful Hints

① If θ were expressed as a function of the vertical displacement y instead of the time t, the acceleration would become a function of the displacement and we would use $v\,dv = a\,dy$.

② We see that the results do not depend on k, the rate at which the force changes direction.

PROBLEMS

Introductory Problems

3/1 The 50-kg crate is projected along the floor with an initial speed of 7 m/s at $x = 0$. The coefficient of kinetic friction is 0.40. Calculate the time required for the crate to come to rest and the corresponding distance x traveled.

Ans. $t = 1.784$ s, $x = 6.24$ m

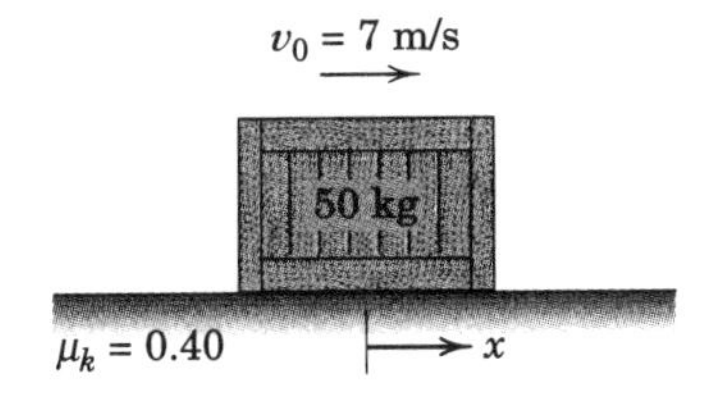

Problem 3/1

3/2 The 50-kg crate of Prob. 3/1 is now projected down an incline as shown with an initial speed of 7 m/s. Investigate the time t required for the crate to come to rest and the corresponding distance x traveled if (*a*) $\theta = 15°$ and (*b*) $\theta = 30°$.

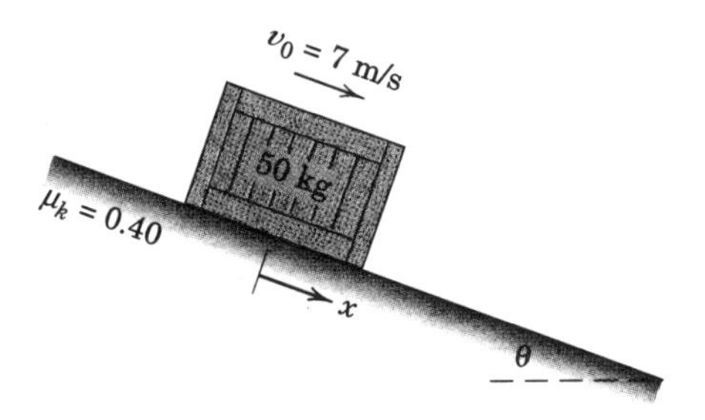

Problem 3/2

3/3 The 100-kg crate is carefully placed with zero velocity on the incline. Describe what happens if (*a*) $\theta = 15°$ and (*b*) $\theta = 20°$.

Ans. (*a*) $a = 0$; no motion
(*b*) $a = 1.051$ m/s^2 down incline

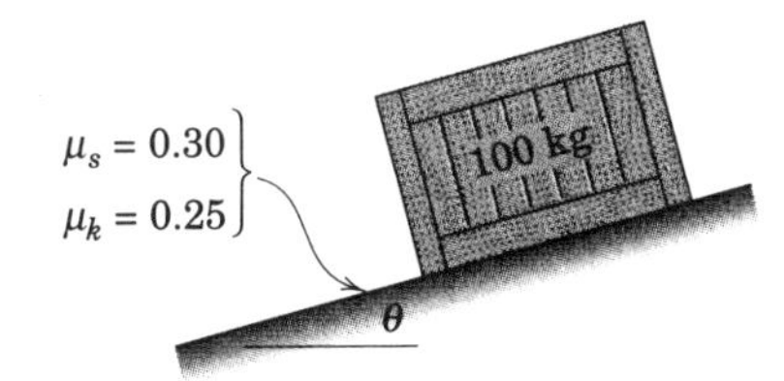

Problem 3/3

3/4 During a brake test, the rear-engine car is stopped from an initial speed of 100 km/h in a distance of 50 m. If it is known that all four wheels contribute equally to the braking force, determine the braking force F at each wheel. Assume a constant deceleration for the 1500-kg car.

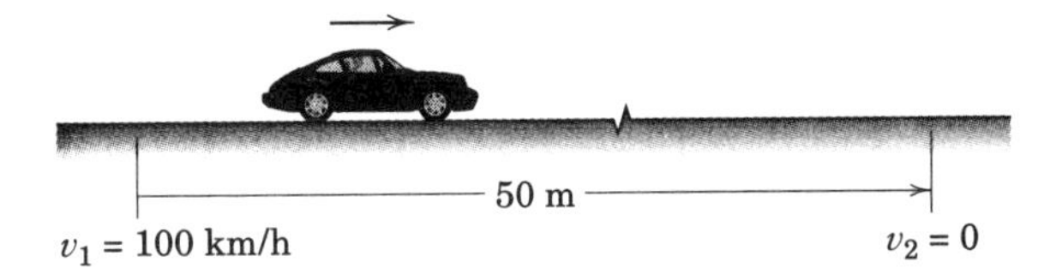

Problem 3/4

3/5 What fraction n of the weight of the jet airplane is the net thrust (nozzle thrust T minus air resistance R) required for the airplane to climb at an angle θ with the horizontal with an acceleration a in the direction of flight?

Ans. $n = \sin\theta + \dfrac{a}{g}$

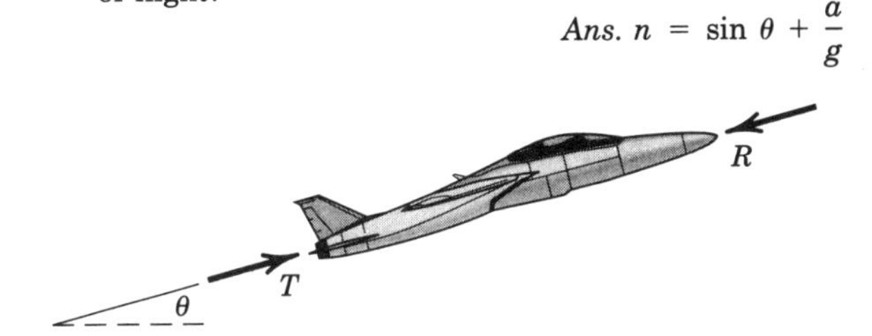

Problem 3/5

3/6 The 300-Mg jet airliner has three engines, each of which produces a nearly constant thrust of 240 kN during the takeoff roll. Determine the length s of runway required if the takeoff speed is 220 km/h. Compute s first for an uphill takeoff direction from A to B and second for a downhill takeoff from B to A on the slightly inclined runway. Neglect air and rolling resistance.

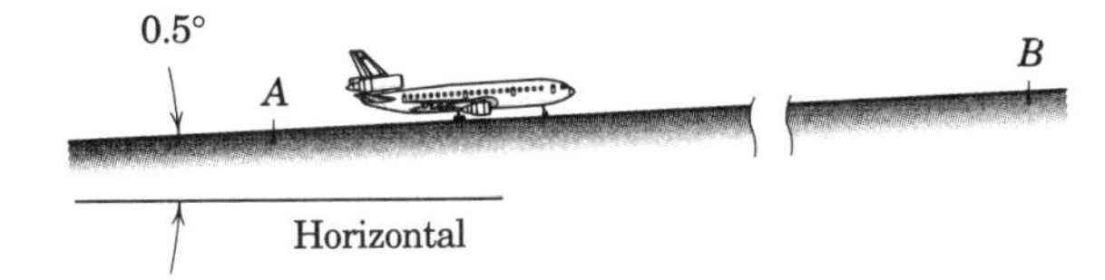

Problem 3/6

3/7 Calculate the vertical acceleration a of the 150-kg cylinder for each of the two cases illustrated. Neglect friction and the mass of the pulleys.

Ans. (*a*) $a = 1.401 \text{ m/s}^2$
(*b*) $a = 3.27 \text{ m/s}^2$

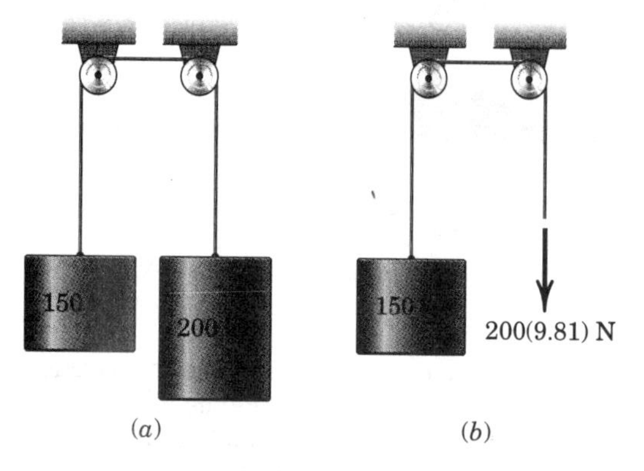

Problem 3/7

3/8 A car is climbing the hill of slope θ_1 at a constant speed v. If the slope decreases abruptly to θ_2 at point A, determine the acceleration a of the car just after passing point A if the driver does not change the throttle setting or shift into a different gear.

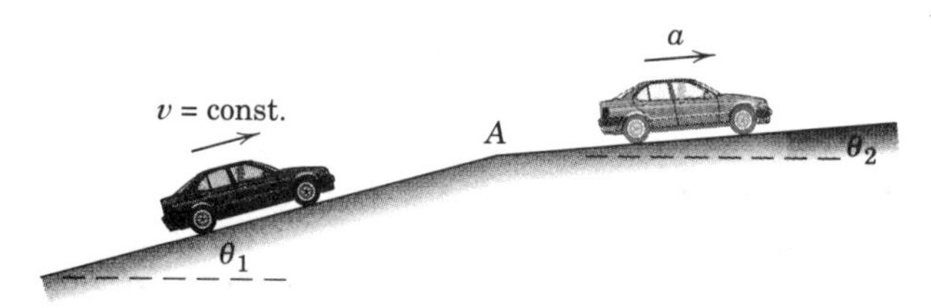

Problem 3/8

3/9 The 340-Mg jetliner A has four engines, each of which produces a nearly constant thrust of 200 kN during the takeoff roll. A small commuter aircraft B taxis toward the end of the runway at a constant speed $v_B = 25$ km/h. Determine the velocity and acceleration which A appears to have relative to an observer in B 10 seconds after A begins its takeoff roll. Neglect air and rolling resistance.

Ans. $\mathbf{v}_{A/B} = -29.5\mathbf{i} - 3.47\mathbf{j}$ m/s
$\mathbf{a}_{A/B} = -2.35\mathbf{i} \text{ m/s}^2$

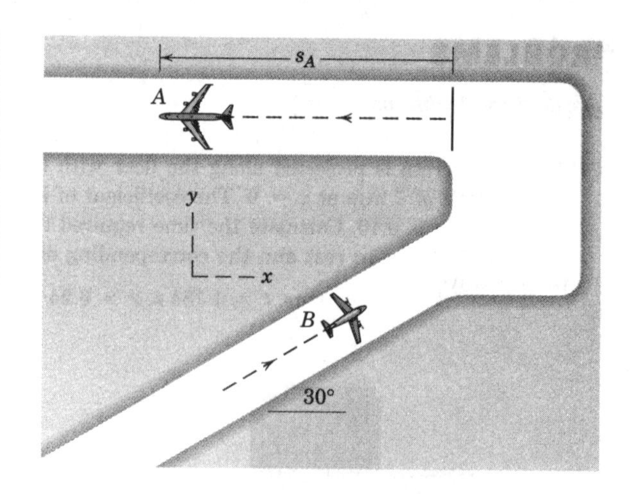

Problem 3/9

3/10 The tractor-trailer unit is moving down the incline with a speed of 8 km/h when the driver brakes the tractor to a stop in a distance of 1.2 m. Estimate the percent increase n in the hitch-force component which is parallel to the incline, compared with the force present at steady speed. The cart and its load have a combined mass of 220 kg. State any assumptions.

Problem 3/10

3/11 The block-and-tackle system is released from rest with all cables taut. Neglect the mass and friction of all pulleys and determine the acceleration of each cylinder and the tensions T_1 and T_2 in the two cables.

Ans. $a_A = 1.401 \text{ m/s}^2$ up
$a_B = 5.61 \text{ m/s}^2$ down
$T_1 = 16.82$ N, $T_2 = 67.3$ N

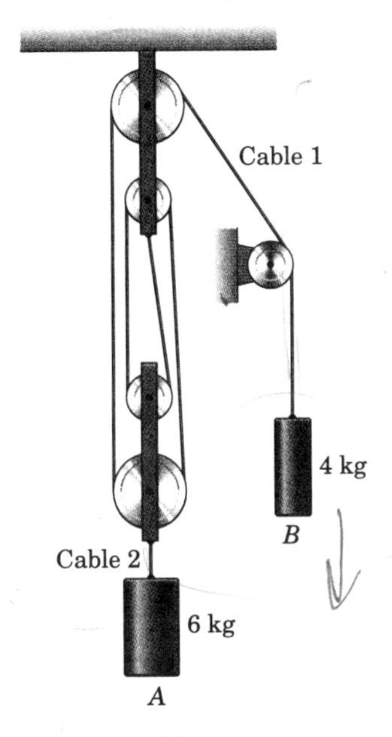

Problem 3/11

3/12 Determine the tension P in the cable which will give the 50-kg block a steady acceleration of 2 m/s^2 up the incline.

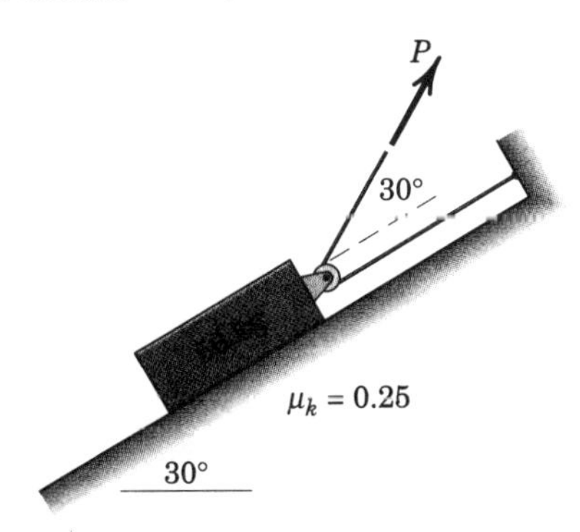

Problem 3/12

3/13 A toy train has magnetic couplers whose maximum attractive force is 0.9 N between adjacent cars. What is the maximum force P with which a child can pull the locomotive and not break the train apart at a coupler? If P is slightly exceeded, which coupler fails? Neglect the mass and friction associated with all wheels.

Ans. $P = 1.8$ N, coupler 1

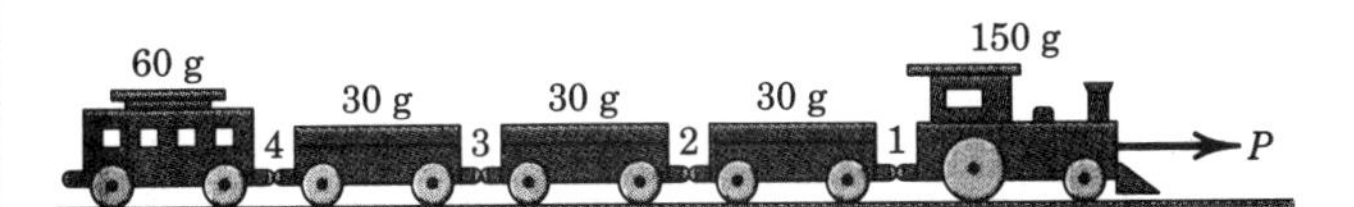

Problem 3/13

Representative Problems

3/14 The steel ball is suspended from the accelerating frame by the two cords A and B. Determine the acceleration a of the frame which will cause the tension in A to be twice that in B.

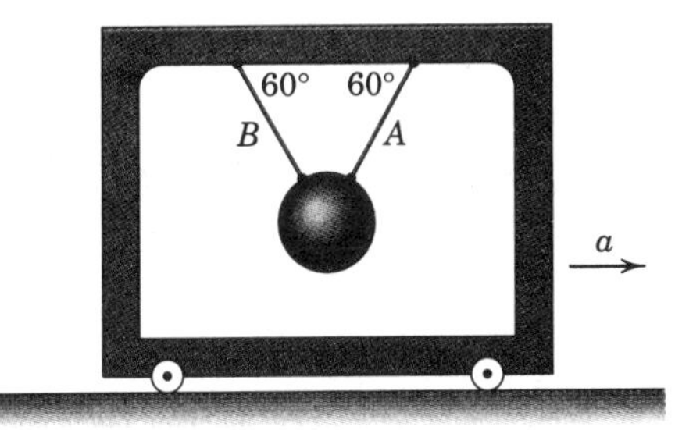

Problem 3/14

3/15 The collar A is free to slide along the smooth shaft B mounted in the frame. The plane of the frame is vertical. Determine the horizontal acceleration a of the frame necessary to maintain the collar in a fixed position on the shaft.

Ans. $a = 5.66 \text{ m/s}^2$

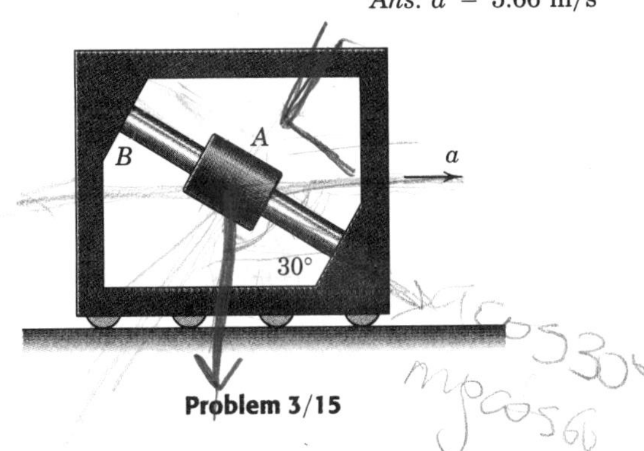

Problem 3/15

3/16 The beam and attached hoisting mechanism have a combined mass of 1200 kg with center of mass at G. If the initial acceleration a of a point P on the hoisting cable is 6 m/s^2, calculate the corresponding reaction at the support A.

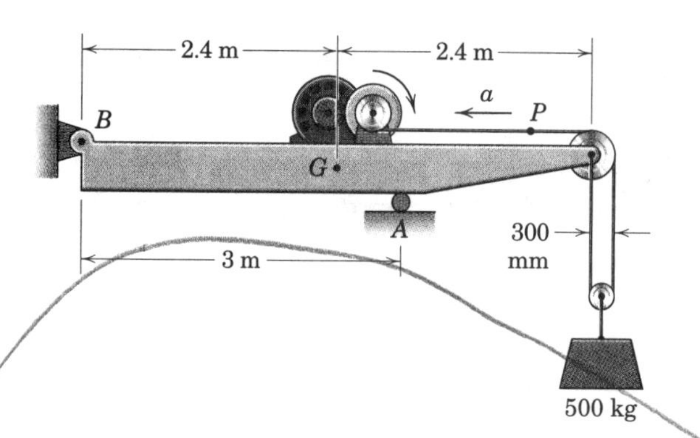

Problem 3/16

3/17 The coefficient of static friction between the flat bed of the truck and the crate it carries is 0.30. Determine the minimum stopping distance s which the truck can have from a speed of 70 km/h with constant deceleration if the crate is not to slip forward.

Ans. $s = 64.3$ m

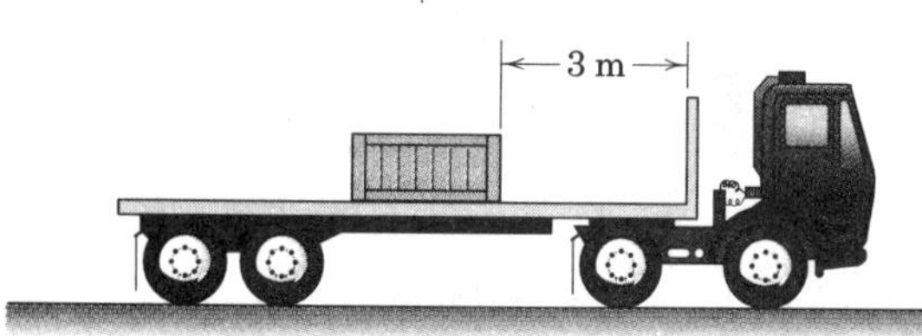

Problem 3/17

3/18 If the truck of Prob. 3/17 comes to a stop from an initial forward speed of 70 km/h in a distance of 50 m with uniform deceleration, determine whether or not the crate strikes the wall at the forward end of the flat bed. If the crate does strike the wall, calculate its speed relative to the truck as the impact occurs. Use the friction coefficients $\mu_s = 0.30$ and $\mu_k = 0.25$.

3/19 The block shown is observed to have a velocity $v_1 = 20$ m/s as it passes point A and a velocity $v_2 = 10$ m/s as it passes point B on the incline. Calculate the coefficient of kinetic friction μ_k between the block and the incline if $x = 75$ m and $\theta = 15°$.

Ans. $\mu_k = 0.479$

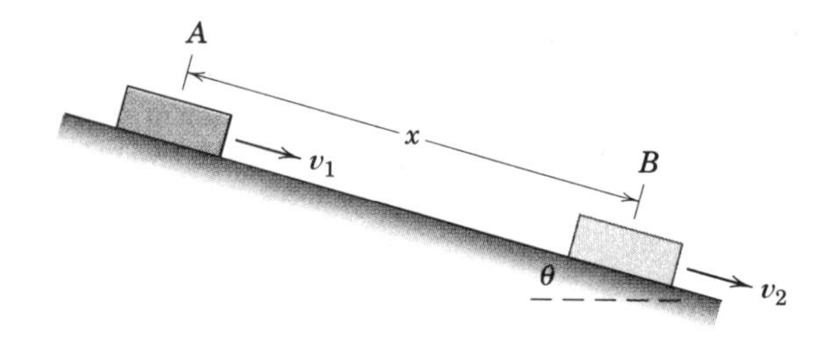

Problem 3/19

3/20 Small objects are delivered to the 2-m inclined chute by a conveyor belt A which moves at a speed $v_1 = 0.4$ m/s. If the conveyor belt B has a speed $v_2 = 0.9$ m/s and the objects are delivered to this belt with no slipping, calculate the coefficient of friction μ_k between the objects and the chute.

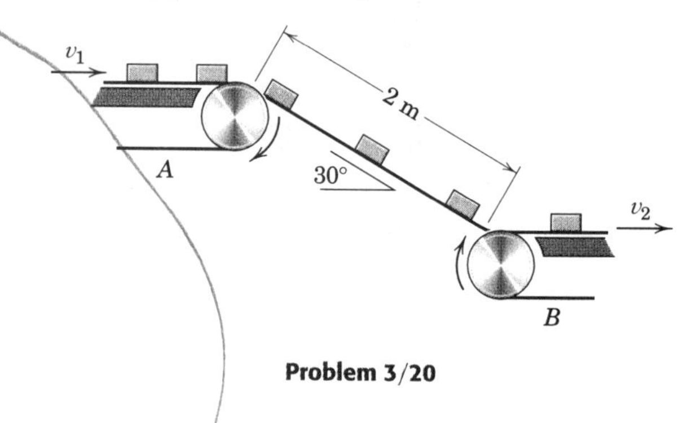

Problem 3/20

3/21 During a reliability test, a circuit board of mass m is attached to an electromagnetic shaker and subjected to a harmonic displacement $x = X \sin \omega t$, where X is the motion amplitude, ω is the motion frequency in radians per second, and t is time. Determine the magnitude F_{max} of the maximum horizontal force which the shaker exerts on the circuit board.

Ans. $F_{max} = mX\omega^2$

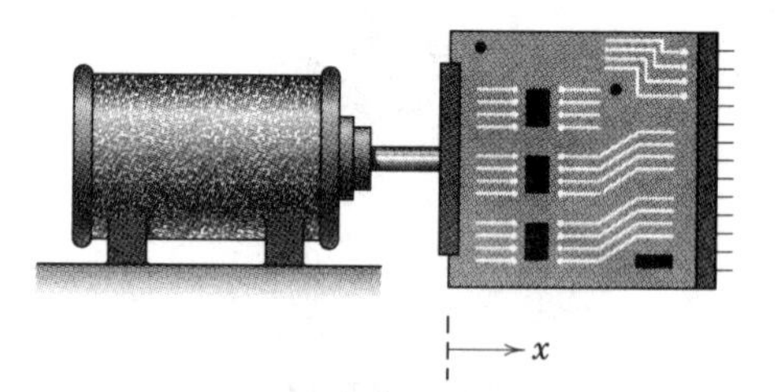

Problem 3/21

3/22 A cesium-ion engine for deep-space propulsion is designed to produce a constant thrust of 2.5 N for long periods of time. If the engine is to propel a 70-Mg spacecraft on an interplanetary mission, compute the time t required for a speed increase from 40 000 km/h to 65 000 km/h. Also find the distance s traveled during this interval. Assume that the spacecraft is moving in a remote region of space where the thrust from its ion engine is the only force acting on the spacecraft in the direction of its motion.

3/23 If the coefficients of static and kinetic friction between the 20-kg block A and the 100-kg cart B are both essentially the same value of 0.50, determine the acceleration of each part for (a) $P = 60$ N and (b) $P = 40$ N.

Ans. (a) $a_A = 1.095\ \text{m/s}^2$, $a_B = 0.981\ \text{m/s}^2$
(b) $a_A = a_B = 0.667\ \text{m/s}^2$

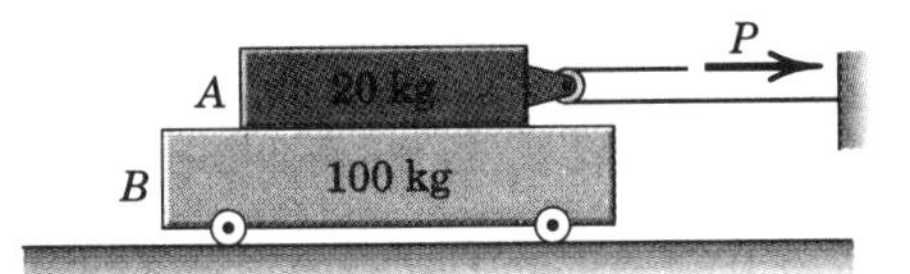

Problem 3/23

3/24 A simple pendulum is pivoted at O and is free to swing in the vertical plane of the plate. If the plate is given a constant acceleration a up the incline θ, write an expression for the steady angle β assumed by the pendulum after all initial start-up oscillations have ceased. Neglect the mass of the slender supporting rod.

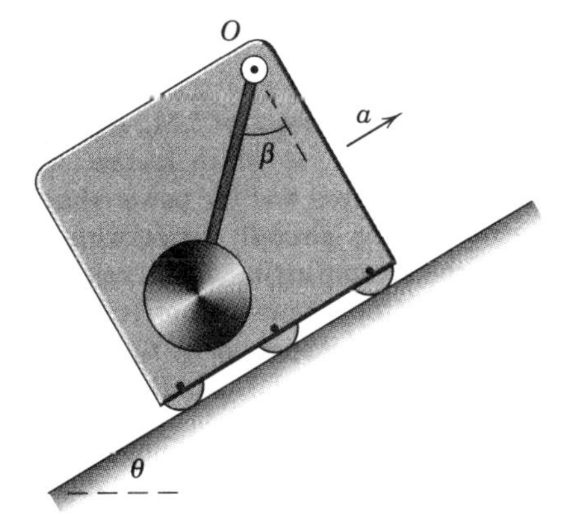

Problem 3/24

3/25 The device shown is used as an accelerometer and consists of a 100-g plunger A which deflects the spring as the housing of the unit is given an upward acceleration a. Specify the necessary spring stiffness k which will permit the plunger to deflect 6 mm beyond the equilibrium position and touch the electrical contact when the steadily but slowly increasing upward acceleration reaches $5g$. Friction may be neglected.

Ans. $k = 818$ N/m

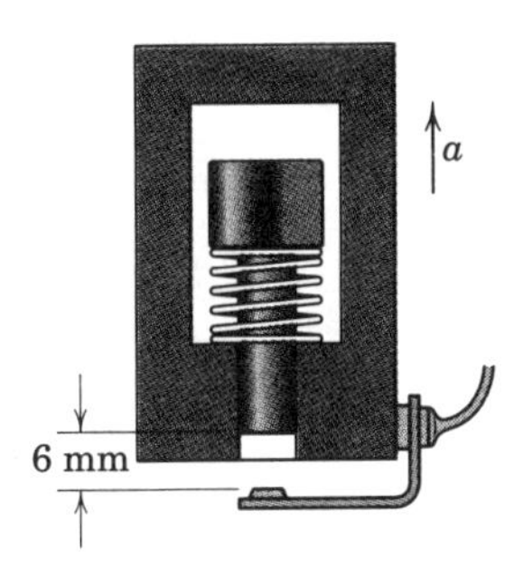

Problem 3/25

3/26 A cylinder of mass m rests in a supporting carriage as shown. If $\beta = 45°$ and $\theta = 30°$, calculate the maximum acceleration a which the carriage may be given up the incline so that the cylinder does not lose contact at B.

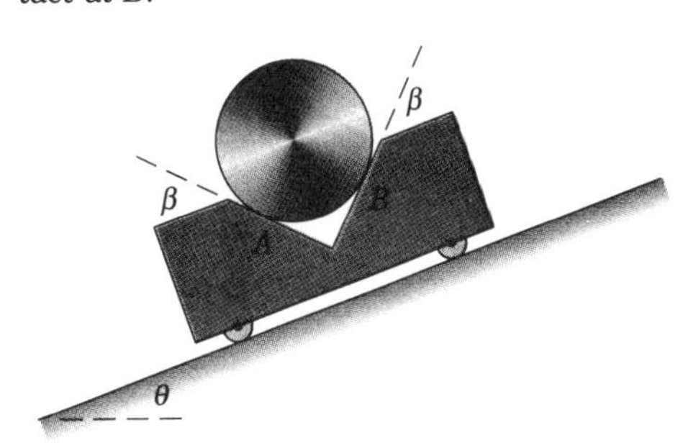

Problem 3/26

3/27 The system of Prob. 2/208 is repeated here with additional mass information specified. Neglect all friction and the mass of the pulleys and determine the accelerations of bodies A and B upon release from rest.

Ans. $a_A = 1.024\ \text{m/s}^2$ down the incline
$a_B = 0.682\ \text{m/s}^2$ up

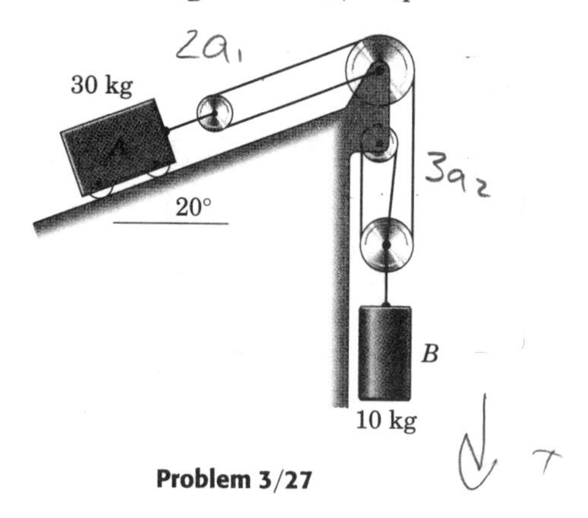

Problem 3/27

3/28 The system is released from rest with the cable taut. For the friction coefficients $\mu_s = 0.25$ and $\mu_k = 0.20$, calculate the acceleration of each body and the tension T in the cable. Neglect the small mass and friction of the pulleys.

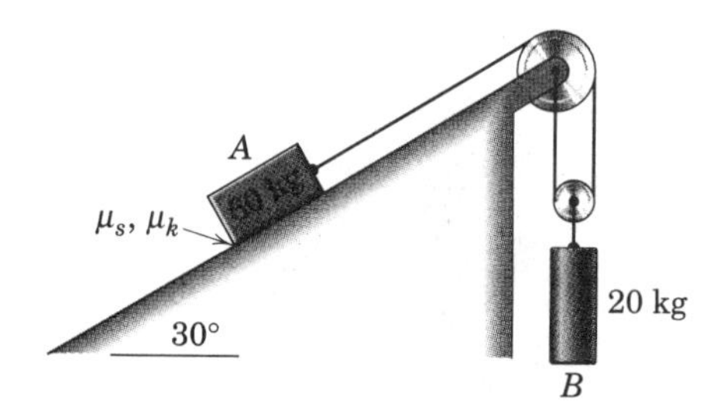

Problem 3/28

3/29 The acceleration of the 50-kg carriage A in its smooth vertical guides is controlled by the tension T exerted on the control cable which passes around the two circular pegs fixed to the carriage. Determine the value of T required to limit the downward acceleration of the carriage to $1.2\ \text{m/s}^2$ if the coefficient of friction between the cable and the pegs is 0.20. (Recall the relation between the tensions in a flexible cable which is slipping on a fixed peg: $T_2 = T_1e^{\mu\beta}$.)

Ans. $T = 171.3$ N

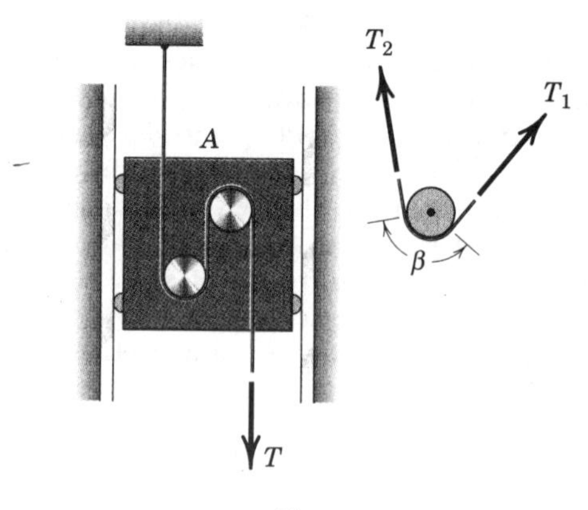

Problem 3/29

3/30 A player pitches a baseball horizontally toward a speed-sensing radar gun. The baseball has a mass of 146 g and has a circumference of 232 mm. If the speed at $x = 0$ is $v_0 = 150$ km/h, estimate the speed as a function of x. Assume that the horizontal aerodynamic drag on the baseball is given by $D = C_D(\frac{1}{2}\rho v^2)S$, where C_D is the drag coefficient, ρ is the air density, v is the speed, and S is the cross-sectional area of the baseball. Use a value of 0.3 for C_D. Neglect the vertical component of the motion but comment on the validity of this assumption. Evaluate your answer for $x = 18$ m, which is the approximate distance between a pitcher's hand and home plate.

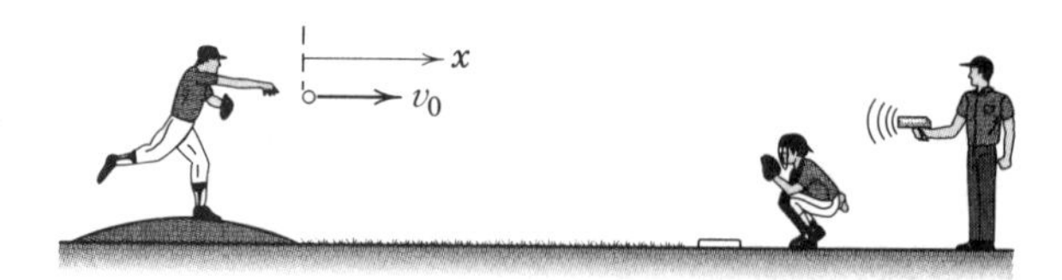

Problem 3/30

3/31 A jet airplane with a mass of 5 Mg has a touchdown speed of 300 km/h, at which instant the braking parachute is deployed and the power shut off. If the total drag on the aircraft varies with velocity as shown in the accompanying graph, calculate the distance x along the runway required to reduce the speed to 150 km/h. Approximate the variation of the drag by an equation of the form $D = kv^2$, where k is a constant.

Ans. $x = 201$ m

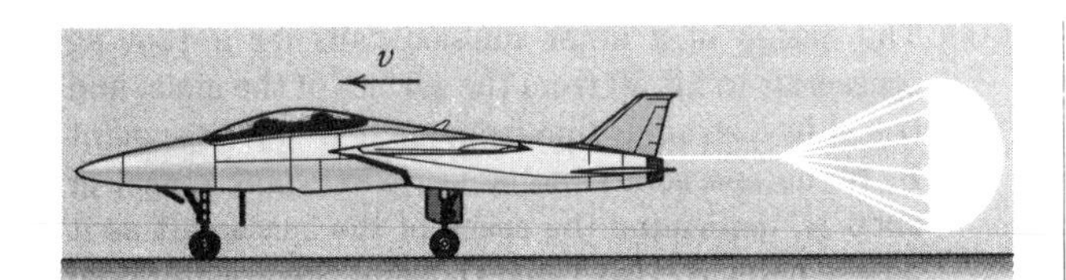

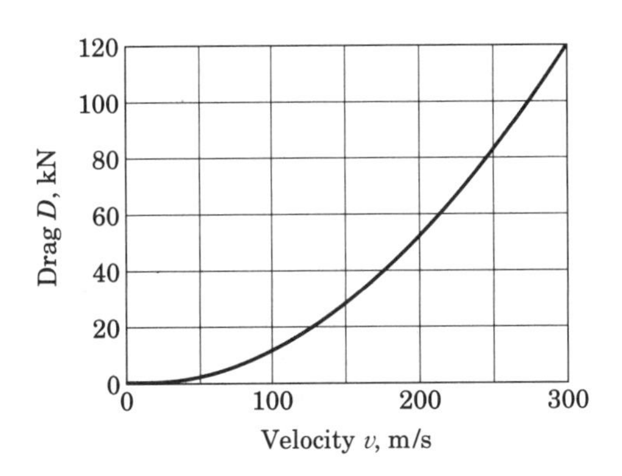

Problem 3/31

3/32 A heavy chain with a mass ρ per unit length is pulled by the constant force P along a horizontal surface consisting of a smooth section and a rough section. The chain is initially at rest on the rough surface with $x = 0$. If the coefficient of kinetic friction between the chain and the rough surface is μ_k, determine the velocity v of the chain when $x = L$. The force P is greater than $\mu_k \rho g L$ in order to initiate motion.

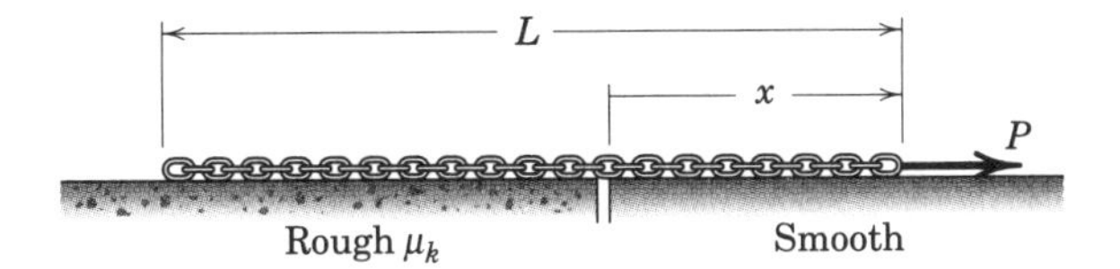

Problem 3/32

3/33 The 1.8-kg collar is released from rest against the light elastic spring, which has a stiffness of 1750 N/m and has been compressed a distance of 150 mm. Determine the acceleration a of the collar as a function of the vertical displacement x of the collar measured in meters from the point of release. Find the velocity v of the collar when $x = 0.15$ m. Friction is negligible.

Ans. $a_x = 136.0 - 972x$, $v = 4.35$ m/s

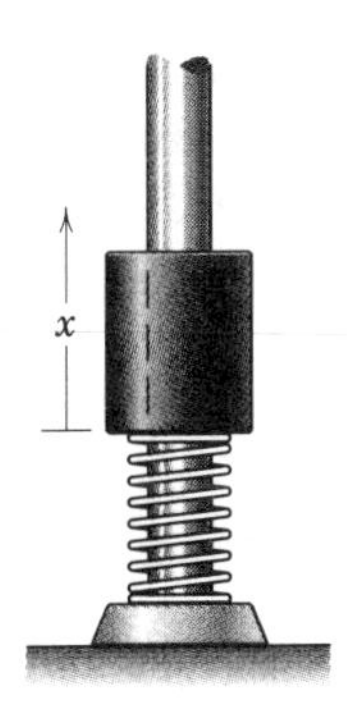

Problem 3/33

3/34 A force P is applied to the initially stationary cart. Determine the velocity and displacement at time $t = 5$ s for each of the force histories P_1 and P_2. Neglect friction.

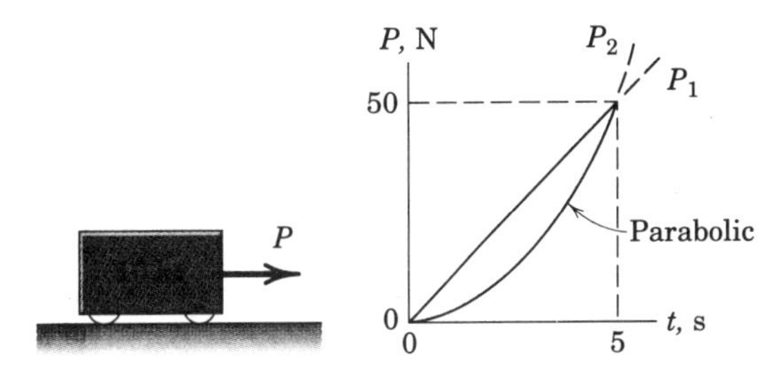

Problem 3/34

3/35 A bar of length l and negligible mass connects the cart of mass M and the particle of mass m. If the cart is subjected to a constant acceleration a to the right, what is the resulting steady-state angle θ which the freely pivoting bar makes with the vertical? Determine the net force P (not shown) which must be applied to the cart to cause the specified acceleration.

Ans. $\theta = \tan^{-1}\left(\frac{a}{g}\right)$, $P = (M + m)a$

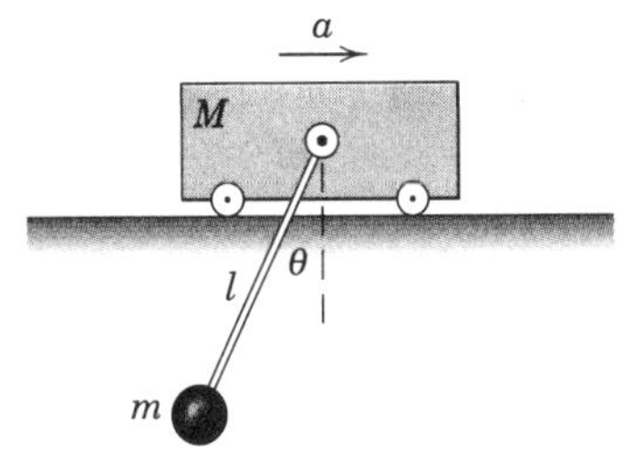

Problem 3/35

3/36 Determine the accelerations of bodies A and B and the tension in the cable due to the application of the 300-N force. Neglect all friction and the masses of the pulleys.

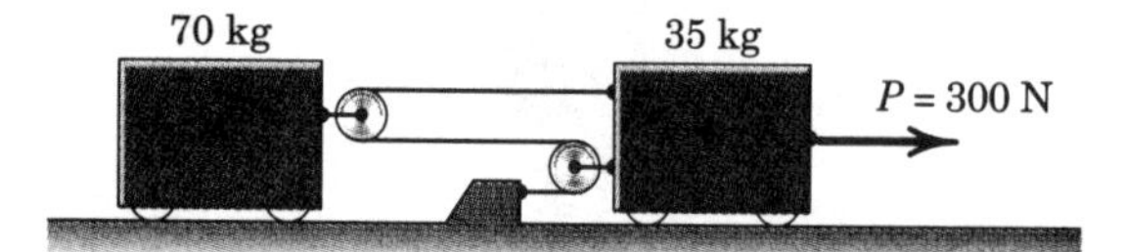

Problem 3/36

3/37 Two configurations for raising an elevator are shown. Elevator A with attached hoisting motor and drum has a total mass of 900 kg. Elevator B without motor and drum also has a mass of 900 kg. If the motor supplies a constant torque of 600 N·m to its 250-mm-diameter drum for 2 s in each case, select the configuration which results in the greater upward acceleration and determine the corresponding velocity v of the elevator 1.2 s after it starts from rest. The mass of the motorized drum is small, thus permitting it to be analyzed as though it were in equilibrium. Neglect the mass of cables and pulleys and all friction.

Ans. Case (a) has the higher acceleration
$v = 7.43$ m/s

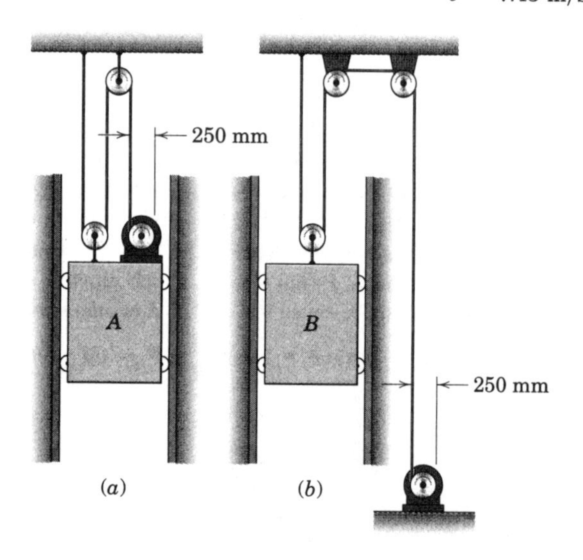

Problem 3/37

3/38 The design of a lunar mission calls for a 1200-kg spacecraft to lift off from the surface of the moon and travel in a straight line from point A and pass point B. If the spacecraft motor has a constant thrust of 2500 N, determine the speed of the spacecraft as it passes point B. Use Table D/2 and the gravitational law from Chapter 1 as needed.

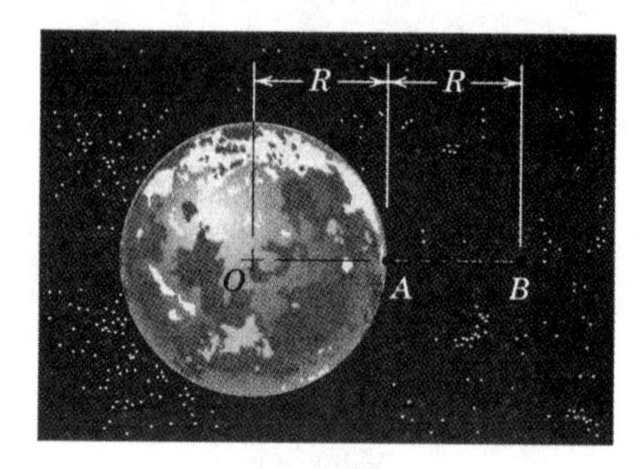

Problem 3/38

3/39 In a test of resistance to motion in an oil bath, a small steel ball of mass m is released from rest at the surface ($y = 0$). If the resistance to motion is given by $R = kv$ where k is a constant, derive an expression for the depth h required for the ball to reach a velocity v.

Ans. $h = \frac{m^2g}{k^2} \ln \left(\frac{1}{1 - kv/(mg)} \right) - \frac{mv}{k}$

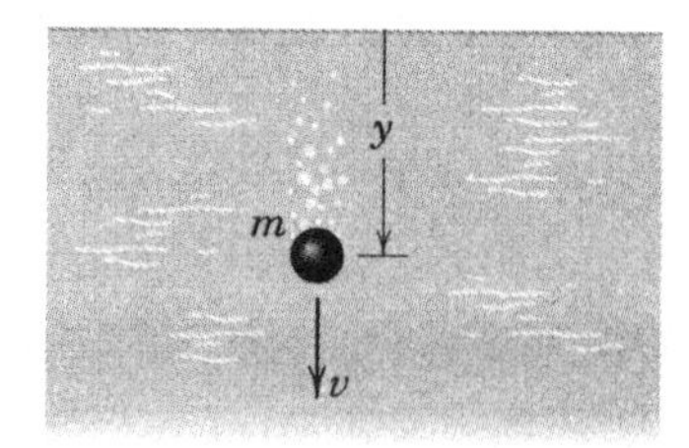

Problem 3/39

3/40 If the steel ball of Prob. 3/39 is released from rest at the surface of a liquid in which the resistance to motion is $R = cv^2$, where c is a constant and v is the downward velocity of the ball, determine the depth h required for the ball to reach a velocity v.

3/41 The inclined block A is given a constant rightward acceleration a. Determine the range of values of θ for which block B will not slip relative to block A, regardless of how large the acceleration a is. The coefficient of static friction between the blocks is μ_s.

Ans. $\tan^{-1}(1/\mu_s) \leq \theta \leq \frac{\pi}{2}$

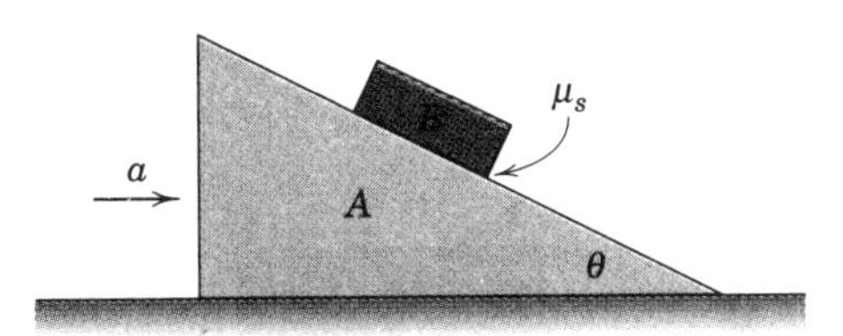

Problem 3/41

3/42 A shock absorber is a mechanical device which provides resistance to compression or extension given by $R = cv$, where c is a constant and v is the time rate of change of the length of the absorber. An absorber of constant $c = 3000$ N·s/m is shown being tested with a 100-kg cylinder suspended from it. The system is released with the cable taut at $y = 0$ and allowed to extend. Determine (a) the steady-state velocity v_s of the lower end of the absorber and (b) the time t and displacement y of the lower end when the cylinder has reached 90 percent of its steady-state speed. Neglect the mass of the piston and attached rod.

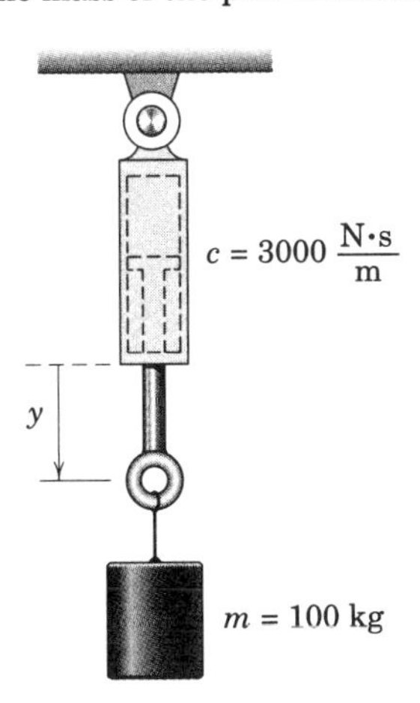

Problem 3/42

3/43 The sliders A and B are connected by a light rigid bar of length $l = 0.5$ m and move with negligible friction in the horizontal slots shown. For the position where $x_A = 0.4$ m, the velocity of A is $v_A = 0.9$ m/s to the right. Determine the acceleration of each slider and the force in the bar at this instant.

Ans. $a_A = 1.364$ m/s² right
$a_B = 9.32$ m/s² down
$T = 46.6$ N

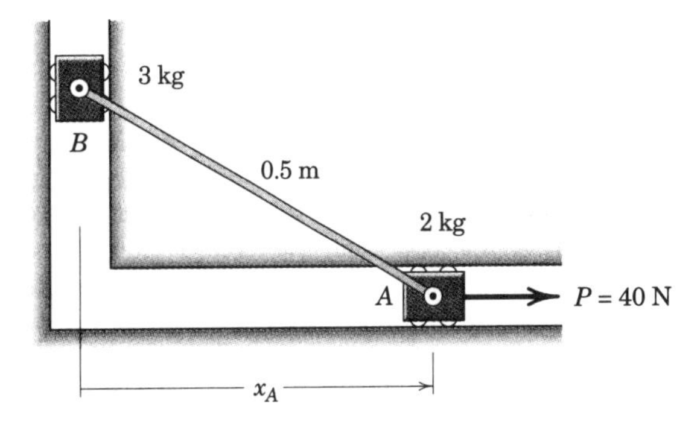

Problem 3/43

3/44 The sliders A and B are connected by a light rigid bar and move with negligible friction in the slots, both of which lie in a horizontal plane. For the position shown, the velocity of A is 0.4 m/s to the right. Determine the acceleration of each slider and the force in the bar at this instant.

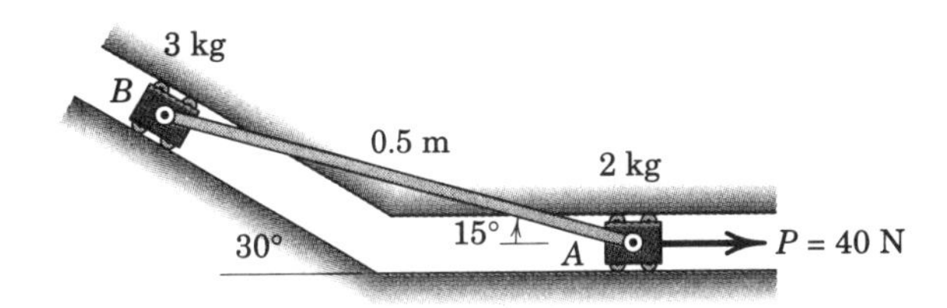

Problem 3/44

3/45 For what value(s) of the angle θ will the acceleration of the 35-kg block be 9 m/s² to the right?

Ans. $\theta = 11.88°, 41.3°$

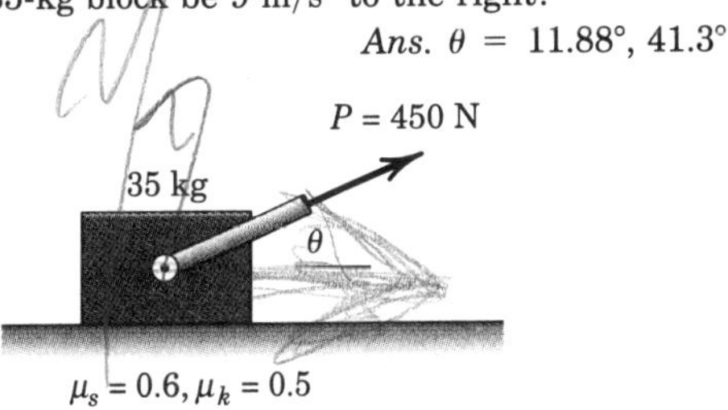

Problem 3/45

▶ **3/46** With the blocks initially at rest, the force P is increased slowly from zero to 260 N. Plot the accelerations of both masses as functions of P.

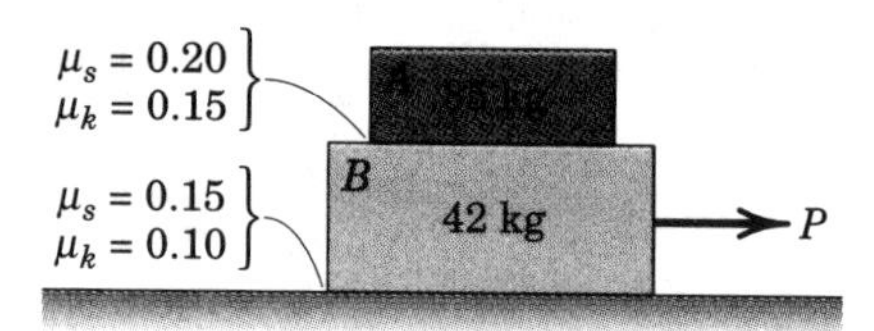

Problem 3/46

▶ **3/47** Two iron spheres, each of which is 100 mm in diameter, are released from rest with a center-to-center separation of 1 m. Assume an environment in space with no forces other than the force of mutual gravitational attraction and calculate the time t required for the spheres to contact each other and the absolute speed v of each sphere upon contact.

Ans. t = 13 h 33 min
$v = 4.76(10^{-5})$ m/s

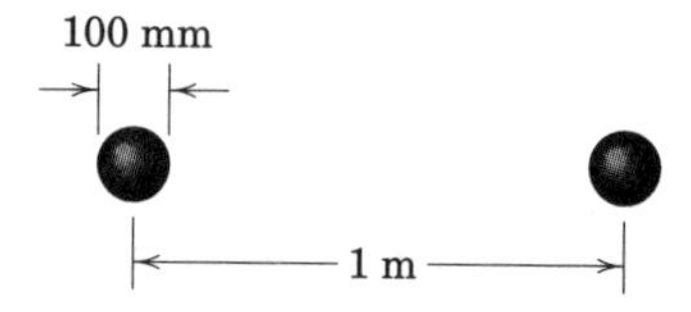

Problem 3/47

▶ **3/48** The system is released from rest in the position shown. Calculate the tension T in the cord and the acceleration a of the 30-kg block. The small pulley attached to the block has negligible mass and friction. (*Suggestion:* First establish the kinematic relationship between the accelerations of the two bodies.)

Ans. T = 138.0 N, a = 0.766 m/s^2

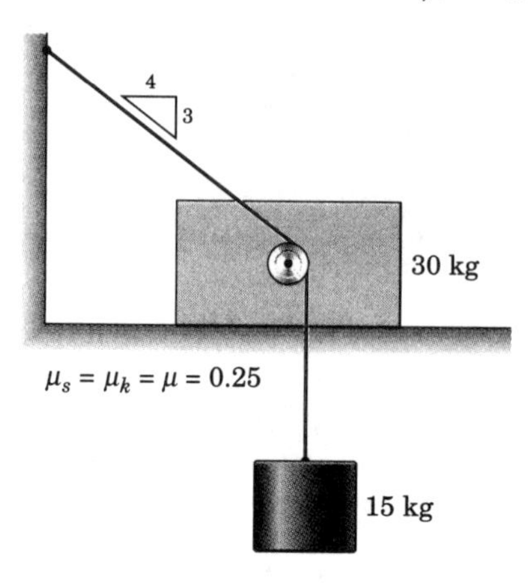

Problem 3/48

3/5 CURVILINEAR MOTION

We turn our attention now to the kinetics of particles which move along plane curvilinear paths. In applying Newton's second law, Eq. 3/3, we will make use of the three coordinate descriptions of acceleration in curvilinear motion which we developed in Arts. 2/4, 2/5, and 2/6.

The choice of an appropriate coordinate system depends on the conditions of the problem and is one of the basic decisions to be made in solving curvilinear-motion problems. We now rewrite Eq. 3/3 in three ways, the choice of which depends on which coordinate system is most appropriate.

Rectangular coordinates (Art. 2/4, Fig. 2/7)

$$\Sigma F_x = ma_x \qquad \Sigma F_y = ma_y \tag{3/6}$$

where $a_x = \ddot{x}$ and $a_y = \ddot{y}$

Normal and tangential coordinates (Art. 2/5, Fig. 2/10)

$$\Sigma F_n = ma_n \qquad \Sigma F_t = ma_t \tag{3/7}$$

where $a_n = \rho\dot{\beta}^2 = v^2/\rho = v\dot{\beta}$, $a_t = \dot{v}$, and $v = \rho\dot{\beta}$

Polar coordinates (Art. 2/6, Fig. 2/15)

$$\Sigma F_r = ma_r \qquad \Sigma F_\theta = ma_\theta \tag{3/8}$$

where $a_r = \ddot{r} - r\dot{\theta}^2$ and $a_\theta = r\ddot{\theta} + 2\dot{r}\dot{\theta}$

In applying these motion equations to a body treated as a particle, you should follow the general procedure established in the previous article on rectilinear motion. After you identify the motion and choose the coordinate system, draw the free-body diagram of the body. Then obtain the appropriate force summations from this diagram in the usual way. The free-body diagram should be complete to avoid incorrect force summations.

Once you assign reference axes, you must use the expressions for both the forces and the acceleration which are consistent with that assignment. In the first of Eqs. 3/7, for example, the positive sense of the n-axis is *toward* the center of curvature, and so the positive sense of our force summation ΣF_n must also be *toward* the center of curvature to agree with the positive sense of the acceleration $a_n = v^2/\rho$.

Sample Problem 3/6

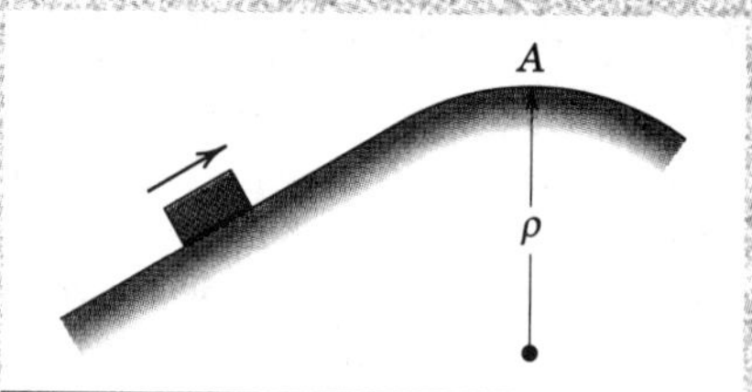

Determine the maximum speed v which the sliding block may have as it passes point A without losing contact with the surface.

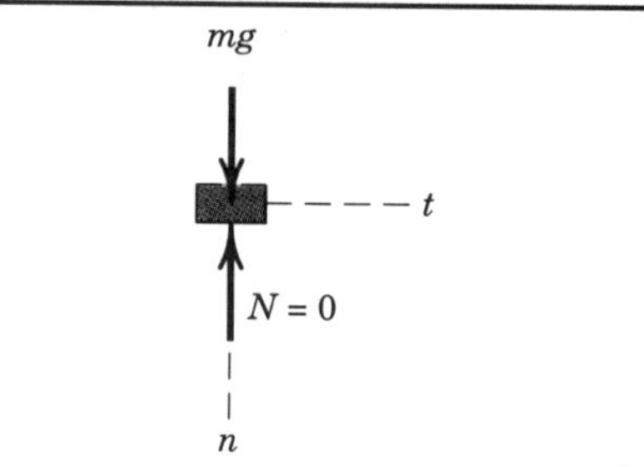

Solution. The condition for loss of contact is that the normal force N which the surface exerts on the block goes to zero. Summing forces in the normal direction gives

$$[\Sigma F_n = ma_n] \qquad mg = m\frac{v^2}{\rho} \qquad v = \sqrt{g\rho} \qquad \textit{Ans.}$$

If the speed at A were less than $\sqrt{g\rho}$, then an upward normal force exerted by the surface on the block would exist. In order for the block to have a speed at A which is greater than $\sqrt{g\rho}$, some type of constraint, such as a second curved surface above the block, would have to be introduced to provide additional downward force.

Sample Problem 3/7

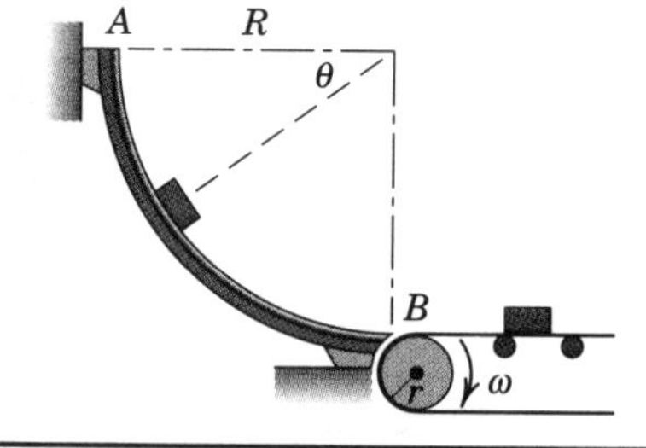

Small objects are released from rest at A and slide down the smooth circular surface of radius R to a conveyor B. Determine the expression for the normal contact force N between the guide and each object in terms of θ and specify the correct angular velocity ω of the conveyor pulley of radius r to prevent any sliding on the belt as the objects transfer to the conveyor.

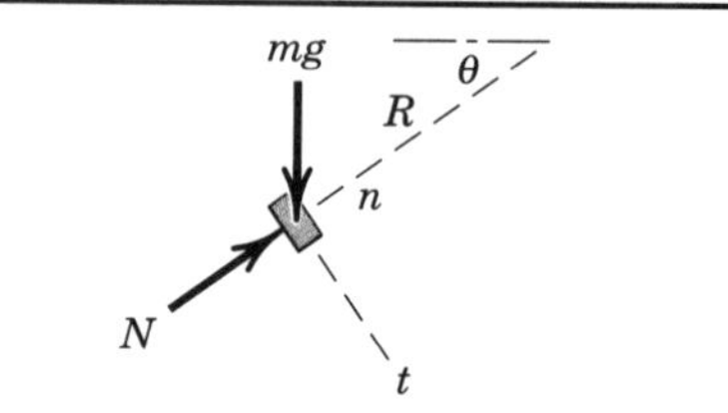

Solution. The free-body diagram of the object is shown together with the coordinate directions n and t. The normal force N depends on the n-component of the acceleration which, in turn, depends on the velocity. The velocity will be cumulative according to the tangential acceleration a_t. Hence, we will find a_t first for any general position.

$$[\Sigma F_t = ma_t] \qquad mg \cos\theta = ma_t \qquad a_t = g\cos\theta$$

① Now we can find the velocity by integrating

$$[v\,dv = a_t\,ds] \qquad \int_0^v v\,dv = \int_0^\theta g\cos\theta\,d(R\theta) \qquad v^2 = 2gR\sin\theta$$

We obtain the normal force by summing forces in the positive n-direction, which is the direction of the n-component of acceleration.

$$[\Sigma F_n = ma_n] \quad N - mg\sin\theta = m\frac{v^2}{R} \qquad N = 3mg\sin\theta \qquad \textit{Ans.}$$

The conveyor pulley must turn at the rate $v = r\omega$ for $\theta = \pi/2$, so that

$$\omega = \sqrt{2gR}/r \qquad \textit{Ans.}$$

Helpful Hint

① It is essential here that we recognize the need to express the tangential acceleration as a function of position so that v may be found by integrating the kinematical relation $v\,dv = a_t\,ds$, in which all quantities are measured along the path.

Sample Problem 3/8

A 1500-kg car enters a section of curved road in the horizontal plane and slows down at a uniform rate from a speed of 100 km/h at A to a speed of 50 km/h as it passes C. The radius of curvature ρ of the road at A is 400 m and at C is 80 m. Determine the total horizontal force exerted by the road on the tires at positions A, B, and C. Point B is the inflection point where the curvature changes direction.

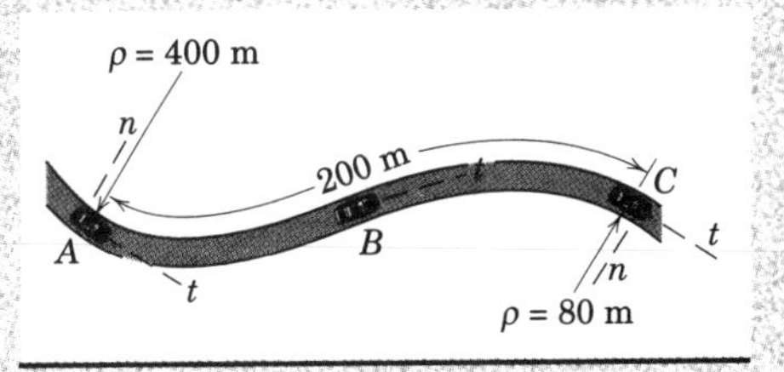

Solution. The car will be treated as a particle so that the effect of all forces exerted by the road on the tires will be treated as a single force. Since the motion is described along the direction of the road, normal and tangential coordinates will be used to specify the acceleration of the car. We will then determine the forces from the accelerations.

The constant tangential acceleration is in the negative t-direction and its magnitude is given by

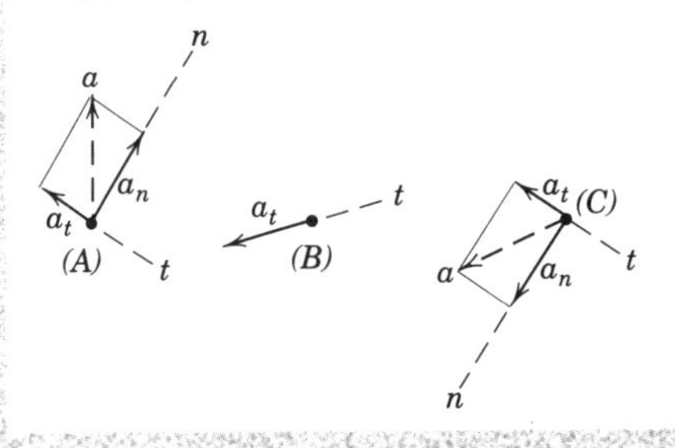

① $[v_C{}^2 = v_A{}^2 + 2a_t\,\Delta s]$ $\quad a_t = \left|\dfrac{(50/3.6)^2 - (100/3.6)^2}{2(200)}\right| = 1.447 \text{ m/s}^2$

The normal components of acceleration at A, B, and C are

② $[a_n = v^2/\rho]$ $\quad$ At A, $\quad a_n = \dfrac{(100/3.6)^2}{400} = 1.929 \text{ m/s}^2$

At B, $\quad a_n = 0$

At C, $\quad a_n = \dfrac{(50/3.6)^2}{80} = 2.41 \text{ m/s}^2$

Application of Newton's second law in both the n- and t-directions to the free-body diagrams of the car gives

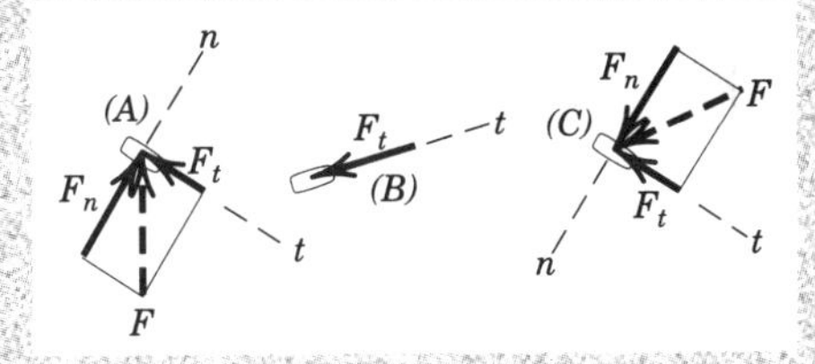

$[\Sigma F_t = ma_t]$ $\quad F_t = 1500(1.447) = 2170 \text{ N}$

③ $[\Sigma F_n = ma_n]$ $\quad$ At A, $\quad F_n = 1500(1.929) = 2890 \text{ N}$

At B, $\quad F_n = 0$

At C, $\quad F_n = 1500(2.41) = 3620 \text{ N}$

Thus, the total horizontal force acting on the tires becomes

At A, $\quad F = \sqrt{F_n{}^2 + F_t{}^2} = \sqrt{(2890)^2 + (2170)^2} = 3620 \text{ N}$ *Ans.*

At B, $\quad F = F_t = 2170 \text{ N}$ *Ans.*

④ At C, $\quad F = \sqrt{F_n{}^2 + F_t{}^2} = \sqrt{(3620)^2 + (2170)^2} = 4220 \text{ N}$ *Ans.*

Helpful Hints

① Recognize the numerical value of the conversion factor from km/h to m/s as 1000/3600 or 1/3.6.

② Note that a_n is always directed toward the center of curvature.

③ Note that the direction of F_n must agree with that of a_n.

④ The angle made by **a** and **F** with the direction of the path can be computed if desired.

Sample Problem 3/9

Compute the magnitude v of the velocity required for the spacecraft S to maintain a circular orbit of altitude 320 km above the surface of the earth.

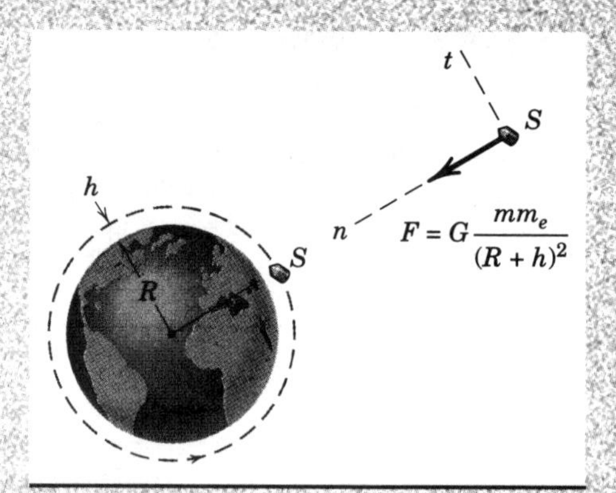

Solution. The only external force acting on the spacecraft is the force of gravitational attraction to the earth (i.e., its weight), as shown in the free-body diagram. Summing forces in the normal direction yields ①

$$[\Sigma F_n = ma_n] \quad G\frac{mm_e}{(R+h)^2} = m\frac{v^2}{(R+h)}, \quad v = \sqrt{\frac{Gm_e}{(R+h)}} = R\sqrt{\frac{g}{(R+h)}}$$

where the substitution $gR^2 = Gm_e$ has been made. Substitution of numbers gives

$$v = (6371)(1000)\sqrt{\frac{9.825}{(6371 + 320)(1000)}} = 7720 \text{ m/s} \qquad \textit{Ans.}$$

Helpful Hint

① Note that, for observations made within an inertial frame of reference, there is no such quantity as "centrifugal force" acting in the minus n-direction. Note also that neither the spacecraft nor its occupants are "weightless," because the weight in each case is given by Newton's law of gravitation. For this altitude, the weights are only about 10 percent less than the earth-surface values. Finally, the term "zero-g" is also misleading. It is only when we make our observations with respect to a coordinate system which has an acceleration equal to the gravitational acceleration (such as in an orbiting spacecraft) that we appear to be in a "zero-g" environment. The quantity which does go to zero aboard orbiting spacecraft is the familiar normal force associated with, for example, an object in contact with a horizontal surface within the spacecraft.

Sample Problem 3/10

Tube A rotates about the vertical O-axis with a constant angular rate $\dot{\theta} = \omega$ and contains a small cylindrical plug B of mass m whose radial position is controlled by the cord which passes freely through the tube and shaft and is wound around the drum of radius b. Determine the tension T in the cord and the horizontal component F_θ of force exerted by the tube on the plug if the constant angular rate of rotation of the drum is ω_0 first in the direction for case (a) and second in the direction for case (b). Neglect friction.

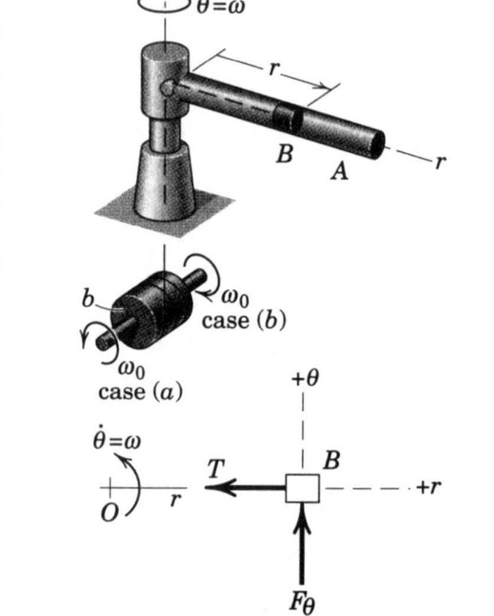

Solution. With r a variable, we use the polar-coordinate form of the equations of motion, Eqs. 3/8. The free-body diagram of B is shown in the horizontal plane and discloses only T and F_θ. The equations of motion are

$$[\Sigma F_r = ma_r] \qquad -T = m(\ddot{r} - r\dot{\theta}^2)$$

$$[\Sigma F_\theta = ma_\theta] \qquad F_\theta = m(r\ddot{\theta} + 2\dot{r}\dot{\theta})$$

Case (a). With $\dot{r} = +b\omega_0$, $\ddot{r} = 0$, and $\ddot{\theta} = 0$, the forces become

$$T = mr\omega^2 \qquad F_\theta = 2mb\omega_0\omega \qquad \textit{Ans.}$$

① ***Case (b).*** With $\dot{r} = -b\omega_0$, $\ddot{r} = 0$, and $\ddot{\theta} = 0$, the forces become

$$T = mr\omega^2 \qquad F_\theta = -2mb\omega_0\omega \qquad \textit{Ans.}$$

Helpful Hint

① The minus sign shows that F_θ is in the direction opposite to that shown on the free-body diagram.

PROBLEMS

Introductory Problems

3/49 The small 0.6-kg block slides with a small amount of friction on the circular path of radius 3 m in the vertical plane. If the speed of the block is 5 m/s as it passes point A and 4 m/s as it passes point B, determine the normal force exerted on the block by the surface at each of these two locations.

Ans. $N_A = 10.89$ N, $N_B = 8.30$ N

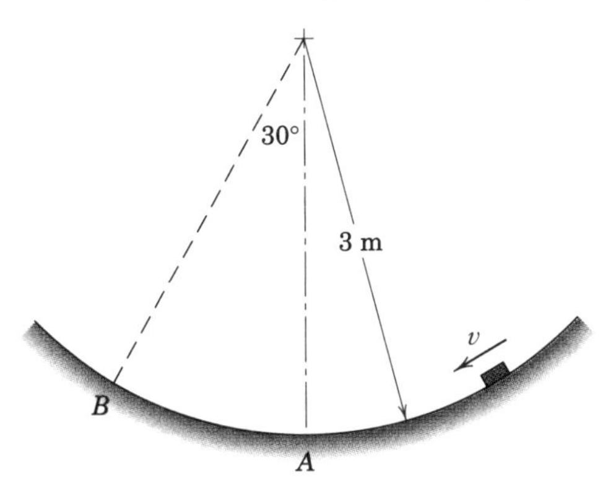

Problem 3/49

3/50 If the 2-kg block passes over the top B of the circular portion of the path with a speed of 3.5 m/s, calculate the magnitude N_B of the normal force exerted by the path on the block. Determine the maximum speed v which the block can have at A without losing contact with the path.

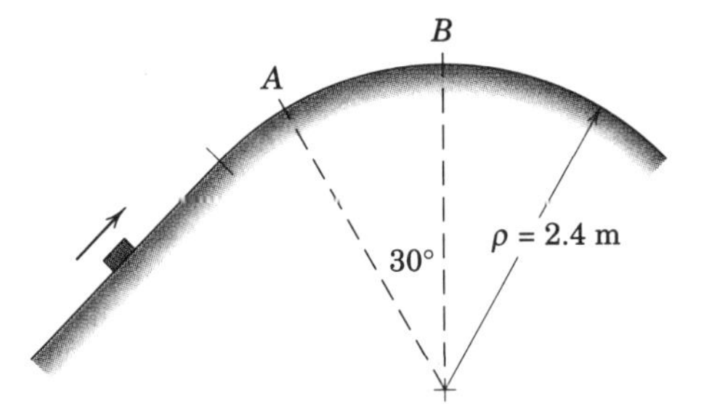

Problem 3/50

3/51 If the 80-kg ski-jumper attains a speed of 25 m/s as he approaches the takeoff position, calculate the magnitude N of the normal force exerted by the snow on his skis just before he reaches A.

Ans. $N = 1791$ N

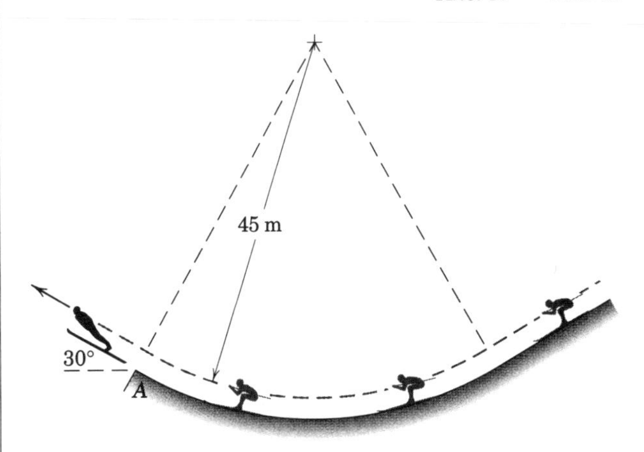

Problem 3/51

3/52 A 0.8-kg slider is propelled upward at A along the fixed curved bar which lies in a vertical plane. If the slider is observed to have a speed of 4 m/s as it passes position B, determine (*a*) the magnitude N of the force exerted by the fixed rod on the slider and (*b*) the rate at which the speed of the slider is decreasing. Assume that friction is negligible.

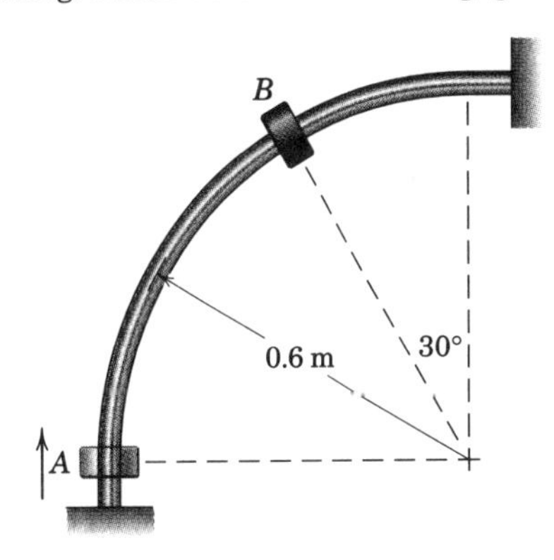

Problem 3/52

3/53 The 120-g slider has a speed $v = 1.4$ m/s as it passes point A of the smooth guide, which lies in a horizontal plane. Determine the magnitude R of the force which the guide exerts on the slider (a) just before it passes point A of the guide and (b) as it passes point B.

Ans. (a) $R = 1.177$ N, (b) $R = 1.664$ N

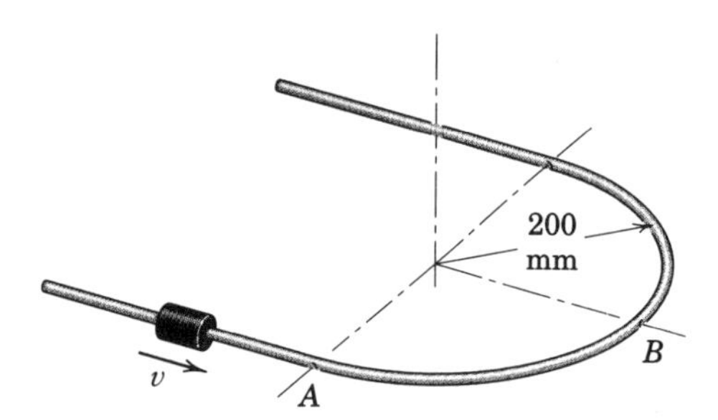

Problem 3/53

3/54 The hollow tube is pivoted about a horizontal axis through point O and is made to rotate in the vertical plane with a constant counterclockwise angular velocity $\dot{\theta} = 3$ rad/s. If a 0.1-kg particle is sliding in the tube toward O with a velocity of 1.2 m/s relative to the tube when the position $\theta = 30°$ is passed, calculate the magnitude N of the normal force exerted by the wall of the tube on the particle at this instant.

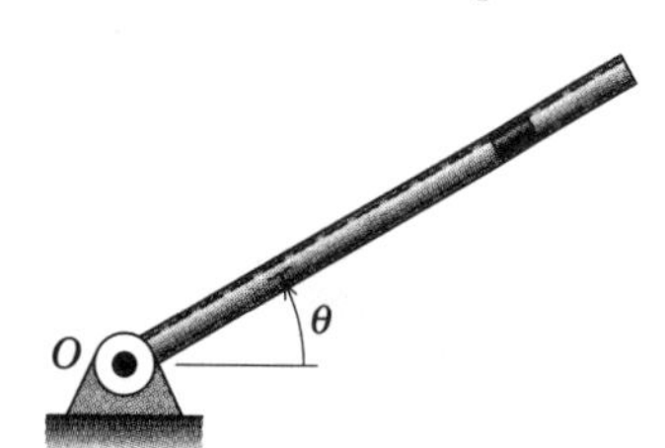

Problem 3/54

3/55 The member OA rotates about a horizontal axis through O with a constant counterclockwise velocity $\omega = 3$ rad/s. As it passes the position $\theta = 0$, a small block of mass m is placed on it at a radial distance $r = 450$ mm. If the block is observed to slip at $\theta = 50°$, determine the coefficient of static friction μ_s between the block and the member.

Ans. $\mu_s = 0.549$

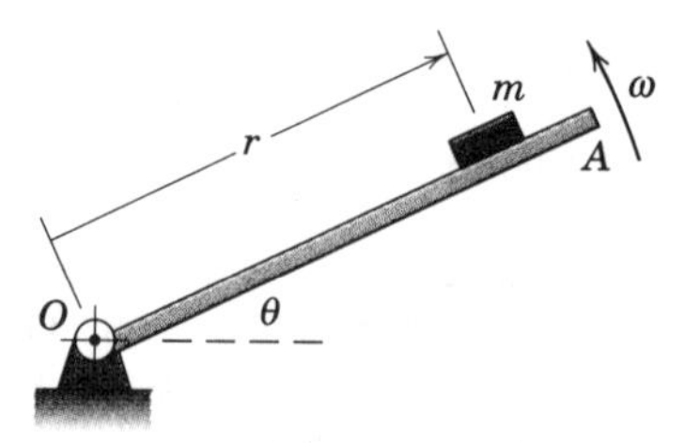

Problem 3/55

3/56 The standard test to determine the maximum lateral acceleration of a car is to drive it around a 60-m-diameter circle painted on a level asphalt surface. The driver slowly increases the vehicle speed until he is no longer able to keep both wheel pairs straddling the line. If this maximum speed is 55 km/h for a 1400-kg car, determine its lateral acceleration capability a_n in g's and compute the magnitude F of the total friction force exerted by the pavement on the car tires.

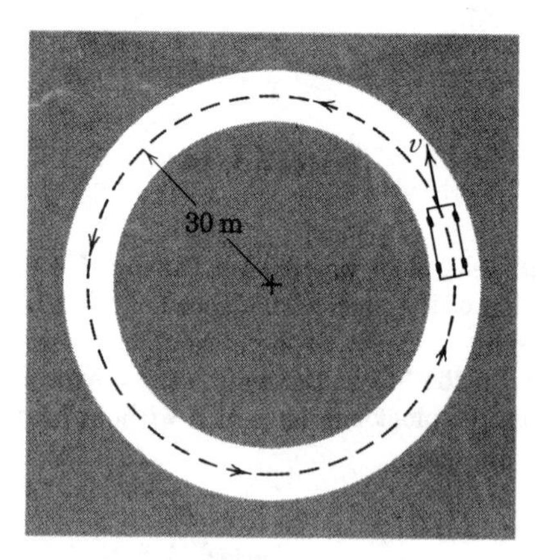

Problem 3/56

3/57 The car of Prob. 3/56 is traveling at 40 km/h when the driver applies the brakes, and the car continues to move along the circular path. What is the maximum deceleration possible if the tires are limited to a total horizontal friction force of 10.6 kN?

Ans. $a_t = -6.36$ m/s^2

3/58 The slotted arm rotates about its center in a horizontal plane at the constant angular rate $\dot{\theta} = 10$ rad/s and carries a 1.5-kg spring-mounted slider which oscillates freely in the slot. If the slider has a speed of 600 mm/s relative to the slot as it crosses the center, calculate the horizontal side thrust P exerted by the slotted arm on the slider at this instant. Determine which side, A or B, of the slot is in contact with the slider.

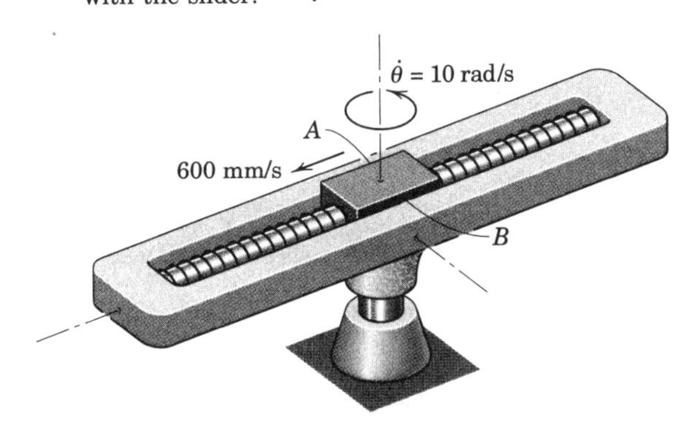

Problem 3/58

3/59 In the design of a space station to operate outside the earth's gravitational field, it is desired to give the structure a rotational speed N which will simulate the effect of the earth's gravity for members of the crew. If the centers of the crew's quarters are to be located 12 m from the axis of rotation, calculate the necessary rotational speed N of the space station in revolutions per minute.

Ans. $N = 8.63$ rev/min

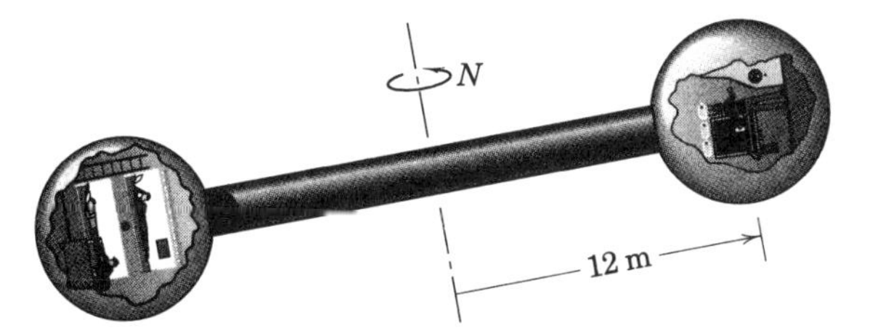

Problem 3/59

3/60 Calculate the necessary rotational speed N for the aerial ride in an amusement park in order that the arms of the gondolas will assume an angle $\theta = 60°$ with the vertical. Neglect the mass of the arms to which the gondolas are attached and treat each gondola as a particle.

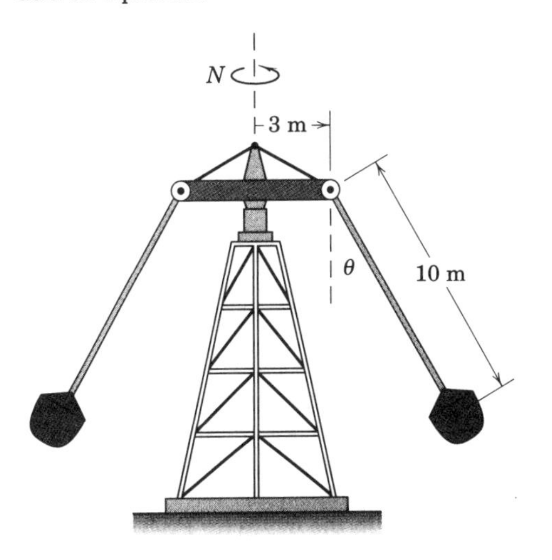

Problem 3/60

3/61 The barrel of a rifle is rotating in a horizontal plane about the vertical z-axis at the constant angular rate $\dot{\theta} = 0.5$ rad/s when a 60-g bullet is fired. If the velocity of the bullet relative to the barrel is 600 m/s just before it reaches the muzzle A, determine the resultant horizontal side thrust P exerted by the barrel on the bullet just before it emerges from A. On which side of the barrel does P act?

Ans. $P = 36$ N, right-hand side

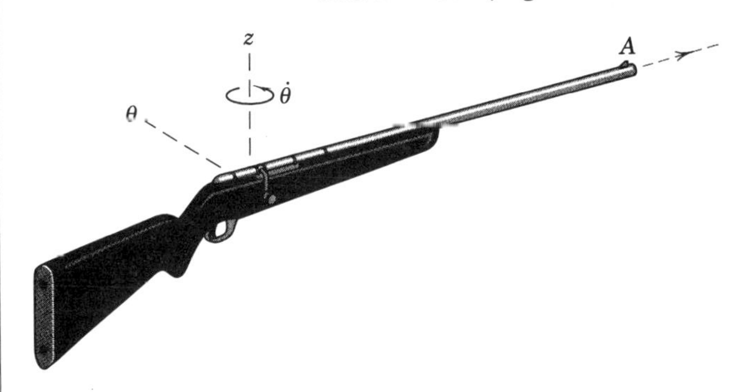

Problem 3/61

3/62 In order to simulate a condition of apparent "weightlessness" experienced by astronauts in an orbiting spacecraft, a jet transport can change its direction at the top of its flight path by dropping its flight-path direction at a prescribed rate $\dot{\theta}$ for a short interval of time. Specify $\dot{\theta}$ if the aircraft has a speed $v = 600$ km/h.

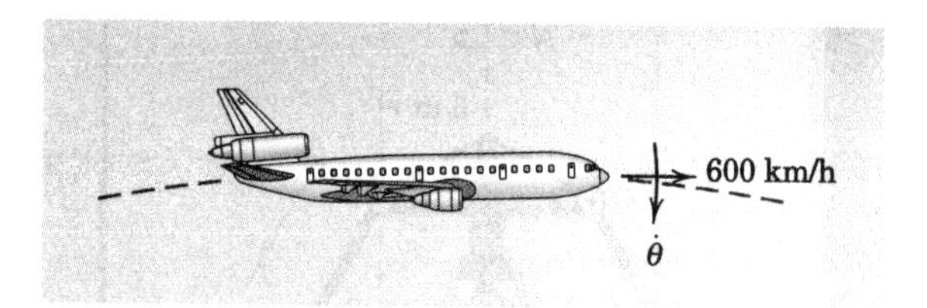

Problem 3/62

3/63 The weight of an object as measured by a spring scale at rest on the ground is 20 N. The object is weighed by the same spring scale placed on the cabin floor of an aircraft which is at the top of a vertical circular arc as shown in the figure for Prob. 3/62. If the airplane is flying with a speed $v = 800$ km/h at an altitude of 6000 m and is dropping its forward line of sight at the rate of 1 degree per second, what reading W' is registered by the scale?

Ans. $W' = 12.09$ N

3/64 The small ball of mass m and its supporting wire become a simple pendulum when the horizontal cord is severed. Determine the ratio k of the tension T in the supporting wire immediately after the cord is cut to that in the wire before the cord is cut.

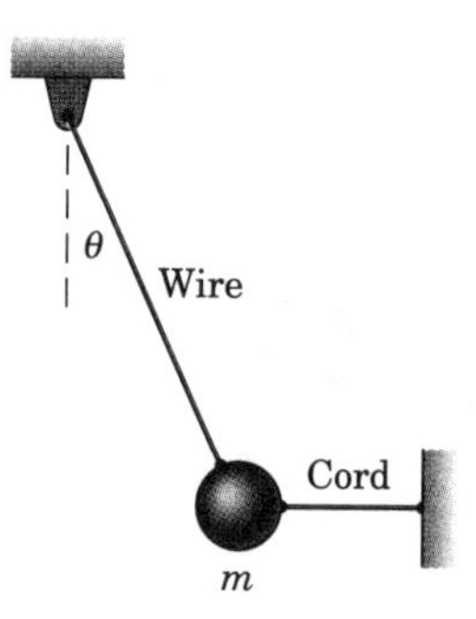

Problem 3/64

3/65 A 1500-kg car enters an S-curve at A with a speed of 96 km/h with brakes applied to reduce the speed to 72 km/h at a uniform rate in a distance of 90 m measured along the curve from A to B. The radius of curvature of the path of the car at B is 180 m. Calculate the total friction force exerted by the road on the tires at B. The road at B lies in a horizontal plane.

Ans. $F = 4220$ N

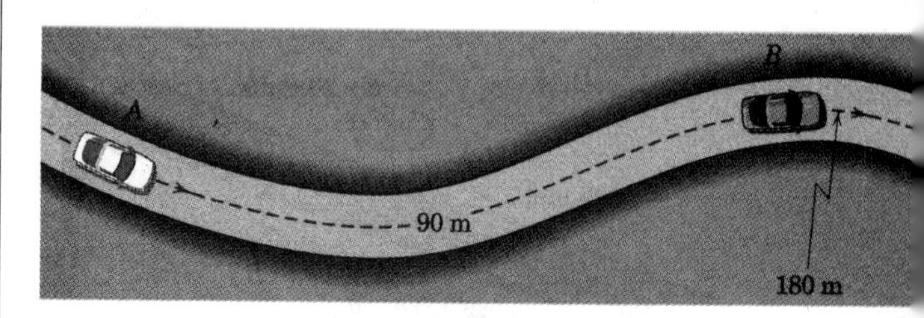

Problem 3/65

3/66 A pilot flies an airplane at a constant speed of 600 km/h in the vertical circle of radius 1000 m. Calculate the force exerted by the seat on the 90-kg pilot at point A and at point B.

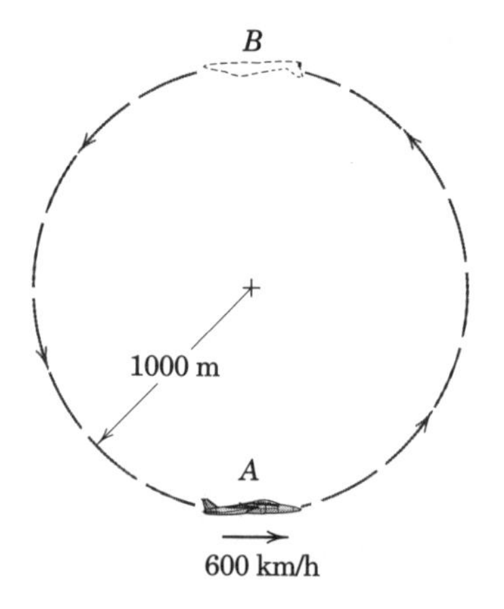

Problem 3/66

Representative Problems

3/67 A 2-kg sphere S is being moved in a vertical plane by a robotic arm. When the angle θ is 30°, the angular velocity of the arm about a horizontal axis through O is 50 deg/s clockwise and its angular acceleration is 200 deg/s^2 counterclockwise. In addition, the hydraulic element is being shortened at the constant rate of 500 mm/s. Determine the necessary minimum gripping force P if the coefficient of static friction between the sphere and the gripping surfaces is 0.5. Compare P with the minimum gripping force P_s required to hold the sphere in static equilibrium in the 30° position.

Ans. $P = 27.0$ N, $P_s = 19.62$ N

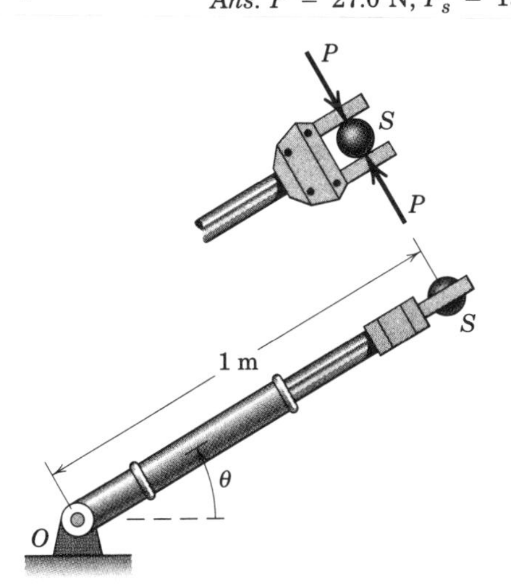

Problem 3/67

3/68 Determine the altitude h (in kilometers) above the surface of the earth at which a satellite in a circular orbit has the same period, 23.9344 h, as the earth's absolute rotation. If such an orbit lies in the equatorial plane of the earth, it is said to be geosynchronous, because the satellite does not appear to move relative to an earth-fixed observer.

3/69 A flatbed truck going 100 km/h rounds a horizontal curve of 300-m radius inwardly banked at 10°. The coefficient of static friction between the truck bed and the 200-kg crate it carries is 0.70. Calculate the friction force F acting on the crate.

Ans. $F = 165.9$ N

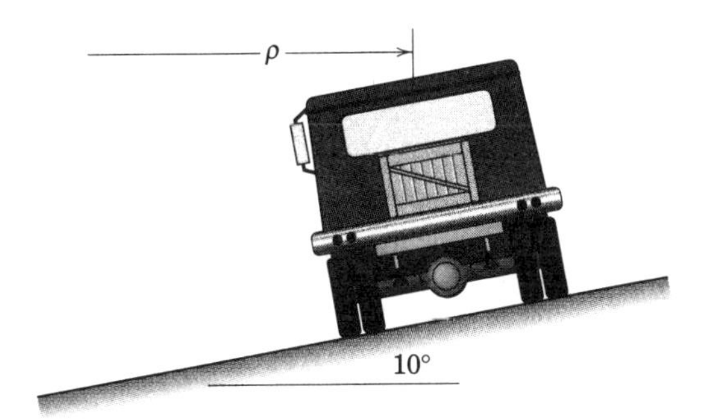

Problem 3/69

3/70 A small bead of mass m is carried by a circular hoop of radius r which rotates about a fixed vertical axis. Show how one might determine the angular speed ω of the hoop by observing the angle θ which locates the bead. Neglect friction in your analysis, but assume that a small amount of friction is present to damp out any motion of the bead relative to the hoop once a constant angular speed has been established. Note any restrictions on your solution.

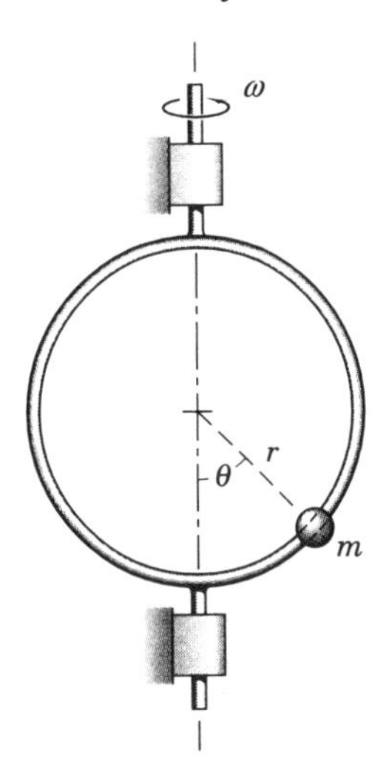

Problem 3/70

3/71 A small object A is held against the vertical side of the rotating cylindrical container of radius r by centrifugal action. If the coefficient of static friction between the object and the container is μ_s, determine the expression for the minimum rotational rate $\dot{\theta} = \omega$ of the container which will keep the object from slipping down the vertical side.

Ans. $\omega = \sqrt{\dfrac{g}{\mu_s r}}$

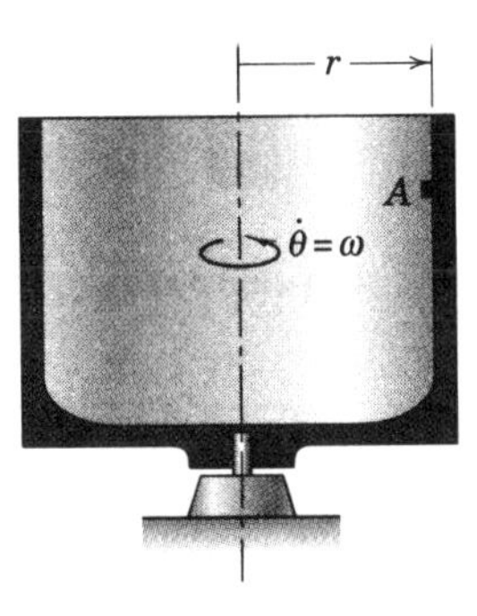

Problem 3/71

3/72 The small object is placed on the inner surface of the conical dish at the radius shown. If the coefficient of static friction between the object and the conical surface is 0.30, for what range of angular velocities ω about the vertical axis will the block remain on the dish without slipping? Assume that speed changes are made slowly so that any angular acceleration may be neglected.

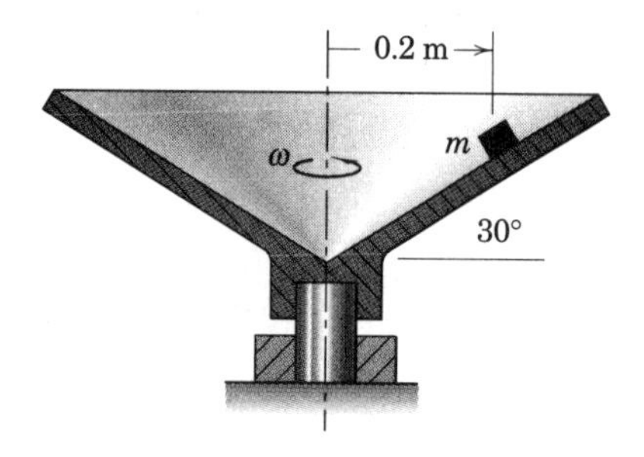

Problem 3/72

3/73 The small object of mass m is placed on the rotating conical surface at the radius shown. If the coefficient of static friction between the object and the rotating surface is 0.8, calculate the maximum angular velocity ω of the cone about the vertical axis for which the object will not slip. Assume very gradual angular-velocity changes.

Ans. $\omega = 2.73$ rad/s

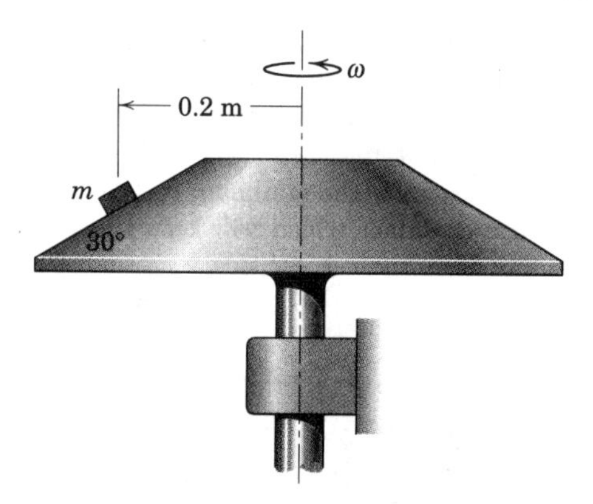

Problem 3/73

3/74 The 2-kg slider fits loosely in the smooth slot of the disk, which rotates about a vertical axis through point O. The slider is free to move slightly along the slot before one of the wires becomes taut. If the disk starts from rest at time $t = 0$ and has a constant clockwise angular acceleration of 0.5 rad/s^2, plot the tensions in wires 1 and 2 and the magnitude N of the force normal to the slot as functions of time t for the interval $0 \le t \le 5$ s.

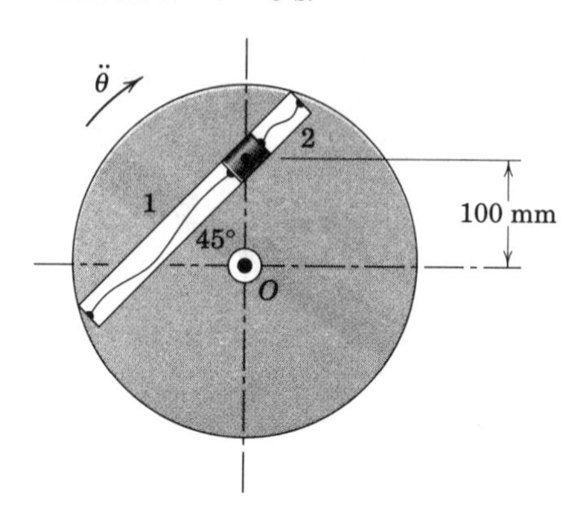

Problem 3/74

3/75 Beginning from rest when $\theta = 20°$, a 35-kg child slides with negligible friction down the sliding board which is in the shape of a 2.5-m circular arc. Determine the tangential acceleration and speed of the child, and the normal force exerted on her (*a*) when $\theta = 30°$ and (*b*) when $\theta = 90°$.

Ans. (*a*) $a_t = 8.50$ m/s^2, $v = 2.78$ m/s
$N = 280$ N
(*b*) $a_t = 0$, $v = 5.68$ m/s
$N = 795$ N

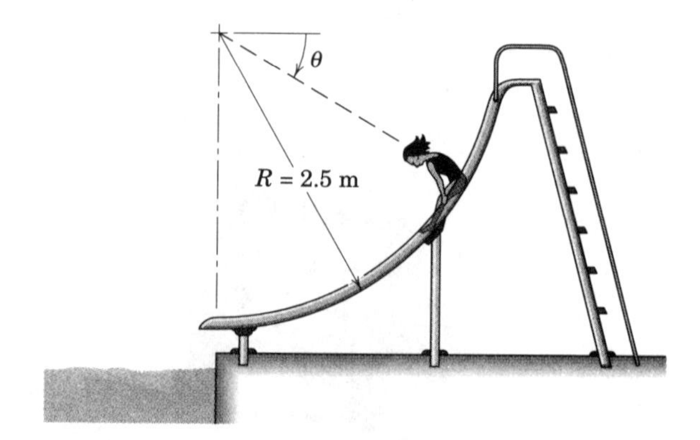

Problem 3/75

3/76 Determine the speed v at which the racecar will have no tendency to slip sideways on the banked track, i.e., the speed at which there is no reliance on friction. In addition, determine the minimum and maximum speeds, using the coefficient of static friction $\mu_s = 0.90$. State any assumptions.

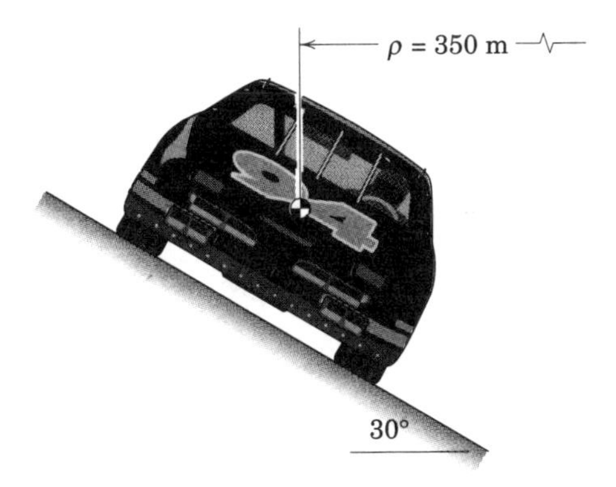

Problem 3/76

3/77 The rocket moves in a vertical plane and is being propelled by a thrust T of 32 kN. It is also subjected to an atmospheric resistance R of 9.6 kN. If the rocket has a velocity of 3 km/s and if the gravitational acceleration is 6 m/s^2 at the altitude of the rocket, calculate the radius of curvature ρ of its path for the position described and the time-rate-of-change of the magnitude v of the velocity of the rocket. The mass of the rocket at the instant considered is 2000 kg.

Ans. $\rho = 3000$ km, $\dot{v} = 6.00\ \text{m/s}^2$

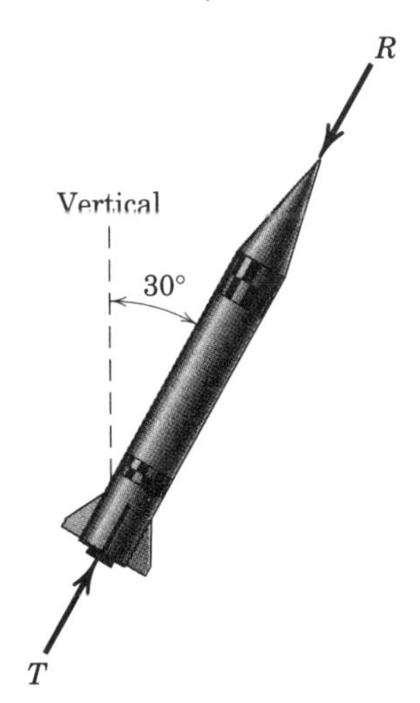

Problem 3/77

3/78 A stretch of highway includes a succession of evenly spaced dips and humps, the contour of which may be represented by the relation $y = b \sin (2\pi x/L)$. What is the maximum speed at which the car A can go over a hump and still maintain contact with the road? If the car maintains this critical speed, what is the total reaction N under its wheels at the bottom of a dip? The mass of the car is m.

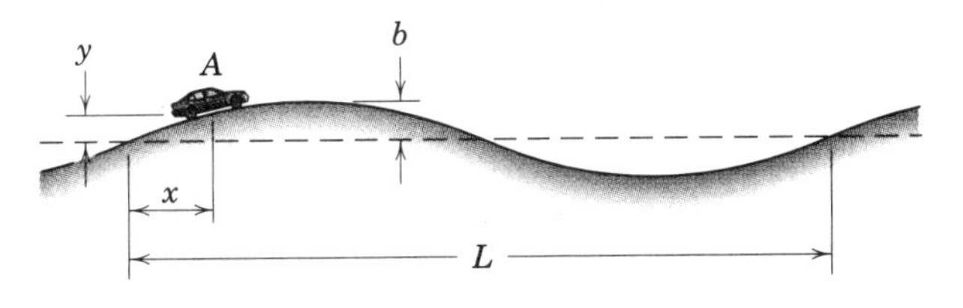

Problem 3/78

3/79 The small ball of mass m is attached to a light cord of length L and moves as a conical pendulum in a horizontal circle with a tangential velocity v. Locate the plane of motion by determining h, and find the tension T in the cord. (*Note:* Use the relation $v = r\dot{\theta} = r\omega$, where ω is the angular velocity about the vertical axis.)

Ans. $h = g/\omega^2$, $T = mL\omega^2$

Problem 3/79

3/80 Small steel balls, each with a mass of 65 g, enter the semicircular trough in the vertical plane with a horizontal velocity of 4.1 m/s at A. Find the force R exerted by the trough on each ball in terms of θ and the velocity v_B of the balls at B. Friction is negligible.

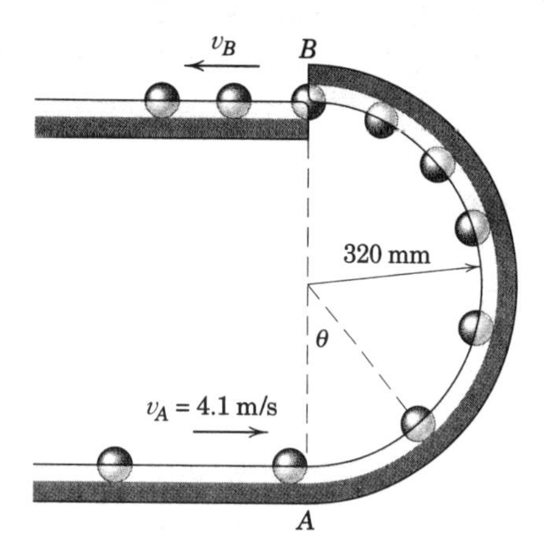

Problem 3/80

3/81 The slotted arm revolves in the horizontal plane about the fixed vertical axis through point O. The 2-kg slider C is drawn toward O at the constant rate of 50 mm/s by pulling the cord S. At the instant for which $r = 225$ mm, the arm has a counterclockwise angular velocity $\omega = 6$ rad/s and is slowing down at the rate of 2 rad/s^2. For this instant, determine the tension T in the cord and the magnitude N of the force exerted on the slider by the sides of the smooth radial slot. Indicate which side, A or B, of the slot contacts the slider.

Ans. $T = 16.20$ N, $N = 2.10$ N, side B

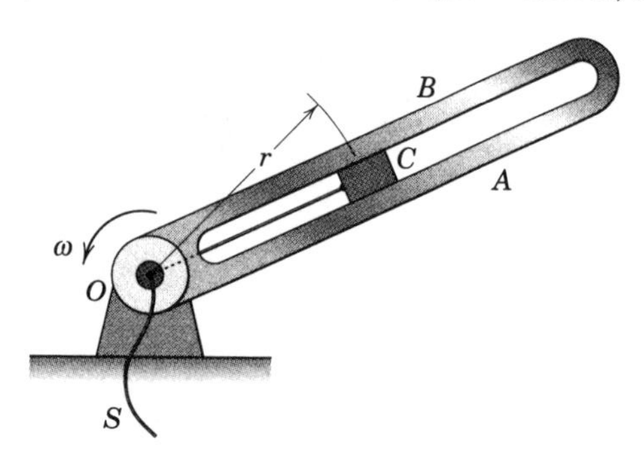

Problem 3/81

3/82 The rotating drum of a clothes dryer is shown in the figure. Determine the angular velocity Ω of the drum which results in loss of contact between the clothes and the drum at $\theta = 50°$. Assume that the small vanes prevent slipping until loss of contact.

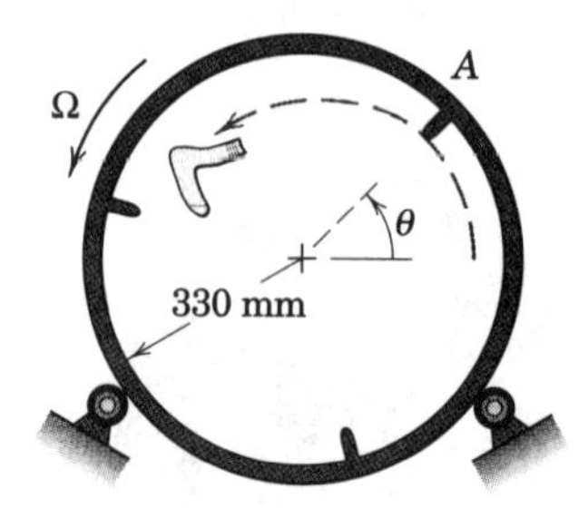

Problem 3/82

3/83 The small 180-g slider A moves without appreciable friction in the hollow tube, which rotates in a horizontal plane with a constant speed $\Omega = 7$ rad/s. The slider is launched with an initial speed $\dot{r}_0 = 20$ m/s relative to the tube at the inertial coordinates $x = 150$ mm and $y = 0$. Determine the magnitude P of the horizontal force exerted on the slider by the tube just before the slider exits the tube.

Ans. $P = 53.3$ N

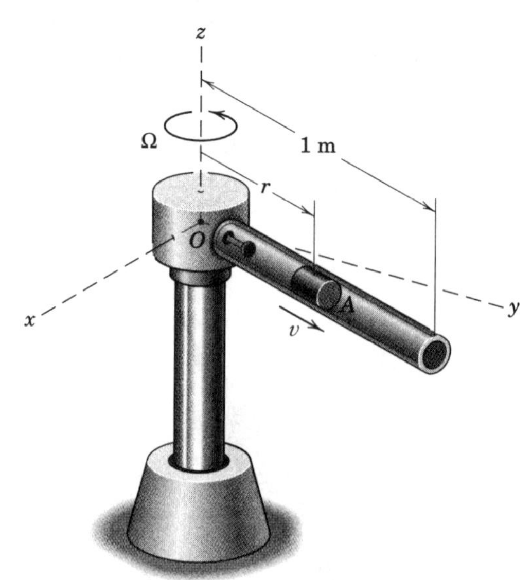

Problem 3/83

3/84 The 1500-kg car is traveling at 100 km/h on the straight portion of the road, and then its speed is reduced uniformly from A to C, at which point it comes to rest. Compute the magnitude F of the total friction force exerted by the road on the car (a) just before it passes point B, (b) just after it passes point B, and (c) just before it stops at point C.

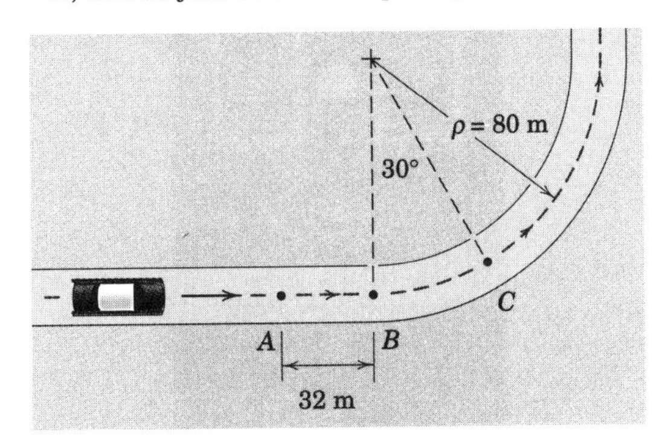

Problem 3/84

3/85 The spring-mounted 0.8-kg collar A oscillates along the horizontal rod, which is rotating at the constant angular rate $\dot{\theta} = 6$ rad/s. At a certain instant, r is increasing at the rate of 800 mm/s. If the coefficient of kinetic friction between the collar and the rod is 0.40, calculate the friction force F exerted by the rod on the collar at this instant.

Ans. $F = 4.39$ N

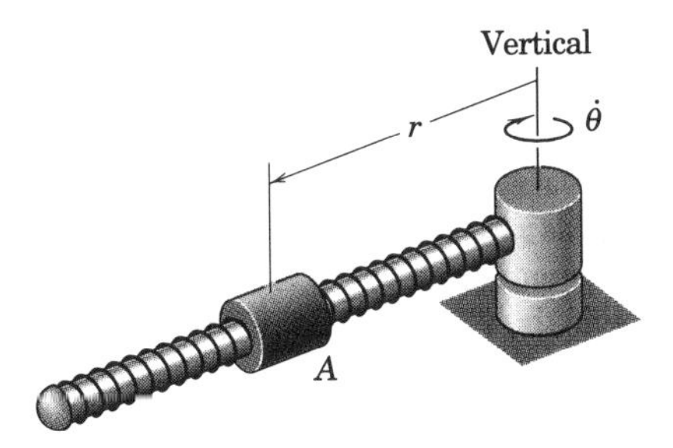

Problem 3/85

3/86 A small rocket-propelled vehicle of mass m travels down the circular path of effective radius r under the action of its weight and a constant thrust T from its rocket motor. If the vehicle starts from rest at A, determine its speed v when it reaches B and the magnitude N of the force exerted by the guide on the wheels just prior to reaching B. Neglect any friction and any loss of mass of the rocket.

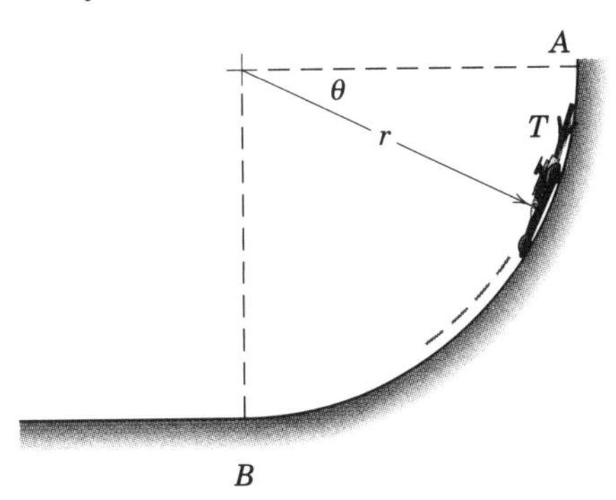

Problem 3/86

3/87 A small vehicle enters the top A of the circular path with a horizontal velocity v_0 and gathers speed as it moves down the path. Determine an expression for the angle β which locates the point where the vehicle leaves the path and becomes a projectile. Evaluate your expression for $v_0 = 0$. Neglect friction and treat the vehicle as a particle.

Ans. $\beta = \cos^{-1}\left(\frac{2}{3} + \frac{v_0^2}{3gR}\right)$, $\beta = 48.2°$

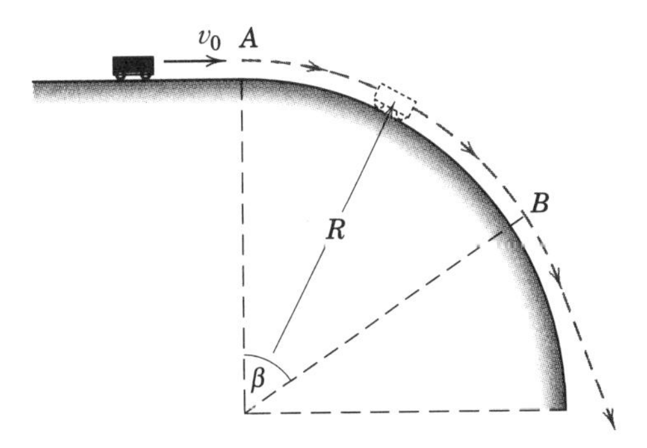

Problem 3/87

3/88 A small coin is placed on the horizontal surface of the rotating disk. If the disk starts from rest and is given a constant angular acceleration $\ddot{\theta} = \alpha$, determine an expression for the number of revolutions N through which the disk turns before the coin slips. The coefficient of static friction between the coin and the disk is μ_s.

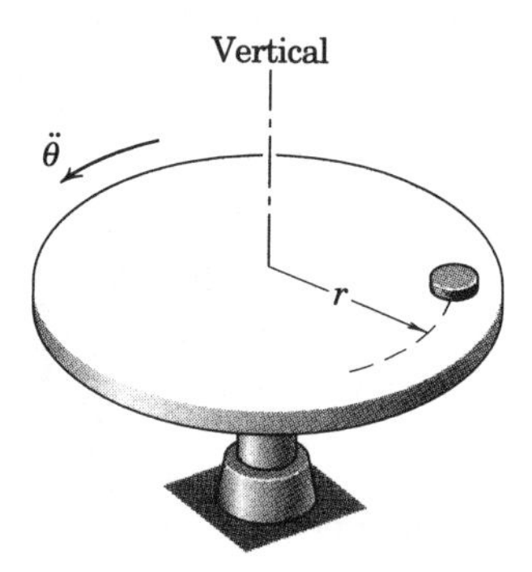

Problem 3/88

3/89 The slotted arm revolves with a constant angular velocity $\dot{\theta} = \omega$ about a vertical axis through the center O of the fixed cam. The cam is cut so that the radius to the path of the center of the pin A varies according to $r = r_0 + b \sin N\omega t$ where N equals the number of lobes—six in this case. If $\omega = 12$ rad/s, $r_0 = 100$ mm, and $b = 10$ mm, and if the spring compression varies from 11.5 N to 19.1 N from valley to crest, calculate the magnitude R of the force between the cam and the 100-g pin A as it passes over the top of the lobe in the position shown. Neglect friction.

Ans. $R = 12.33$ N

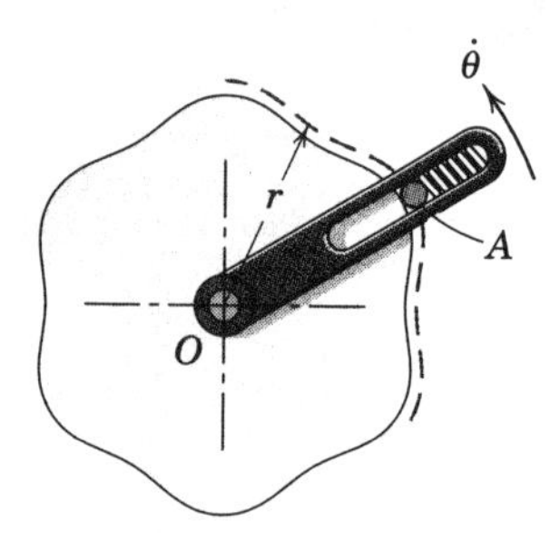

Problem 3/89

3/90 A right-handed baseball pitcher throws a curve ball initially aimed at the right edge of home plate B. It curves so as to "break" 150 mm as shown. Assume that the horizontal velocity component is constant at $v = 38$ m/s, neglect vertical motion, and estimate (*a*) the average radius of curvature ρ of the baseball path and (*b*) the normal force R acting on the 146-g baseball.

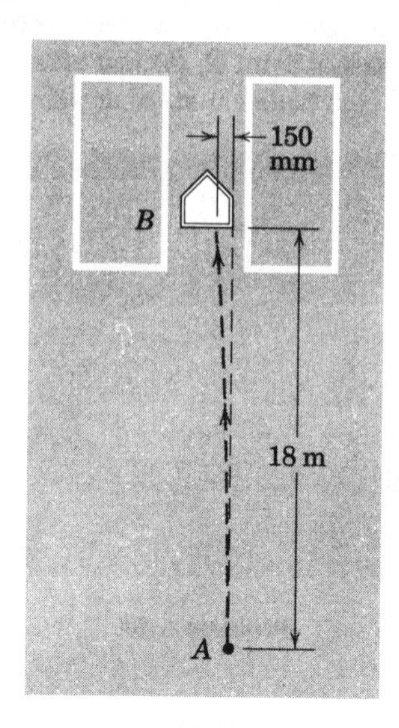

Problem 3/90

3/91 A body at rest relative to the surface of the earth rotates with the earth and therefore moves in a circular path about the polar axis of the earth considered fixed. Derive an expression for the ratio k of the apparent weight of such a body as measured by a spring scale at the equator (calibrated to read the actual force applied) to the true weight of the body, which is the absolute gravitational attraction to the earth. The absolute acceleration due to gravity at the equator is $g = 9.815$ m/s^2. The radius of the earth at the equator is $R = 6378$ km, and the angular velocity of the earth is $\omega = 0.729(10^{-4})$ rad/s. If the true weight is 100 N, what is the apparent measured weight W'?

Ans. $k = 1 - \dfrac{R\omega^2}{g}$

$W' = 99.655$ N

3/92 The particle P is released at time $t = 0$ from the position $r = r_0$ inside the smooth tube with no velocity relative to the tube, which is driven at the constant angular velocity ω_0 about a vertical axis. Determine the radial velocity v_r, the radial position r, and the transverse velocity v_θ as functions of time t. Explain why the radial velocity increases with time in the absence of radial forces. Plot the absolute path of the particle during the time it is inside the tube for $r_0 = 0.1$ m, $l = 1$ m, and $\omega_0 = 1$ rad/s.

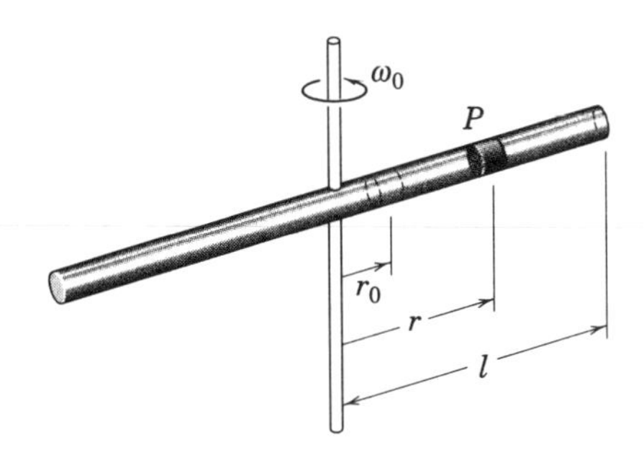

Problem 3/92

3/93 Remove the assumption of smooth surfaces as stated in Prob. 3/92 and assume a coefficient of kinetic friction μ_k between the particle and rotating tube. Determine the radial position r of the particle as a function of time t if it is released with no relative velocity at $r = r_0$ when $t = 0$. Assume that static friction is overcome.

Ans.

$$r = \frac{r_0}{2\sqrt{\mu_k^2 + 1}}[(\mu_k + \sqrt{\mu_k^2 + 1})e^{\omega_0(-\mu_k + \sqrt{\mu_k^2 + 1})t} + (-\mu_k + \sqrt{\mu_k^2 + 1})e^{\omega_0(-\mu_k - \sqrt{\mu_k^2 + 1})t}]$$

3/94 The tangential or hoop stress in the rim of a flywheel can be approximated by simulating the rim as a metal hoop rotating in its plane about its geometric axis. Derive an expression for the hoop stress σ_t in terms of the rim speed v, the mass ρ per unit length of the rim, and the cross-sectional area A of the hoop rim. Apply the equation of motion to the free-body diagram of a differential element of the hoop, treated as a particle.

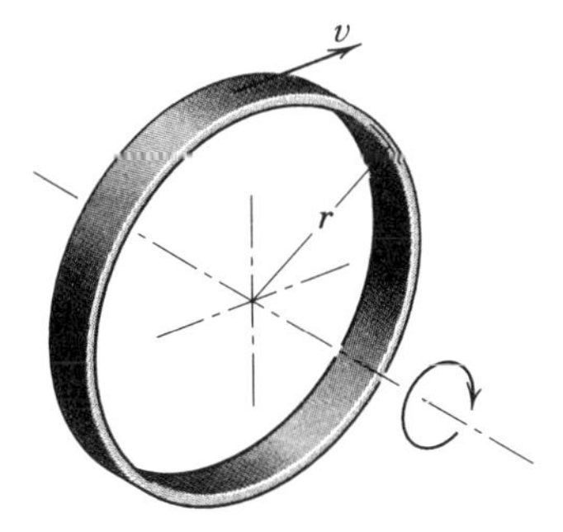

Problem 3/94

3/95 The uniform slender rod of length L, mass m, and cross-sectional area A is rotating in a horizontal plane about the vertical central axis O-O at a constant high angular velocity ω. By analyzing the horizontal forces on the accelerating differential element shown, derive an expression for the tensile stress σ in the rod as a function of r. The stress, commonly referred to as centrifugal stress, equals the tensile force divided by the cross-sectional area A.

$$Ans.\ \sigma = \frac{mL\omega^2}{2A}\left(\frac{1}{4} - \frac{r^2}{L^2}\right)$$

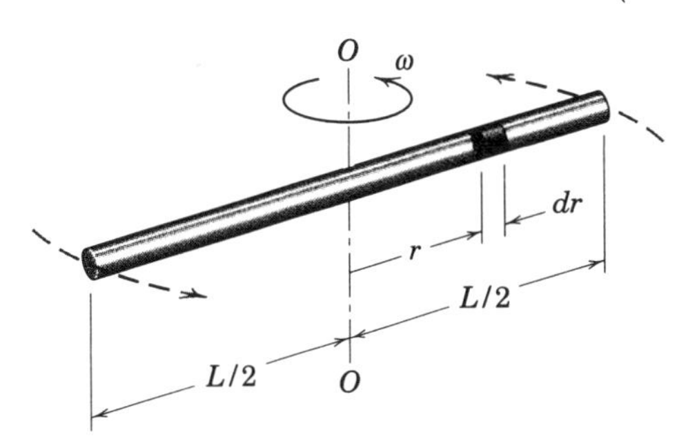

Problem 3/95

3/96 The spacecraft P is in the elliptical orbit shown. At the instant represented, its speed is $v = 4230$ m/s. Determine the corresponding values of $\dot{r}$, $\dot{\theta}$, $\ddot{r}$, and $\ddot{\theta}$. Use $g = 9.825$ m/s^2 as the acceleration of gravity on the surface of the earth and $R = 6371$ km as the radius of the earth.

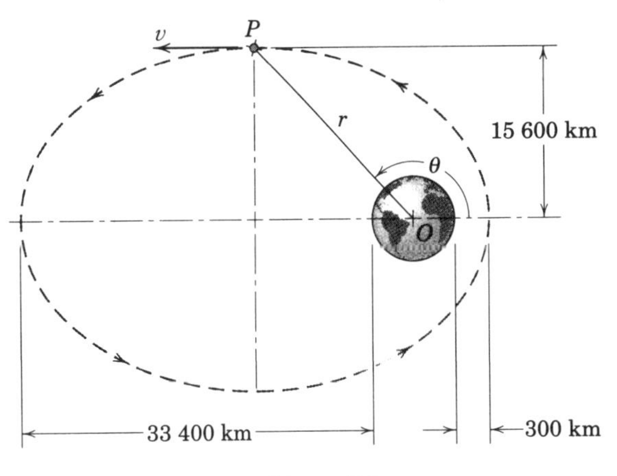

Problem 3/96

▶**3/97** A hollow tube rotates about the horizontal axis through point O with constant angular velocity ω_0. A particle of mass m is introduced with zero relative velocity at $r = 0$ when $\theta = 0$ and slides outward through the smooth tube. Determine r as a function of θ.

Ans. $r = \dfrac{g}{2\omega_0^2}(\sinh\theta - \sin\theta)$

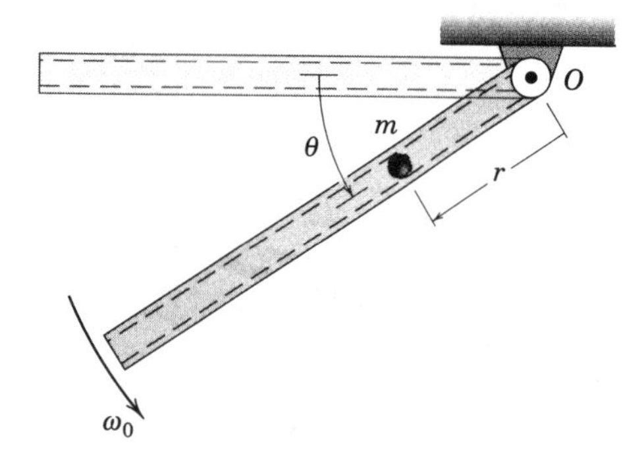

Problem 3/97

▶**3/98** The small pendulum of mass m is suspended from a trolley which runs on a horizontal rail. The trolley and pendulum are initially at rest with $\theta = 0$. If the trolley is given a constant acceleration $a = g$, determine the maximum angle θ_{max} through which the pendulum swings. Also find the tension T in the cord in terms of θ.

Ans. $\theta_{max} = \pi/2$, $T = mg(3\sin\theta + 3\cos\theta - 2)$

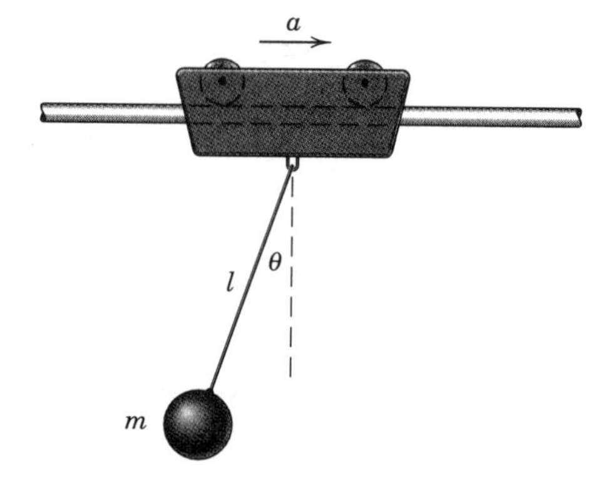

Problem 3/98

▶**3/99** A small object is released from rest at A and slides with friction down the circular path. If the coefficient of friction is 0.2, determine the velocity of the object as it passes B. (*Hint:* Write the equations of motion in the n- and t-directions, eliminate N, and substitute $v\,dv = a_t r\,d\theta$. The resulting equation is a linear nonhomogeneous differential equation of the form $dy/dx + f(x)y = g(x)$, the solution of which is well known.

Ans. $v = 5.52$ m/s

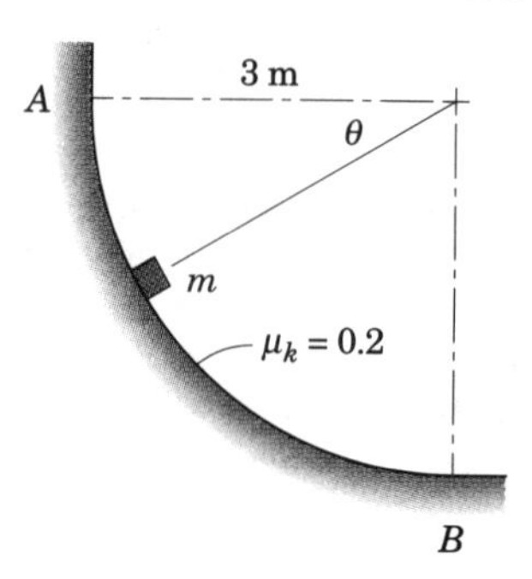

Problem 3/99

▶**3/100** A small collar of mass m is given an initial velocity of magnitude v_0 on the horizontal circular track fabricated from a slender rod. If the coefficient of kinetic friction is μ_k, determine the distance traveled before the collar comes to rest. (*Hint:* Recognize that the friction force depends on the net normal force.)

Ans. $s = \dfrac{r}{2\mu_k}\ln\left[\dfrac{v_0^2 + \sqrt{v_0^4 + r^2g^2}}{rg}\right]$

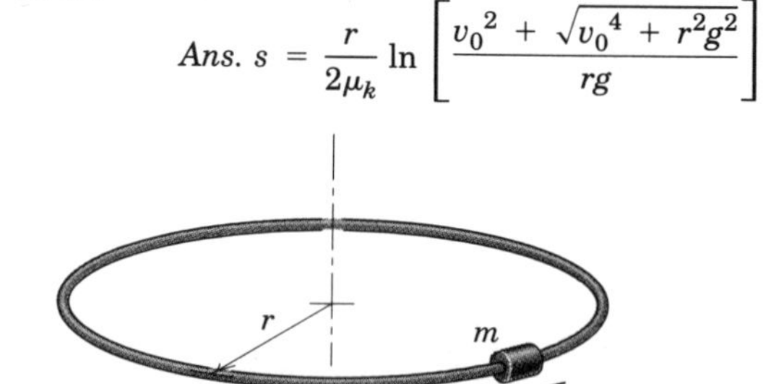

Problem 3/100

▶**3/101** The slotted arm OB rotates in a horizontal plane about point O of the fixed circular cam with constant angular velocity $\dot{\theta} = 15$ rad/s. The spring has a stiffness of 5 kN/m and is uncompressed when $\theta = 0$. The smooth roller A has a mass of 0.5 kg. Determine the normal force N which the cam exerts on A and also the force R exerted on A by the sides of the slot when $\theta = 45°$. All surfaces are smooth. Neglect the small diameter of the roller.

Ans. $N = 81.6$ N, $R = 38.7$ N

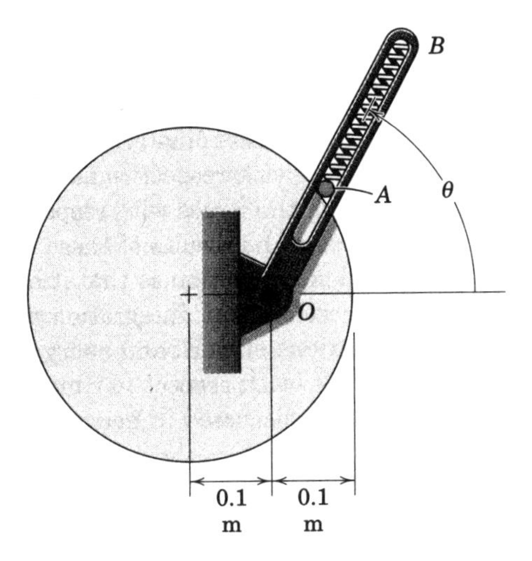

Problem 3/101

▶**3/102** The small cart is nudged with negligible velocity from its horizontal position at A onto the parabolic path, which lies in a vertical plane. Neglect friction and show that the cart maintains contact with the path for all values of k.

Ans. $N = \dfrac{mg}{(1 + 4k^2x^2)^{3/2}} > 0$

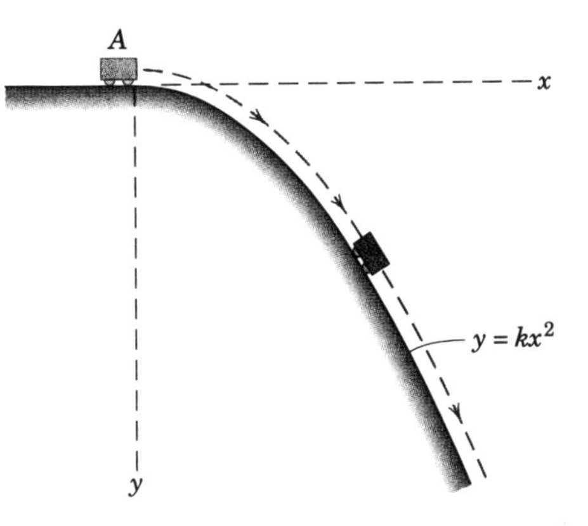

Problem 3/102

SECTION B. WORK AND ENERGY

3/6 Work and Kinetic Energy

In the previous two articles, we applied Newton's second law $\mathbf{F} = m\mathbf{a}$ to various problems of particle motion to establish the instantaneous relationship between the net force acting on a particle and the resulting acceleration of the particle. When we needed to determine the change in velocity or the corresponding displacement of the particle, we integrated the computed acceleration by using the appropriate kinematic equations.

There are two general classes of problems in which the cumulative effects of unbalanced forces acting on a particle are of interest to us. These cases involve (1) integration of the forces with respect to the displacement of the particle and (2) integration of the forces with respect to the time they are applied. We may incorporate the results of these integrations directly into the governing equations of motion so that it becomes unnecessary to solve directly for the acceleration. Integration with respect to displacement leads to the equations of work and energy, which are the subject of this article. Integration with respect to time leads to the equations of impulse and momentum, discussed in Section C.

Definition of Work

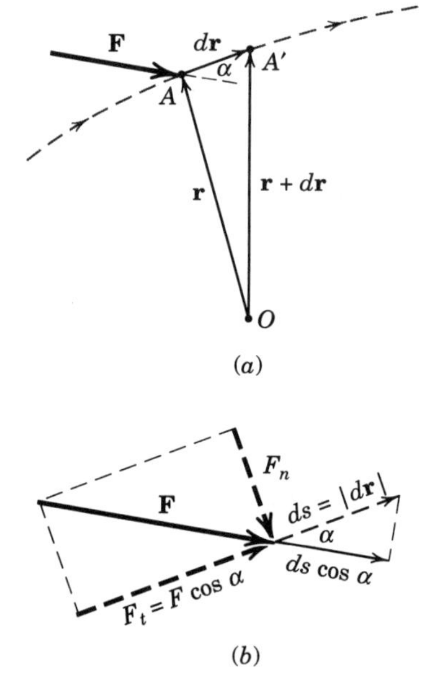

Figure 3/2

We now develop the quantitative meaning of the term "work."* Figure 3/2*a* shows a force $\mathbf{F}$ acting on a particle at A which moves along the path shown. The position vector $\mathbf{r}$ measured from some convenient origin O locates the particle as it passes point A, and $d\mathbf{r}$ is the differential displacement associated with an infinitesimal movement from A to A'. The work done by the force $\mathbf{F}$ during the displacement $d\mathbf{r}$ is defined as

$$dU = \mathbf{F} \cdot d\mathbf{r}$$

The magnitude of this dot product is $dU = F\,ds \cos\alpha$, where α is the angle between $\mathbf{F}$ and $d\mathbf{r}$ and where ds is the magnitude of $d\mathbf{r}$. This expression may be interpreted as the displacement multiplied by the force component $F_t = F\cos\alpha$ in the direction of the displacement, as represented by the dashed lines in Fig. 3/2*b*. Alternatively, the work dU may be interpreted as the force multiplied by the displacement component $ds \cos\alpha$ in the direction of the force, as represented by the full lines in Fig. 3/2*b*.

With this definition of work, it should be noted that the component $F_n = F\sin\alpha$ normal to the displacement does no work. Thus, the work dU may be written as

$$dU = F_t\,ds$$

Work is positive if the working component F_t is in the direction of the displacement and negative if it is in the opposite direction. Forces which

*The concept of work was also developed in the study of virtual work in Chapter 7 of *Vol. 1 Statics*.

do work are termed *active forces.* Constraint forces which do no work are termed *reactive forces.*

Units of Work

The SI units of work are those of force (N) times displacement (m) or N·m. This unit is given the special name *joule* (J), which is defined as the work done by a force of 1 N acting through a distance of 1 m in the direction of the force. Consistent use of the joule for work (and energy) rather than the units N·m will avoid possible ambiguity with the units of moment of a force or torque, which are also written N·m.

In the U.S. customary system, work has the units of ft-lb. Dimensionally, work and moment are the same. In order to distinguish between the two quantities, it is recommended that work be expressed as foot pounds (ft-lb) and moment as pound feet (lb-ft). It should be noted that work is a scalar as given by the dot product and involves the product of a force and a distance, both measured along the same line. Moment, on the other hand, is a vector as given by the cross product and involves the product of force and distance measured at right angles to the force.

Calculation of Work

During a finite movement of the point of application of a force, the force does an amount of work equal to

$$U = \int \mathbf{F} \cdot d\mathbf{r} = \int (F_x\, dx + F_y\, dy + F_z\, dz)$$

or

$$U = \int F_t\, ds$$

In order to carry out this integration, it is necessary to know the relations between the force components and their respective coordinates or the relation between F_t and s. If the functional relationship is not known as a mathematical expression which can be integrated but is specified in the form of approximate or experimental data, then we can compute the work by carrying out a numerical or graphical integration as represented by the area under the curve of F_t versus s, as shown in Fig. 3/3.

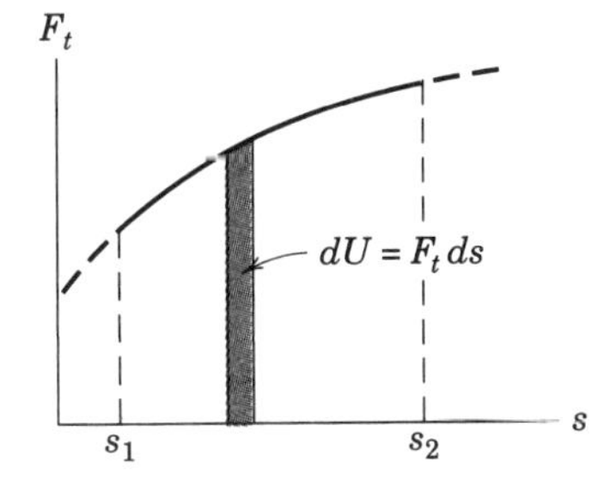

Figure 3/3

Work and Linear Springs

A common example of the work done on a body by a variable force is found in the action of a spring attached to a movable body. We consider here the common linear spring of stiffness k, where the force F in the spring, tension or compression, is proportional to its deformation x, so that $F = kx$. Figure 3/4 shows the two cases where the body is moved by a force P so as to stretch the spring a distance x or to compress the spring a distance x. Because the force exerted by the spring on the body in each case is in the sense *opposite* to the displacement, it does *negative*

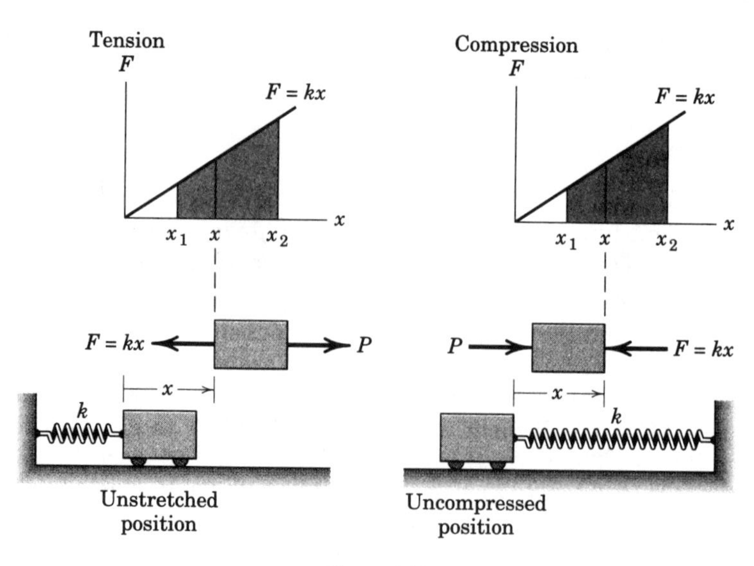

Figure 3/4

work on the body. Thus, for both stretching and compressing the spring, the work done *on* the body is *negative* and is given by

$$U_{1\text{-}2} = -\int_{x_1}^{x_2} F\,dx = -\int_{x_1}^{x_2} kx\,dx = -\tfrac{1}{2}k(x_2^{\,2} - x_1^{\,2})$$

Whether a spring is in tension or compression while being relaxed, we see from both examples in Fig. 3/4 that the deformation changes from x_2 to a lesser deformation x_1. The force exerted on the body by the spring in both cases is in the *same* sense as the displacement, and, therefore, the work done *on* the body is *positive.*

The magnitude of the work, positive or negative, is seen to be equal to the shaded trapezoidal area shown for both cases in Fig. 3/4. In calculating the work done by a spring force, take care to use consistent units for k and x. For example, if x is in meters (or feet), k must be in N/m (or lb/ft).

The expression $F = kx$ is actually a static relationship which is true only when elements of the spring have no acceleration. Dynamic behavior of a spring when its mass is accounted for is a fairly complex problem which will not be treated here. We will assume that the mass of the spring is small compared with the masses of other accelerating parts of the system, in which case, the linear static relationship will not involve appreciable error.

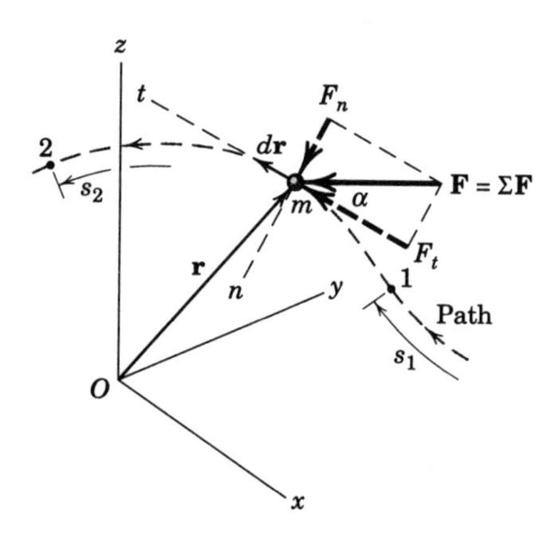

Figure 3/5

Work and Curvilinear Motion

We now consider the work done on a particle of mass m, Fig. 3/5, moving along a curved path under the action of the force $\mathbf{F}$, which stands for the resultant $\Sigma\mathbf{F}$ of all forces acting on the particle. The position of m is specified by the position vector $\mathbf{r}$, and its displacement along its path during time dt is represented by the change $d\mathbf{r}$ in its position vector. The work done by $\mathbf{F}$ during a finite movement of the particle from point 1 to point 2 is

$$U_{1\text{-}2} = \int_1^2 \mathbf{F}\cdot d\mathbf{r} = \int_{s_1}^{s_2} F_t\, ds$$

where the limits specify the initial and final end points of the motion.

When we substitute Newton's second law $\mathbf{F} = m\mathbf{a}$, the expression for the work of all forces becomes

$$U_{1\text{-}2} = \int_1^2 \mathbf{F}\cdot d\mathbf{r} = \int_1^2 m\mathbf{a}\cdot d\mathbf{r}$$

But $\mathbf{a}\cdot d\mathbf{r} = a_t\, ds$, where a_t is the tangential component of the acceleration of m. In terms of the velocity v of the particle, Eq. 2/3 gives $a_t\, ds = v\, dv$. Thus, the expression for the work of $\mathbf{F}$ becomes

$$U_{1\text{-}2} = \int_1^2 \mathbf{F}\cdot d\mathbf{r} = \int_{v_1}^{v_2} mv\, dv = \tfrac{1}{2}m(v_2^2 - v_1^2) \qquad \mathbf{(3/9)}$$

where the integration is carried out between points 1 and 2 along the curve, at which points the velocities have the magnitudes v_1 and v_2, respectively.

Principle of Work and Kinetic Energy

The *kinetic energy* T of the particle is defined as

$$T = \tfrac{1}{2}mv^2 \qquad \mathbf{(3/10)}$$

and is the total work which must be done on the particle to bring it from a state of rest to a velocity v. Kinetic energy T is a scalar quantity with the units of N·m or *joules* (J) in SI units and ft-lb in U.S. customary units. Kinetic energy is *always* positive, regardless of the direction of the velocity.

Equation 3/9 may be restated as

$$U_{1\text{-}2} = T_2 - T_1 = \Delta T \qquad \mathbf{(3/11)}$$

which is the *work-energy equation* for a particle. The equation states that the *total work done* by all forces acting on a particle as it moves from point 1 to point 2 equals the corresponding *change in kinetic energy* of the particle. Although T is always positive, the change ΔT may be positive, negative, or zero. When written in this concise form, Eq. 3/11 tells us that the work always results in a *change* of kinetic energy.

Alternatively, the work-energy relation may be expressed as the initial kinetic energy T_1 plus the work done $U_{1\text{-}2}$ equals the final kinetic energy T_2, or

$$T_1 + U_{1\text{-}2} = T_2 \tag{3/11a}$$

When written in this form, the terms correspond to the natural sequence of events. Clearly, the two forms 3/11 and 3/11*a* are equivalent.

Advantages of the Work-Energy Method

We now see from Eq. 3/11 that a major advantage of the method of work and energy is that it avoids the necessity of computing the acceleration and leads directly to the velocity changes as functions of the forces which do work. Further, the work-energy equation involves only those forces which do work and thus give rise to changes in the magnitude of the velocities.

We consider now a system of two particles joined together by a connection which is frictionless and incapable of any deformation. The forces in the connection are equal and opposite, and their points of application necessarily have identical displacement components in the direction of the forces. Therefore, the net work done by these internal forces is zero during any movement of the system. Thus, Eq. 3/11 is applicable to the entire system, where $U_{1\text{-}2}$ is the total or net work done on the system by forces external to it and ΔT is the change, $T_2 - T_1$, in the total kinetic energy of the system. The total kinetic energy is the sum of the kinetic energies of both elements of the system. We thus see that another advantage of the work-energy method is that it enables us to analyze a system of particles joined in the manner described without dismembering the system.

Application of the work-energy method requires isolation of the particle or system under consideration. For a single particle you should draw a *free-body diagram* showing all externally applied forces. For a system of particles rigidly connected without springs, draw an *active-force diagram* showing only those external forces which do work (active forces) on the entire system.*

Power

The capacity of a machine is measured by the time rate at which it can do work or deliver energy. The total work or energy output is not a measure of this capacity since a motor, no matter how small, can deliver a large amount of energy if given sufficient time. On the other hand, a large and powerful machine is required to deliver a large amount of energy in a short period of time. Thus, the capacity of a machine is rated by its *power*, which is defined as the *time rate of doing work*.

*The active-force diagram was introduced in the method of virtual work in statics. See Chapter 7 of *Vol. 1 Statics*.

Accordingly, the power P developed by a force $\mathbf{F}$ which does an amount of work U is $P = dU/dt = \mathbf{F}\cdot d\mathbf{r}/dt$. Because $d\mathbf{r}/dt$ is the velocity $\mathbf{v}$ of the point of application of the force, we have

$$P = \mathbf{F}\cdot\mathbf{v} \tag{3/12}$$

Power is clearly a scalar quantity, and in SI it has the units of N·m/s = J/s. The special unit for power is the *watt* (W), which equals one joule per second (J/s). In U.S. customary units, the unit for mechanical power is the *horsepower* (hp). These units and their numerical equivalences are

$$1\ \text{W} = 1\ \text{J/s}$$
$$1\ \text{hp} = 550\ \text{ft-lb/sec} = 33{,}000\ \text{ft-lb/min}$$
$$1\ \text{hp} = 746\ \text{W} = 0.746\ \text{kW}$$

Efficiency

The ratio of the work done *by* a machine to the work done *on* the machine during the same time interval is called the *mechanical efficiency* e_m of the machine. This definition assumes that the machine operates uniformly so that there is no accumulation or depletion of energy within it. Efficiency is always less than unity since every device operates with some loss of energy and since energy cannot be created within the machine. In mechanical devices which involve moving parts, there will always be some loss of energy due to the negative work of kinetic friction forces. This work is converted to heat energy which, in turn, is dissipated to the surroundings. The mechanical efficiency at any instant of time may be expressed in terms of mechanical power P by

$$e_m = \frac{P_{\text{output}}}{P_{\text{input}}} \tag{3/13}$$

In addition to energy loss by mechanical friction, there may also be electrical and thermal energy loss, in which case, the *electrical efficiency* e_e and *thermal efficiency* e_t are also involved. The *overall efficiency* e in such instances is

$$e = e_m e_e e_t$$

Sample Problem 3/11

Calculate the velocity v of the 50-kg crate when it reaches the bottom of the chute at B if it is given an initial velocity of 4 m/s down the chute at A. The coefficient of kinetic friction is 0.30.

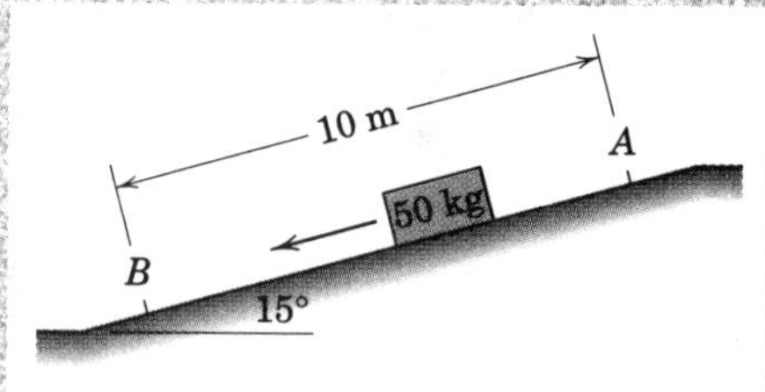

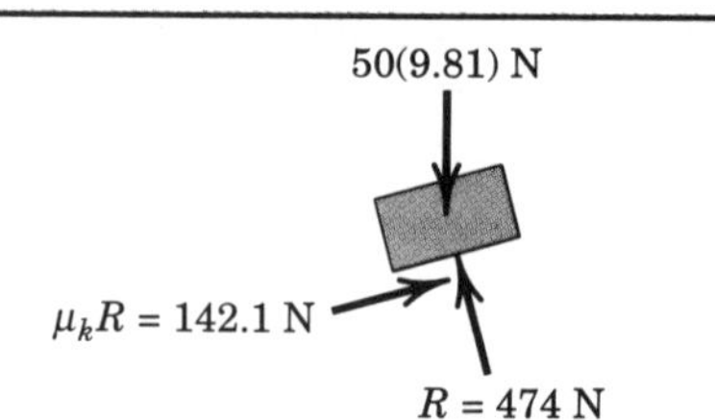

Solution. The free-body diagram of the crate is drawn and includes the normal force R and the kinetic friction force F calculated in the usual manner. The work done by the component of the weight down the plane is positive, whereas that done by the friction force is negative. The total work done on the crate during the motion is

① $[U = Fs]$ $\quad U_{1\text{-}2} = [50(9.81)\sin 15° - 142.1]10 = -151.9 \text{ J}$

The change in kinetic energy is $T_2 - T_1 = \Delta T$

$[T = \frac{1}{2}mv^2] \qquad \Delta T = \frac{1}{2}(50)(v^2 - 4^2)$

The work-energy equation gives

$[U_{1\text{-}2} = \Delta T] \qquad -151.9 = 25(v^2 - 16)$

$v^2 = 9.93 \text{ (m/s)}^2 \qquad v = 3.15 \text{ m/s}$ *Ans.*

Since the net work done is negative, we obtain a decrease in the kinetic energy.

Helpful Hint

① Since the net work done is negative, we obtain a decrease in the kinetic energy.

Sample Problem 3/12

The flatbed truck, which carries an 80-kg crate, starts from rest and attains a speed of 72 km/h in a distance of 75 m on a level road with constant acceleration. Calculate the work done by the friction force acting on the crate during this interval if the static and kinetic coefficients of friction between the crate and the truck bed are (*a*) 0.30 and 0.28, respectively, or (*b*) 0.25 and 0.20, respectively.

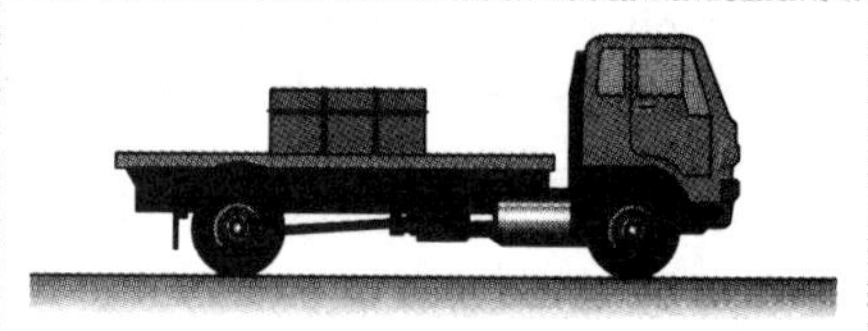

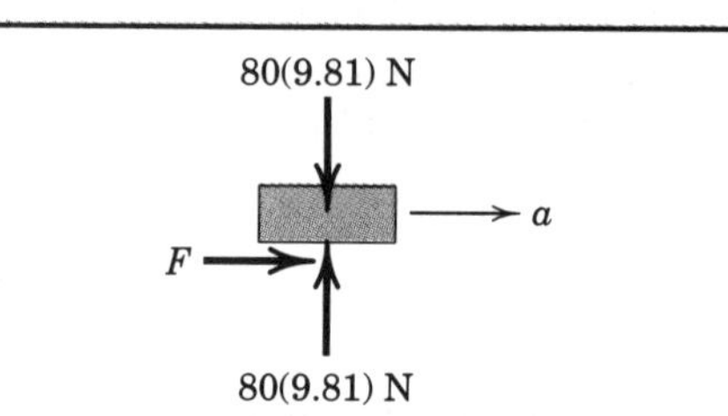

Solution. If the crate does not slip on the bed, its acceleration will be that of the truck, which is

$$[v^2 = 2as] \qquad a = \frac{v^2}{2s} = \frac{(72/3.6)^2}{2(75)} = 2.67 \text{ m/s}^2$$

Case (a). This acceleration requires a friction force on the block of

$[F = ma] \qquad F = 80(2.67) = 213 \text{ N}$

which is less than the maximum possible value of $\mu_s N = 0.30(80)(9.81) = 235$ N. Therefore, the crate does not slip and the work done by the actual static friction force of 213 N is

① $[U = Fs] \qquad U_{1\text{-}2} = 213(75) = 16\,000 \text{ J}$ or 16 kJ

Case (b). For $\mu_s = 0.25$, the maximum possible friction force is 0.25(80)(9.81) = 196.2 N, which is slightly less than the value of 213 N required for no slipping. Therefore, we conclude that the crate slips, and the friction force is governed by the kinetic coefficient and is $F = 0.20(80)(9.81) = 157.0$ N. The acceleration becomes

$[F = ma] \qquad a = F/m = 157.0/80 = 1.962 \text{ m/s}^2$

The distances traveled by the crate and the truck are in proportion to their accelerations. Thus, the crate has a displacement of (1.962/2.67)75 = 55.2 m, and the work done by kinetic friction is

② $[U = Fs] \qquad U_{1\text{-}2} = 157.0(55.2) = 8660 \text{ J}$ or 8.66 kJ *Ans.*

Helpful Hints

① We note that static friction forces do no work when the contacting surfaces are both at rest. When they are in motion, however, as in this problem, the static friction force acting on the crate does positive work and that acting on the truck bed does negative work.

② This problem shows that a kinetic friction force can do positive work when the surface which supports the object and generates the friction force is in motion. If the supporting surface is at rest, then the kinetic friction force acting on the moving part always does negative work.

Sample Problem 3/13

The 50-kg block at A is mounted on rollers so that it moves along the fixed horizontal rail with negligible friction under the action of the constant 300-N force in the cable. The block is released from rest at A, with the spring to which it is attached extended an initial amount $x_1 = 0.233$ m. The spring has a stiffness $k = 80$ N/m. Calculate the velocity v of the block as it reaches position B.

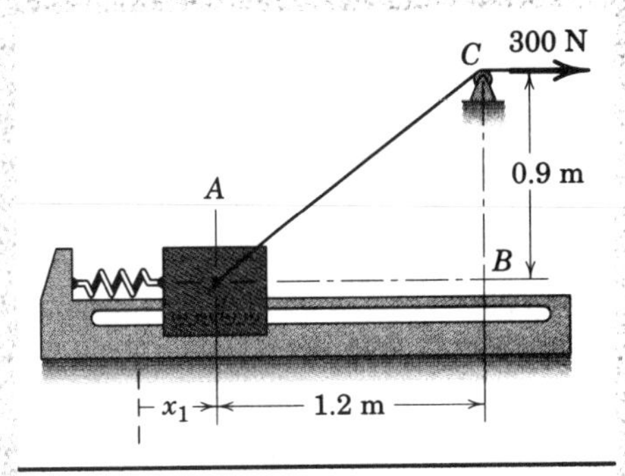

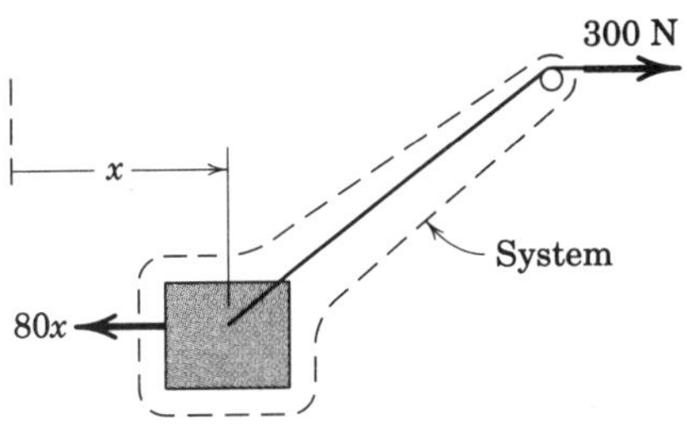

Solution. It will be assumed initially that the stiffness of the spring is small enough to allow the block to reach position B. The active-force diagram for the system composed of both block and cable is shown for a general position. The spring force $80x$ and the 300-N tension are the only forces external to this system which do work on the system. The force exerted on the block by the rail, the weight of the block, and the reaction of the small pulley on the cable do no work on the system and are not included on the active-force diagram.

As the block moves from $x = 0.233$ m to $x = 0.233 + 1.2 = 1.433$ m, the work done by the spring force acting on the block is negative and equals

① $$\left[U = \int F\,dx\right] \quad U_{1\text{-}2} = -\int_{0.233}^{1.433} 80x\,dx = -40x^2\Big|_{0.233}^{1.433} = -80.0 \text{ J}$$

The work done on the system by the constant 300-N force in the cable is the force times the net horizontal movement of the cable over pulley C, which is $\sqrt{(1.2)^2 + (0.9)^2} - 0.9 = 0.6$ m. Thus, the work done is $300(0.6) = 180$ J. We now apply the work-energy equation to the system and get

$$[U_{1\text{-}2} = \Delta T] \quad -80.0 + 180 = \tfrac{1}{2}(50)(v^2 - 0) \qquad v = 2.0 \text{ m/s} \qquad \textit{Ans.}$$

We take special note of the advantage to our choice of system. If the block alone had constituted the system, the horizontal component of the 300-N cable tension acting on the block would have to be integrated over the 1.2-m displacement. This step would require considerably more effort than was needed in the solution as presented. If there had been appreciable friction between the block and its guiding rail, we would have found it necessary to isolate the block alone in order to compute the variable normal force and, hence, the variable friction force. Integration of the friction force over the displacement would then be required to evaluate the negative work which it would do.

Helpful Hint

① If the variable x had been measured from the starting position A, the spring force would be $80(0.233 + x)$, and the limits of integration would be 0 and 1.2 m.

Sample Problem 3/14

The power winch A hoists the 360-kg log up the 30° incline at a constant speed of 1.2 m/s. If the power output of the winch is 4 kW, compute the coefficient of kinetic friction μ_k between the log and the incline. If the power is suddenly increased to 6 kW, what is the corresponding instantaneous acceleration a of the log?

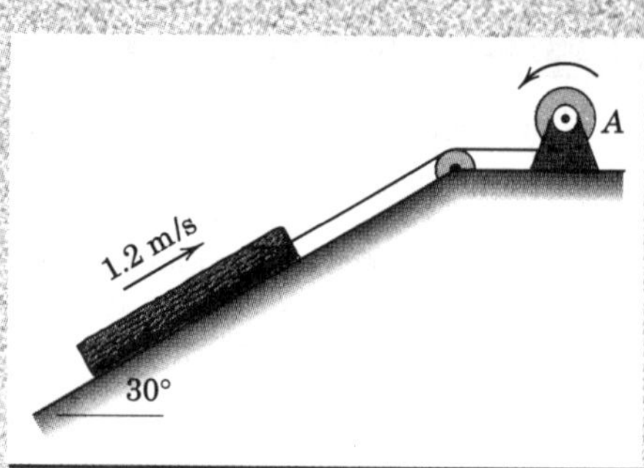

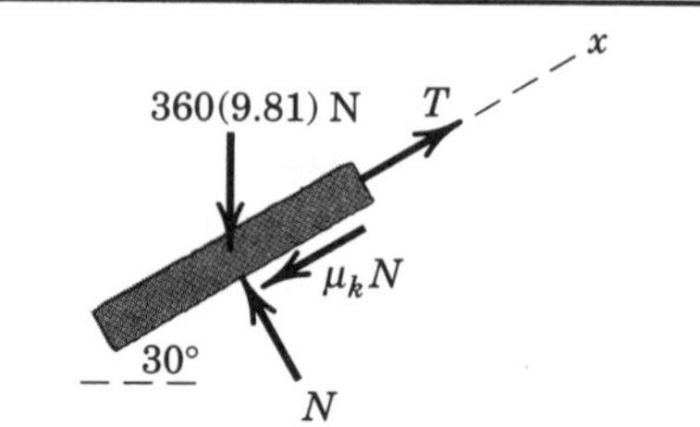

Solution. From the free-body diagram of the log, we get $N = 360(9.81)\cos 30° = 3060$ N, and the kinetic friction force becomes $3060\mu_k$. For constant speed, the forces are in equilibrium so that

$$[\Sigma F_x = 0] \quad T - 3060\mu_k - 360(9.81)\sin 30° = 0 \qquad T = 3060\mu_k + 1766$$

The power output of the winch gives the tension in the cable

① $$[P = Tv] \qquad T = P/v = 4000/1.2 = 3330 \text{ N}$$

Substituting T gives

$$3330 = 3060\mu_k + 1766 \qquad \mu_k = 0.513 \qquad \textit{Ans.}$$

When the power is increased, the tension momentarily becomes

$$[P = Tv] \qquad T = P/v = 6000/1.2 = 5000 \text{ N}$$

and the corresponding acceleration is given by

$$[\Sigma F_x = ma_x] \qquad 5000 - 3060(0.513) - 360(9.81)\sin 30° = 360a$$

② $$a = 4.63 \text{ m/s}^2 \qquad \textit{Ans.}$$

Helpful Hints

① Note the conversion from kilowatts to watts. Also remember to use J/s rather than N · m/s.

② As the speed increases, the acceleration will drop until the speed stabilizes at a value higher than 1.2 m/s.

Sample Problem 3/15

A satellite of mass m is put into an elliptical orbit around the earth. At point A, its distance from the earth is $h_1 = 500$ km and it has a velocity $v_1 = 30\,000$ km/h. Determine the velocity v_2 of the satellite as it reaches point B, a distance $h_2 = 1200$ km from the earth.

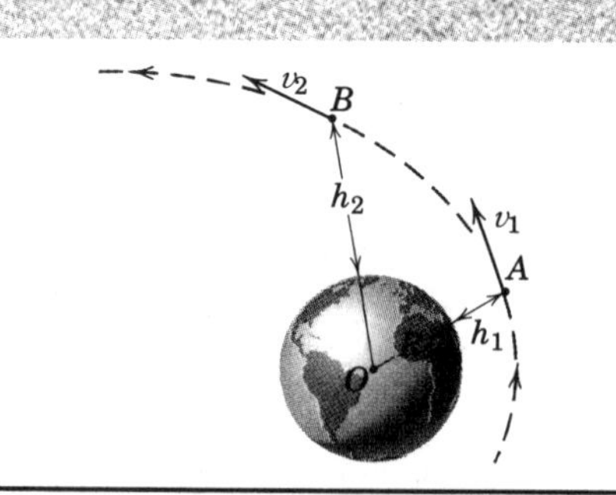

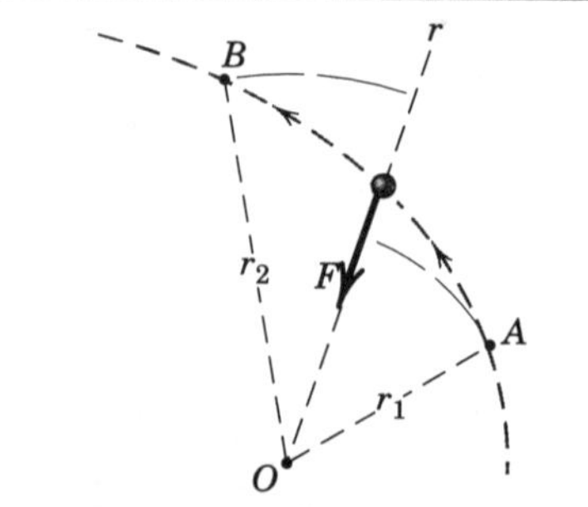

Solution. The satellite is moving outside of the earth's atmosphere so that the only force acting on it is the gravitational attraction of the earth. With the mass and radius of the earth expressed by m_e and R, respectively, the gravitational law of Eq. 1/2 gives $F = Gmm_e/r^2 = gR^2m/r^2$ when the substitution $Gm_e = gR^2$ is made for the surface values $F = mg$ and $r = R$. The work done by F is due only to the radial component of motion along the line of action of F and is negative for increasing r.

$$U_{1\text{-}2} = -\int_{r_1}^{r_2} F\,dr = -mgR^2\int_{r_1}^{r_2}\frac{dr}{r^2} = mgR^2\left(\frac{1}{r_2} - \frac{1}{r_1}\right)$$

The work-energy equation $U_{1\text{-}2} = \Delta T$ gives

①② $$mgR^2\left(\frac{1}{r_2} - \frac{1}{r_1}\right) = \tfrac{1}{2}m(v_2^2 - v_1^2) \qquad v_2^2 = v_1^2 + 2gR^2\left(\frac{1}{r_2} - \frac{1}{r_1}\right)$$

Substituting the numerical values gives

$$v_2^2 = \left(\frac{30\,000}{3.6}\right)^2 + 2(9.81)[(6371)(10^3)]^2\left(\frac{10^{-3}}{6371 + 1200} - \frac{10^{-3}}{6371 + 500}\right)$$

$$= 69.44(10^6) - 10.72(10^6) = 58.73(10^6)\ (\text{m/s})^2$$

$$v_2 = 7663 \text{ m/s} \quad \text{or} \quad v_2 = 7663(3.6) = 27\,590 \text{ km/h} \qquad \textit{Ans.}$$

Helpful Hints

① Note that the result is independent of the mass of the satellite.

② Consult Table D/2, Appendix D, to find the radius R of the earth.

PROBLEMS

Introductory Problems

3/103 The spring is unstretched when $x = 0$. If the body moves from the initial position $x_1 = 100$ mm to the final position $x_2 = 200$ mm, (*a*) determine the work done by the spring on the body and (*b*) determine the work done on the body by its weight.

Ans. (*a*) $U_{1\text{-}2} = -60$ J, (*b*) $U_{1\text{-}2} = 2.35$ J

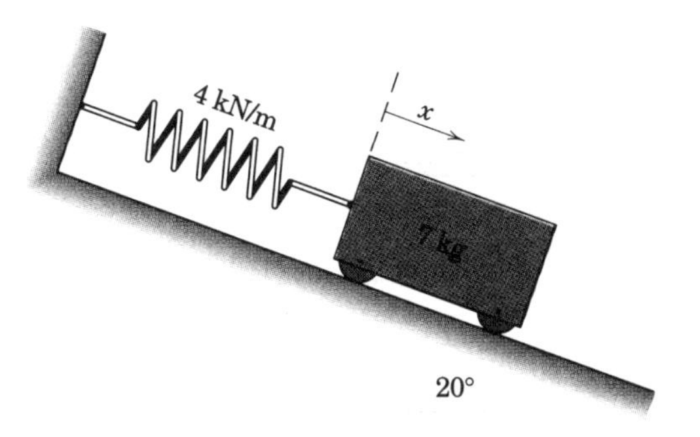

Problem 3/103

3/104 The small body has a speed $v_A = 5$ m/s at point *A*. Neglecting friction, determine its speed v_B at point *B* after it has risen 0.8 m. Is knowledge of the shape of the track necessary?

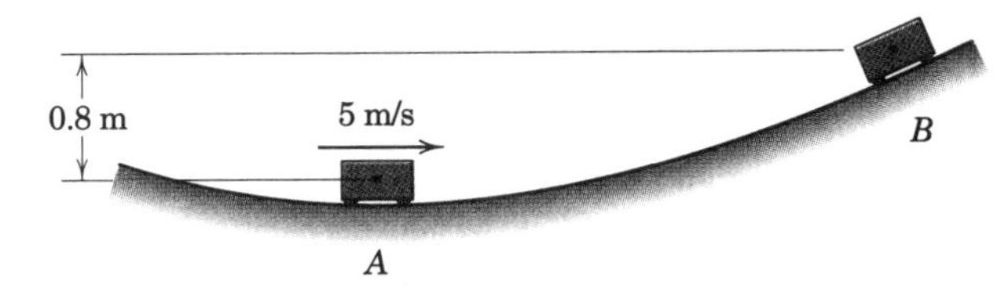

Problem 3/104

3/105 The 30-kg crate slides down the curved path in the vertical plane. If the crate has a velocity of 1.2 m/s down the incline at *A* and a velocity of 8 m/s at *B*, compute the work U_f done on the crate by friction during the motion from *A* to *B*.

Ans. $U_f = -827$ J

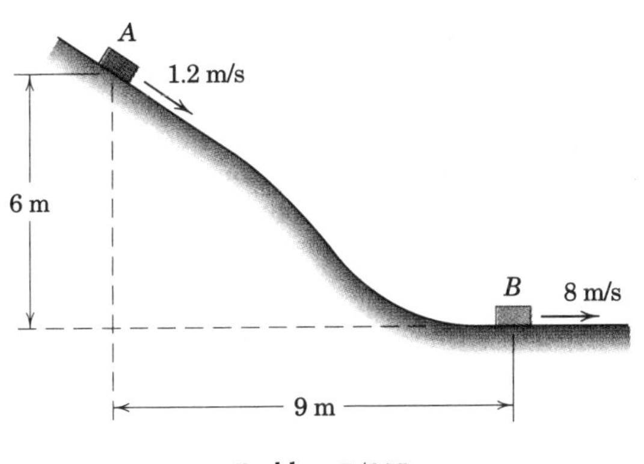

Problem 3/105

3/106 The crawler wrecking crane is moving with a constant speed of 3 km/h when it is suddenly brought to a stop. Compute the maximum angle θ through which the cable of the wrecking ball swings.

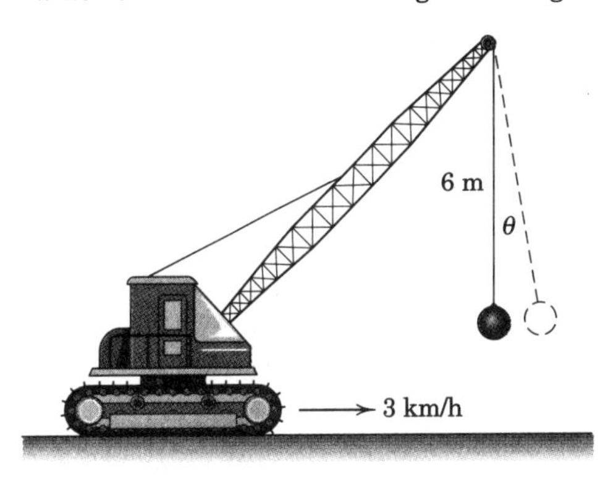

Problem 3/106

3/107 The 0.8-kg collar slides with negligible friction on the fixed rod in the vertical plane. If the collar starts from rest at *A* under the action of the constant 8-N horizontal force, calculate its velocity v as it hits the stop at *B*.

Ans. $v = 4.73$ m/s

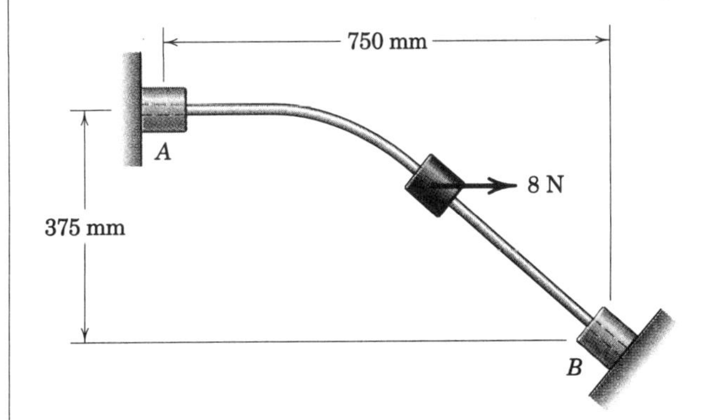

Problem 3/107

3/108 The 15-kg collar A is released from rest in the position shown and slides with negligible friction up the fixed rod inclined 30° from the horizontal under the action of a constant force P = 200 N applied to the cable. Calculate the required stiffness k of the spring so that its maximum deflection equals 180 mm. The position of the small pulley at B is fixed.

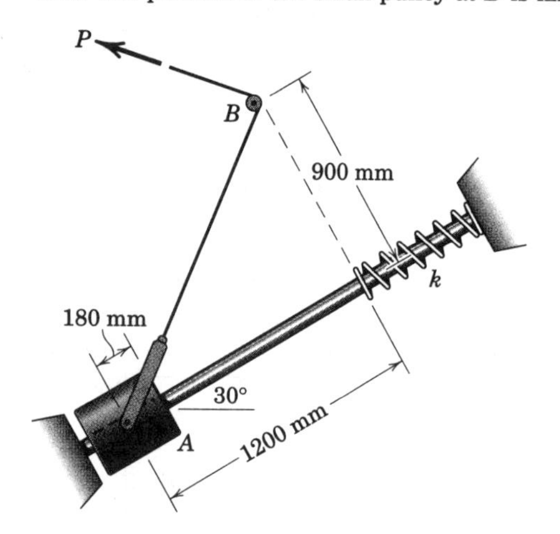

Problem 3/108

3/109 In the design of a spring bumper for a 1500-kg car, it is desired to bring the car to a stop from a speed of 8 km/h in a distance equal to 150 mm of spring deformation. Specify the required stiffness k for each of the two springs behind the bumper. The springs are undeformed at the start of impact.

Ans. k = 164.6 kN/m

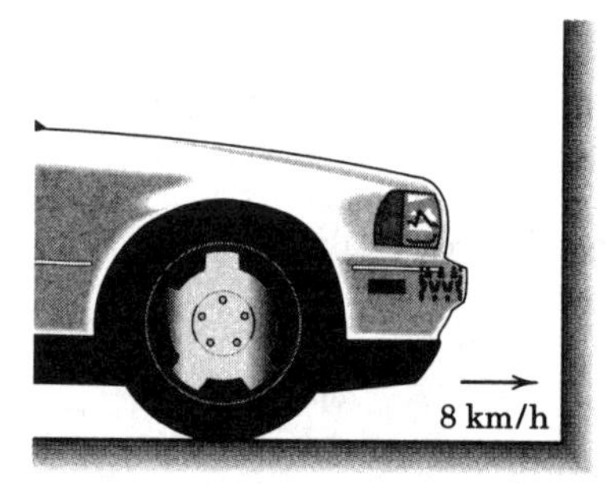

Problem 3/109

3/110 The small collar of mass m is released from rest at A and slides down the curved rod in the vertical plane with negligible friction. Express the velocity v of the collar as it strikes the base at B in terms of the given conditions.

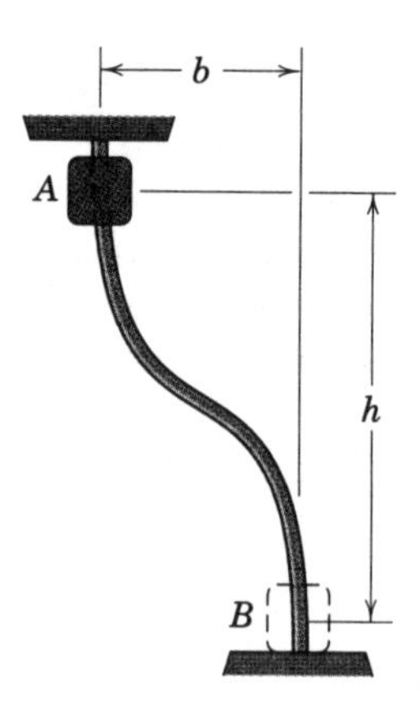

Problem 3/110

3/111 For the sliding collar of Prob. 3/110, if m = 0.5 kg, b = 0.8 m, and h = 1.5 m, and if the velocity of the collar as it strikes the base B is 4.70 m/s after release of the collar from rest at A, calculate the work Q of friction. What happens to the energy which is lost?

Ans. $Q = -1.835$ J

3/112 The position vector of a particle is given by $\mathbf{r} = 8t\mathbf{i} + 1.2t^2\mathbf{j} - 0.5(t^3 - 1)\mathbf{k}$, where t is the time in seconds from the start of the motion and where $\mathbf{r}$ is expressed in meters. For the condition when t = 4 s, determine the power P developed by the force $\mathbf{F} = 40\mathbf{i} - 20\mathbf{j} - 36\mathbf{k}$ N which acts on the particle.

3/113 The man and his bicycle have a combined mass of 95 kg. What power P is the man developing in riding up a 5-percent grade at a constant speed of 20 km/h?

Ans. P = 259 W

Problem 3/113

3/114 The 4-kg ball and the attached light rod rotate in the vertical plane about the fixed axis at O. If the assembly is released from rest at $\theta = 0$ and moves under the action of the 60-N force, which is maintained normal to the rod, determine the velocity v of the ball as θ approaches 90°. Treat the ball as a particle.

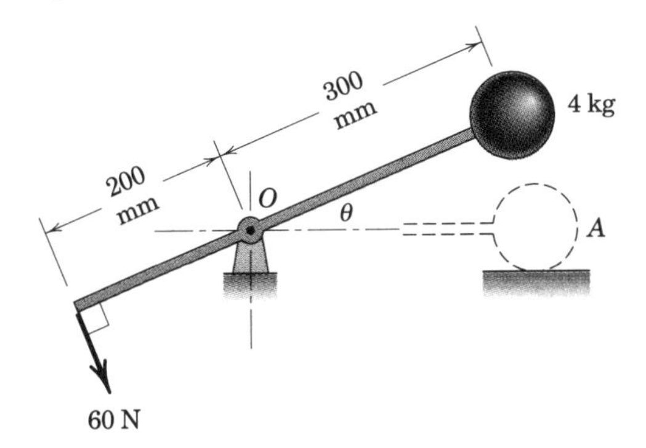

Problem 3/114

3/115 Each of the two systems is released from rest. Calculate the velocity v of each 25-kg cylinder after the 20-kg cylinder has dropped 2 m. The 10-kg cylinder of case (a) is replaced by a 10(9.81)-N force in case (b).

Ans. (a) $v = 1.889$ m/s, (b) $v = 2.09$ m/s

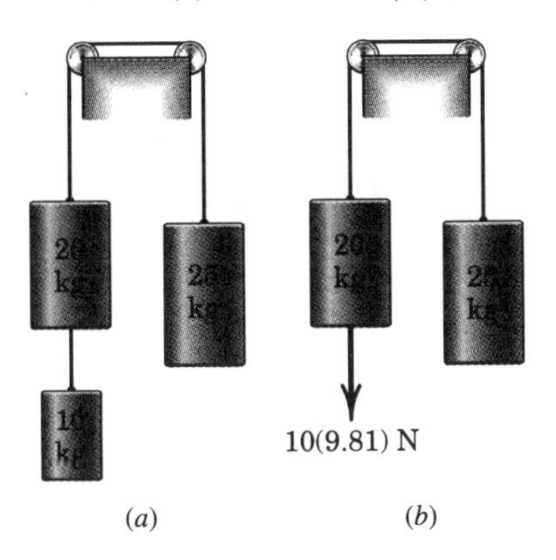

Problem 3/115

3/116 Small metal blocks are discharged with a velocity of 0.45 m/s to a ramp by the upper conveyor shown. If the coefficient of kinetic friction between the blocks and the ramp is 0.30, calculate the angle θ which the ramp must make with the horizontal so that the blocks will transfer without slipping to the lower conveyor moving at the speed of 0.15 m/s.

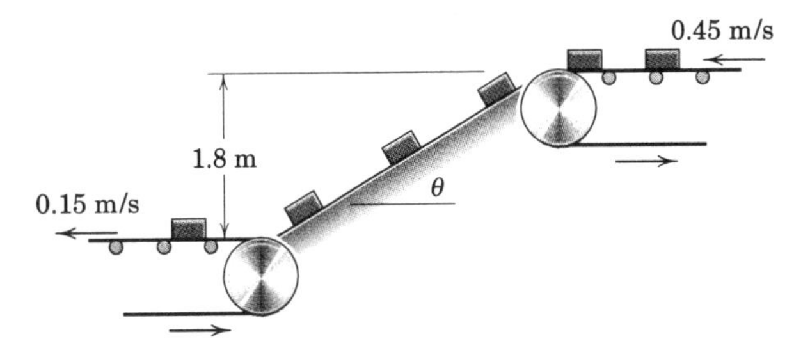

Problem 3/116

3/117 The 2-kg collar is released from rest at A and slides down the inclined fixed rod in the vertical plane. The coefficient of kinetic friction is 0.4. Calculate (a) the velocity v of the collar as it strikes the spring and (b) the maximum deflection x of the spring.

Ans. (a) $v = 2.56$ m/s, (b) $x = 98.9$ mm

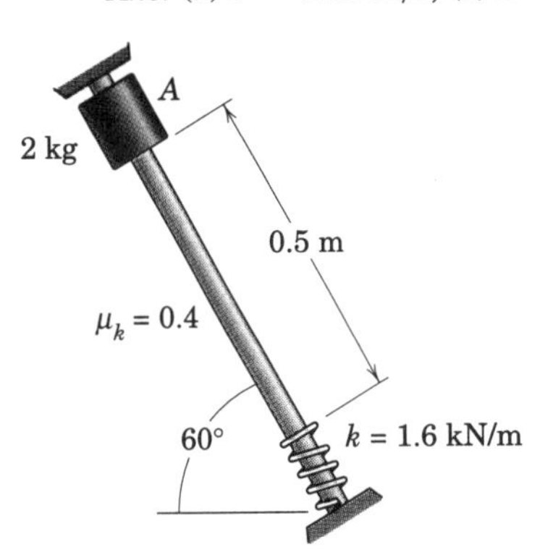

Problem 3/117

3/118 A 1200-kg car enters an 8-percent downhill grade at a speed of 100 km/h. The driver applies her brakes to bring the car to a speed of 25 km/h in a distance of 0.5 km measured along the road. Calculate the energy loss Q dissipated from the brakes in the form of heat. Neglect any friction losses from other causes such as air resistance.

3/119 The 54-kg woman jogs up the flight of stairs in 5 seconds. Determine her average power output.

Ans. $P = 291$ W

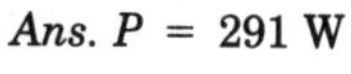

Problem 3/119

Representative Problems

3/120 The 0.8-kg collar slides freely on the fixed circular rod. Calculate the velocity v of the collar as it hits the stop at B if it is elevated from rest at A by the action of the constant 40-N force in the cord. The cord is guided by the small fixed pulleys.

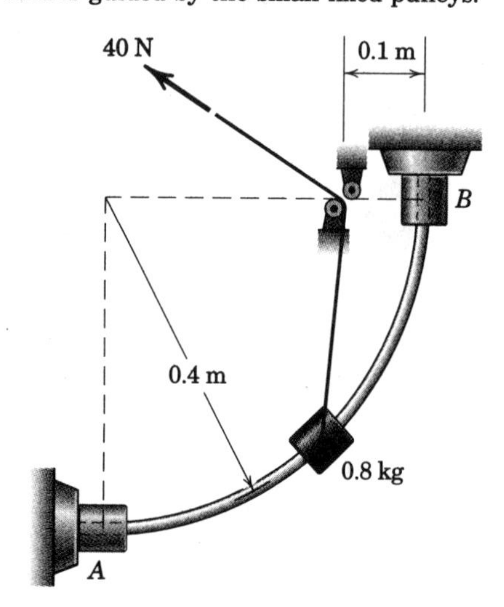

Problem 3/120

3/121 The 0.2-kg slider moves freely along the fixed curved rod from A to B in the vertical plane under the action of the constant 5-N tension in the cord. If the slider is released from rest at A, calculate its velocity v as it reaches B.

Ans. $v = 4.48$ m/s

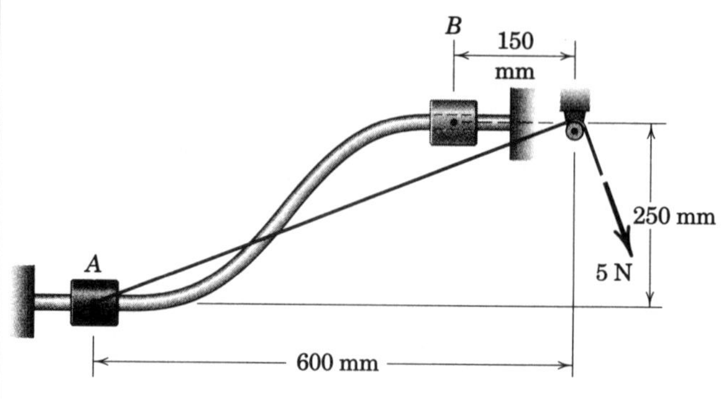

Problem 3/121

3/122 A department-store escalator handles a steady load of 30 people per minute in elevating them from the first to the second floor through a vertical rise of 7 m. The average person has a mass of 65 kg. If the motor which drives the unit delivers 3 kW, calculate the mechanical efficiency e of the system.

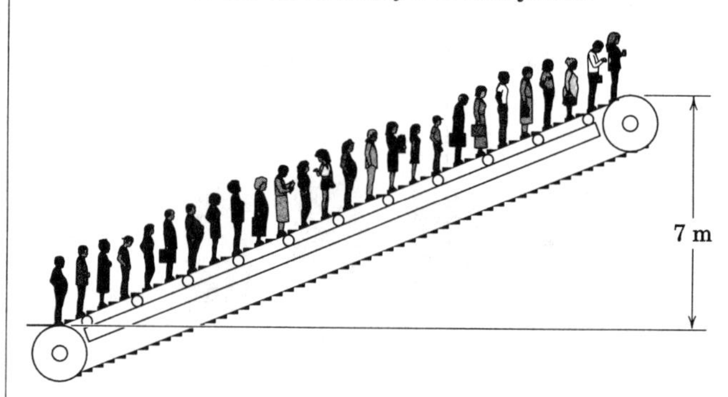

Problem 3/122

3/123 A car with a mass of 1500 kg starts from rest at the bottom of a 10-percent grade and acquires a speed of 50 km/h in a distance of 100 m with constant acceleration up the grade. What is the power P delivered to the drive wheels by the engine when the car reaches this speed?

Ans. $P = 40.4$ kW

3/124 In a test to determine the crushing characteristics of a packing material, a steel cone of mass m is released, falls a distance h, and then penetrates the material. The radius of the cone is proportional to the square of the distance from its tip. The resistance R of the material to penetration depends on the cross-sectional area of the penetrating object and thus is proportional to the fourth power of the cone penetration distance x, or $R = kx^4$. If the cone comes to rest at a distance $x = d$, determine the constant k in terms of the test conditions and results. Utilize a single application of the work-energy equation.

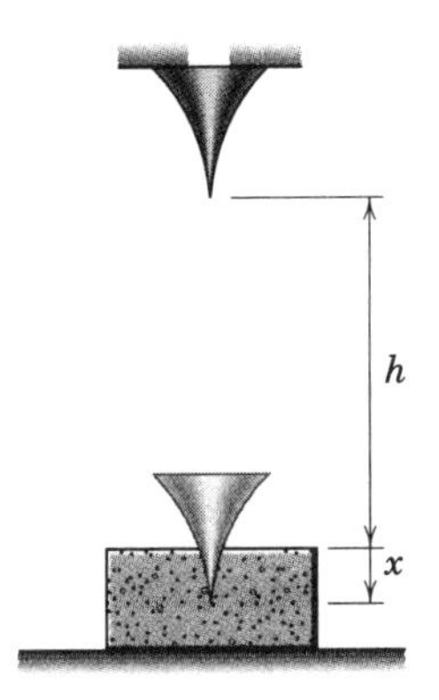

Problem 3/124

3/125 The motor unit A is used to elevate the 300-kg cylinder at a constant rate of 2 m/s. If the power meter B registers an electrical input of 2.20 kW, calculate the combined electrical and mechanical efficiency e of the system.

Ans. $e = 0.892$

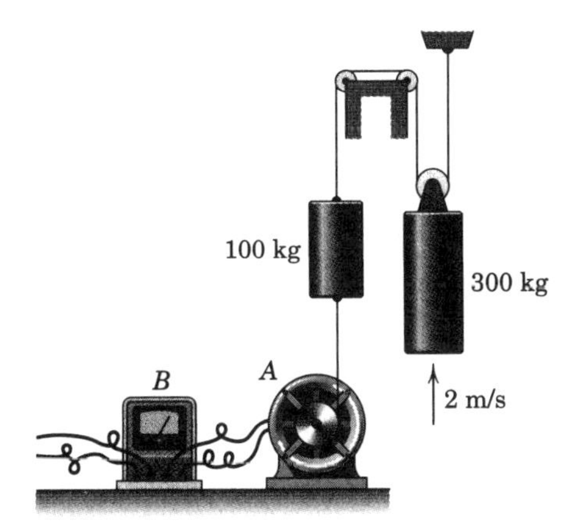

Problem 3/125

3/126 In a railroad classification yard, a 68-Mg freight car moving at 0.5 m/s at A encounters a retarder section of track at B which exerts a retarding force of 32 kN on the car in the direction opposite to motion. Over what distance x should the retarder be activated in order to limit the speed of the car to 3 m/s at C?

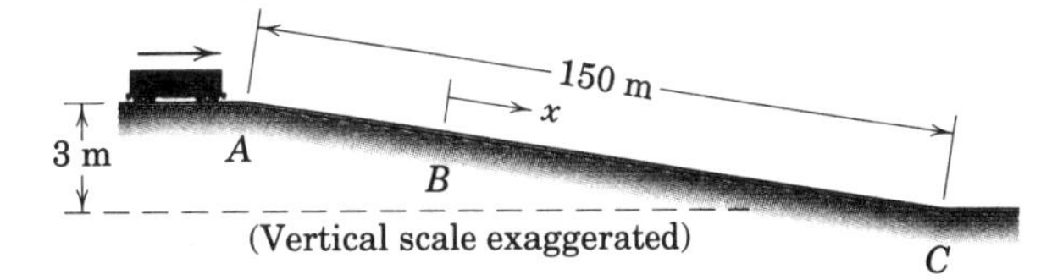

Problem 3/126

3/127 The 0.5-kg collar slides with negligible friction along the fixed spiral rod, which lies in the vertical plane. The rod has the shape of the spiral $r = 0.3\theta$, where r is in meters and θ is in radians. The collar is released from rest at A and slides to B under the action of a constant radial force $T = 10$ N. Calculate the velocity v of the slider as it reaches B.

Ans. $v = 5.30$ m/s

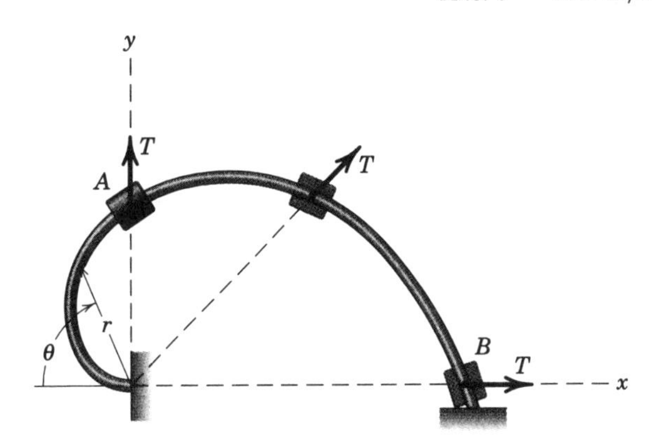

Problem 3/127

3/128 A small rocket-propelled test vehicle with a total mass of 100 kg starts from rest at A and moves with negligible friction along the track in the vertical plane as shown. If the propelling rocket exerts a constant thrust T of 1.5 kN from A to position B where it is shut off, determine the distance s which the vehicle rolls up the incline before stopping. The loss of mass due to the expulsion of gases by the rocket is small and may be neglected.

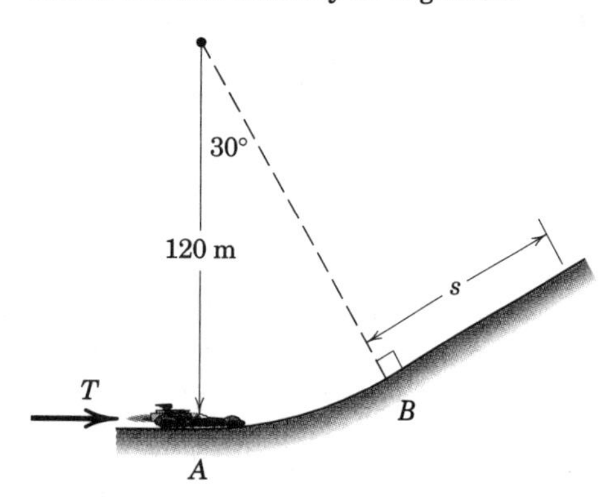

Problem 3/128

3/129 The small slider of mass m is released from rest while in position A and then slides along the vertical-plane track. The track is smooth from A to D and rough (coefficient of kinetic friction μ_k) from point D on. Determine (*a*) the normal force N_B exerted by the track on the slider just after it passes point B, (*b*) the normal force N_C exerted by the track on the slider as it passes the bottom point C, and (*c*) the distance s traveled along the incline past point D before the slider stops.

Ans. (*a*) $N_B = 4mg$, (*b*) $N_C = 7mg$

$$(c)\ s = \frac{4R}{1 + \mu_k\sqrt{3}}$$

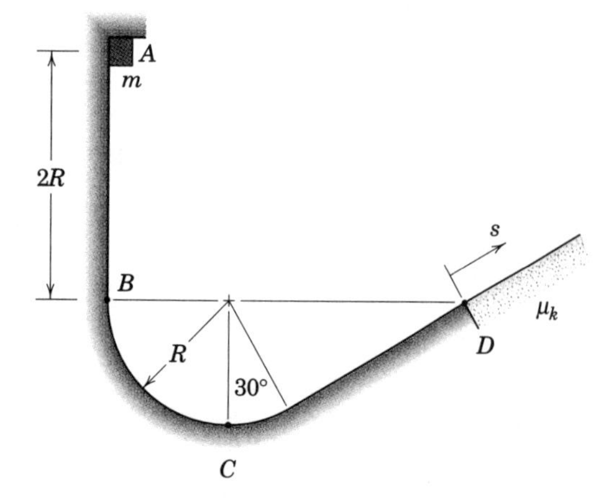

Problem 3/129

3/130 Each of the sliders A and B has a mass of 2 kg and moves with negligible friction in its respective guide, with y being in the vertical direction. A 20-N horizontal force is applied to the midpoint of the connecting link of negligible mass, and the assembly is released from rest with $\theta = 0$. Calculate the velocity v_A with which A strikes the horizontal guide when $\theta = 90°$.

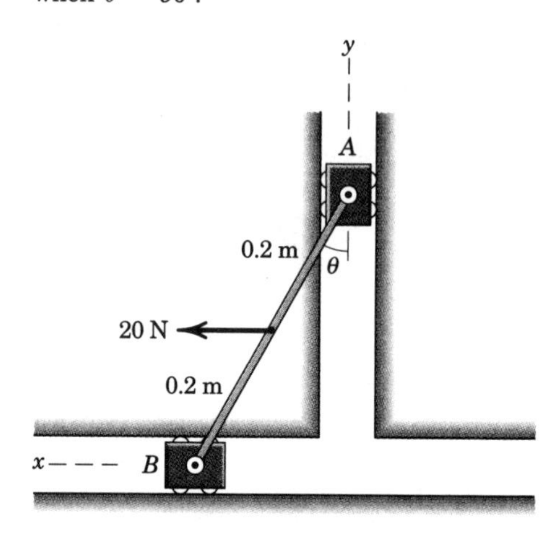

Problem 3/130

3/131 The ball is released from position A with a velocity of 3 m/s and swings in a vertical plane. At the bottom position, the cord strikes the fixed bar at B, and the ball continues to swing in the dashed arc. Calculate the velocity v_C of the ball as it passes position C.

Ans. $v_C = 3.59$ m/s

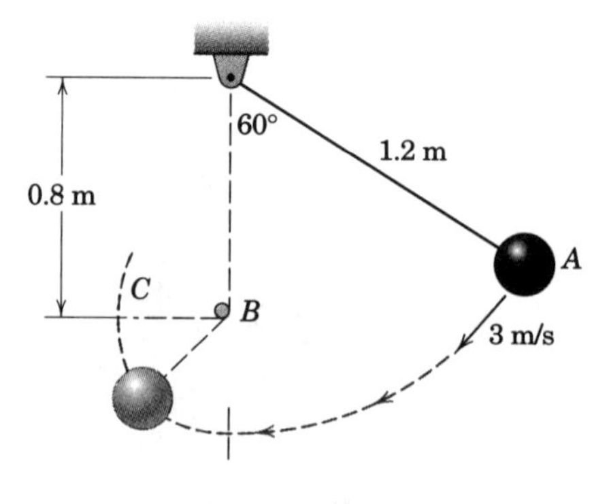

Problem 3/131

3/132 Calculate the horizontal velocity v with which the 20-kg carriage must strike the spring in order to compress it a maximum of 100 mm. The spring is known as a "hardening" spring, since its stiffness increases with deflection as shown in the accompanying graph.

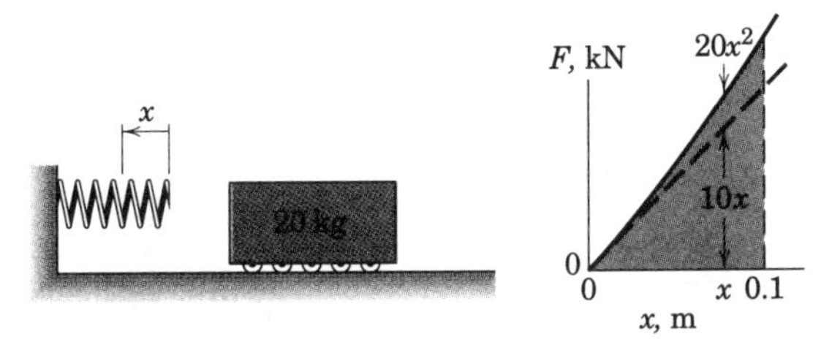

Problem 3/132

3/133 It is experimentally determined that the drive wheels of a car must exert a tractive force of 560 N on the road surface in order to maintain a steady vehicle speed of 90 km/h on a horizontal road. If it is known that the overall drivetrain efficiency is $e_m = 0.70$, determine the required motor power output P.

Ans. $P = 20$ kW

3/134 The 6-kg cylinder is released from rest in the position shown and falls on the spring, which has been initially precompressed 50 mm by the light strap and restraining wires. If the stiffness of the spring is 4 kN/m, compute the additional deflection δ of the spring produced by the falling cylinder before it rebounds.

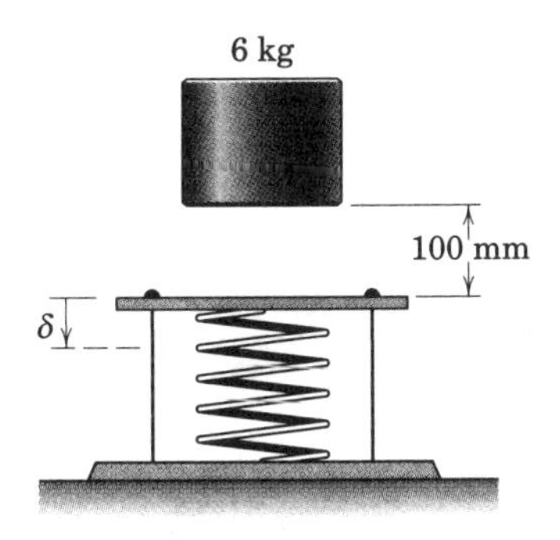

Problem 3/134

3/135 The 7-kg collar A slides with negligible friction on the fixed vertical shaft. When the collar is released from rest at the bottom position shown, it moves up the shaft under the action of the constant force $F = 200$ N applied to the cable. Calculate the stiffness k which the spring must have if its maximum compression is to be limited to 75 mm. The position of the small pulley at B is fixed.

Ans. $k = 8.79$ kN/m

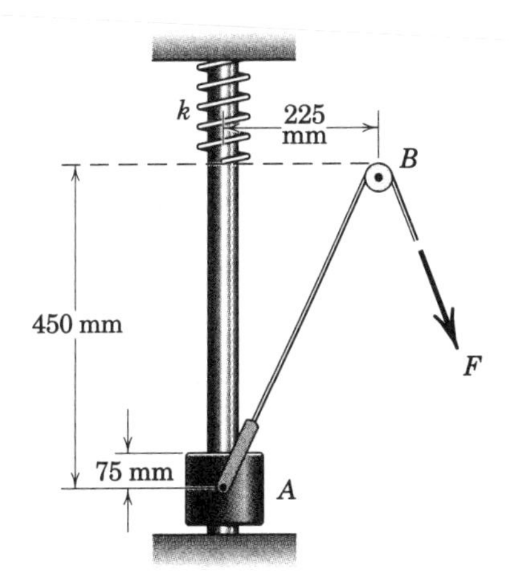

Problem 3/135

3/136 The vertical motion of the 20-kg block is controlled by the two forces P applied to the ends A and B of the linkage, where A and B are constrained to move in the horizontal guide. If forces $P = 1100$ N are applied with the linkage initially at rest with $\theta = 60°$, determine the upward velocity v of the block as θ approaches 180°. Neglect friction and the weight of the links and note that P is greater than its equilibrium value of $(5W/2) \cot 30° = 850$ N.

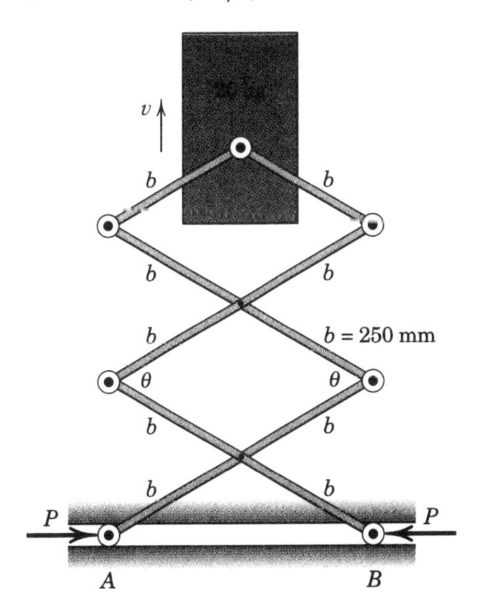

Problem 3/136

3/137 Extensive testing of an experimental 900-kg automobile reveals the aerodynamic drag force F_D and the total nonaerodynamic rolling-resistance force F_R to be as shown in the plot. Determine (*a*) the power required for steady speeds of 50 and 100 km/h on a level road, (*b*) the power required for a steady speed of 100 km/h both up and down a 6-percent incline, and (*c*) the steady speed at which no power is required going down the 6-percent incline.

Ans. (*a*) $P_{50} = 4.34$ kW, $P_{100} = 13.89$ kW
(*b*) $P_{up} = 28.6$ kW, $P_{down} = -800$ W
(*c*) $v = 105.6$ km/h

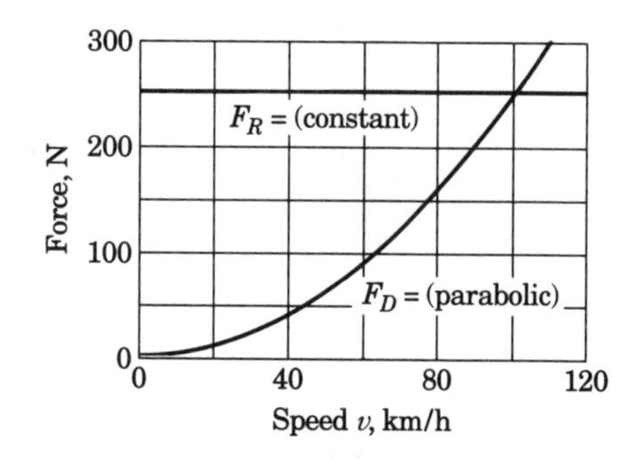

Problem 3/137

3/138 The 25-kg slider in the position shown has an initial velocity $v_0 = 0.6$ m/s on the inclined rail and slides under the influence of gravity and friction. The coefficient of kinetic friction between the slider and the rail is 0.5. Calculate the velocity of the slider as it passes the position for which the spring is compressed a distance $x = 100$ mm. The spring offers a compressive resistance C and is known as a "hardening" spring, since its stiffness increases with deflection as shown in the accompanying graph.

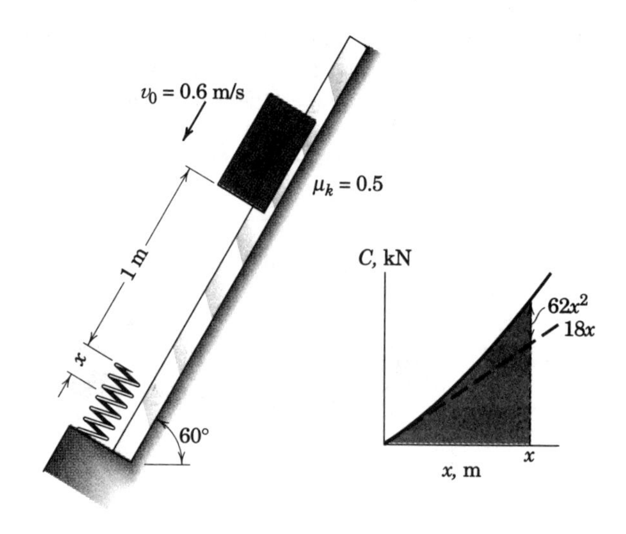

Problem 3/138

3/139 The spring attached to the 10-kg mass is nonlinear, having the force–deflection relationship shown in the figure, and is unstretched when $x = 0$. If the mass is moved to the position $x = 100$ mm and released from rest, determine its velocity v when $x = 0$. Determine the corresponding velocity v' if the spring were linear according to $F = 4x$, where x is in meters and the force F is in kilonewtons.

Ans. $v' = 1.899$ m/s

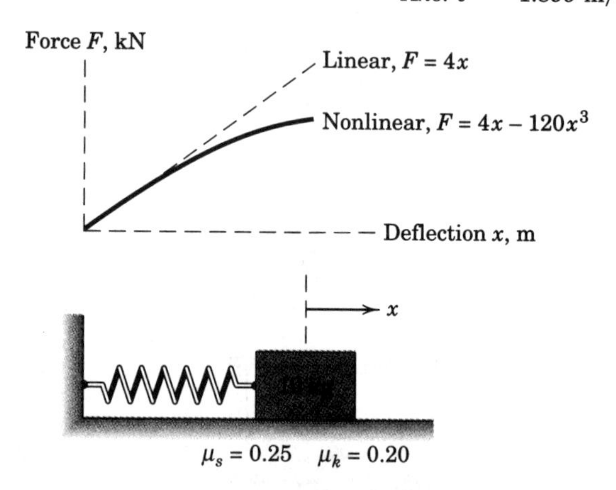

Problem 3/139

3/140 The car of mass m accelerates on a level road under the action of the driving force F from a speed v_1 to a higher speed v_2 in a distance s. If the engine develops a constant power output P, determine v_2. Treat the car as a particle under the action of the single horizontal force F.

Problem 3/140

3/141 The carriage of mass m is released from rest in its horizontal guide against the spring of stiffness k which has been compressed an amount x_0. Derive an expression for the power P developed by the spring in terms of the deflection x of the spring. Also determine the maximum power P_{max} and the corresponding value of x.

$$Ans.\ P = kx\sqrt{\frac{k}{m}(x_0{}^2 - x^2)}$$

$$P_{max} = \frac{k}{2}\sqrt{\frac{k}{m}}\,x_0{}^2 \text{ at } x = \frac{x_0}{\sqrt{2}}$$

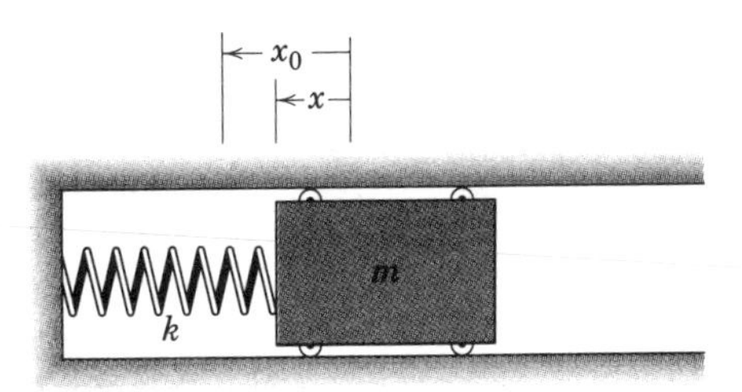

Problem 3/141

3/142 The three springs of equal moduli are unstretched when the cart is released from rest in the position $x = 0$. If $k = 120$ N/m and $m = 10$ kg, determine (*a*) the speed v of the cart when $x = 50$ mm, (*b*) the maximum displacement x_{max} of the cart, and (*c*) the steady-state displacement x_{ss} that would exist after all oscillations cease.

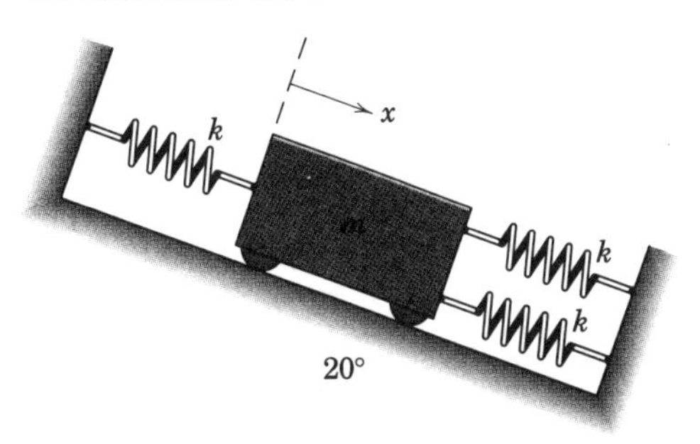

Problem 3/142

3/7 Potential Energy

In the previous article on work and kinetic energy, we isolated a particle or a combination of joined particles and determined the work done by gravity forces, spring forces, and other externally applied forces acting on the particle or system. We did this to evaluate U in the work-energy equation. In the present article we will introduce the concept of *potential energy* to treat the work done by gravity forces and by spring forces. This concept will simplify the analysis of many problems.

Gravitational Potential Energy

We consider first the motion of a particle of mass m in close proximity to the surface of the earth, where the gravitational attraction (weight) mg is essentially constant, Fig. 3/6*a*. The *gravitational potential energy* V_g of the particle is defined as the work mgh done *against* the gravitational field to elevate the particle a distance h above some arbitrary reference plane, where V_g is taken to be zero. Thus, we write the potential energy as

$$V_g = mgh \qquad \textbf{(3/14)}$$

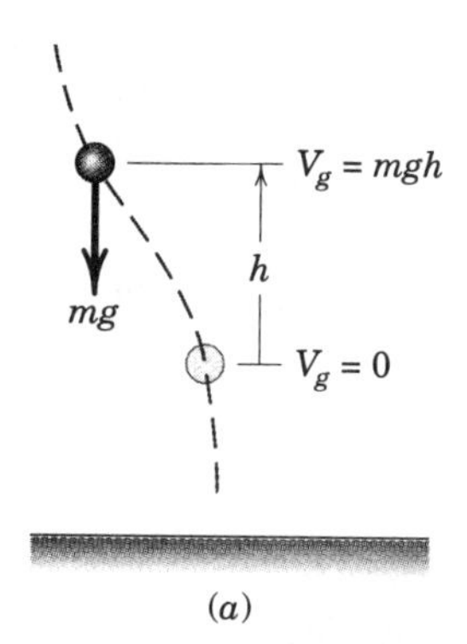

(*a*)

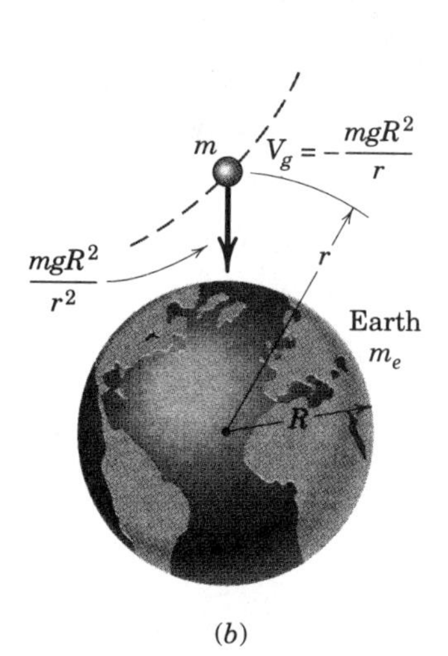

(*b*)

Figure 3/6

This work is called potential energy because it may be converted into energy if the particle is allowed to do work on a supporting body while it returns to its lower original datum plane. In going from one level at $h = h_1$ to a higher level at $h = h_2$, the *change* in potential energy becomes

$$\Delta V_g = mg(h_2 - h_1) = mg\Delta h$$

The corresponding work done *by* the gravitational force on the particle is $-mg\Delta h$. Thus, the work done by the gravitational force is the negative of the change in potential energy.

When large changes in altitude in the field of the earth are encountered, Fig. 3/6*b*, the gravitational force $Gmm_e/r^2 = mgR^2/r^2$ is no longer constant. The work done *against* this force to change the radial position of the particle from r to r' is the change $V_g' - V_g$ in gravitational potential energy, which is

$$\int_r^{r'} mgR^2 \frac{dr}{r^2} = mgR^2\left(\frac{1}{r} - \frac{1}{r'}\right) = V_g' - V_g$$

It is customary to take $V_g' = 0$ when $r' = \infty$, so that with this datum we have

$$V_g = -\frac{mgR^2}{r} \qquad \textbf{(3/15)}$$

In going from r_1 to r_2, the corresponding change in potential energy is

$$\Delta V_g = mgR^2\left(\frac{1}{r_1} - \frac{1}{r_2}\right)$$

which, again, is the *negative* of the work done *by* the gravitational force. We note that the potential energy of a given particle depends only on its position, h or r, and not on the particular path it followed in reaching that position.

Elastic Potential Energy

The second example of potential energy occurs in the deformation of an elastic body, such as a spring. The work which is done on the spring to deform it is stored in the spring and is called its *elastic potential energy* V_e. This energy is recoverable in the form of work done by the spring on the body attached to its movable end during the release of the deformation of the spring. For the one-dimensional linear spring of stiffness k, which we discussed in Art. 3/6 and illustrated in Fig. 3/4, the force supported by the spring at any deformation x, tensile or compressive, from its undeformed position is $F = kx$. Thus, we define the elastic potential energy of the spring as the work done on it to deform it an amount x, and we have

$$V_e = \int_0^x kx\,dx = \frac{1}{2}kx^2 \qquad \textbf{(3/16)}$$

If the deformation, either tensile or compressive, of a spring increases from x_1 to x_2 during the motion, then the change in potential energy of the spring is its final value minus its initial value or

$$\Delta V_e = \frac{1}{2}k(x_2{}^2 - x_1{}^2)$$

which is positive. Conversely, if the deformation of a spring decreases during the motion interval, then the change in potential energy of the spring becomes negative. The magnitude of these changes is represented by the shaded trapezoidal area in the F-x diagram of Fig. 3/4.

Because the force exerted *on* the spring *by* the moving body is equal and opposite to the force F exerted *by* the spring *on* the body (Fig. 3/4), it follows that the work done on the spring is the negative of the work done on the body. Therefore, we may replace the work U done by the spring on the body by $-\Delta V_e$, the negative of the potential energy change for the spring, provided the spring is now included within the system.

Work-Energy Equation

With the elastic member included in the system, we now modify the work-energy equation to account for the potential-energy terms. If $U'_{1\text{-}2}$ stands for the work of all external forces *other than* gravitational forces and spring forces, we may write Eq. 3/11 as $U'_{1\text{-}2} + (-\Delta V_g) + (-\Delta V_e) = \Delta T$ or

$$U'_{1\text{-}2} = \Delta T + \Delta V_g + \Delta V_e \qquad \textbf{(3/17)}$$

This alternative form of the work-energy equation is often far more convenient to use than Eq. 3/11, since the work of both gravity and spring forces is accounted for by focusing attention on the end-point positions of the particle and on the end-point lengths of the elastic spring. The path followed between these end-point positions is of no consequence in the evaluation of ΔV_g and ΔV_e.

Note that Eq. 3/17 may be rewritten in the equivalent form

$$T_1 + V_{g_1} + V_{e_1} + U'_{1\text{-}2} = T_2 + V_{g_2} + V_{e_2} \qquad \textbf{(3/17a)}$$

To help clarify the difference between the use of Eqs. 3/11 and 3/17, Fig. 3/7 shows schematically a particle of mass m constrained to move along a fixed path under the action of forces F_1 and F_2, the gravitational force $W = mg$, the spring force F, and the normal reaction N. In Fig. 3/7b, the particle is isolated with its free-body diagram. The work done by each of the forces F_1, F_2, W, and the spring force $F = kx$ is evaluated, say, from A to B, and equated to the change ΔT in kinetic energy using Eq. 3/11. The constraint reaction N, if normal to the path, will do no work. The alternative approach is shown in Fig. 3/7c, where the spring is included as a part of the isolated system. The work done during the interval by F_1 and F_2 is the $U'_{1\text{-}2}$-term of Eq. 3/17 with the changes in elastic and gravitational potential energies included on the energy side of the equation.

We note with the first approach that the work done by $F = kx$ could require a somewhat awkward integration to account for the changes in

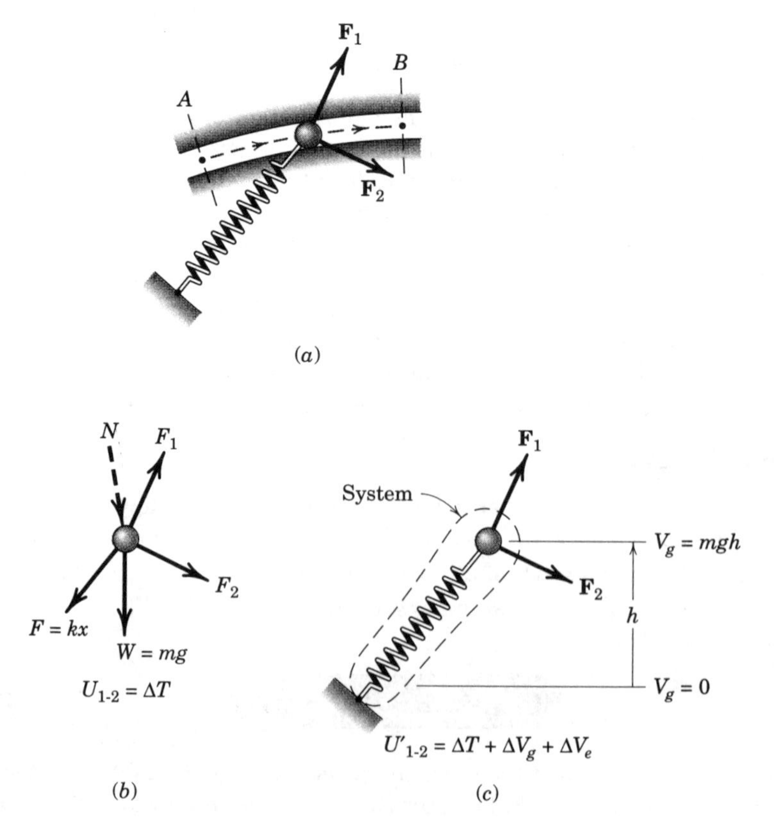

Figure 3/7

magnitude and direction of F as the particle moves from A to B. With the second approach, however, only the initial and final lengths of the spring are required to evaluate ΔV_e. This greatly simplifies the calculation.

We may rewrite the alternative work-energy relation, Eq. 3/17, for a particle-and-spring system as

$$U'_{1\text{-}2} = \Delta(T + V_g + V_e) = \Delta E \tag{3/17b}$$

where $E = T + V_g + V_e$ is the total mechanical energy of the particle and its attached linear spring. Equation 3/17*b* states that the net work done on the system by all forces other than gravitational forces and elastic forces equals the change in the total mechanical energy of the system. For problems where the only forces are gravitational, elastic, and nonworking constraint forces, the U'-term is zero, and the energy equation becomes merely

$$\Delta E = 0 \qquad \text{or} \qquad E = \text{constant} \tag{3/18}$$

When E is constant, we see that transfers of energy between kinetic and potential may take place as long as the total mechanical energy $T + V_g + V_e$ does not change. Equation 3/18 expresses the *law of conservation of dynamical energy*.

Conservative Force Fields*

We have observed that the work done against a gravitational or an elastic force depends only on the net change of position and not on the particular path followed in reaching the new position. Forces with this characteristic are associated with *conservative force fields*, which possess an important mathematical property.

Consider a force field where the force $\mathbf{F}$ is a function of the coordinates, Fig. 3/8. The work done by $\mathbf{F}$ during a displacement $d\mathbf{r}$ of its point of application is $dU = \mathbf{F}\cdot d\mathbf{r}$. The total work done along its path from 1 to 2 is

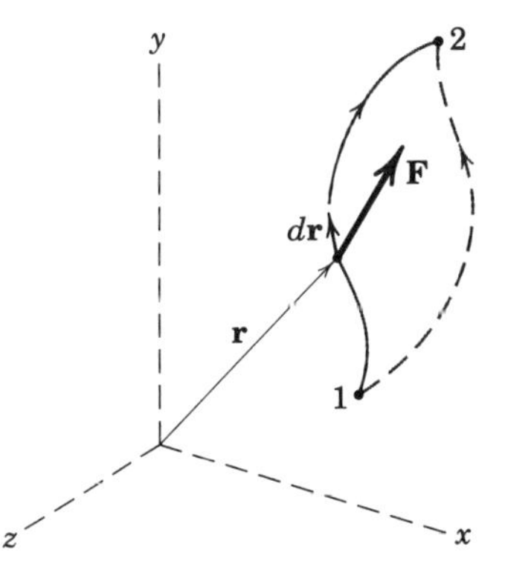

Figure 3/8

$$U = \int \mathbf{F}\cdot d\mathbf{r} = \int (F_x\,dx + F_y\,dy + F_z\,dz)$$

The integral $\int \mathbf{F}\cdot d\mathbf{r}$ is a line integral which depends, in general, on the particular path followed between any two points 1 and 2 in space. If, however, $\mathbf{F}\cdot d\mathbf{r}$ is an *exact differential*† $-dV$ of some scalar function V of the coordinates, then

$$U_{1\text{-}2} = \int_{V_1}^{V_2} -dV = -(V_2 - V_1) \tag{3/19}$$

*Optional.

†Recall that a function $d\phi = P\,dx + Q\,dy + R\,dz$ is an exact differential in the coordinates x-y-z if

$$\frac{\partial P}{\partial y} = \frac{\partial Q}{\partial x} \qquad \frac{\partial P}{\partial z} = \frac{\partial R}{\partial x} \qquad \frac{\partial Q}{\partial z} = \frac{\partial R}{\partial y}$$

which depends only on the end points of the motion and which is thus *independent* of the path followed. The minus sign before dV is arbitrary but is chosen to agree with the customary designation of the sign of potential energy change in the gravity field of the earth.

If V exists, the differential change in V becomes

$$dV = \frac{\partial V}{\partial x}\,dx + \frac{\partial V}{\partial y}\,dy + \frac{\partial V}{\partial z}\,dz$$

Comparison with $-dV = \mathbf{F}\cdot d\mathbf{r} = F_x\,dx + F_y\,dy + F_z\,dz$ gives us

$$F_x = -\frac{\partial V}{\partial x} \qquad F_y = -\frac{\partial V}{\partial y} \qquad F_z = -\frac{\partial V}{\partial z}$$

The force may also be written as the vector

$$\mathbf{F} = -\boldsymbol{\nabla} V \tag{3/20}$$

where the symbol $\boldsymbol{\nabla}$ stands for the vector operator "del", which is

$$\boldsymbol{\nabla} = \mathbf{i}\frac{\partial}{\partial x} + \mathbf{j}\frac{\partial}{\partial y} + \mathbf{k}\frac{\partial}{\partial z}$$

The quantity V is known as the *potential function*, and the expression $\boldsymbol{\nabla} V$ is known as the *gradient of the potential function.*

When force components are derivable from a potential as described, the force is said to be *conservative*, and the work done by $\mathbf{F}$ between any two points is independent of the path followed.

Sample Problem 3/16

The 10-kg slider A moves with negligible friction up the inclined guide. The attached spring has a stiffness of 60 N/m and is stretched 0.6 m in position A, where the slider is released from rest. The 250-N force is constant and the pulley offers negligible resistance to the motion of the cord. Calculate the velocity v of the slider as it passes point C.

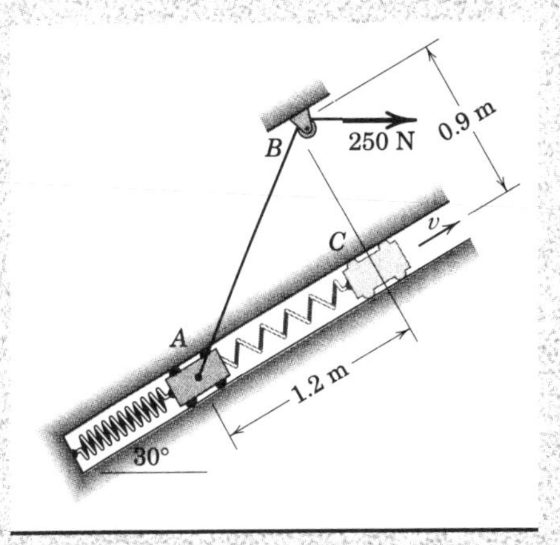

Solution. The slider and inextensible cord together with the attached spring will be analyzed as a system, which permits the use of Eq. 3/17. The only nonpotential force doing work on this system is the 250-N tension applied to the cord. While the slider moves from A to C, the point of application of the 250-N force moves a distance of $\overline{AB} - \overline{BC}$ or $1.5 - 0.9 = 0.6$ m.

① $$U'_{1\text{-}2} = 250(0.6) = 150 \text{ J}$$

The change in kinetic energy of the slider is

$$\Delta T = \tfrac{1}{2}m(v^2 - v_0^2) = \tfrac{1}{2}(10)v^2$$

where the initial velocity v_0 is zero. The change in gravitational potential energy is

② $$\Delta V_g = mg(\Delta h) = 10(9.81)(1.2 \sin 30°) = 58.9 \text{ J}$$

The change in elastic potential energy is

③ $$\Delta V_e = \tfrac{1}{2}k(x_2^2 - x_1^2) = \tfrac{1}{2}(60)([1.2 + 0.6]^2 - [0.6]^2) = 86.4 \text{ J}$$

Substitution into the alternative work-energy equation gives

$$[U'_{1\text{-}2} = \Delta T + \Delta V_g + \Delta V_e] \quad 150 = \tfrac{1}{2}(10)v^2 + 58.9 + 86.4$$

$$v = 0.974 \text{ m/s} \qquad \textit{Ans.}$$

Helpful Hints

① The reactions of the guides on the slider are normal to the direction of motion and do no work.

② Since the center of mass of the slider has an upward component of displacement, ΔV_g is positive.

③ Be very careful not to make the mistake of using $\frac{1}{2}k(x_2 - x_1)^2$ for ΔV_e. We want the difference of the squares and not the square of the difference.

Sample Problem 3/17

The 3-kg slider is released from rest at point A and slides with negligible friction in a vertical plane along the circular rod. The attached spring has a stiffness of 350 N/m and has an unstretched length of 0.6 m. Determine the velocity of the slider as it passes position B.

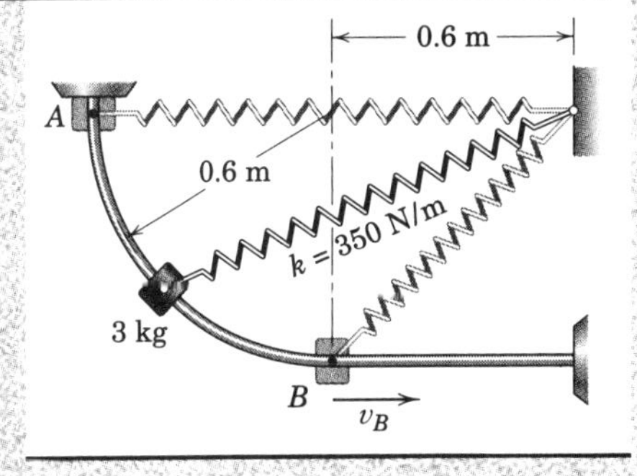

Solution. The work done by the weight and the spring force on the slider will be treated as changes in the potential energies, and the reaction of the rod on the slider is normal to the motion and does no work. Hence, $U'_{1\text{-}2} = 0$. The changes in the potential and kinetic energies for the system of slider and spring are

① $$\Delta V_e = \tfrac{1}{2}k(x_B^2 - x_A^2) = \tfrac{1}{2}(350)\{(0.6[\sqrt{2} - 1])^2 - (0.6)^2\} = -52.2 \text{ J}$$

$$\Delta V_g = W\Delta h = 3(9.81)(-0.6) = -17.66 \text{ J}$$

$$\Delta T = \frac{1}{2}m(v_B^2 - v_A^2) = \frac{1}{2}3(v_B^2 - 0) = 1.5v_B^2$$

$$[\Delta T + \Delta V_g + \Delta V_e = 0] \quad 1.5v_B^2 - 17.66 - 52.2 = 0$$

$$v_B = 6.82 \text{ m/s} \qquad \textit{Ans.}$$

① Note that if we evaluated the work done by the spring force acting on the slider by means of the integral $\int \mathbf{F} \cdot d\mathbf{r}$, it would necessitate a lengthy computation to account for the change in the magnitude of the force, along with the change in the angle between the force and the tangent to the path. Note further that v_B depends only on the end conditions of the motion and does not require knowledge of the shape of the path.

PROBLEMS

Introductory Problems

3/143 The 0.9-kg collar is released from rest at A and slides freely up the inclined rod, striking the stop at B with a velocity v. The spring of stiffness k = 24 N/m has an unstretched length of 375 mm. Calculate v.

Ans. $v = 1.156$ m/s

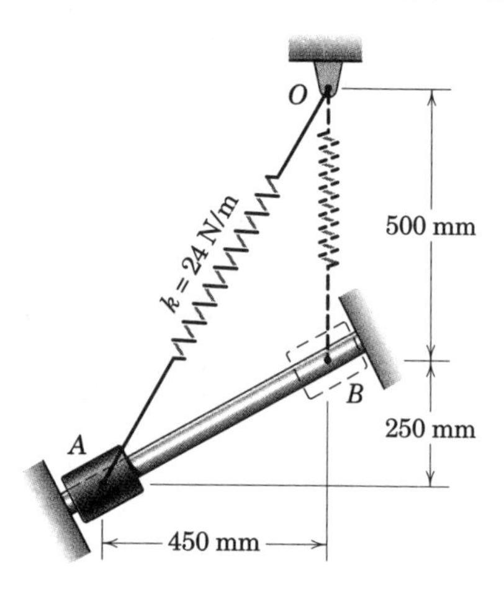

Problem 3/143

3/144 The 4-kg slider is released from rest at A and slides with negligible friction down the circular rod in the vertical plane. Determine (a) the velocity v of the slider as it reaches the bottom at B and (b) the maximum deformation x of the spring.

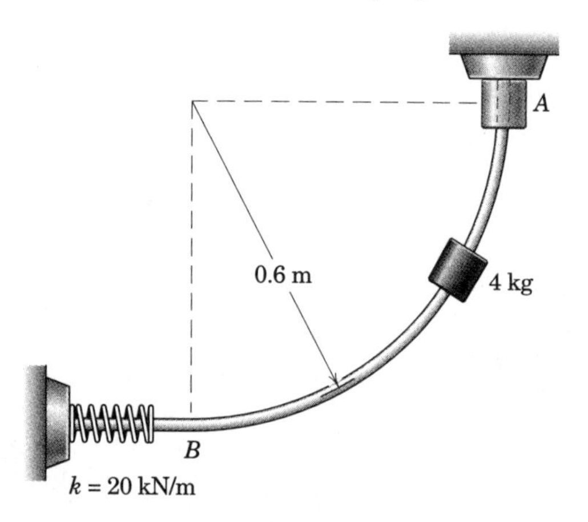

Problem 3/144

3/145 Point P on the 2-kg cylinder has an initial velocity v_0 = 0.8 m/s as it passes position A. Neglect the mass of the pulleys and cable and determine the distance y of point P below A when the 3-kg cylinder has acquired an upward velocity of 0.6 m/s.

Ans. $y = 0.224$ m

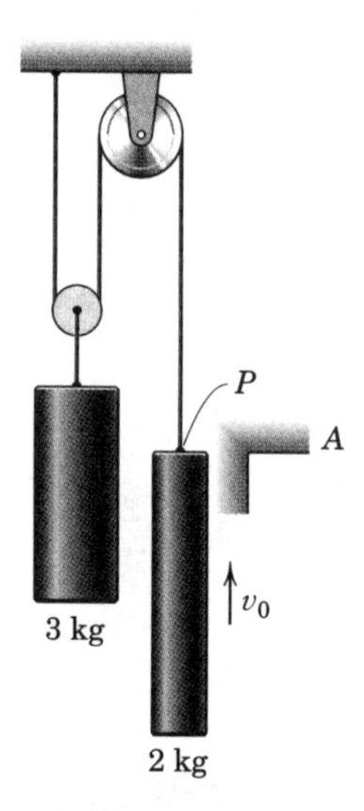

Problem 3/145

3/146 The 1.2-kg slider is released from rest in position A and slides without friction along the vertical-plane guide shown. Determine (a) the speed v_B of the slider as it passes position B and (b) the maximum deflection δ of the spring.

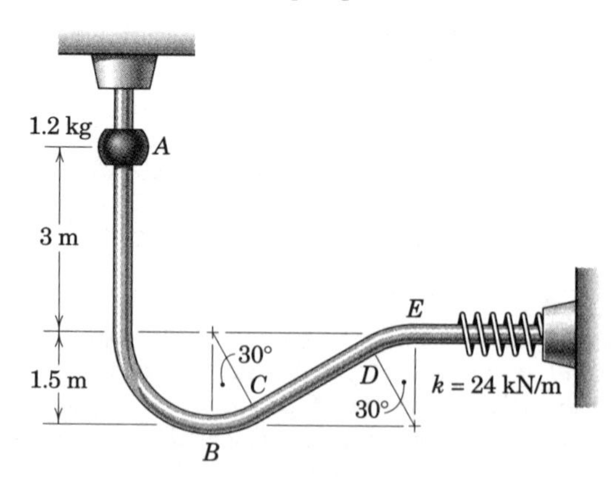

Problem 3/146

3/147 The 1.2-kg slider of the system of Prob. 3/146 is released from rest in position A and slides without friction along the vertical-plane guide. Determine the normal force exerted by the guide on the slider (*a*) just before it passes point C, (*b*) just after it passes point C, and (*c*) just before it passes point E.

Ans. (*a*) $N_C = 77.7$ N, (*b*) $N_C = 10.19$ N
(*c*) $N_E = 35.3$ N (down)

3/148 A bead with a mass of 0.25 kg is released from rest at A and slides down and around the fixed smooth wire. Determine the force N between the wire and the bead as it passes point B.

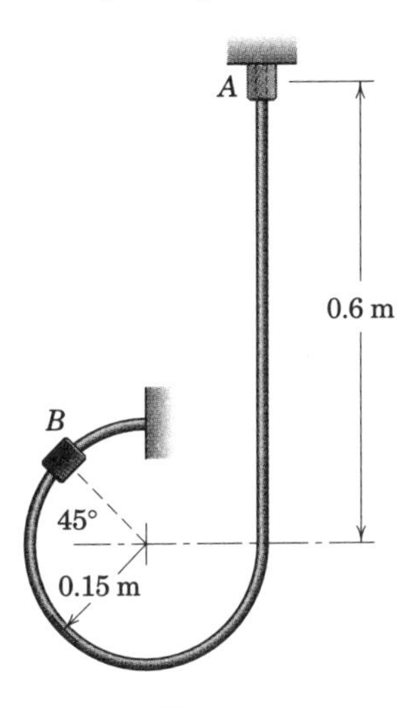

Problem 3/148

3/149 The spring of constant k is unstretched when the slider of mass m passes position B. If the slider is released from rest in position A, determine its speed as it passes points B and C. What is the normal force exerted by the guide on the slider at position C? Neglect friction between the mass and the circular guide, which lies in a vertical plane.

$$\textit{Ans. } v_B = \sqrt{2gR + \frac{kR^2}{m}(3 - 2\sqrt{2})}$$

$$v_C = \sqrt{4gR + \frac{kR^2}{m}(3 - 2\sqrt{2})}$$

$$N = m\left[5g + \frac{kR}{m}(3 - 2\sqrt{2})\right]$$

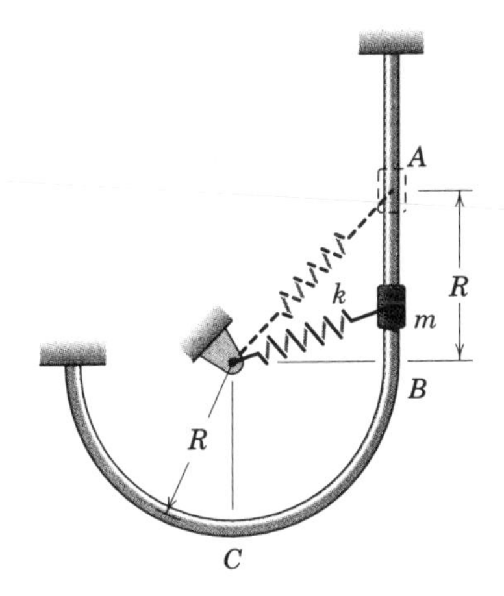

Problem 3/149

3/150 It is desired that the 45-kg container, when released from rest in the position shown, have no velocity after dropping 2 m to the platform below. Specify the proper mass m of the counterbalancing cylinder.

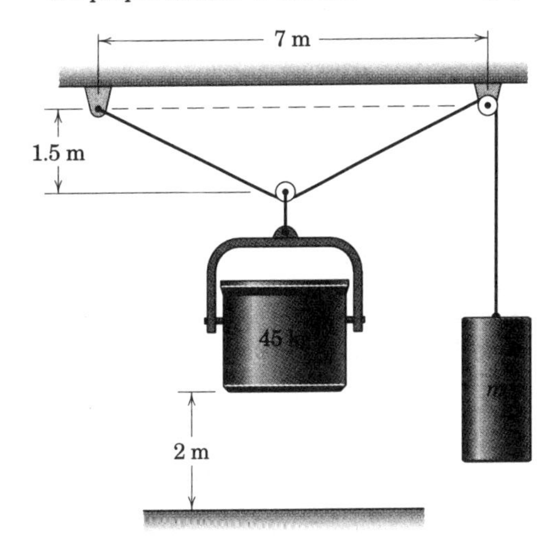

Problem 3/150

3/151 The light rod is pivoted at O and carries the 2- and 4-kg particles. If the rod is released from rest at $\theta = 60°$ and swings in the vertical plane, calculate (*a*) the velocity v of the 2-kg particle just before it hits the spring in the dashed position and (*b*) the maximum compression x of the spring. Assume that x is small so that the position of the rod when the spring is compressed is essentially horizontal.

Ans. (*a*) $v = 1.162$ m/s, (*b*) $x = 12.07$ mm

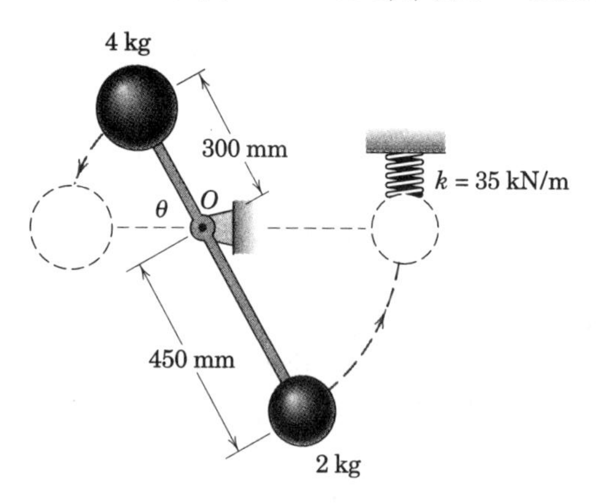

Problem 3/151

3/153 In the design of an inside loop for an amusement park ride, it is desired to maintain the same centripetal acceleration throughout the loop. Assume negligible loss of energy during the motion and determine the radius of curvature ρ of the path as a function of the height y above the low point A, where the velocity and radius of curvature are v_0 and ρ_0, respectively. For a given value of ρ_0, what is the minimum value of v_0 for which the vehicle will not leave the track at the top of the loop?

Ans. $\rho = \rho_0\left(1 - \dfrac{2gy}{v_0^2}\right)$, $v_{0_{min}} = \sqrt{\rho_0 g}$

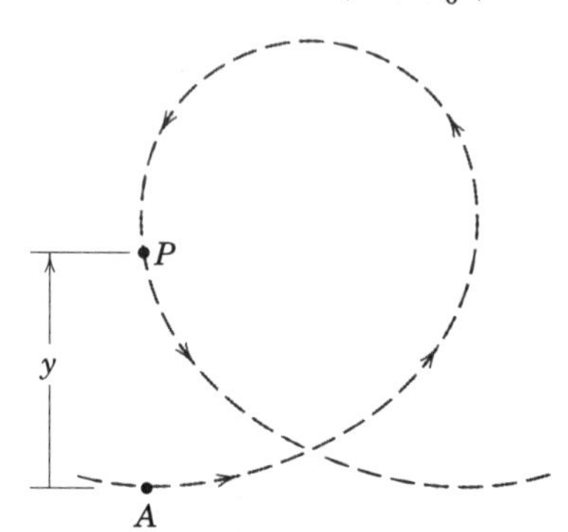

Problem 3/153

Representative Problems

3/152 The 1.5-kg slider C moves along the fixed rod under the action of the spring whose unstretched length is 0.3 m. If the velocity of the slider is 2 m/s at point A and 3 m/s at point B, calculate the work U_f done by friction between these two points. Also, determine the average friction force acting on the slider between A and B if the length of the path is 0.70 m. The x-y plane is horizontal.

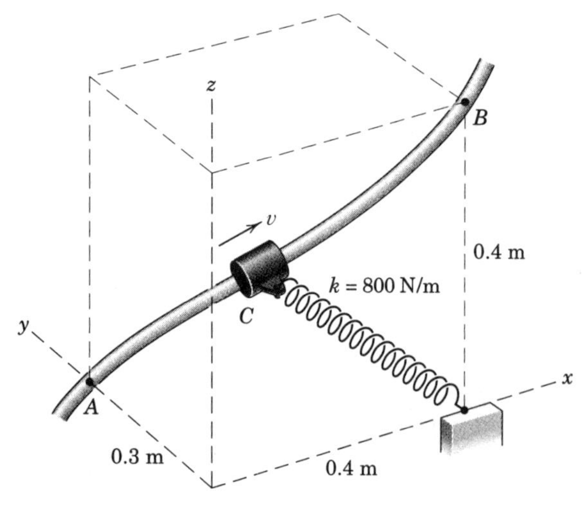

Problem 3/152

3/154 The springs are undeformed in the position shown. If the 6-kg collar is released from rest in the position where the lower spring is compressed 125 mm, determine the maximum compression x_B of the upper spring.

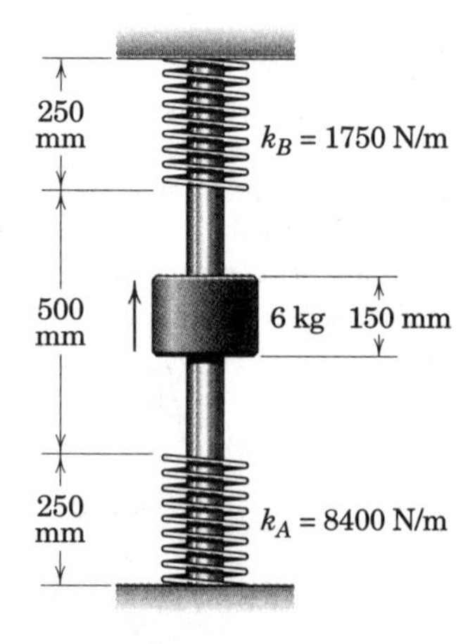

Problem 3/154

3/155 The two springs, each of stiffness $k = 1.2$ kN/m, are of equal length and undeformed when $\theta = 0$. If the mechanism is released from rest in the position $\theta = 20°$, determine its angular velocity $\dot{\theta}$ when $\theta = 0$. The mass m of each sphere is 3 kg. Treat the spheres as particles and neglect the masses of the light rods and springs.

Ans. $\dot{\theta} = 4.22$ rad/s

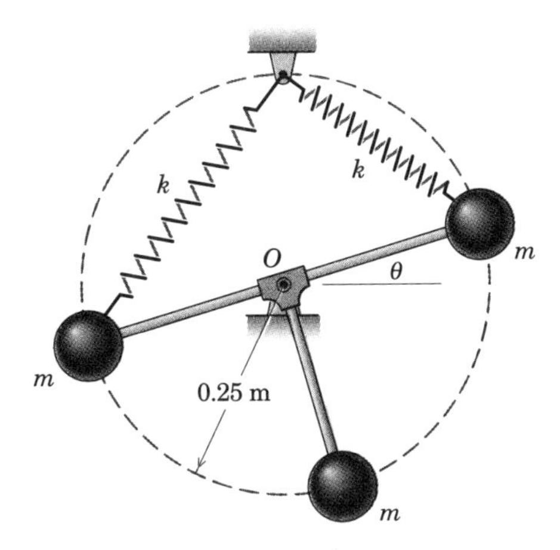

Problem 3/155

3/156 The 1.5-kg ball is given an initial velocity $v_A = 2.5$ m/s in the vertical plane at position A, where the two horizontal attached springs are unstretched. The ball follows the dashed path shown and crosses point B, which is 125 mm directly below A. Calculate the velocity v_B of the ball at B. Each spring has a stiffness of 1800 N/m.

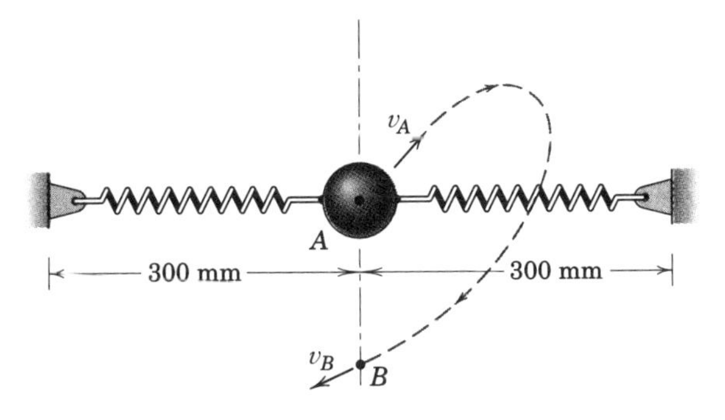

Problem 3/156

3/157 The spring has an unstretched length of 625 mm. If the system is released from rest in the position shown, determine the speed v of the ball (*a*) when it has dropped a vertical distance of 250 mm and (*b*) when the rod has rotated 35°.

Ans. (*a*) $v = 0.919$ m/s
(*b*) $v = 0.470$ m/s

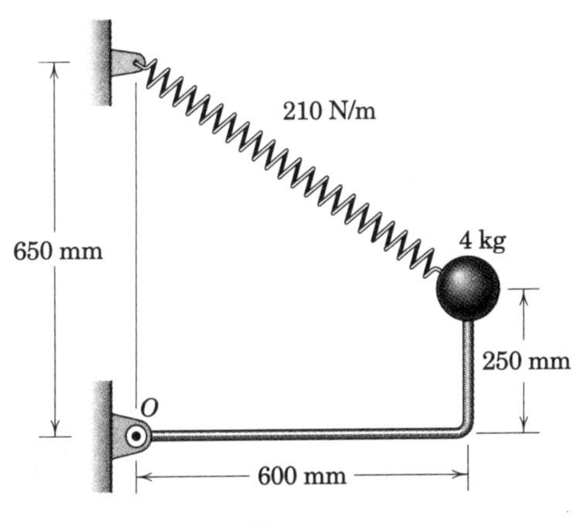

Problem 3/157

3/158 If the system is released from rest, determine the speeds of both masses after B has moved 1 m. Neglect friction and the masses of the pulleys.

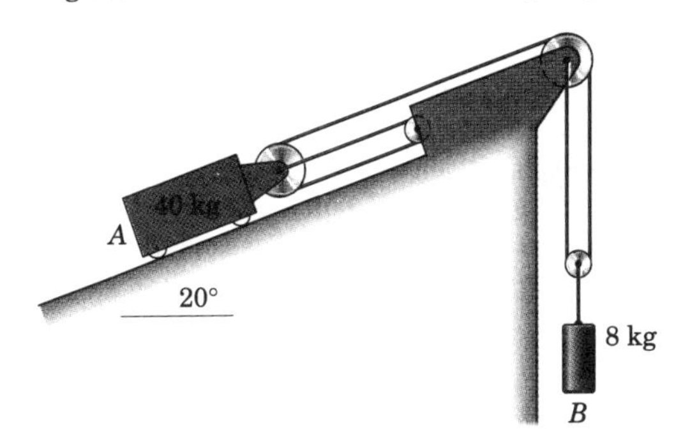

Problem 3/158

3/159 The two 1.5-kg spheres are released from rest and gently nudged outward from the position $\theta = 0$ and then rotate in a vertical plane about the fixed centers of their attached gears, thus maintaining the same angle θ for both rods. Determine the velocity v of each sphere as the rods pass the position $\theta = 30°$. The spring is unstretched when $\theta = 0$, and the masses of the two identical rods and the two gear wheels may be neglected.

Ans. $v = 0.331$ m/s

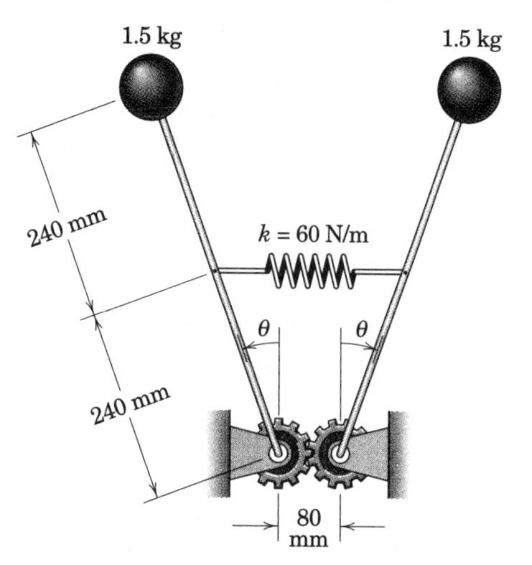

Problem 3/159

3/160 A projectile is fired vertically upward from the north pole with a velocity v_0. Calculate the least value of v_0 which will allow the projectile to escape from the earth's gravitational pull assuming no atmospheric resistance. The earth's radius is 6371 km. Use the absolute value $g = 9.825$ m/s^2.

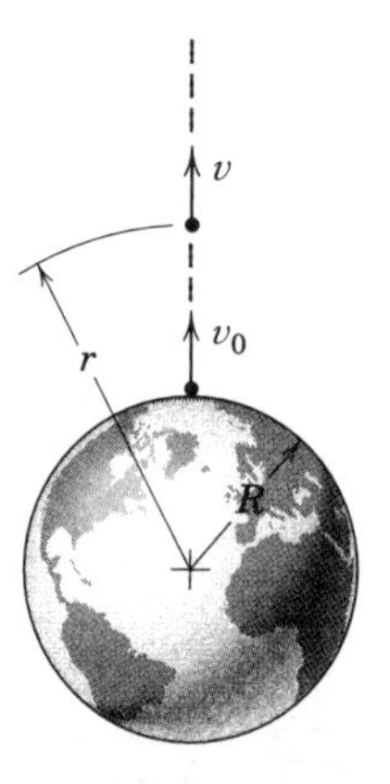

Problem 3/160

3/161 A satellite is put into an elliptical orbit around the earth and has a velocity v_P at the perigee position P. Determine the expression for the velocity v_A at the apogee position A. The radii to A and P are, respectively, r_A and r_P. Note that the total energy remains constant.

Ans. $v_A = \sqrt{v_P{}^2 - 2gR^2\left(\frac{1}{r_P} - \frac{1}{r_A}\right)}$

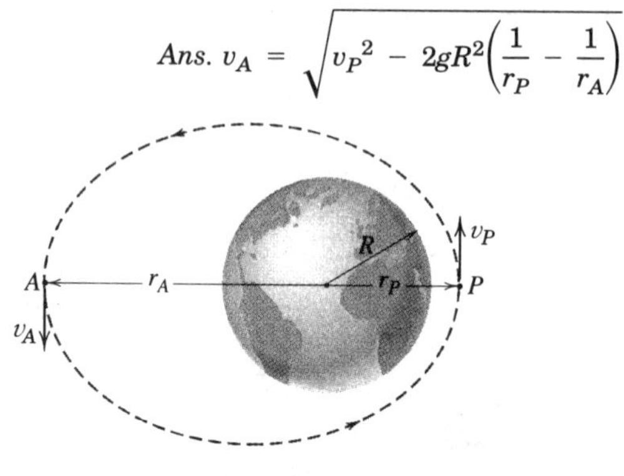

Problem 3/161

3/162 A rocket launches an unpowered space capsule at point A with an absolute velocity $v_A = 13\,000$ km/h at an altitude of 40 km. After the capsule has traveled a distance of 400 km measured along its absolute space trajectory, its velocity at B is 12 400 km/h and its altitude is 80 km. Determine the average resistance P to motion in the rarified atmosphere. The mass of the capsule is 22 kg, and the mean radius of the earth is 6371 km. Consider the center of the earth fixed in space.

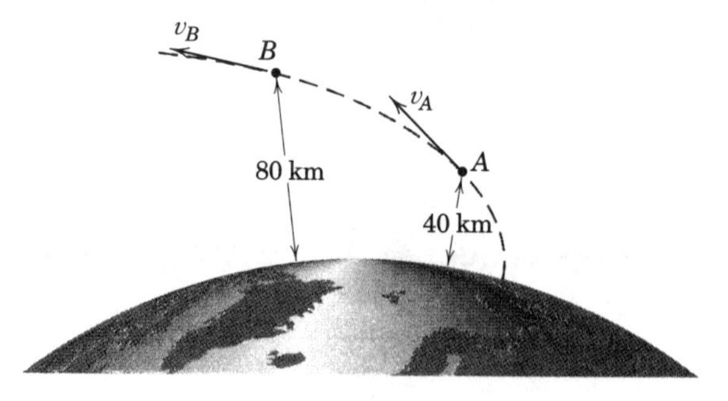

Problem 3/162

3/163 Upon its return voyage from a space mission, the spacecraft has a velocity of 24 000 km/h at point A, which is 7000 km from the center of the earth. Determine the velocity of the spacecraft when it reaches point B, which is 6500 km from the center of the earth. The trajectory between these two points is outside the effect of the earth's atmosphere.

Ans. $v_B = 26\ 300$ km/h

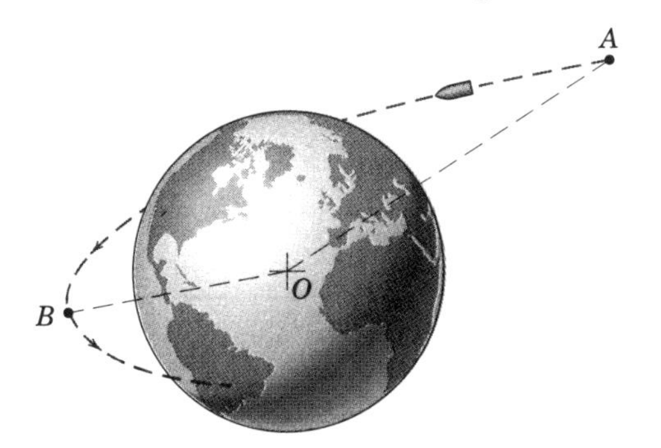

Problem 3/163

3/164 The shank of the 2-kg vertical plunger occupies the dashed position when resting in equilibrium against the spring of stiffness $k = 1.6$ kN/m. The upper end of the spring is welded to the plunger, and the lower end is welded to the base plate. If the plunger is lifted 40 mm above its equilibrium position and released from rest, calculate its velocity v as it strikes the button A. Friction is negligible.

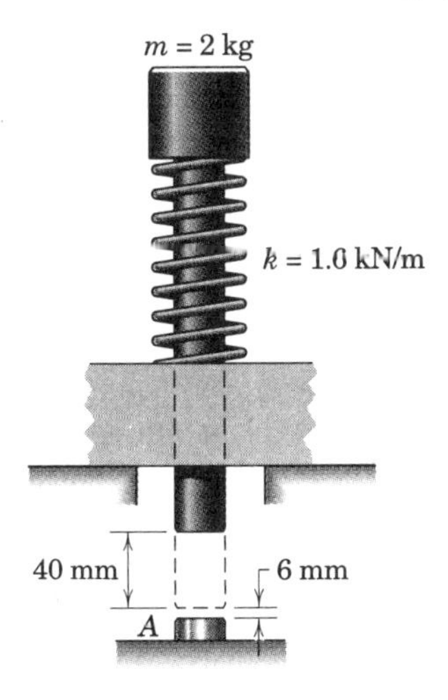

Problem 3/164

3/165 The 5-kg cylinder is released from rest in the position shown and compresses the spring of stiffness $k = 1.8$ kN/m. Determine the maximum compression x_{max} of the spring and the maximum velocity v_{max} of the cylinder along with the corresponding deflection x of the spring.

Ans. $x_{max} = 105.9$ mm, $v_{max} = 1.493$ m/s
$x = 27.2$ mm

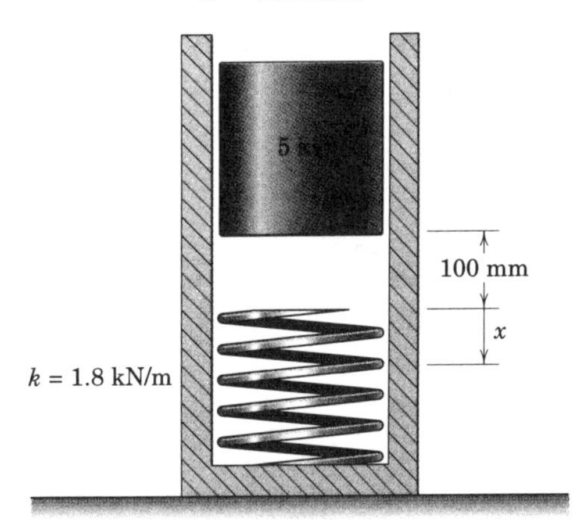

Problem 3/165

3/166 The cylinder of mass m is attached to the collar bracket at A by a spring of stiffness k. The collar fits loosely on the vertical shaft, which is lowering both the collar and the suspended cylinder with a constant velocity v. When the collar strikes the base B, it stops abruptly with essentially no rebound. Determine the maximum additional deflection δ of the spring after the impact.

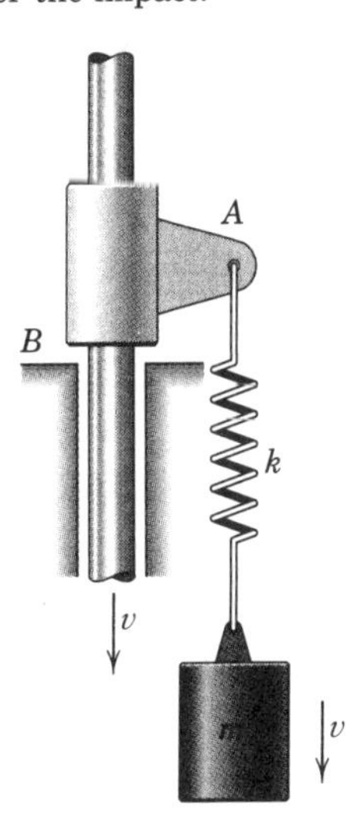

Problem 3/166

3/167 The two right-angle rods with attached spheres are released from rest in the position $\theta = 0$. If the system is observed to momentarily come to rest when $\theta = 45°$, determine the spring constant k. The spring is unstretched when $\theta = 0$. Treat the spheres as particles and neglect friction.

Ans. $k = 155.1$ N/m

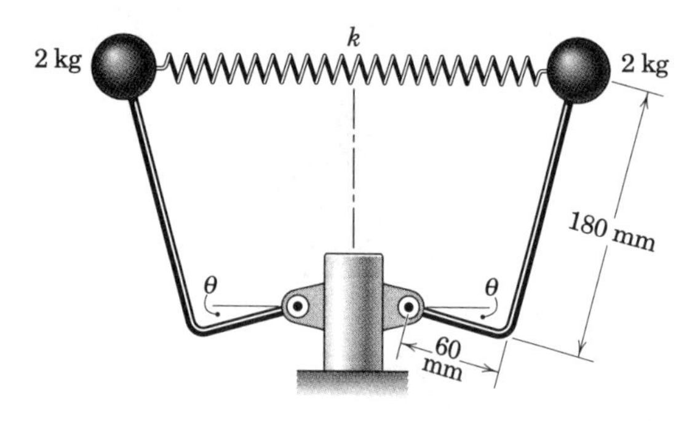

Problem 3/167

3/168 The mechanism is released from rest with $\theta = 180°$, where the uncompressed spring of stiffness $k = 900$ N/m is just touching the underside of the 4-kg collar. Determine the angle θ corresponding to the maximum compression of the spring. Motion is in the vertical plane, and the mass of the links may be neglected.

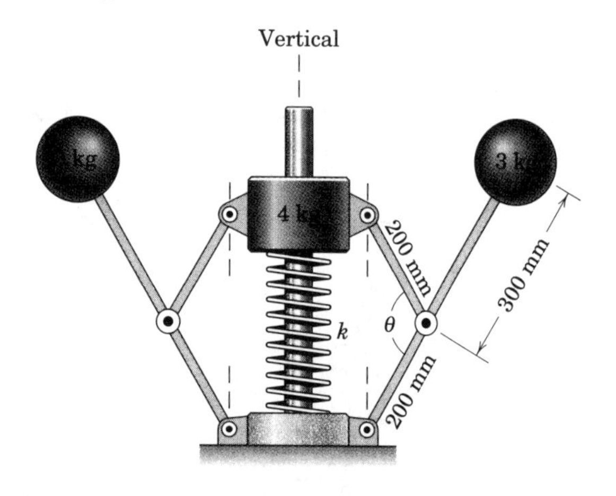

Problem 3/168

3/169 The fixed point O is located at one of the two foci of the elliptical guide. The spring has a stiffness of 3 N/m and is unstretched when the slider is at A. If the speed v_A is such that the speed of the 0.4-kg slider approaches zero at C, determine its speed at point B. The smooth guide lies in a horizontal plane. (If necessary, refer to Eqs. 3/39 for elliptical geometry.)

Ans. $v_B = 2.51$ m/s

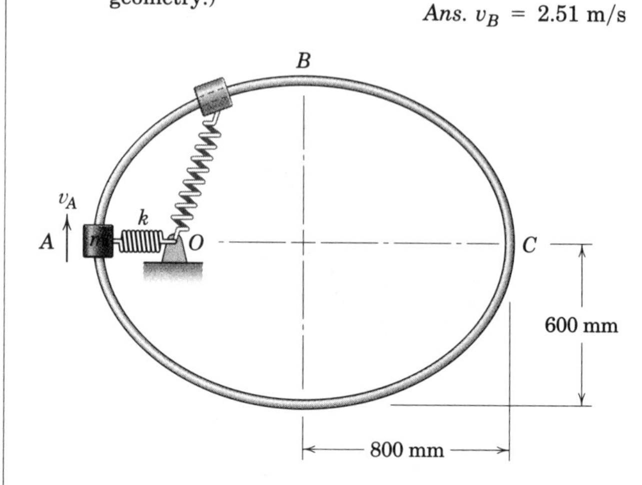

Problem 3/169

3/170 An 80-kg pole vaulter carrying a uniform 4.9-m, 4.5-kg pole approaches the jump with a velocity v and manages to barely clear the bar set at a height of 5.5 m. As he clears the bar, his velocity and that of the pole are essentially zero. Calculate the minimum possible value of v required for him to make the jump. Both the horizontal pole and the center of gravity of the vaulter are 1.1 m above the ground during the approach.

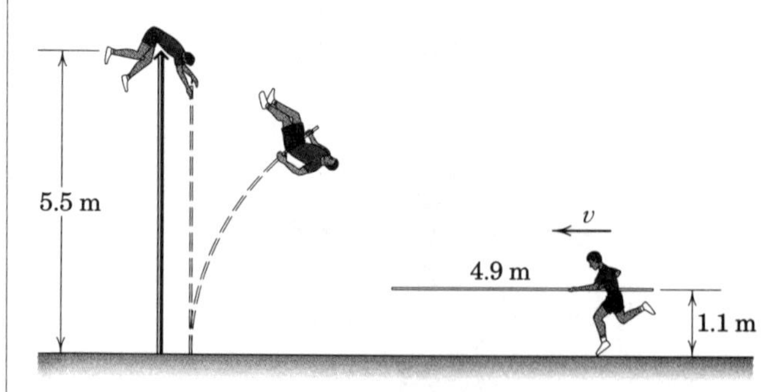

Problem 3/170

3/171 The 3-kg sphere is carried by the parallelogram linkage where the spring is unstretched when $\theta = 90°$. If the mechanism is released from rest at $\theta = 90°$, calculate the velocity v of the sphere when the position $\theta = 135°$ is passed. The links are in the vertical plane, and their mass is small and may be neglected.

Ans. $v = 1.143$ m/s

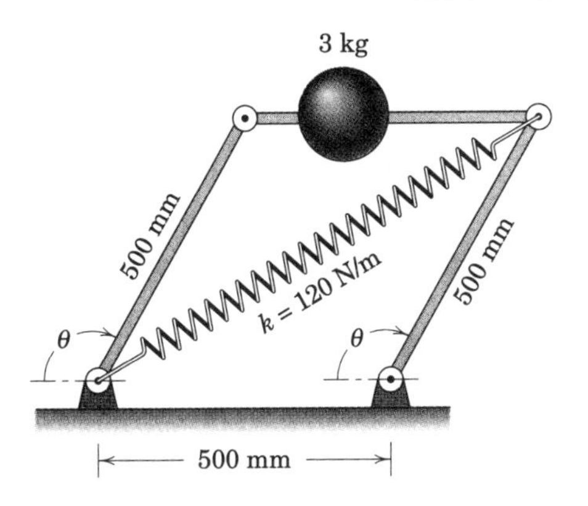

Problem 3/171

3/172 The cars of an amusement-park ride have a speed $v_1 = 90$ km/h at the lowest part of the track. Determine their speed v_2 at the highest part of the track. Neglect energy loss due to friction. (*Caution:* Give careful thought to the change in potential energy of the system of cars.)

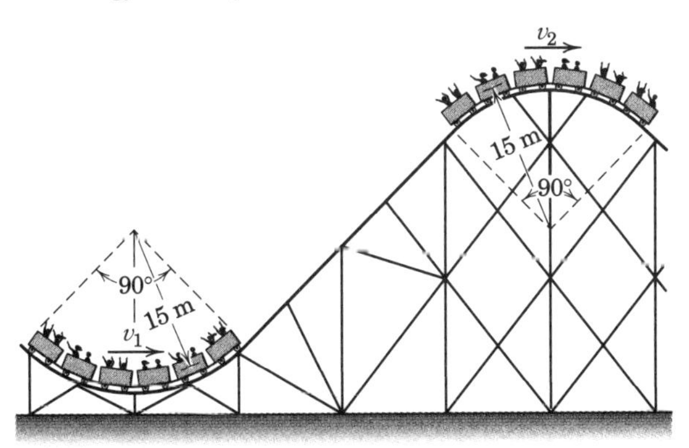

Problem 3/172

3/173 The 0.6-kg slider is released from rest at A and slides down the smooth parabolic guide (which lies in a vertical plane) under the influence of its own weight and of the spring of constant 120 N/m. Determine the speed of the slider as it passes point B and the corresponding normal force exerted on it by the guide. The unstretched length of the spring is 200 mm.

Ans. $v_B = 5.92$ m/s, $N = 84.1$ N

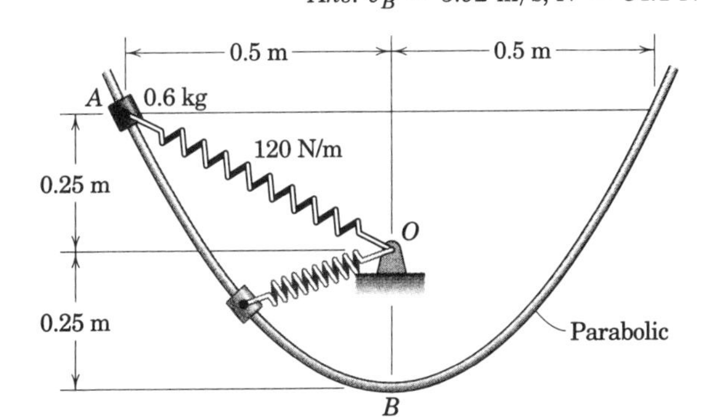

Problem 3/173

3/174 Calculate the maximum velocity of slider B if the system is released from rest with $x = y$. Motion is in the vertical plane. Assume friction is negligible. The sliders have equal masses.

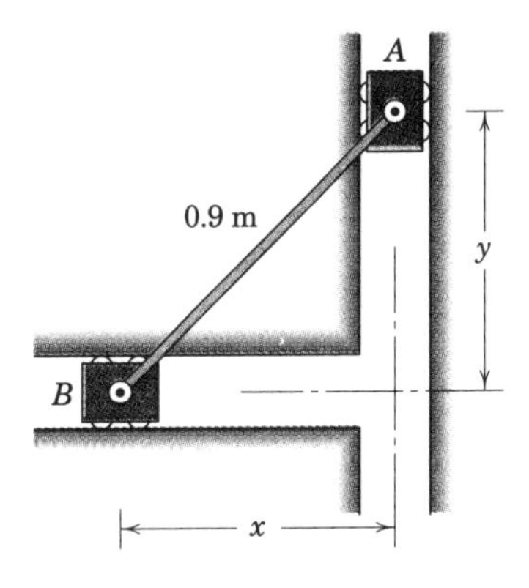

Problem 3/174

3/175 The chain of length L is released from rest on the smooth incline with $x = 0$. Determine the velocity v of the links in terms of x.

$$Ans.\ v = \sqrt{2gx\left[\sin\theta - \frac{x}{2L}(1 - \sin\theta)\right]}$$

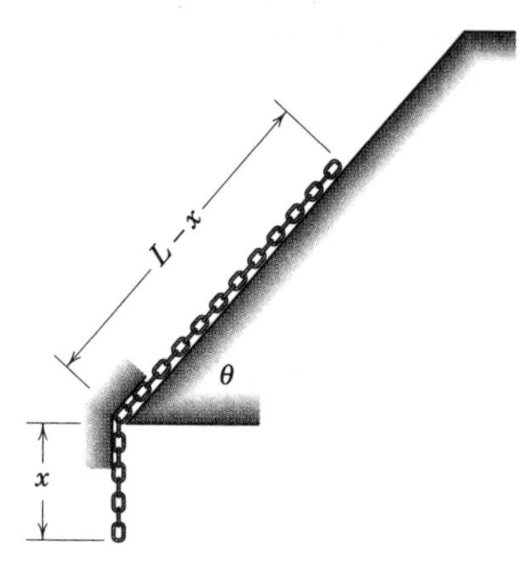

Problem 3/175

3/176 The flexible bicycle-type chain of length $\pi r/2$ and mass per unit length ρ is released from rest with $\theta = 0$ in the smooth circular channel and falls through the hole in the supporting surface. Determine the velocity v of the chain as the last link leaves the slot.

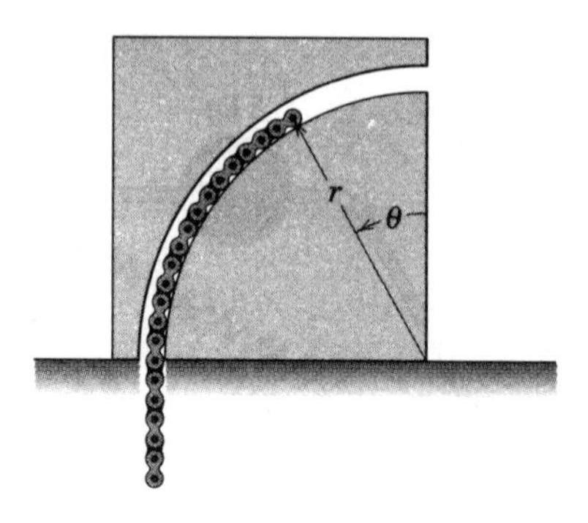

Problem 3/176

SECTION C. IMPULSE AND MOMENTUM

3/8 Introduction

In the previous two articles, we focused attention on the equations of work and energy, which are obtained by integrating the equation of motion $\mathbf{F} = m\mathbf{a}$ with respect to the displacement of the particle. We found that the velocity changes could be expressed directly in terms of the work done or in terms of the overall changes in energy. In the next two articles, we will integrate the equation of motion with respect to time rather than displacement. This approach leads to the equations of impulse and momentum. These equations greatly facilitate the solution of many problems in which the applied forces act during extremely short periods of time (as in impact problems) or over specified intervals of time.

3/9 Linear Impulse and Linear Momentum

Consider again the general curvilinear motion in space of a particle of mass m, Fig. 3/9, where the particle is located by its position vector $\mathbf{r}$ measured from a fixed origin O. The velocity of the particle is $\mathbf{v} = \dot{\mathbf{r}}$ and is tangent to its path (shown as a dashed line). The resultant $\Sigma\mathbf{F}$ of all forces on m is in the direction of its acceleration $\dot{\mathbf{v}}$. We may now write the basic equation of motion for the particle, Eq. 3/3, as

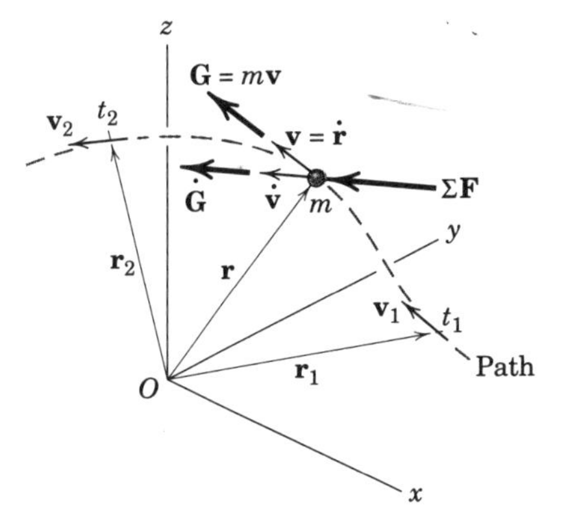

Figure 3/9

$$\Sigma\mathbf{F} = m\dot{\mathbf{v}} = \frac{d}{dt}(m\mathbf{v}) \quad \text{or} \quad \boxed{\Sigma\mathbf{F} = \dot{\mathbf{G}}} \tag{3/21}$$

where the product of the mass and velocity is defined as the *linear momentum* $\mathbf{G} = m\mathbf{v}$ of the particle. Equation 3/21 states that *the resultant of all forces acting on a particle equals its time rate of change of linear momentum*. In SI the units of linear momentum $m\mathbf{v}$ are seen to be kg·m/s, which also equals N·s. In U.S. customary units, the units of linear momentum $m\mathbf{v}$ are [lb/(ft/sec^2)][ft/sec] = lb-sec.

Because Eq. 3/21 is a vector equation, we recognize that, in addition to the equality of the magnitudes of $\Sigma\mathbf{F}$ and $\dot{\mathbf{G}}$, the direction of the resultant force coincides with the direction of the rate of change in linear momentum, which is the direction of the rate of change in velocity. Equation 3/21 is one of the most useful and important relationships in dynamics, and it is valid as long as the mass m of the particle is not changing with time. The case where m changes with time is discussed in Art. 4/7 of Chapter 4.

We now write the three scalar components of Eq. 3/21 as

$$\Sigma F_x = \dot{G}_x \qquad \Sigma F_y = \dot{G}_y \qquad \Sigma F_z = \dot{G}_z \tag{3/22}$$

These equations may be applied independently of one another.

The Linear Impulse-Momentum Principle

All that we have done so far in this article is to rewrite Newton's second law in an alternative form in terms of momentum. But we are now able to describe the effect of the resultant force $\Sigma\mathbf{F}$ on the linear momentum of the particle over a finite period of time simply by integrating Eq. 3/21 with respect to the time t. Multiplying the equation by dt gives $\Sigma\mathbf{F}\ dt = d\mathbf{G}$, which we integrate from time t_1 to time t_2 to obtain

$$\int_{t_1}^{t_2} \Sigma\mathbf{F}\ dt = \mathbf{G}_2 - \mathbf{G}_1 = \Delta\mathbf{G} \qquad (3/23)$$

Here the linear momentum at time t_2 is $\mathbf{G}_2 = m\mathbf{v}_2$ and the linear momentum at time t_1 is $\mathbf{G}_1 = m\mathbf{v}_1$. The product of force and time is defined as the *linear impulse* of the force, and Eq. 3/23 states that *the total linear impulse on m equals the corresponding change in linear momentum of m.*

Alternatively, we may write Eq. 3/23 as

$$\mathbf{G}_1 + \int_{t_1}^{t_2} \Sigma\mathbf{F}\ dt = \mathbf{G}_2 \qquad (3/23a)$$

which says that the initial linear momentum of the body plus the linear impulse applied to it equals its final linear momentum.

The impulse integral is a vector which, in general, may involve changes in both magnitude and direction during the time interval. Under these conditions, it will be necessary to express $\Sigma\mathbf{F}$ and $\mathbf{G}$ in component form and then combine the integrated components. The components of Eq. 3/23 become the scalar equations

$$\int_{t_1}^{t_2} \Sigma F_x\ dt = (mv_x)_2 - (mv_x)_1$$

$$\int_{t_1}^{t_2} \Sigma F_y\ dt = (mv_y)_2 - (mv_y)_1$$

$$\int_{t_1}^{t_2} \Sigma F_z\ dt = (mv_z)_2 - (mv_z)_1$$

These three scalar impulse-momentum equations are completely independent. The scalar expressions corresponding to the vector equations 3/23*a* are simply rearrangements of these equations.

There are cases where a force acting on a particle varies with the time in a manner determined by experimental measurements or by other approximate means. In this case a graphical or numerical integration must be performed. If, for example, a force F acting on a particle in a given direction varies with the time t as indicated in Fig. 3/10, then the impulse, $\int_{t_1}^{t_2} F\,dt$, of this force from t_1 to t_2 is the shaded area under the curve.

In evaluating the resultant impulse, it is necessary to include the effect of *all* forces acting on m, except those whose magnitudes are negligible. By now you should be aware that the only reliable method of accounting for the effects of *all* forces is to isolate the particle in question by drawing its *free-body diagram*.

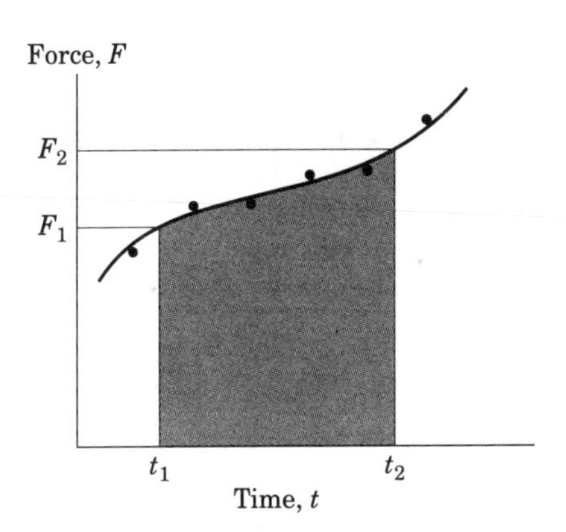

Figure 3/10

Conservation of Linear Momentum

If the resultant force on a particle is zero during an interval of time, we see that Eq. 3/21 requires that its linear momentum **G** remain constant. In this case, the linear momentum of the particle is said to be *conserved*. Linear momentum may be conserved in one coordinate direction, such as x, but not necessarily in the y- or z-direction. A careful examination of the free-body diagram of the particle will disclose whether the total linear impulse on the particle in a particular direction is zero. If it is, the corresponding linear momentum is unchanged (conserved) in that direction.

Consider now the motion of two particles a and b which interact during an interval of time. If the interactive forces **F** and $-$**F** between them are the only unbalanced forces acting on the particles during the interval, it follows that the linear impulse on particle a is the negative of the linear impulse on particle b. Therefore, from Eq. 3/23, the change in linear momentum $\Delta\mathbf{G}_a$ of particle a is the negative of the change $\Delta\mathbf{G}_b$ in linear momentum of particle b. So we have $\Delta\mathbf{G}_a = -\Delta\mathbf{G}_b$ or $\Delta(\mathbf{G}_a + \mathbf{G}_b) = \mathbf{0}$. Thus, the total linear momentum $\mathbf{G} = \mathbf{G}_a + \mathbf{G}_b$ for the system of the two particles remains constant during the interval, and we write

$$\Delta\mathbf{G} = \mathbf{0} \quad \text{or} \quad \mathbf{G}_1 = \mathbf{G}_2 \tag{3/24}$$

Equation 3/24 expresses the *principle of conservation of linear momentum*.

Sample Problem 3/18

A 0.2-kg particle moves in the vertical y-z plane (z up, y horizontal) under the action of its weight and a force **F** which varies with time. The linear momentum of the particle in newton-seconds is given by the expression $\mathbf{G} = \frac{3}{2}(t^2 + 3)\mathbf{j} - \frac{2}{3}(t^3 - 4)\mathbf{k}$, where t is the time in seconds. Determine the force **F** and its magnitude for the instant when $t = 2$ s.

Solution. The weight expressed as a vector is $-0.2(9.81)\mathbf{k}$ N. Thus, the force-momentum equation becomes

① $[\Sigma\mathbf{F} = \dot{\mathbf{G}}]$ $\quad \mathbf{F} - 0.2(9.81)\mathbf{k} = \dfrac{d}{dt}[\frac{3}{2}(t^2 + 3)\mathbf{j} - \frac{2}{3}(t^3 - 4)\mathbf{k}]$

$$= 3t\mathbf{j} - 2t^2\mathbf{k}$$

For $t = 2$ s, $\quad \mathbf{F} = 0.2(9.81)\mathbf{k} + 3(2)\mathbf{j} - 2(2^2)\mathbf{k}$

$$= 6\mathbf{j} - 6.04\mathbf{k}\ \text{N} \qquad \textit{Ans.}$$

Thus, $\quad F = \sqrt{6^2 + 6.04^2} = 8.51$ N $\qquad$ *Ans.*

Helpful Hint

① Don't forget that $\Sigma\mathbf{F}$ includes *all* forces on the particle, of which the weight is one.

Sample Problem 3/19

A particle with a mass of 0.5 kg has a velocity $u = 10$ m/s in the x-direction at time $t = 0$. Forces $\mathbf{F}_1$ and $\mathbf{F}_2$ act on the particle, and their magnitudes change with time according to the graphical schedule shown. Determine the velocity **v** of the particle at the end of the 3 s.

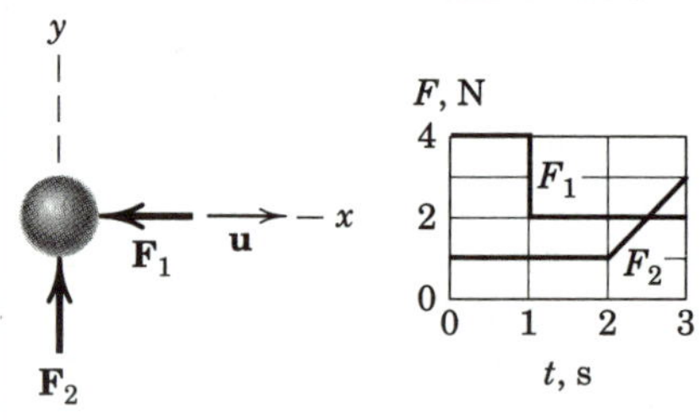

Solution. The impulse-momentum equation is applied in component form and gives for the x- and y-directions, respectively,

① $\left[\int \Sigma F_x\, dt = m\Delta v_x\right]$ $\quad -[4(1) + 2(3 - 1)] = 0.5(v_x - 10)$

$$v_x = -6 \text{ m/s}$$

$\left[\int \Sigma F_y\, dt = m\Delta v_y\right]$ $\quad [1(2) + 2(3 - 2)] = 0.5(v_y - 0)$

$$v_y = 8 \text{ m/s}$$

Thus,

$$\mathbf{v} = -6\mathbf{i} + 8\mathbf{j} \text{ m/s} \quad \text{and} \quad v = \sqrt{6^2 + 8^2} = 10 \text{ m/s}$$

$$\theta_x = \tan^{-1}\frac{8}{-6} = 126.9° \qquad \textit{Ans.}$$

② Although not called for, the path of the particle for the first 3 s is plotted in the figure. The velocity at $t = 3$ s is shown together with its components.

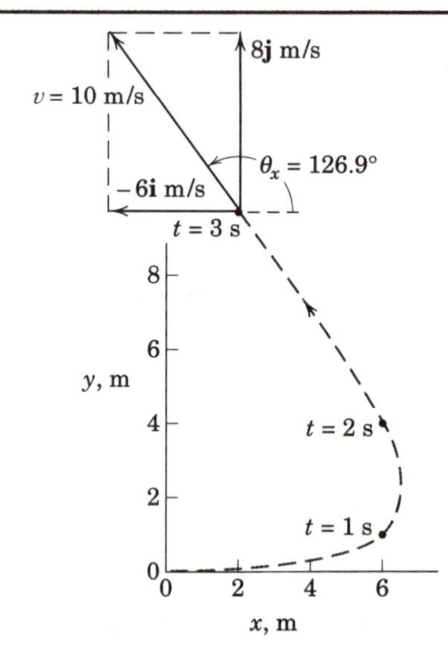

Helpful Hints

① The impulse in each direction is the corresponding area under the force-time graph. Note that F_1 is in the negative x-direction so its impulse is negative.

② It is important to note that the algebraic signs must be carefully respected in applying the momentum equations. Also we must recognize that impulse and momentum are vector quantities in contrast to work and energy, which are scalars.

Sample Problem 3/20

The loaded 150-kg skip is rolling down the incline at 4 m/s when a force P is applied to the cable as shown at time $t = 0$. The force P is increased uniformly with the time until it reaches 600 N at $t = 4$ s, after which time it remains constant at this value. Calculate (*a*) the time t_1 at which the skip reverses its direction and (*b*) the velocity v of the skip at $t = 8$ s. Treat the skip as a particle.

Solution. The stated variation of P with the time is plotted, and the free-body diagram of the skip is drawn.

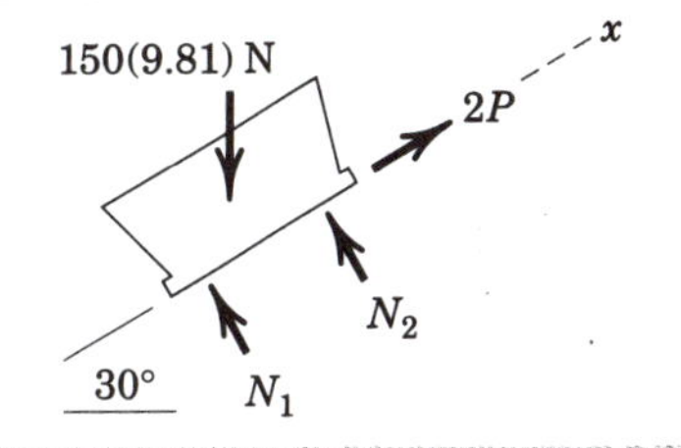

Part (a). The skip reverses direction when its velocity becomes zero. We will assume that this condition occurs at $t = 4 + \Delta t$ s. The impulse-momentum equation applied consistently in the positive x-direction gives

$$\left[\int \Sigma F_x \, dt = m\Delta v_x\right]$$

① $$\tfrac{1}{2}(4)(2)(600) + 2(600)\Delta t - 150(9.81)\sin 30°(4 + \Delta t) = 150(0 - [-4])$$

$$464\Delta t = 1143 \qquad \Delta t = 2.46 \text{ s} \qquad t = 4 + 2.46 = 6.46 \text{ s} \qquad \textit{Ans.}$$

Part (b). Applying the impulse-momentum equation to the entire interval gives

$$\left[\int \Sigma F_x \, dt = m\Delta v_x\right]$$

$$\tfrac{1}{2}(4)(2)(600) + 4(2)(600) - 150(9.81)\sin 30°(8) = 150(v - [-4])$$

$$150v = 714 \qquad v = 4.76 \text{ m/s} \qquad \textit{Ans.}$$

The same result is obtained by analyzing the interval from t_1 to 8 s.

Helpful Hint

① The free-body diagram keeps us from making the error of using the impulse of P rather than $2P$ or of forgetting the impulse of the component of the weight. The first term in the equation is the triangular area of the P-t relation for the first 4 s, doubled for the force of $2P$.

Sample Problem 3/21

The 50-g bullet traveling at 600 m/s strikes the 4-kg block centrally and is embedded within it. If the block slides on a smooth horizontal plane with a velocity of 12 m/s in the direction shown prior to impact, determine the velocity **v** of the block and embedded bullet immediately after impact.

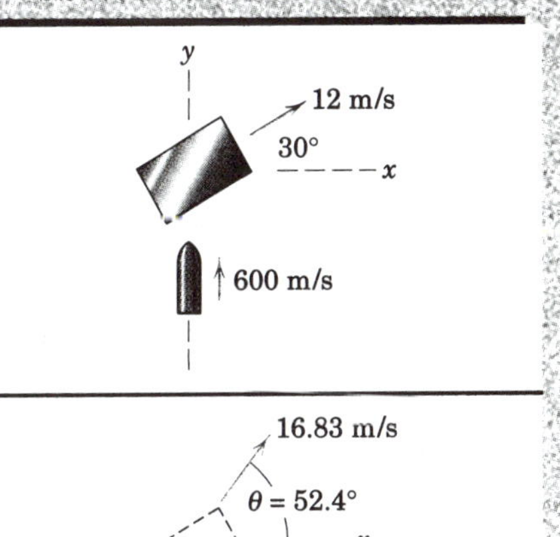

Solution. Since the force of impact is internal to the system composed of the block and bullet and since there are no other external forces acting on the system in the plane of motion, it follows that the linear momentum of the system is conserved. Thus,

① $[\mathbf{G}_1 = \mathbf{G}_2]$ $$0.050(600\mathbf{j}) + 4(12)(\cos 30°\mathbf{i} + \sin 30°\mathbf{j}) = (4 + 0.050)\mathbf{v}$$

$$\mathbf{v} = 10.26\mathbf{i} + 13.33\mathbf{j} \text{ m/s} \qquad \textit{Ans.}$$

The final velocity and its direction are given by

$\left[v = \sqrt{v_x^2 + v_y^2}\right]$ $$v = \sqrt{(10.26)^2 + (13.33)^2} = 16.83 \text{ m/s} \qquad \textit{Ans.}$$

$[\tan\theta = v_y/v_x]$ $$\tan\theta = \frac{13.33}{10.26} = 1.299 \qquad \theta = 52.4° \qquad \textit{Ans.}$$

Helpful Hint

① Working with the vector form of the principle of conservation of linear momentum is clearly equivalent to working with the component form.

PROBLEMS

Introductory Problems

3/177 The 1500-kg car has a velocity of 30 km/h up the 10-percent grade when the driver applies more power for 8 s to bring the car up to a speed of 60 km/h. Calculate the time average F of the total force tangent to the road exerted on the tires during the 8 s. Treat the car as a particle and neglect air resistance.

Ans. $F = 3.03$ kN

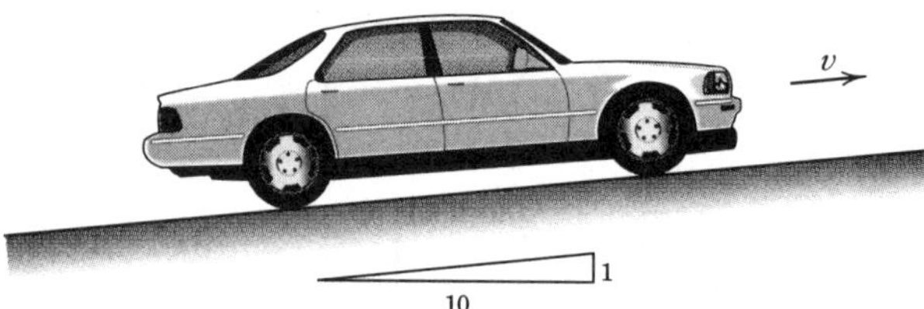

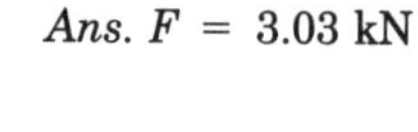

Problem 3/177

3/178 The velocity of a 1.2-kg particle is given by $\mathbf{v} = 1.5t^3\mathbf{i} + (2.4 - 3t^2)\mathbf{j} + 5\mathbf{k}$, where $\mathbf{v}$ is in meters per second and the time t is in seconds. Determine the linear momentum $\mathbf{G}$ of the particle, its magnitude G, and the net force $\mathbf{R}$ which acts on the particle when $t = 2$ s.

3/179 A 75-g projectile traveling at 600 m/s strikes and becomes embedded in the 50-kg block, which is initially stationary. Compute the energy lost during the impact. Express your answer as an absolute value $|\Delta E|$ and as a percentage n of the original system energy E.

Ans. $|\Delta E| = 13\,480$ J, $n = 99.9\%$

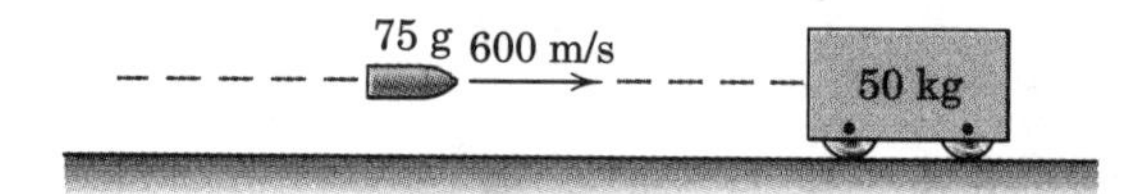

Problem 3/179

3/180 A jet-propelled airplane with a mass of 10 Mg is flying horizontally at a constant speed of 1000 km/h under the action of the engine thrust T and the equal and opposite air resistance R. The pilot ignites two rocket-assist units, each of which develops a forward thrust T_0 of 8 kN for 9 s. If the velocity of the airplane in its horizontal flight is 1050 km/h at the end of the 9 s, calculate the time-average increase ΔR in air resistance. The mass of the rocket fuel used is negligible compared with that of the airplane.

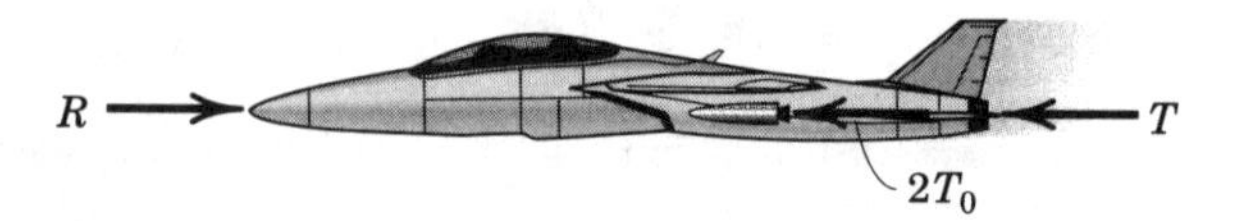

Problem 3/180

3/181 Freight car A with a total mass of 80 Mg is moving along the horizontal track in a switching yard at 3 km/h. Freight car B with a total mass of 60 Mg and moving at 5 km/h overtakes car A and is coupled to it. Determine (*a*) the common velocity v of the two cars as they move together after being coupled and (*b*) the loss of energy $|\Delta E|$ due to the impact.

Ans. (*a*) $v = 3.86$ km/h, (*b*) $|\Delta E| = 5290$ J

Problem 3/181

3/182 A railroad car of mass m and initial speed v collides with and becomes coupled with the two identical cars. Compute the final speed v' of the group of three cars and the fractional loss n of energy if (*a*) the initial separation distance $d = 0$ (that is, the two stationary cars are initially coupled together with no slack in the coupling) and (*b*) the distance $d \neq 0$ so that the cars are uncoupled and slightly separated. Neglect rolling resistance.

Problem 3/182

3/183 The 200-kg lunar lander is descending onto the moon's surface with a velocity of 6 m/s when its retro-engine is fired. If the engine produces a thrust T for 4 s which varies with the time as shown and then cuts off, calculate the velocity of the lander when $t = 5$ s, assuming that it has not yet landed. Gravitational acceleration at the moon's surface is 1.62 m/s^2.

Ans. $v = 2.10$ m/s

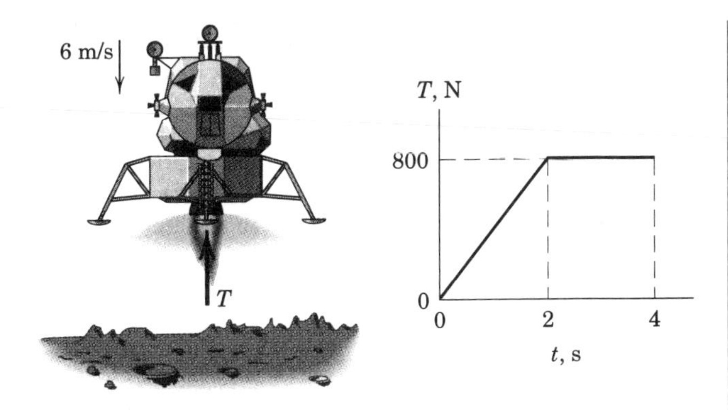

Problem 3/183

3/184 A 3-Mg experimental rocket sled is propelled by six rocket motors each with an impulse rating of 100 kN·s. The rockets are fired at $\frac{1}{4}$-s intervals, and the duration of each rocket firing is 1.5 s. If the velocity of the sled 3 s from the start is 150 m/s, determine the time average R of the total aerodynamic and mechanical resistance to motion. Neglect the loss of mass due to exhausted fuel compared with the mass of the sled.

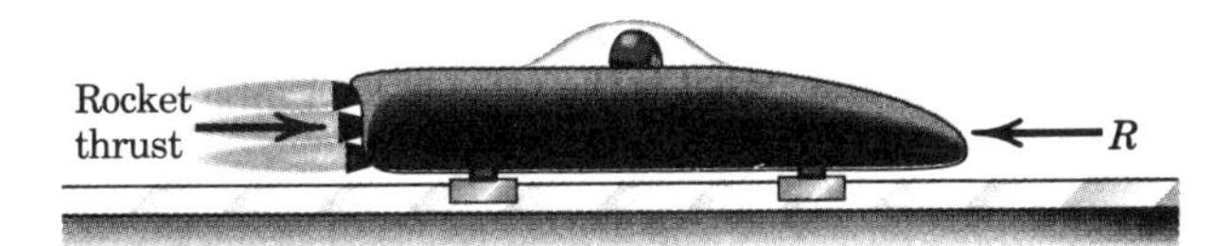

Problem 3/184

3/185 The 9-kg block is moving to the right with a velocity of 0.6 m/s on a horizontal surface when a force P is applied to it at time $t = 0$. Calculate the velocity v of the block when $t = 0.4$ s. The kinetic coefficient of friction is $\mu_k = 0.3$.

Ans. $v = 1.823$ m/s

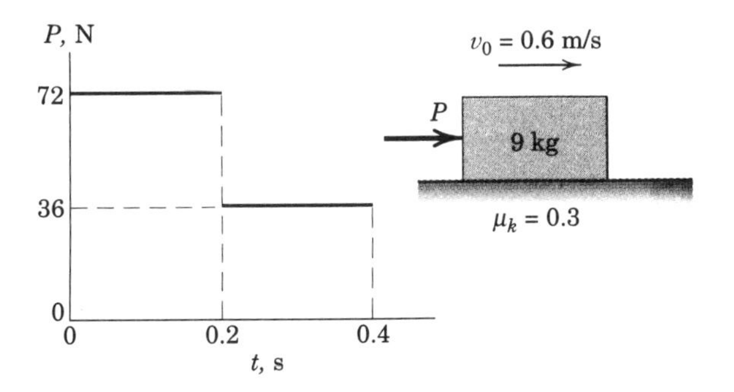

Problem 3/185

3/186 The resistance to motion of a certain racing toboggan is 2 percent of the normal force on its runners. Calculate the time t required for the toboggan to reach a speed of 100 km/h down the slope if it starts from rest.

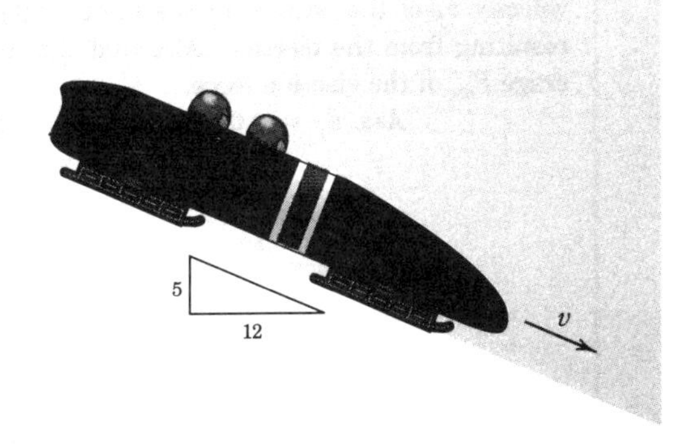

Problem 3/186

3/187 The pilot of a 40-Mg airplane which is originally flying horizontally at a speed of 650 km/h cuts off all engine power and enters a 5° glide path as shown. After 120 s the airspeed is 600 km/h. Calculate the time-average drag force D (air resistance to motion along the flight path).

Ans. $D = 38.8$ kN

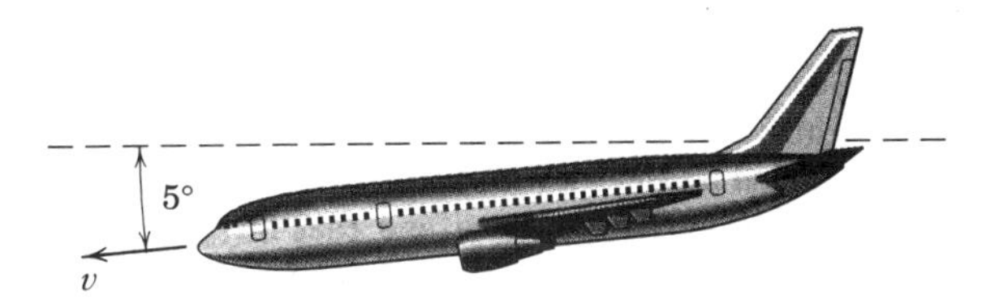

Problem 3/187

Representative Problems

3/188 The supertanker has a total displacement (mass) of $150(10^3)$ metric tons (one metric ton equals 1000 kg) and is lying still in the water when the tug commences a tow. If a constant tension of 200 kN is developed in the tow cable, compute the time required to bring the tanker to a speed of 1 knot from rest. At this low speed, hull resistance to motion through the water is very small and may be neglected. (1 knot = 1.852 km/h)

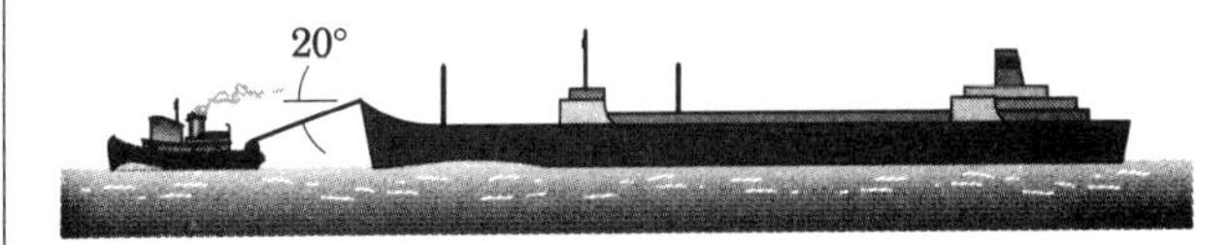

Problem 3/188

3/189 The space shuttle launches an 800-kg satellite by ejecting it from the cargo bay as shown. The ejection mechanism is activated and is in contact with the satellite for 4 s to give it a velocity of 0.3 m/s in the z-direction relative to the shuttle. The mass of the shuttle is 90 Mg. Determine the component of velocity v_f of the shuttle in the minus z-direction resulting from the ejection. Also find the time average F_{av} of the ejection force.

Ans. $v_f = 0.00264$ m/s, $F_{av} = 59.5$ N

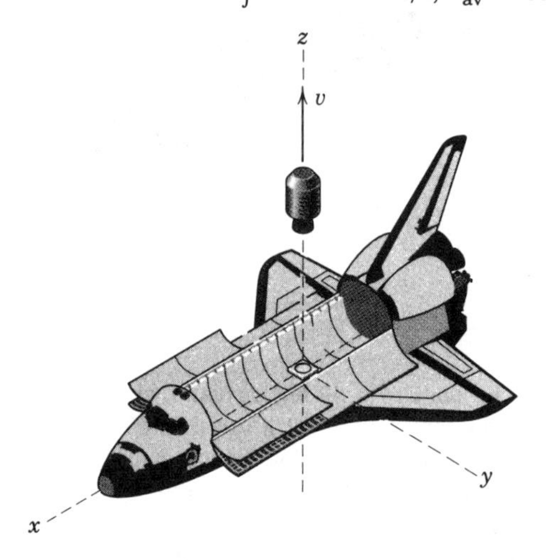

Problem 3/189

3/190 Car B is initially stationary and is struck by car A, which is moving with speed v. The cars become entangled and move with speed v' after the collision. The mass of car A is m and that of car B is pm, where p is the ratio of the mass of car B to that of car A. If the duration of the collision is Δt, express the common speed v' after the collision and the average acceleration of each car during the collision in terms of v, p, and Δt. Evaluate your expressions for $p = 0.5$.

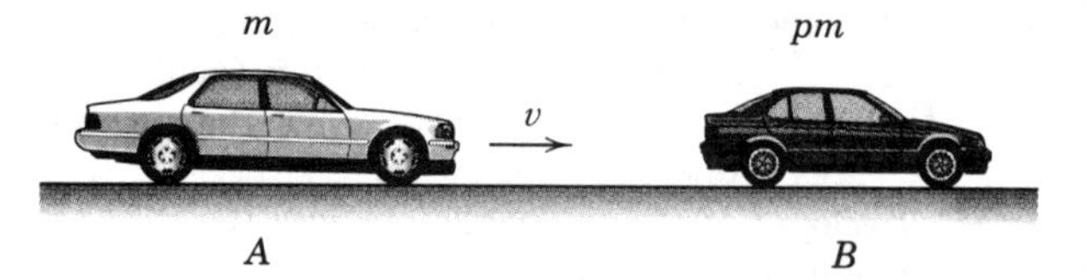

Problem 3/190

3/191 The hydraulic braking system for the truck and trailer is set to produce equal braking forces for the two units. If the brakes are applied uniformly for 5 s to bring the rig to a stop from a speed of 30 km/h down the 10-percent grade, determine the force P in the coupling between the trailer and the truck. The mass of the truck is 10 Mg and that of the trailer is 7.5 Mg.

Ans. $P = 3.30$ kN (tension)

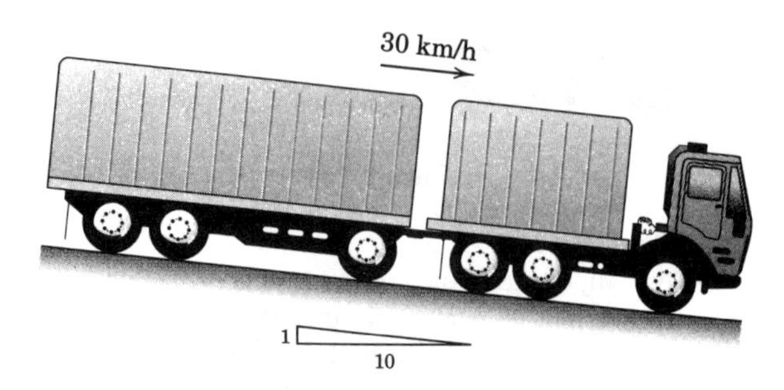

Problem 3/191

3/192 The cart of mass m is subjected to the exponentially decreasing force F, which represents a shock or blast loading. If the cart is stationary at time $t = 0$, determine its velocity v and displacement s as functions of time. What is the value of v for large values of t?

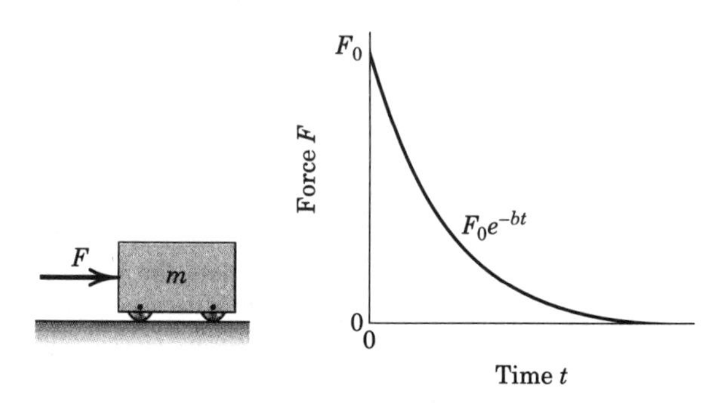

Problem 3/192

3/193 The 4-kg cart, at rest at time $t = 0$, is acted on by a horizontal force which varies with time t as shown. Neglect friction and determine the velocity of the cart at $t = 1$ s and at $t = 3$ s.

Ans. $v_1 = 0.417$ m/s, $v_3 = 8.96$ m/s

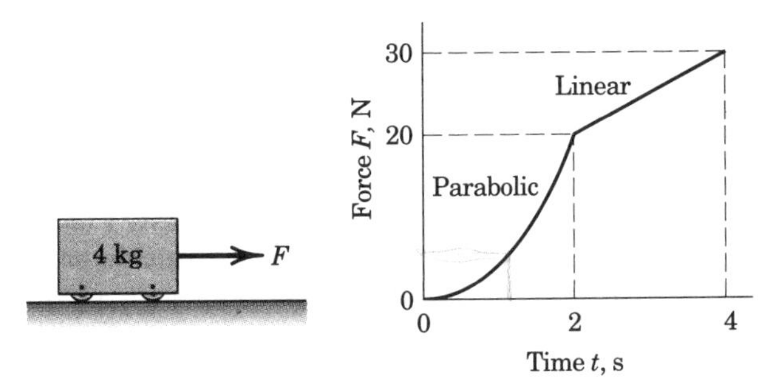

Problem 3/193

3/194 The tow truck with attached 1200-kg car accelerates uniformly from 30 km/h to 70 km/h over a 15-s interval. The average rolling resistance for the car over this speed interval is 500 N. Assume that the 60° angle shown represents the time average configuration and determine the average tension in the tow cable.

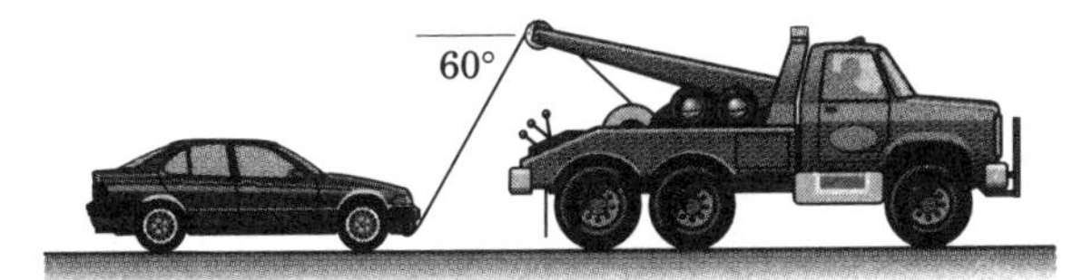

Problem 3/194

3/195 The 450-kg ram of a pile driver falls 1.4 m from rest and strikes the top of a 240-kg pile embedded 0.9 m in the ground. Upon impact the ram is seen to move with the pile with no noticeable rebound. Determine the velocity v of the pile and ram immediately after impact. Can you justify using the principle of conservation of momentum even though the weights act during the impact?

Ans. $v = 3.42$ m/s

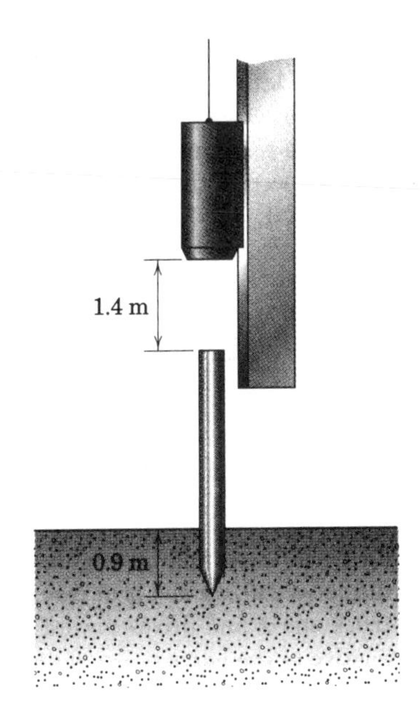

Problem 3/195

3/196 The 900-kg motorized unit A is designed to raise and lower the 600-kg bucket B of concrete. Determine the average force R which supports unit A during the 6 seconds required to slow the descent of the bucket from 3 m/s to 0.5 m/s. Analyze the entire system as a unit without finding the tension in the cable.

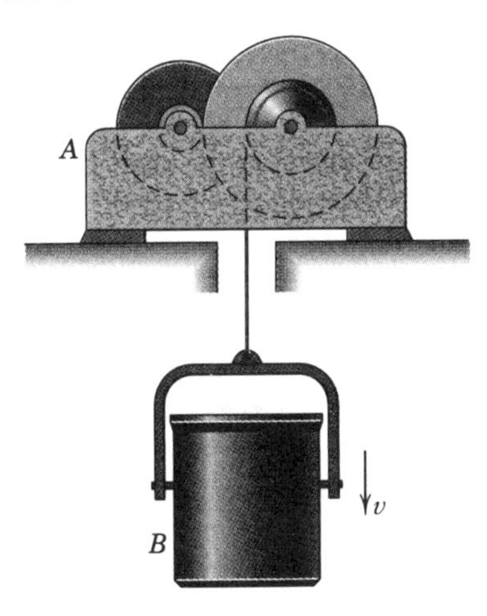

Problem 3/196

3/197 The 12-Mg truck drives onto the 350-Mg barge from the dock at 20 km/h and brakes to a stop on the deck. The barge is free to move in the water, which offers negligible resistance to motion at low speeds. Calculate the speed of the barge after the truck has come to rest on it.

Ans. $v = 0.663$ km/h

Problem 3/197

3/198 An 8-Mg truck is resting on the deck of a barge which displaces 240 Mg and is at rest in still water. If the truck starts and drives toward the bow at a speed relative to the barge $v_{rel} = 6$ km/h, calculate the speed v of the barge. The resistance to the motion of the barge through the water is negligible at low speeds.

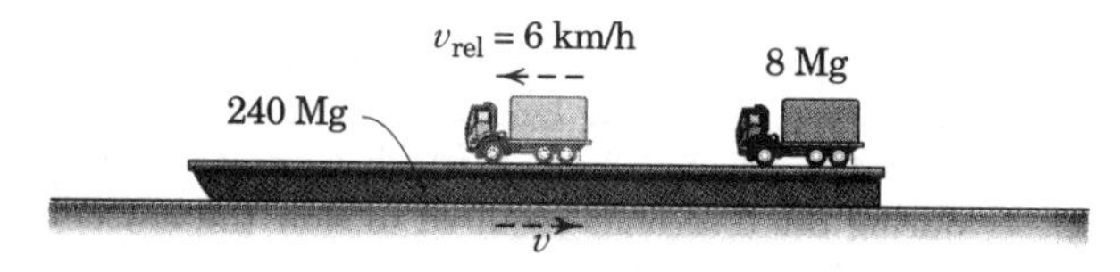

Problem 3/198

3/199 Car B (1500 kg) traveling west at 48 km/h collides with car A (1600 kg) traveling north at 32 km/h as shown. If the two cars become entangled and move together as a unit after the crash, compute the magnitude v of their common velocity immediately after the impact and the angle θ made by the velocity vector with the north direction.

Ans. $v = 28.5$ km/h, $\theta = 54.6°$

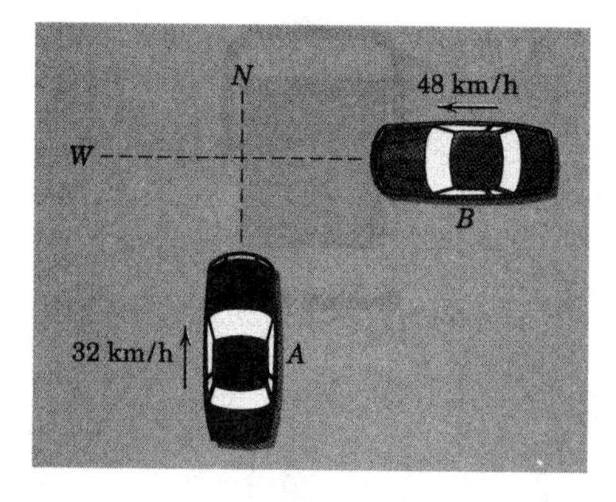

Problem 3/199

3/200 Determine the time required by a diesel-electric locomotive, which produces a constant drawbar pull of 270 kN, to increase the speed of a 1600-Mg freight train from 32 km/h to 48 km/h up a 1-percent grade. Train resistance is 50 N per megagram.

3/201 A 1000-kg spacecraft is traveling in deep space with a speed of $v_s = 2000$ m/s when a 10-kg meteor moving with a velocity $\mathbf{v}_m$ of magnitude 5000 m/s in the direction shown strikes and becomes embedded in the spacecraft. Determine the final velocity $\mathbf{v}$ of the mass center G of the spacecraft. Calculate the angle β between $\mathbf{v}$ and the initial velocity $\mathbf{v}_s$ of the spacecraft.

Ans. $\mathbf{v} = 36.9\mathbf{i} + 1951\mathbf{j} - 14.76\mathbf{k}$ m/s
$\beta = 1.167°$

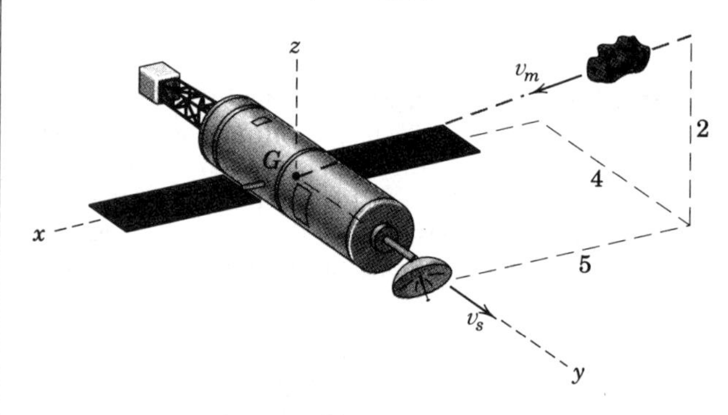

Problem 3/201

3/202 A spacecraft in deep space is programmed to increase its speed by a desired amount Δv by burning its engine for a specified time duration t. Twenty-five percent of the way through the burn, the engine suddenly malfunctions and thereafter produces only half of its normal thrust. What percent n of Δv is achieved if the rocket motor is fired for the planned time t? How much extra time t' would the rocket need to operate in order to compensate for the failure?

3/203 The force P, which is applied to the 10-kg block initially at rest, varies linearly with the time as indicated. If the coefficients of static and kinetic friction between the block and the horizontal surface are 0.6 and 0.4, respectively, determine the velocity of the block when $t = 4$ s.

Ans. $v = 6.61$ m/s

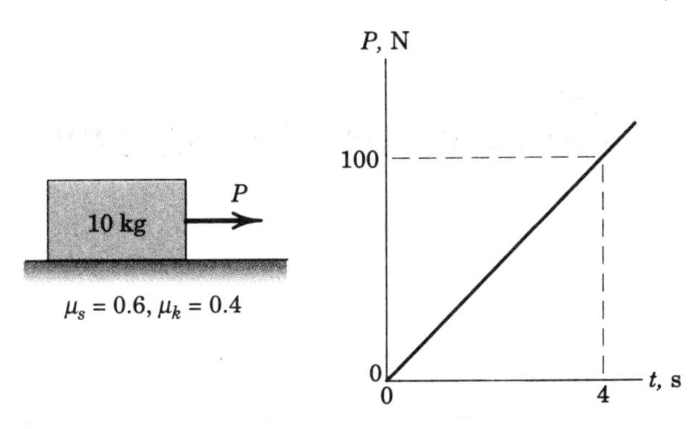

Problem 3/203

3/204 Careful measurements made during the impact of the 200-g metal cylinder with the spring-loaded plate reveal a semielliptical relation between the contact force F and the time t of impact as shown. Determine the rebound velocity v of the cylinder if it strikes the plate with a velocity of 6 m/s.

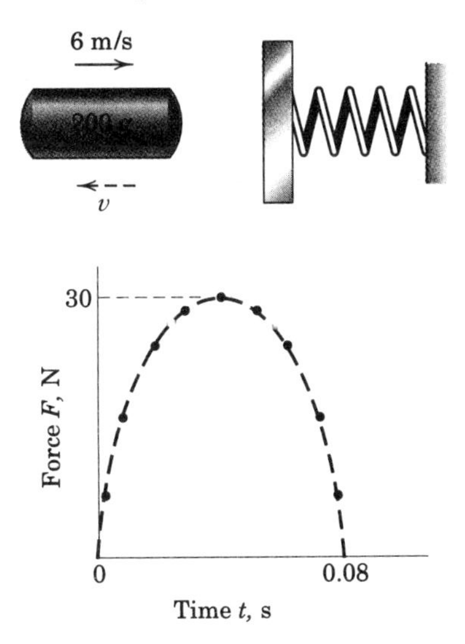

Problem 3/204

3/205 A 4-kg object, which is moving on a smooth horizontal surface with a velocity of 10 m/s in the $-x$-direction, is subjected to a force F_x which varies with the time as shown. Approximate the experimental data by the dashed line and determine the velocity of the object (*a*) at $t = 0.6$ s and (*b*) at $t = 0.9$ s.

Ans. (*a*) $v = 2$ m/s, (*b*) $v = -2.5$ m/s

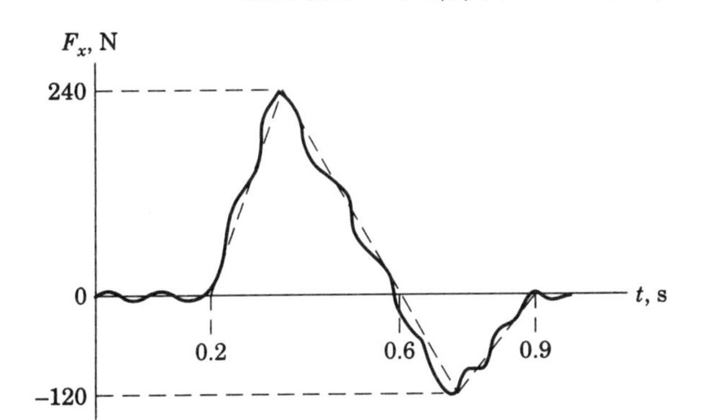

Problem 3/205

3/206 The 45.9-g golf ball is struck by the five-iron and acquires the velocity shown in a time period of 0.001 s. Determine the magnitude R of the average force exerted by the club on the ball. What acceleration magnitude a does this force cause, and what is the distance d over which the launch velocity is achieved, assuming constant acceleration?

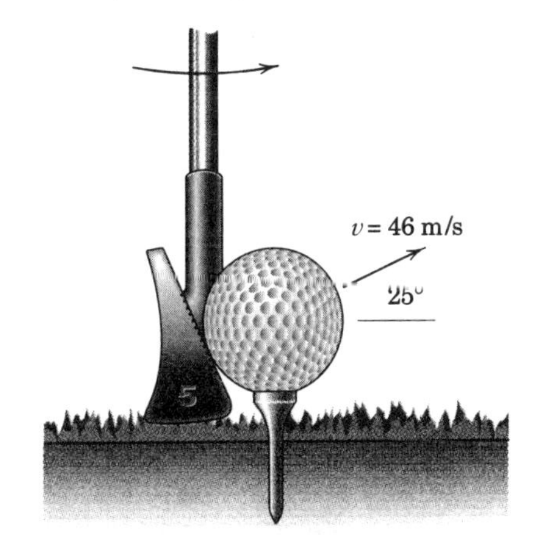

Problem 3/206

3/207 The cart is moving down the incline with a velocity $v_0 = 20$ m/s at $t = 0$, at which time the force P begins to act as shown. After 5 seconds the force continues at the 50-N level. Determine the velocity of the cart at time $t = 8$ s and calculate the time t at which the cart velocity is zero.

Ans. $v_2 = 1.423$ m/s down incline, $t = 8.25$ s

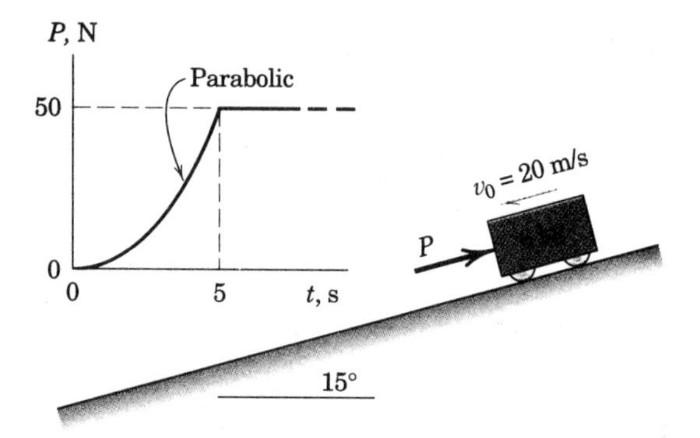

Problem 3/207

3/208 The ballistic pendulum is a simple device to measure projectile velocity v by observing the maximum angle θ to which the box of sand with embedded projectile swings. Calculate the angle θ if the 60-g projectile is fired horizontally into the suspended 20-kg box of sand with a velocity $v = 600$ m/s. Also find the percentage of energy lost during impact.

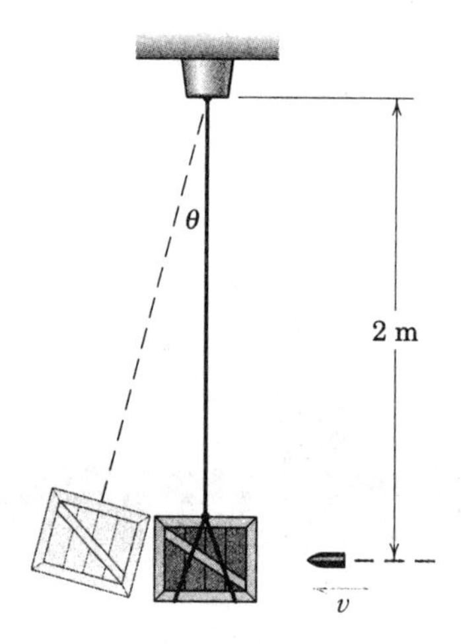

Problem 3/208

3/209 The 500-ton tug is towing the 900-ton coal barge at a steady speed of 6 knots. For a short period of time, the stern winch takes in the towing cable at the rate of 0.5 m/s. Calculate the reduced speed v of the tug during this interval. Assume the tow cable to be horizontal. (Recall 1 knot = 1.852 km and 1 ton = 1000 kg)

Ans. $v = 5.38$ knots

Problem 3/209

3/210 The cylindrical plug A of mass m_A is released from rest at B and slides down the smooth circular guide. The plug strikes the block C and becomes embedded in it. Write the expression for the distance s which the block and plug slide before coming to rest. The coefficient of kinetic friction between the block and the horizontal surface is μ_k.

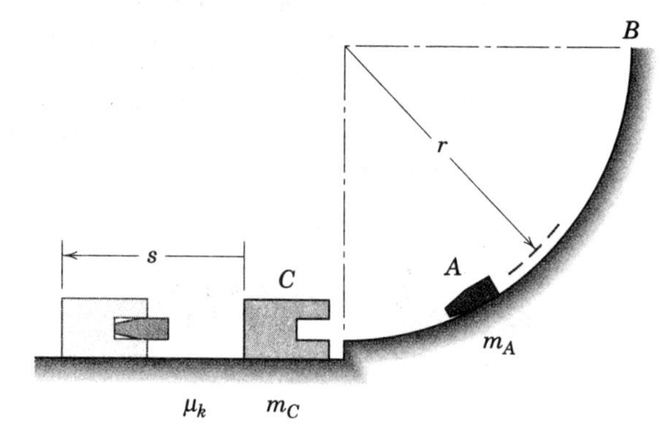

Problem 3/210

3/211 A 10-kg body is traveling in a horizontal straight line with a velocity of 3 m/s when a horizontal force P is applied to it at right angles to the initial direction of motion. If P varies according to the accompanying graph, remains constant in direction, and is the only force acting on the body in its plane of motion, find the magnitude of the velocity of the body when $t = 2$ s and the angle θ which the velocity makes with the direction of P.

Ans. $v = 3.91$ m/s, $\theta = 50.2°$

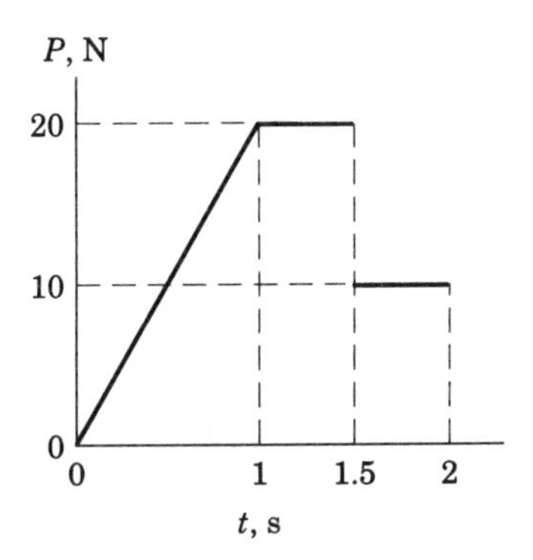

Problem 3/211

3/212 The baseball is traveling with a horizontal velocity of 135 km/h just before impact with the bat. Just after the impact, the velocity of the 146-g ball is 210 km/h directed at 35° to the horizontal as shown. Determine the x- and y-components of the average force **R** exerted by the bat on the baseball during the 0.005-s impact. Comment on treatment of the weight of the baseball (a) during the impact and (b) over the first few seconds after impact.

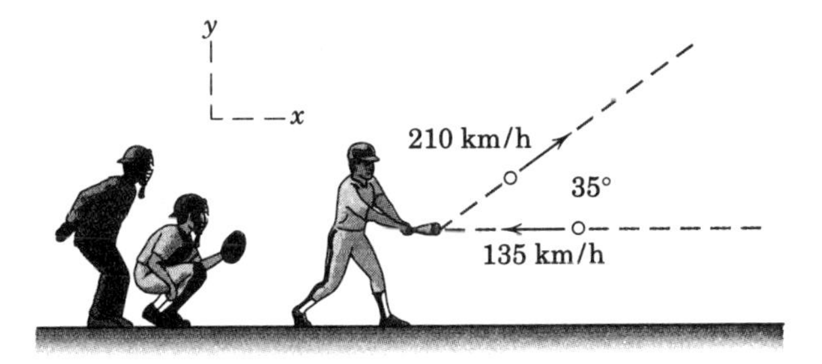

Problem 3/212

3/213 A tennis player strikes the tennis ball with her racket while the ball is still rising. The ball speed before impact with the racket is $v_1 = 15$ m/s and after impact its speed is $v_2 = 22$ m/s, with directions as shown in the figure. If the 60-g ball is in contact with the racket for 0.05 s, determine the magnitude of the average force **R** exerted by the racket on the ball. Find the angle β made by **R** with the horizontal. Comment on the treatment of the ball weight during impact.

Ans. $R = 43.0$ N, $\beta = 8.68°$

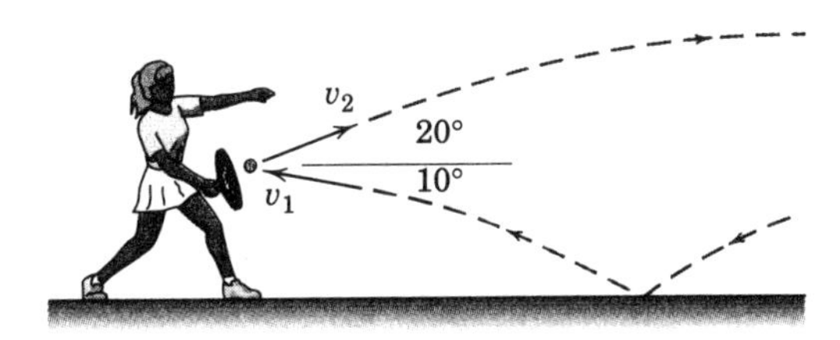

Problem 3/213

3/214 A 0.5-kg body oscillates along the horizontal x-axis under the sole action of an alternating force in the x-direction which decreases in amplitude with time t as shown and is given by $F_x = 4e^{-t} \cos 2\pi t$, where F_x is in newtons and t is in seconds. If the body has a velocity of 1.2 m/s in the negative x-direction at time $t = 0$, determine its velocity v_x at time $t = 2$ s.

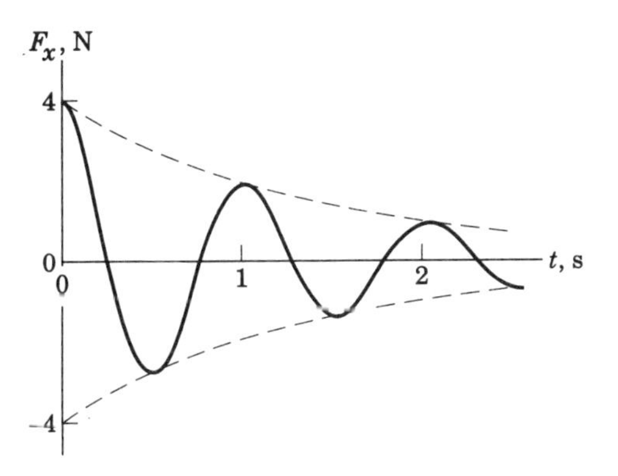

Problem 3/214

3/215 The 40-kg boy has taken a running jump from the upper surface and lands on his 5-kg skateboard with a velocity of 5 m/s in the plane of the figure as shown. If his impact with the skateboard has a time duration of 0.05 s, determine the final speed v along the horizontal surface and the total normal force N exerted by the surface on the skateboard wheels during the impact.

Ans. $v = 3.85$ m/s, $N = 2.44$ kN

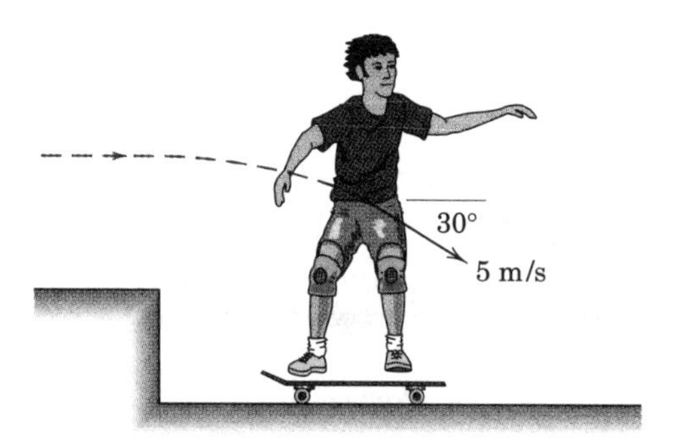

Problem 3/215

3/216 The loaded mine skip has a mass of 3 Mg. The hoisting drum produces a tension T in the cable according to the time schedule shown. If the skip is at rest against A when the drum is activated, determine the speed v of the skip when $t = 6$ s. Friction loss may be neglected.

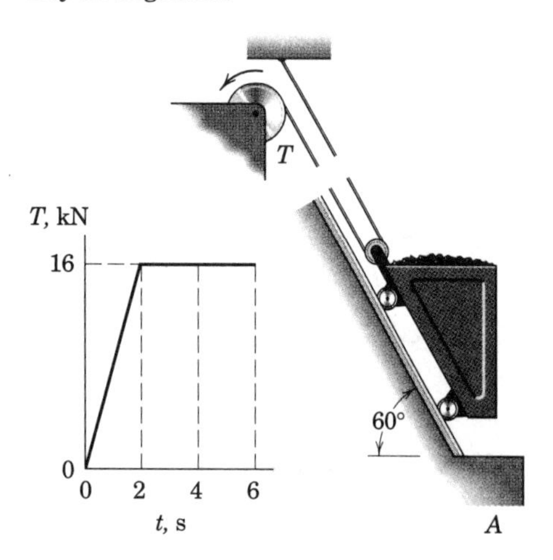

Problem 3/216

▶3/217 The 400-kg ram of a pile driver is designed to fall 1.5 m from rest and strike the top of a 300-kg pile partially embedded in the ground. The deeper the penetration, the greater is the tendency for the ram to rebound as a result of the impact. Calculate the velocity v of the pile immediately after impact for the following three conditions: (a) initial resistance to penetration is small at the outset, and the ram is observed to move with the pile immediately after impact; (b) resistance to penetration has increased, and the ram is seen to have zero velocity immediately after impact; (c) resistance to penetration is high, and the ram is seen to rebound to a height of 100 mm above the point of impact. Why is it permissible to neglect the impulse of the weight of the ram during impact?

Ans. (a) $v = 3.10$ m/s, (b) $v = 7.23$ m/s
(c) $v = 9.10$ m/s

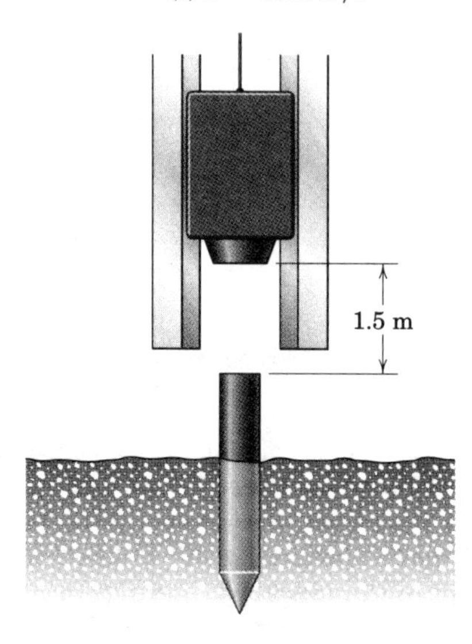

Problem 3/217

▶3/218 Two barges, each with a displacement (mass) of 500 Mg, are loosely moored in calm water. A stunt driver starts his 1500-kg car from rest at A, drives along the deck, and leaves the end of the 15° ramp at a speed of 50 km/h relative to the barge and ramp. The driver successfully jumps the gap and brings his car to rest relative to barge 2 at B. Calculate the velocity v_2 imparted to barge 2 just after the car has come to rest on the barge. Neglect the resistance of the water to motion at the low velocities involved.

Ans. $v_2 = 40.0$ mm/s

Problem 3/218

▶**3/219** The simple pendulum A of mass m_A and length l is suspended from the trolley B of mass m_B. If the system is released from rest at $\theta = 0$, determine the velocity v_B of the trolley when $\theta = 90°$. Friction is negligible.

Ans. $v_B = \dfrac{m_A}{m_B}\sqrt{\dfrac{2gl}{1 + m_A/m_B}}$

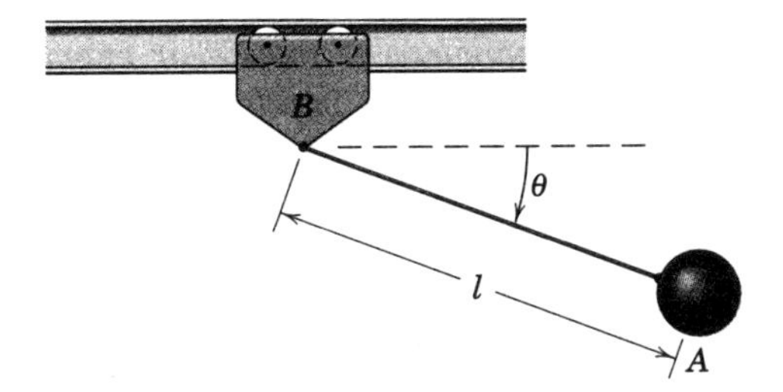

Problem 3/219

3/220 A torpedo boat with a displacement of 60 metric tons is moving at 18 knots when it fires a 140-kg torpedo horizontally with the launch tube at the 30° angle shown. If the torpedo has a velocity relative to the boat of 6 m/s as it leaves the tube, determine the momentary reduction Δv in forward speed of the boat. Recall that 1 knot = 1.852 km/h and 1 metric ton = 1000 kg. Solve the problem also by referring the motion to a coordinate system moving with the initial speed of the boat.

Ans. $\Delta v = 0.01210$ m/s

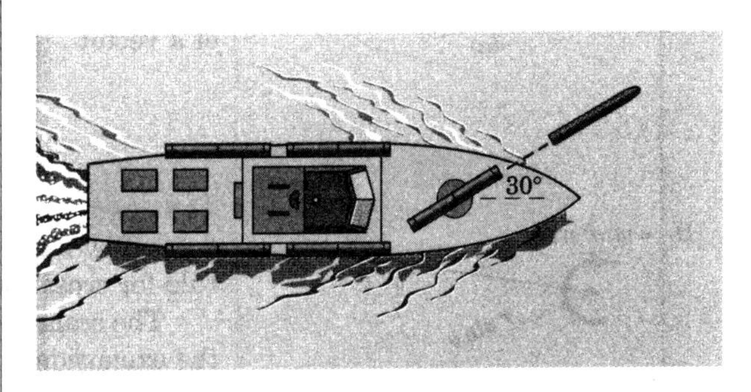

Problem 3/220

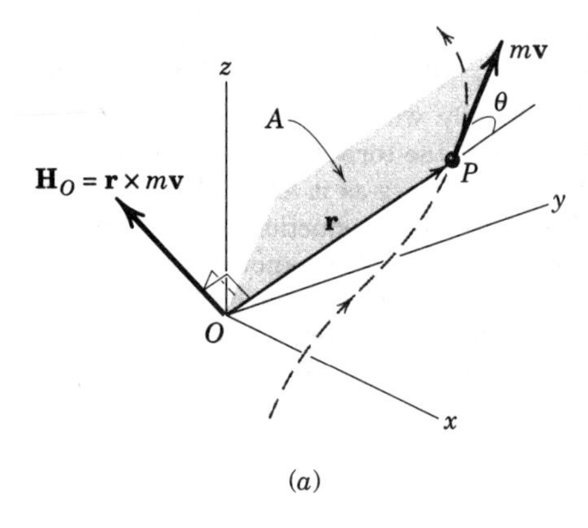

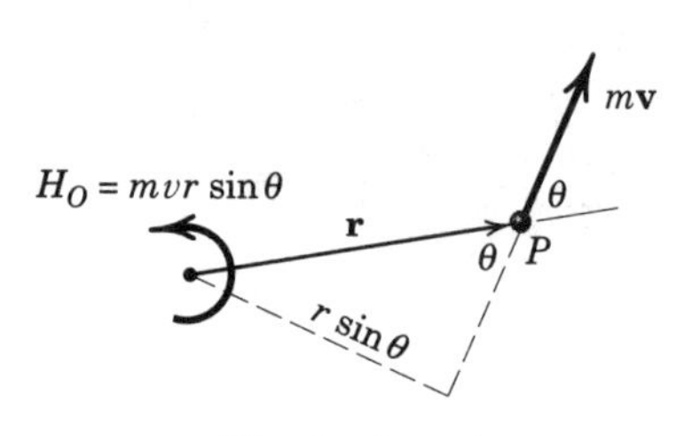

Figure 3/11

3/10 Angular Impulse and Angular Momentum

In addition to the equations of linear impulse and linear momentum, there exists a parallel set of equations for angular impulse and angular momentum. First, we define the term *angular momentum*. Figure 3/11*a* shows a particle P of mass m moving along a curve in space. The particle is located by its position vector $\mathbf{r}$ with respect to a convenient origin O of fixed coordinates x-y-z. The velocity of the particle is $\mathbf{v} = \dot{\mathbf{r}}$, and its linear momentum is $\mathbf{G} = m\mathbf{v}$. The *moment* of the *linear momentum vector* $m\mathbf{v}$ about the origin O is defined as the *angular momentum* $\mathbf{H}_O$ of P about O and is given by the cross-product relation for the moment of a vector

$$\mathbf{H}_O = \mathbf{r} \times m\mathbf{v} \tag{3/25}$$

The angular momentum then is a vector perpendicular to the plane A defined by $\mathbf{r}$ and $\mathbf{v}$. The sense of $\mathbf{H}_O$ is clearly defined by the right-hand rule for cross products.

The scalar components of angular momentum may be obtained from the expansion

$$\mathbf{H}_O = \mathbf{r} \times m\mathbf{v} = m(v_z y - v_y z)\mathbf{i} + m(v_x z - v_z x)\mathbf{j} + m(v_y x - v_x y)\mathbf{k}$$

$$\mathbf{H}_O = m\begin{vmatrix} \mathbf{i} & \mathbf{j} & \mathbf{k} \\ x & y & z \\ v_x & v_y & v_z \end{vmatrix} \tag{3/26}$$

so that

$$H_x = m(v_z y - v_y z) \qquad H_y = m(v_x z - v_z x) \qquad H_z = m(v_y x - v_x y)$$

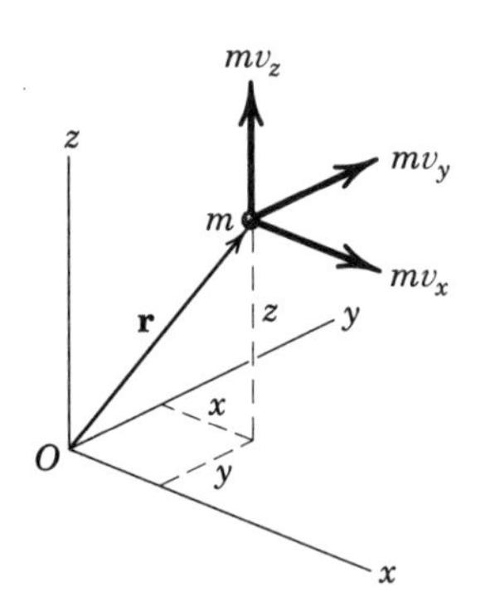

Figure 3/12

Each of these expressions for angular momentum may be checked easily from Fig. 3/12, which shows the three linear-momentum components, by taking the moments of these components about the respective axes.

To help visualize angular momentum, we show in Fig. 3/11*b* a two-dimensional representation in plane A of the vectors shown in part *a* of the figure. The motion is viewed in plane A defined by $\mathbf{r}$ and $\mathbf{v}$. The magnitude of the moment of $m\mathbf{v}$ about O is simply the linear momentum mv times the moment arm $r \sin\theta$ or $mvr \sin\theta$, which is the magnitude of the cross product $\mathbf{H}_O = \mathbf{r} \times m\mathbf{v}$.

Angular momentum is the moment of linear momentum and must not be confused with linear momentum. In SI units, angular momentum has the units $\text{kg}\cdot(\text{m/s})\cdot\text{m} = \text{kg}\cdot\text{m}^2/\text{s} = \text{N}\cdot\text{m}\cdot\text{s}$. In the U.S. customary system, angular momentum has the units $[\text{lb}/(\text{ft/sec}^2)][\text{ft/sec}][\text{ft}] =$ ft-lb-sec.

Rate of Change of Angular Momentum

We are now ready to relate the moment of the forces acting on the particle P to its angular momentum. If $\Sigma\mathbf{F}$ represents the resultant of *all* forces acting on the particle P of Fig. 3/11, the moment $\mathbf{M}_O$ about the origin O is the vector cross product

$$\Sigma \mathbf{M}_O = \mathbf{r} \times \Sigma \mathbf{F} = \mathbf{r} \times m\dot{\mathbf{v}}$$

where Newton's second law $\Sigma \mathbf{F} = m\dot{\mathbf{v}}$ has been substituted. We now differentiate Eq. 3/25 with time, using the rule for the differentiation of a cross product (see item 9, Art. C/7, Appendix C) and obtain

$$\dot{\mathbf{H}}_O = \dot{\mathbf{r}} \times m\mathbf{v} + \mathbf{r} \times m\dot{\mathbf{v}} = \mathbf{v} \times m\mathbf{v} + \mathbf{r} \times m\dot{\mathbf{v}}$$

The term $\mathbf{v} \times m\mathbf{v}$ is zero since the cross product of parallel vectors is identically zero. Substitution into the expression for $\Sigma \mathbf{M}_O$ gives

$$\Sigma \mathbf{M}_O = \dot{\mathbf{H}}_O \qquad (3/27)$$

Equation 3/27 states that the *moment about the fixed point O of all forces acting on m equals the time rate of change of angular momentum of m about O.* This relation, particularly when extended to a system of particles, rigid or nonrigid, provides one of the most powerful tools of analysis in dynamics.

Equation 3/27 is a vector equation with scalar components

$$\Sigma M_{O_x} = \dot{H}_{O_x} \qquad \Sigma M_{O_y} = \dot{H}_{O_y} \qquad \Sigma M_{O_z} = \dot{H}_{O_z} \qquad (3/28)$$

The Angular Impulse-Momentum Principle

Equation 3/27 gives the instantaneous relation between the moment and the time rate of change of angular momentum. To obtain the effect of the moment $\Sigma \mathbf{M}_O$ on the angular momentum of the particle over a finite period of time, we integrate Eq. 3/27 from time t_1 to time t_2. Multiplying the equation by dt gives $\Sigma \mathbf{M}_O \, dt = d\mathbf{H}_O$, which we integrate to obtain

$$\int_{t_1}^{t_2} \Sigma \mathbf{M}_O \, dt = \mathbf{H}_{O_2} - \mathbf{H}_{O_1} = \Delta \mathbf{H}_O \qquad (3/29)$$

where $\mathbf{H}_{O_2} = \mathbf{r}_2 \times m\mathbf{v}_2$ and $\mathbf{H}_{O_1} = \mathbf{r}_1 \times m\mathbf{v}_1$. The product of moment and time is defined as *angular impulse*, and Eq. 3/29 states that the *total angular impulse on m about the fixed point O equals the corresponding change in angular momentum of m about O.*

Alternatively, we may write Eq. 3/29 as

$$\mathbf{H}_{O_1} + \int_{t_1}^{t_2} \Sigma \mathbf{M}_O \, dt = \mathbf{H}_{O_2} \qquad (3/29a)$$

which states that the initial angular momentum of the particle plus the angular impulse applied to it equals its final angular momentum. The units of angular impulse are clearly those of angular momentum, which are N·m·s or kg·m^2/s in SI units and ft-lb-sec in U.S. customary units.

As in the case of linear impulse and linear momentum, the equation of angular impulse and angular momentum is a vector equation where changes in direction as well as magnitude may occur during the interval of integration. Under these conditions, it is necessary to express $\Sigma \mathbf{M}_O$ and $\mathbf{H}_O$ in component form and then combine the integrated components. Thus, the x-component of Eq. 3/29 becomes

$$\int_{t_1}^{t_2} \Sigma M_{O_x}\, dt = (H_{O_x})_2 - (H_{O_x})_1$$
$$= m[(v_z y - v_y z)_2 - (v_z y - v_y z)_1]$$

where the subscripts 1 and 2 refer to the values of the respective quantities at times t_1 and t_2. Similar expressions exist for the y- and z-components of the angular-momentum integral.

Plane-Motion Applications

The foregoing angular-impulse and angular-momentum relations have been developed in their general three-dimensional forms. Most of the applications of interest to us, however, can be analyzed as plane-motion problems where moments are taken about a single axis normal to the plane of motion. In this case, the angular momentum may change magnitude and sense, but the direction of the vector remains unaltered.

Thus, for a particle of mass m moving along a curved path in the x-y plane, Fig. 3/13, the angular momenta about O at points 1 and 2 have the magnitudes $H_{O_1} = |\mathbf{r}_1 \times m\mathbf{v}_1| = mv_1 d_1$ and $H_{O_2} = |\mathbf{r}_2 \times m\mathbf{v}_2| = mv_2 d_2$, respectively. In the illustration both H_{O_1} and H_{O_2} are represented in the counterclockwise sense in accord with the direction of the

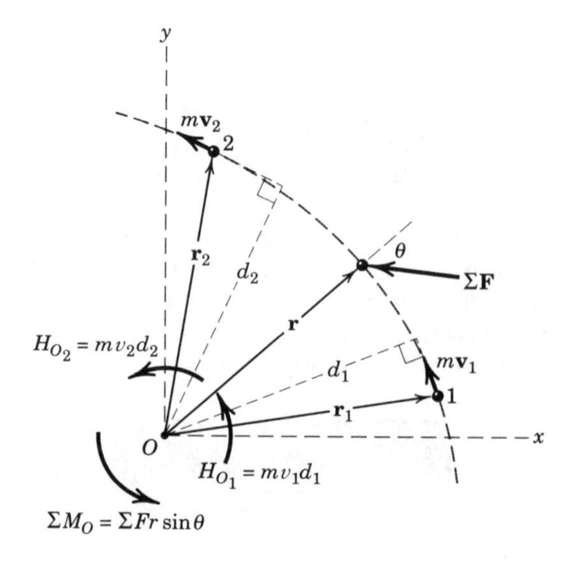

Figure 3/13

moment of the linear momentum. The scalar form of Eq. 3/29 applied to the motion between points 1 and 2 during the time interval t_1 to t_2 becomes

$$\int_{t_1}^{t_2} \Sigma M_O \, dt = H_{O_2} - H_{O_1} \quad \text{or} \quad \int_{t_1}^{t_2} \Sigma Fr \sin \theta \, dt = mv_2 d_2 - mv_1 d_1$$

This example should help clarify the relation between the scalar and vector forms of the angular impulse-momentum relations.

Equations 3/21 and 3/27 add no new basic information since they are merely alternative forms of Newton's second law. We will discover in subsequent chapters, however, that the motion equations expressed in terms of the time rate of change of momentum are applicable to the motion of rigid and nonrigid bodies and provide a very general and powerful approach to many problems. The full generality of Eq. 3/27 is usually not required to describe the motion of a single particle or the plane motion of rigid bodies, but it does have important use in the analysis of the space motion of rigid bodies introduced in Chapter 7.

Conservation of Angular Momentum

If the resultant moment about a fixed point O of all forces acting on a particle is zero during an interval of time, Eq. 3/27 requires that its angular momentum $\mathbf{H}_O$ about that point remain constant. In this case, the angular momentum of the particle is said to be *conserved*. Angular momentum may be conserved about one axis but not about another axis. A careful examination of the free-body diagram of the particle will disclose whether the moment of the resultant force on the particle about a fixed point is zero, in which case, the angular momentum about that point is unchanged (conserved).

Consider now the motion of two particles a and b which interact during an interval of time. If the interactive forces $\mathbf{F}$ and $-\mathbf{F}$ between them are the only unbalanced forces acting on the particles during the interval, it follows that the moments of the equal and opposite forces about any fixed point O not on their line of action are equal and opposite. If we apply Eq. 3/29 to particle a and then to particle b and add the two equations, we obtain $\Delta\mathbf{H}_a + \Delta\mathbf{H}_b = \mathbf{0}$ (where all angular momenta are referred to point O). Thus, the total angular momentum for the system of the two particles remains constant during the interval, and we write

$$\Delta\mathbf{H}_O = \mathbf{0} \quad \text{or} \quad \mathbf{H}_{O_1} = \mathbf{H}_{O_2} \tag{3/30}$$

which expresses the *principle of conservation of angular momentum*.

Sample Problem 3/22

The small 2-kg block slides on a smooth horizontal surface under the action of the force in the spring and a force F. The angular momentum of the block about O varies with time as shown in the graph. When $t = 6.5$ s, it is known that $r = 150$ mm and $\beta = 60°$. Determine F for this instant.

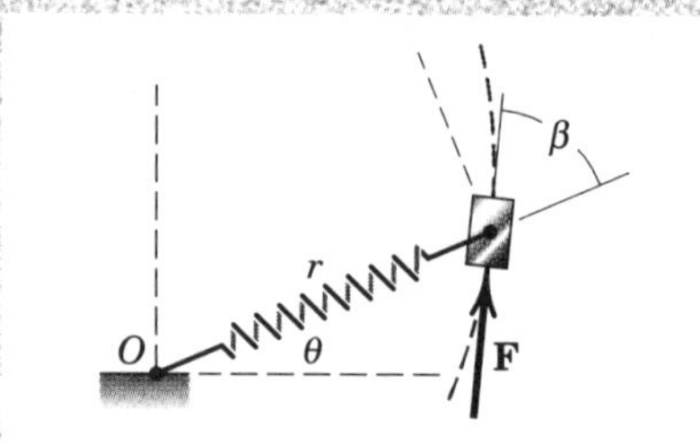

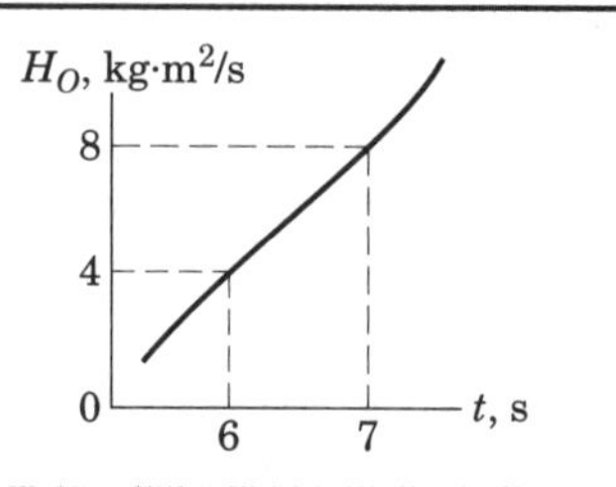

Solution. The only moment of the forces about O is due to F since the spring force passes through O. Thus, $\Sigma M_O = Fr \sin \beta$. From the graph the time rate of change of H_O for $t = 6.5$ s is very nearly $(8 - 4)/(7 - 6)$ or $\dot{H}_O = 4 \text{ kg}\cdot\text{m}^2/\text{s}^2$. The moment-angular momentum relation gives

① $[\Sigma M_O = \dot{H}_O]$ $\quad F(0.150) \sin 60° = 4 \quad F = 30.8 \text{ N}$ *Ans.*

Helpful Hint

① We do not need vector notation here since we have plane motion where the direction of the vector $\mathbf{H}_O$ does not change.

Sample Problem 3/23

A small mass particle is given an initial velocity $\mathbf{v}_0$ tangent to the horizontal rim of a smooth hemispherical bowl at a radius r_0 from the vertical centerline, as shown at point A. As the particle slides past point B, a distance h below A and a distance r from the vertical centerline, its velocity $\mathbf{v}$ makes an angle θ with the horizontal tangent to the bowl through B. Determine θ.

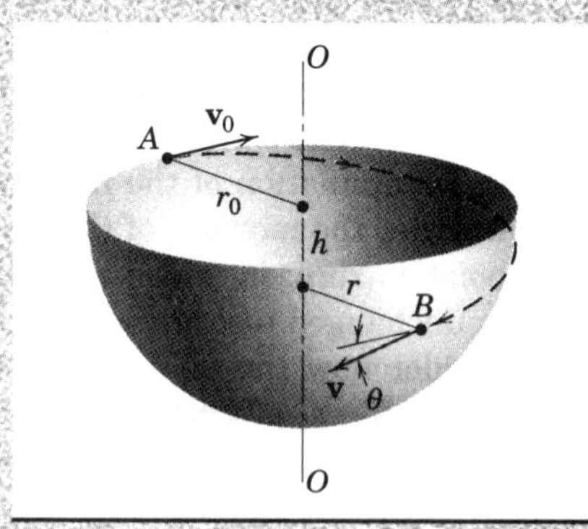

Solution. The forces on the particle are its weight and the normal reaction exerted by the smooth surface of the bowl. Neither force exerts a moment about the axis O–O, so that angular momentum is conserved about that axis. Thus,

① $[(H_O)_1 = (H_O)_2]$ $\quad mv_0r_0 = mvr\cos\theta$

Also, energy is conserved so that $E_1 = E_2$. Thus

$[T_1 + V_{g_1} = T_2 + V_{g_2}]$ $\quad \frac{1}{2}mv_0^2 + mgh = \frac{1}{2}mv^2 + 0$

$$v = \sqrt{v_0^2 + 2gh}$$

Eliminating v and substituting $r^2 = r_0^2 - h^2$ give

$$v_0r_0 = \sqrt{v_0^2 + 2gh}\sqrt{r_0^2 - h^2}\cos\theta$$

$$\theta = \cos^{-1}\frac{1}{\sqrt{1 + \dfrac{2gh}{v_0^2}}\sqrt{1 - \dfrac{h^2}{r_0^2}}} \qquad \textit{Ans.}$$

Helpful Hint

① The angle θ is measured in the plane tangent to the hemispherical surface at B.

PROBLEMS

Introductory Problems

3/221 Determine the magnitude H_O of the angular momentum of the 2-kg sphere about point O (*a*) by using the vector definition of angular momentum and (*b*) by using an equivalent scalar approach. The center of the sphere lies in the *x*-*y* plane.

Ans. $H_O = 69.3\ \text{kg}\cdot\text{m}^2/\text{s}$

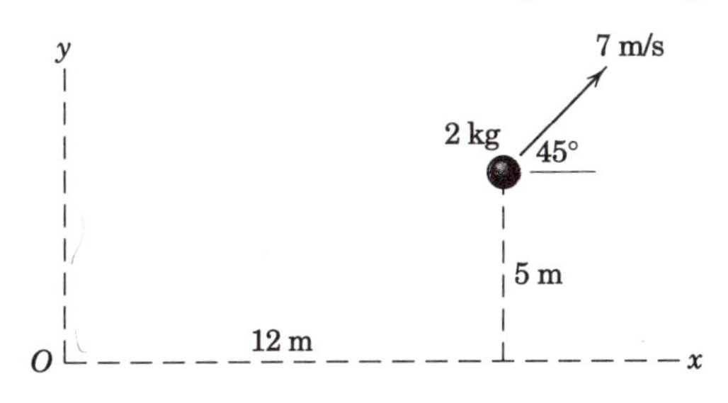

Problem 3/221

3/222 The 3-kg sphere moves in the *x*-*y* plane and has the indicated velocity at a particular instant. Determine its (*a*) linear momentum, (*b*) angular momentum about point O, and (*c*) kinetic energy.

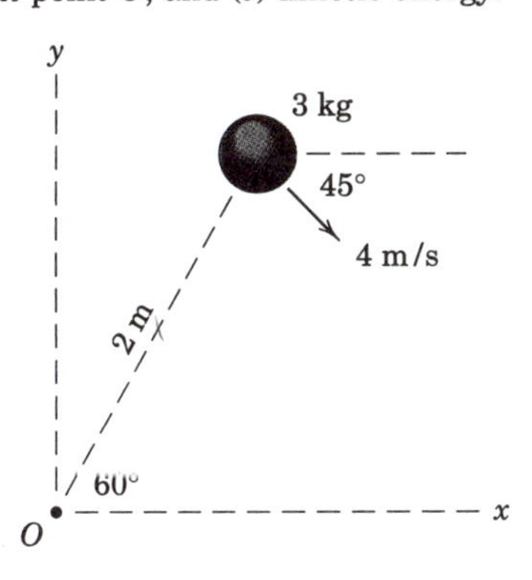

Problem 3/222

3/223 At a certain instant, the linear momentum of a particle is given by $\mathbf{G} = -3\mathbf{i} - 2\mathbf{j} + 3\mathbf{k}$ kg·m/s and its position vector is $\mathbf{r} = 3\mathbf{i} + 4\mathbf{j} - 3\mathbf{k}$ m. Determine the magnitude H_O of the angular momentum of the particle about the origin of coordinates.

Ans. $H_O = 8.49\ \text{kg}\cdot\text{m}^2/\text{s}$

3/224 A particle with a mass of 4 kg has a position vector in meters given by $\mathbf{r} = 3t^2\mathbf{i} - 2t\mathbf{j} - 3t\mathbf{k}$, where t is the time in seconds. For $t = 3$ s determine the magnitude of the angular momentum of the particle and the magnitude of the moment of all forces on the particle, both about the origin of coordinates.

3/225 At a certain instant, the particle of mass m has the position and velocity shown in the figure, and it is acted upon by the force $\mathbf{F}$. Determine its angular momentum about point O and the time rate of change of this angular momentum.

Ans. $\mathbf{H}_O = mv(-c\mathbf{i} + a\mathbf{k})$, $\dot{\mathbf{H}}_O = F(b\mathbf{i} - a\mathbf{j})$

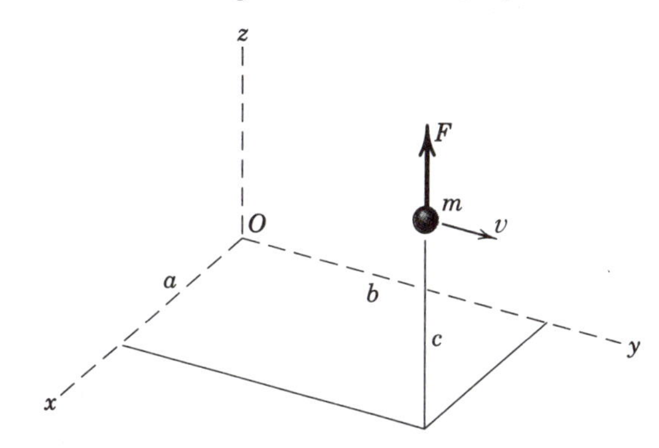

Problem 3/225

3/226 The small spheres, which have the masses and initial velocities shown in the figure, strike and become attached to the spiked ends of the rod, which is freely pivoted at O and is initially at rest. Determine the angular velocity ω of the assembly after impact. Neglect the mass of the rod.

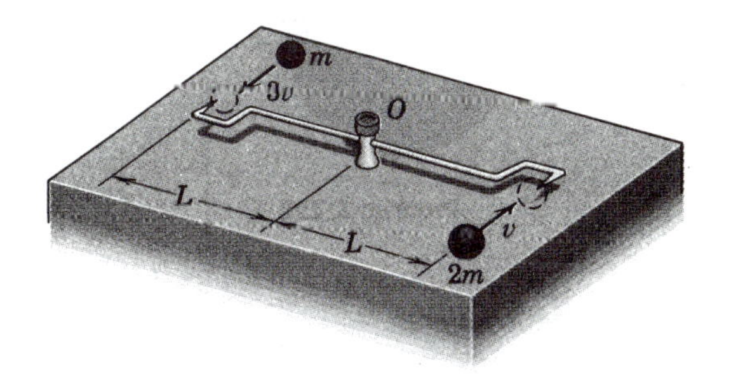

Problem 3/226

3/227 The assembly starts from rest and reaches an angular speed of 150 rev/min under the action of a 20-N force T applied to the string for t seconds. Determine t. Neglect friction and all masses except those of the four 3-kg spheres, which may be treated as particles.

Ans. $t = 15.08$ s

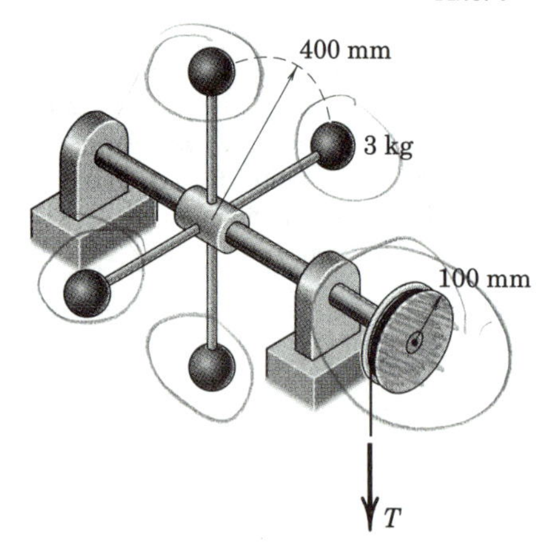

Problem 3/227

3/228 The only force acting on an earth satellite traveling outside of the earth's atmosphere is the radial gravitational attraction. The moment of this force is zero about the earth's center taken as a fixed point. Prove that $r^2\dot{\theta}$ remains constant for the motion of the satellite.

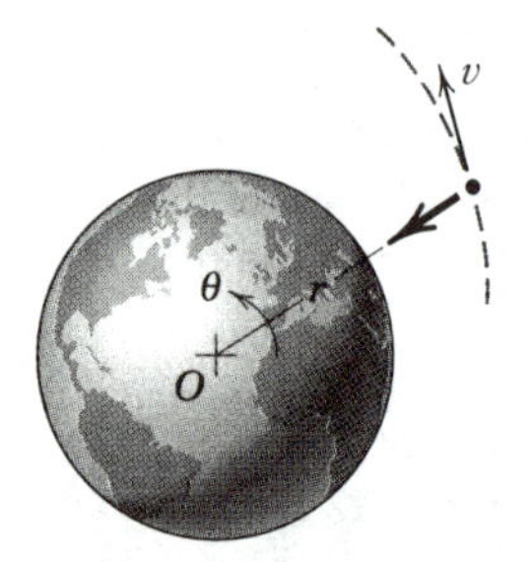

Problem 3/228

Representative Problems

3/229 A small 110-g particle is projected with a horizontal velocity of 2 m/s into the top A of the smooth circular guide fixed in the vertical plane. Calculate the time rate of change $\dot{\mathbf{H}}_B$ of angular momentum about point B when the particle passes the bottom of the guide at C.

Ans. $\dot{\mathbf{H}}_B = 1.519\mathbf{k}$ N·m

Problem 3/229

3/230 The 6-kg sphere and 4-kg block (shown in section) are secured to the arm of negligible mass which rotates in the vertical plane about a horizontal axis at O. The 2-kg plug is released from rest at A and falls into the recess in the block when the arm has reached the horizontal position. An instant before engagement, the arm has an angular velocity $\omega_0 = 2$ rad/s. Determine the angular velocity ω of the arm immediately after the plug has wedged itself in the block.

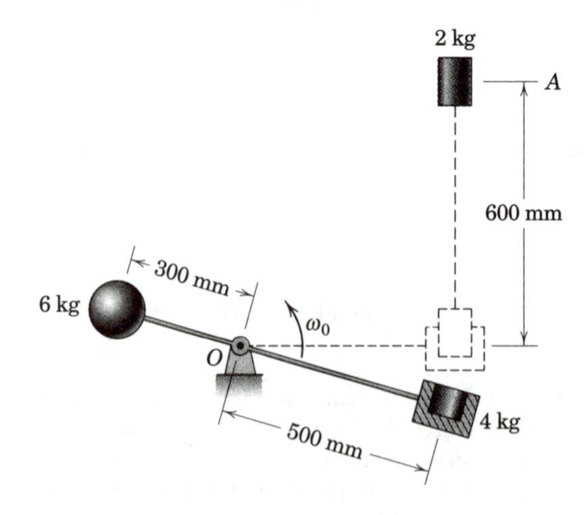

Problem 3/230

3/231 The central attractive force F on an earth satellite can have no moment about the center O of the earth. For the particular elliptical orbit with major and minor axes as shown, a satellite will have a velocity of 33 880 km/h at the perigee altitude of 390 km. Determine the velocity of the satellite at point B and at apogee A. The radius of the earth is 6371 km.

Ans. $v_B = 19\,540$ km/h
$v_A = 11\,300$ km/h

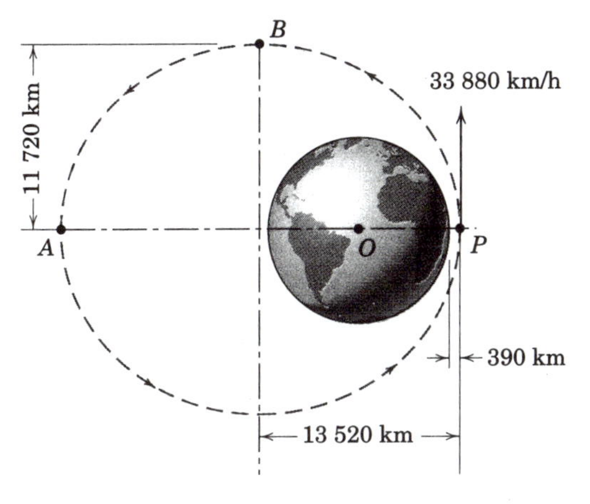

Problem 3/231

3/232 A particle moves on the inside surface of a smooth conical shell and is given an initial velocity $\mathbf{v}_0$ tangent to the horizontal rim of the surface at A. As the particle slides past point B, a distance z below A, its velocity $\mathbf{v}$ makes an angle θ with the horizontal tangent to the surface through B. Determine expressions for θ and the speed v.

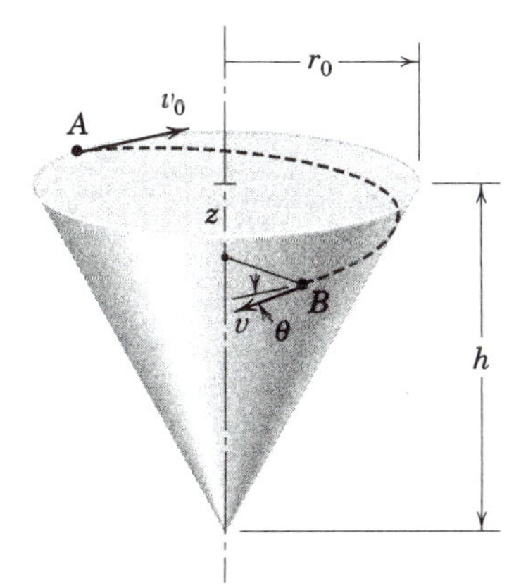

Problem 3/232

3/233 The two spheres of equal mass m are able to slide along the horizontal rotating rod. If they are initially latched in position a distance r from the rotating axis with the assembly rotating freely with an angular velocity ω_0, determine the new angular velocity ω after the spheres are released and finally assume positions at the ends of the rod at a radial distance of $2r$. Also find the fraction n of the initial kinetic energy of the system which is lost. Neglect the small mass of the rod and shaft.

Ans. $\omega = \omega_0/4$, $n = 3/4$

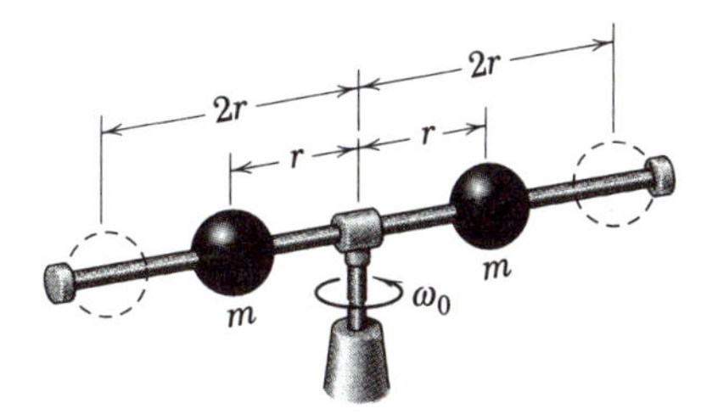

Problem 3/233

3/234 The two balls, each of mass m, are mounted on the light rods which rotate in a horizontal plane at the angular rate $\dot{\theta}$ about the fixed vertical axis. If an internal mechanism lowers the rods to the dashed positions shown without interfering with the free rotation about the vertical axis, determine the new rate of rotation $\dot{\theta}'$ in the lowered position.

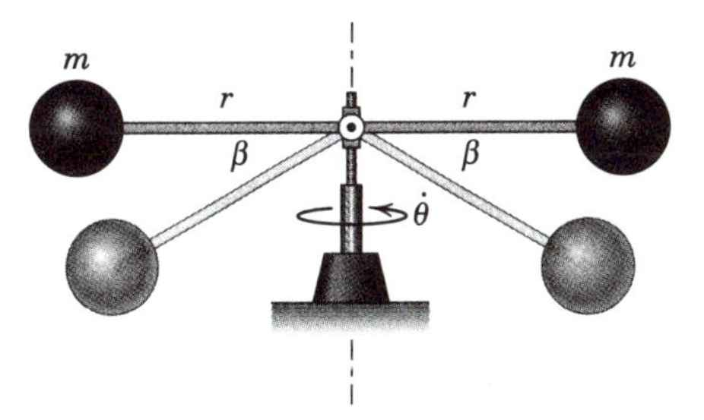

Problem 3/234

3/235 The 0.2-kg ball and its supporting cord are revolving about the vertical axis on the fixed smooth conical surface with an angular velocity of 4 rad/s. The ball is held in the position $b = 300$ mm by the tension T in the cord. If the distance b is reduced to the constant value of 200 mm by increasing the tension T in the cord, compute the new angular velocity ω and the work $U'_{1\text{-}2}$ done on the system by T.

Ans. $\omega = 9$ rad/s, $U'_{1\text{-}2} = 0.233$ J

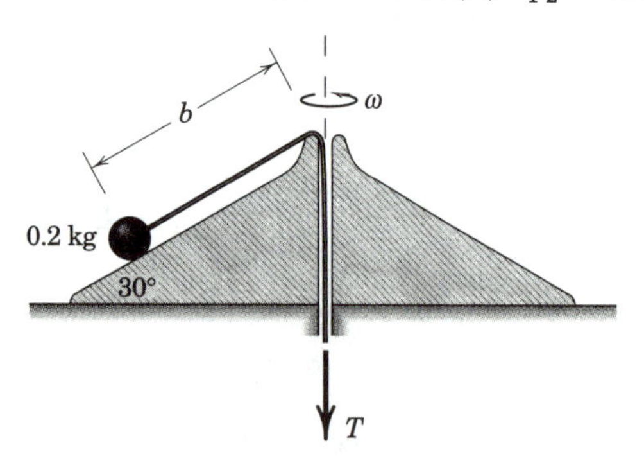

Problem 3/235

3/236 Determine the magnitude H_O of the angular momentum about the launch point O of the projectile of mass m which is launched with speed v_0 at the angle θ as shown (*a*) at the instant of launch and (*b*) at the instant of impact. Qualitatively account for the two results. Neglect atmospheric resistance.

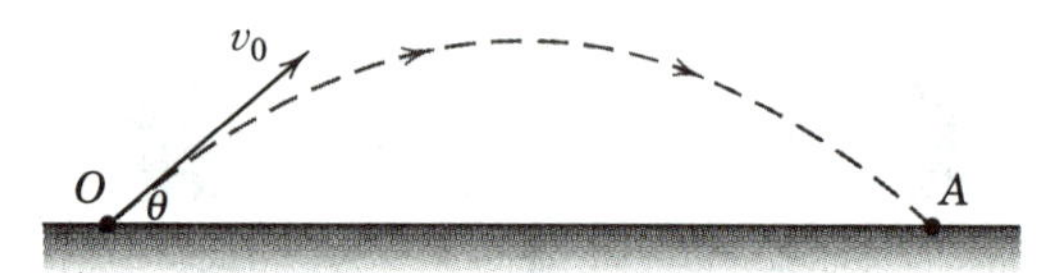

Problem 3/236

3/237 The particle of mass m is launched from point O with a horizontal velocity $\mathbf{u}$ at time $t = 0$. Determine its angular momentum $\mathbf{H}_O$ relative to point O as a function of time.

Ans. $\mathbf{H}_O = -\frac{1}{2}mgut^2\mathbf{k}$

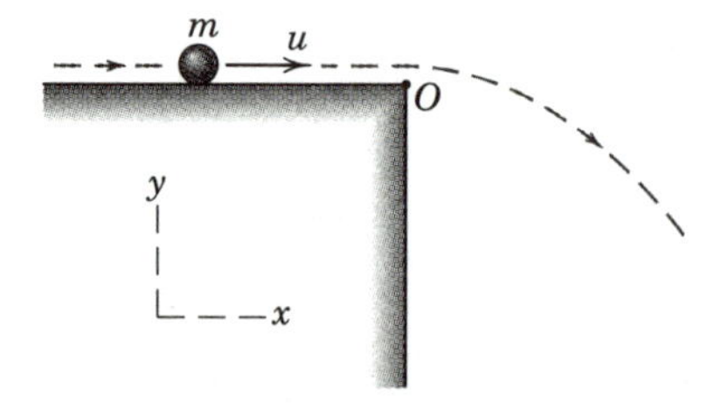

Problem 3/237

3/238 At the point A of closest approach to the sun, a comet has a velocity $v_A = 57.45(10^3)$ m/s. Determine the radial and transverse components of its velocity v_B at point B, where the radial distance from the sun is $120.7(10^6)$ km.

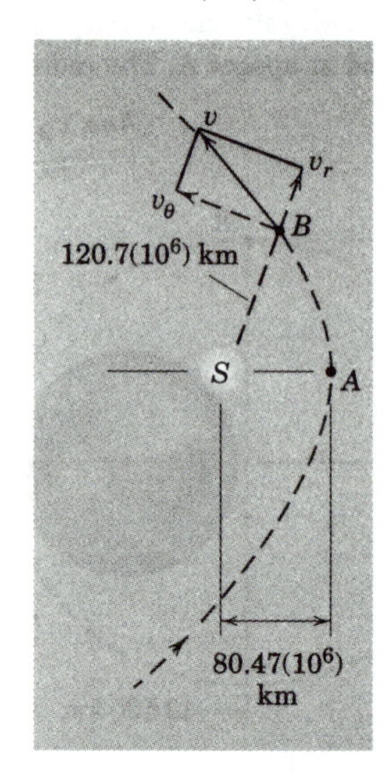

Problem 3/238

3/239 A particle is released on the smooth inside wall of a cylindrical tank at A with a velocity v_0 which makes an angle β with the horizontal tangent. When the particle reaches a point B a distance h below A, determine the expression for the angle θ made by its velocity with the horizontal tangent at B.

Ans. $\theta = \cos^{-1} \dfrac{\cos\beta}{\sqrt{1 + \dfrac{2gh}{v_0^2}}}$

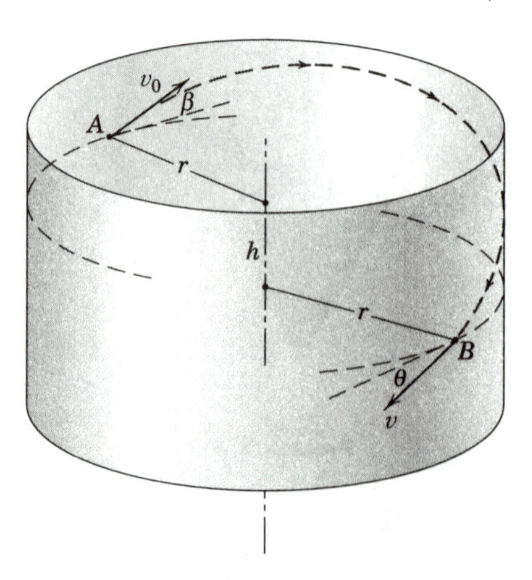

Problem 3/239

3/240 A pendulum consists of two 3.2-kg concentrated masses positioned as shown on a light but rigid bar. The pendulum is swinging through the vertical position with a clockwise angular velocity $\omega = 6$ rad/s when a 50-g bullet traveling with velocity $v = 300$ m/s in the direction shown strikes the lower mass and becomes embedded in it. Calculate the angular velocity ω' which the pendulum has immediately after impact and find the maximum angular deflection θ of the pendulum.

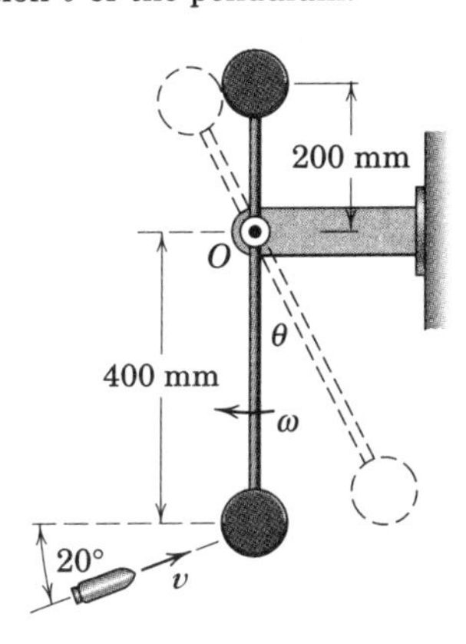

Problem 3/240

3/241 The small particle of mass m and its restraining cord are spinning with an angular velocity ω on the horizontal surface of a smooth disk, shown in section. As the force F is slightly relaxed, r increases and ω changes. Determine the rate of change of ω with respect to r and show that the work done by F during a movement dr equals the change in kinetic energy of the particle.

Ans. $\dfrac{d\omega}{dr} = -\dfrac{2\omega}{r}$

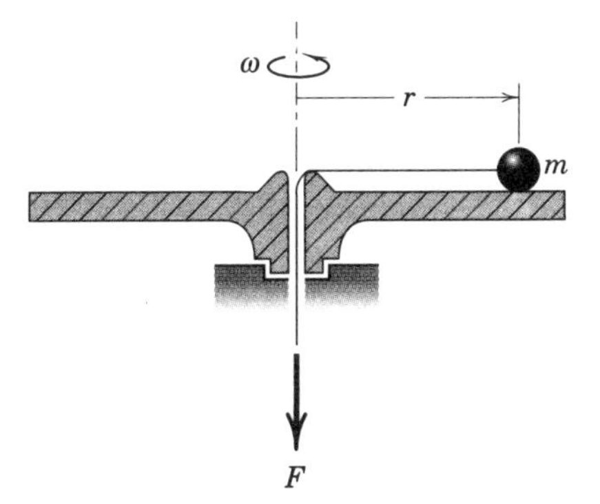

Problem 3/241

3/242 The 0.7-kg sphere moves in a horizontal plane and is controlled by a cord which is reeled in and out below the table in such a way that the center of the sphere is confined to the path given by $(x^2/2.25) + (y^2/1.44) = 1$ where x and y are in meters. If the speed of the sphere is $v_A = 2$ m/s as it passes point A, determine the tension T_B in the cord as the sphere passes point B. Friction is negligible.

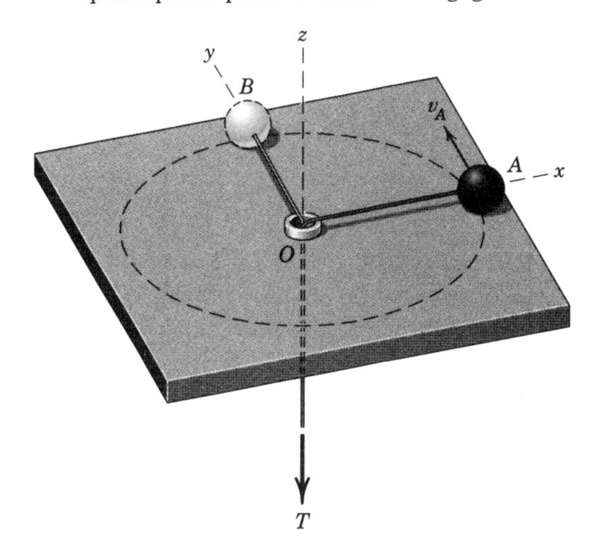

Problem 3/242

▶**3/243** The assembly of two 5-kg spheres is rotating freely about the vertical axis at 40 rev/min with $\theta = 90°$. If the force F which maintains the given position is increased to raise the base collar and reduce θ to 60°, determine the new angular velocity ω. Also determine the work U done by F in changing the configuration of the system. Assume that the mass of the arms and collars is negligible.

Ans. $\omega = 3.00$ rad/s, $U = 5.34$ J

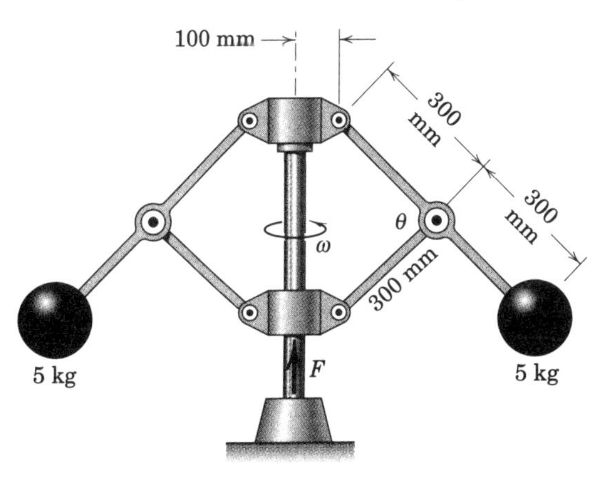

Problem 3/243

▶**3/244** The small particle of mass m is given an initial high velocity in the horizontal plane and winds its cord around the fixed vertical shaft of radius a. All motion occurs essentially in the horizontal plane. If the angular velocity of the cord is ω_0 when the distance from the particle to the tangency point is r_0, determine the angular velocity ω of the cord and its tension T after it has turned through an angle θ. Does either of the principles of conservation of momentum apply?

Ans. $\omega = \dfrac{\omega_0}{1 - a\theta/r_0}$, $T = mr_0\omega_0\omega$

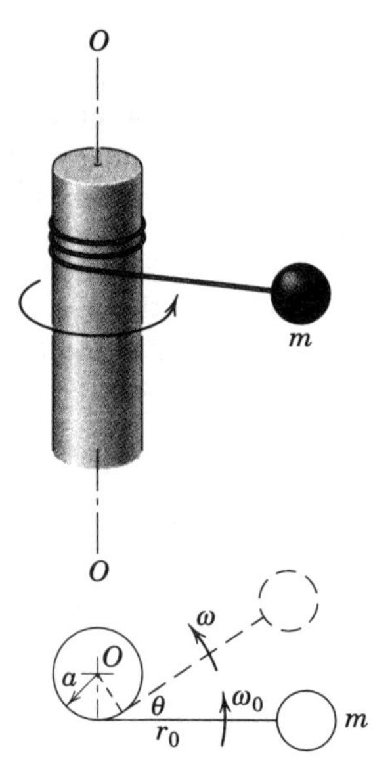

Problem 3/244

Chapter

5

PLANE KINEMATICS OF RIGID BODIES

CHAPTER OUTLINE

5/1 INTRODUCTION

In Chapter 2 on particle kinematics, we developed the relationships governing the displacement, velocity, and acceleration of points as they moved along straight or curved paths. In rigid-body kinematics we use these same relationships but must also account for the rotational motion of the body. Thus rigid-body kinematics involves both linear and angular displacements, velocities, and accelerations.

We need to describe the motion of rigid bodies for two important reasons. First, we frequently need to generate, transmit, or control certain desired motions by the use of cams, gears, and linkages of various types. Here we must analyze the displacement, velocity, and acceleration of the motion to determine the design geometry of the mechanical parts. Furthermore, as a result of the motion generated, forces may be developed which must be accounted for in the design of the parts.

Second, we must often determine the motion of a rigid body caused by the forces applied to it. Calculation of the motion of a rocket under the influence of its thrust and gravitational attraction is an example of such a problem.

We need to apply the principles of rigid-body kinematics in both situations. This chapter covers the kinematics of rigid-body motion which may be analyzed as occurring in a single plane. In Chapter 7 we will present an introduction to the kinematics of motion in three dimensions.

Rigid-Body Assumption

In the previous chapter we defined a *rigid body* as a system of particles for which the distances between the particles remain unchanged. Thus, if each particle of such a body is located by a position vector from reference axes attached to and rotating with the body, there will be no change in any position vector as measured from these axes. This is, of course, an ideal case since all solid materials change shape to some extent when forces are applied to them.

Nevertheless, if the movements associated with the changes in shape are very small compared with the movements of the body as a whole, then the assumption of rigidity is usually acceptable. The displacements due to the flutter of an aircraft wing, for instance, do not affect the description of the flight path of the aircraft as a whole, and thus the rigid-body assumption is clearly acceptable. On the other hand, if the problem is one of describing, as a function of time, the internal wing stress due to wing flutter, then the relative motions of portions of the wing cannot be neglected, and the wing may not be considered a rigid body. In this and the next two chapters, almost all of the material is based on the assumption of rigidity.

Plane Motion

A rigid body executes plane motion when all parts of the body move in parallel planes. For convenience, we generally consider the *plane of motion* to be the plane which contains the mass center, and we treat the body as a thin slab whose motion is confined to the plane of the slab. This idealization adequately describes a very large category of rigid-body motions encountered in engineering.

The plane motion of a rigid body may be divided into several categories, as represented in Fig. 5/1.

Translation is defined as any motion in which every line in the body remains parallel to its original position at all times. In translation there is *no rotation of any line in the body*. In *rectilinear translation*, part *a* of Fig. 5/1, all points in the body move in parallel straight lines. In *curvilinear translation*, part *b*, all points move on congruent curves. We note that in each of the two cases of translation, the motion of the body is completely specified by the motion of any point in the body, since all points have the same motion. Thus, our earlier study of the motion of a point (particle) in Chapter 2 enables us to describe completely the translation of a rigid body.

Rotation about a fixed axis, part *c* of Fig. 5/1, is the angular motion about the axis. It follows that all particles in a rigid body move in circular paths about the axis of rotation, and all lines in the body which are perpendicular to the axis of rotation (including those which do not pass through the axis) rotate through the same angle in the same time.

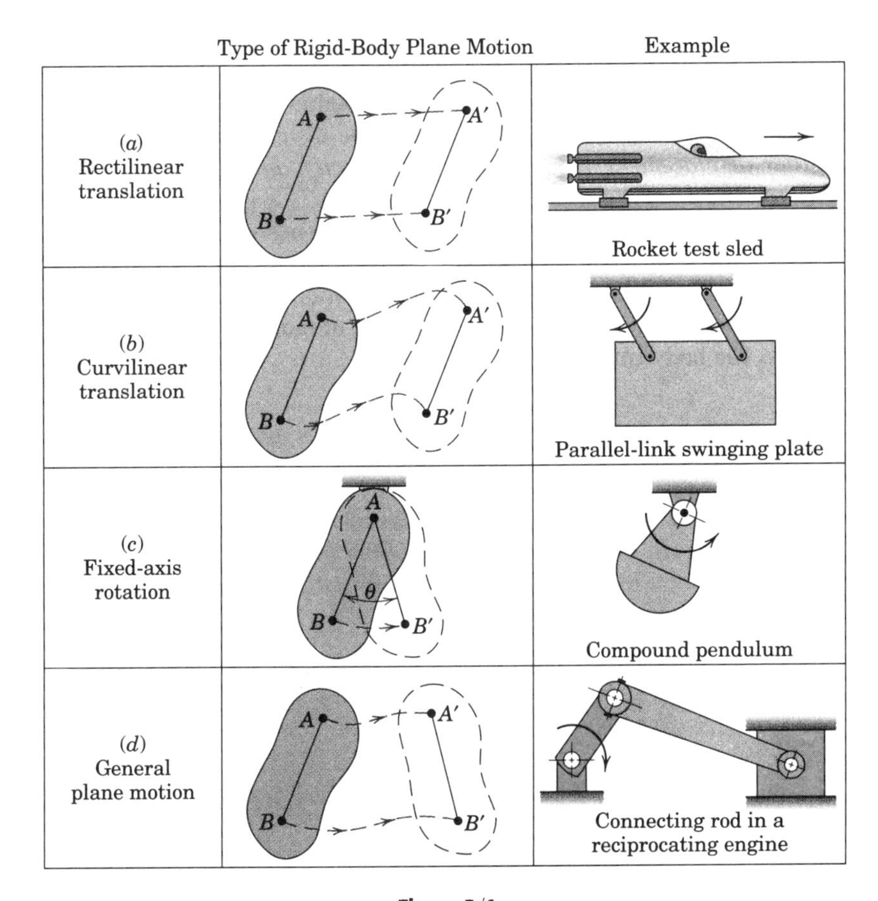

Figure 5/1

Again, our discussion in Chapter 2 on the circular motion of a point enables us to describe the motion of a rotating rigid body, which is treated in the next article.

General plane motion of a rigid body, part d of Fig. 5/1, is a combination of translation and rotation. We will utilize the principles of relative motion covered in Art. 2/8 to describe general plane motion.

Note that in each of the examples cited, the actual paths of all particles in the body are projected onto the single plane of motion as represented in each figure.

Analysis of the plane motion of rigid bodies is accomplished either by directly calculating the absolute displacements and their time derivatives from the geometry involved or by utilizing the principles of relative motion. Each method is important and useful and will be covered in turn in the articles which follow.

5/2 Rotation

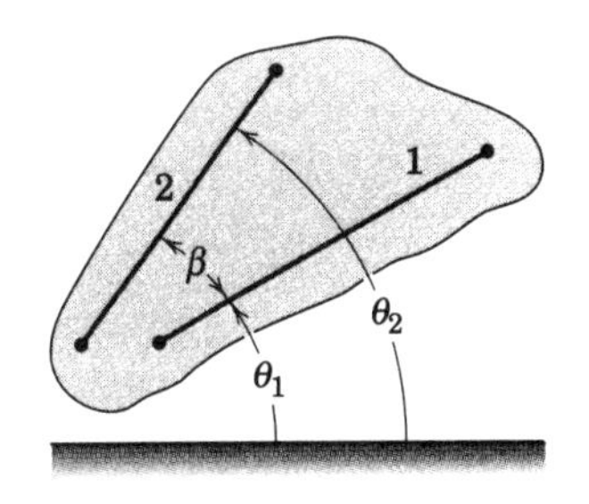

Figure 5/2

The rotation of a rigid body is described by its angular motion. Figure 5/2 shows a rigid body which is rotating as it undergoes plane motion in the plane of the figure. The angular positions of any two lines 1 and 2 attached to the body are specified by θ_1 and θ_2 measured from any convenient fixed reference direction. Because the angle β is invariant, the relation $\theta_2 = \theta_1 + \beta$ upon differentiation with respect to time gives $\dot{\theta}_2 = \dot{\theta}_1$ and $\ddot{\theta}_2 = \ddot{\theta}_1$ or, during a finite interval, $\Delta\theta_2 = \Delta\theta_1$. Thus, *all lines on a rigid body in its plane of motion have the same angular displacement, the same angular velocity, and the same angular acceleration.*

Note that the angular motion of a line depends only on its angular position with respect to any arbitrary fixed reference and on the time derivatives of the displacement. Angular motion does not require the presence of a fixed axis, normal to the plane of motion, about which the line and the body rotate.

Angular-Motion Relations

The angular velocity ω and angular acceleration α of a rigid body in plane rotation are, respectively, the first and second time derivatives of the angular position coordinate θ of any line in the plane of motion of the body. These definitions give

$$\omega = \frac{d\theta}{dt} = \dot{\theta}$$

$$\alpha = \frac{d\omega}{dt} = \dot{\omega} \quad \text{or} \quad \alpha = \frac{d^2\theta}{dt^2} = \ddot{\theta} \tag{5/1}$$

$$\omega\, d\omega = \alpha\, d\theta \quad \text{or} \quad \dot{\theta}\, d\dot{\theta} = \ddot{\theta}\, d\theta$$

The third relation is obtained by eliminating dt from the first two. In each of these relations, the positive direction for ω and α, clockwise or counterclockwise, is the same as that chosen for θ. Equations 5/1 should be recognized as analogous to the defining equations for the rectilinear motion of a particle, expressed by Eqs. 2/1, 2/2, and 2/3. In fact, all relations which were described for rectilinear motion in Art. 2/2 apply to the case of rotation in a plane if the linear quantities s, v, and a are replaced by their respective equivalent angular quantities θ, ω, and α. As we proceed further with rigid-body dynamics, we will find that the analogies between the relationships for linear and angular motion are almost complete throughout kinematics and kinetics. These relations are important to recognize, as they help to demonstrate the symmetry and unity found throughout mechanics.

For rotation with *constant* angular acceleration, the integrals of Eqs. 5/1 become

$$\omega = \omega_0 + \alpha t$$

$$\omega^2 = \omega_0{}^2 + 2\alpha(\theta - \theta_0)$$

$$\theta = \theta_0 + \omega_0 t + \tfrac{1}{2}\alpha t^2$$

Here θ_0 and ω_0 are the values of the angular position coordinate and angular velocity, respectively, at $t = 0$, and t is the duration of the motion considered. You should be able to carry out these integrations easily, as they are completely analogous to the corresponding equations for rectilinear motion with constant acceleration covered in Art. 2/2.

The graphical relationships described for s, v, a, and t in Figs. 2/3 and 2/4 may be used for θ, ω, and α merely by substituting the corresponding symbols. You should sketch these graphical relations for plane rotation. The mathematical procedures for obtaining rectilinear velocity and displacement from rectilinear acceleration may be applied to rotation by merely replacing the linear quantities by their corresponding angular quantities.

Rotation about a Fixed Axis

When a rigid body rotates about a fixed axis, all points other than those on the axis move in concentric circles about the fixed axis. Thus, for the rigid body in Fig. 5/3 rotating about a fixed axis normal to the plane of the figure through O, any point such as A moves in a circle of radius r. From the previous discussion in Art. 2/5, you should already be familiar with the relationships between the linear motion of A and the angular motion of the line normal to its path, which is also the angular motion of the rigid body. With the notation $\omega = \dot{\theta}$ and $\alpha = \dot{\omega} = \ddot{\theta}$ for the angular velocity and angular acceleration, respectively, of the body we have Eqs. 2/11, rewritten as

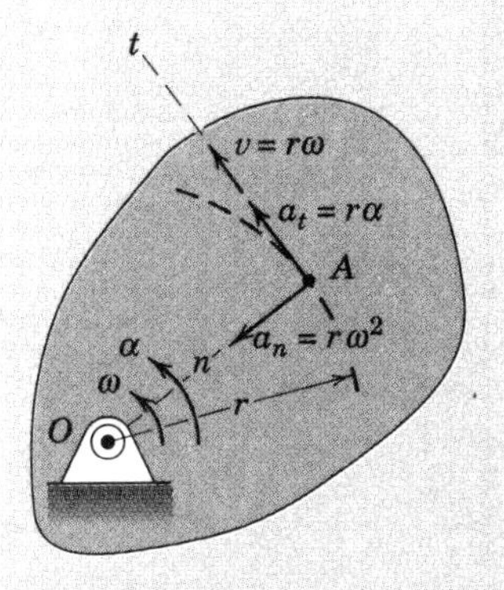

Figure 5/3

$$
\begin{aligned}
v &= r\omega \\
a_n &= r\omega^2 = v^2/r = v\omega \\
a_t &= r\alpha
\end{aligned}
\tag{5/2}
$$

These quantities may be expressed alternatively using the cross-product relationship of vector notation. The vector formulation is especially important in the analysis of three-dimensional motion. The angular velocity of the rotating body may be expressed by the vector $\boldsymbol{\omega}$ normal to the plane of rotation and having a sense governed by the right-hand rule, as shown in Fig. 5/4a. From the definition of the vector cross product, we see that the vector $\mathbf{v}$ is obtained by crossing $\boldsymbol{\omega}$ into $\mathbf{r}$. This cross product gives the correct magnitude and direction for $\mathbf{v}$ and we write

$$\mathbf{v} = \dot{\mathbf{r}} = \boldsymbol{\omega} \times \mathbf{r}$$

The order of the vectors to be crossed must be retained. The reverse order gives $\mathbf{r} \times \boldsymbol{\omega} = -\mathbf{v}$.

The acceleration of point A is obtained by differentiating the cross-product expression for $\mathbf{v}$, which gives

$$
\begin{aligned}
\mathbf{a} = \dot{\mathbf{v}} &= \boldsymbol{\omega} \times \dot{\mathbf{r}} + \dot{\boldsymbol{\omega}} \times \mathbf{r} \\
&= \boldsymbol{\omega} \times (\boldsymbol{\omega} \times \mathbf{r}) + \dot{\boldsymbol{\omega}} \times \mathbf{r} \\
&= \boldsymbol{\omega} \times \mathbf{v} + \boldsymbol{\alpha} \times \mathbf{r}
\end{aligned}
$$

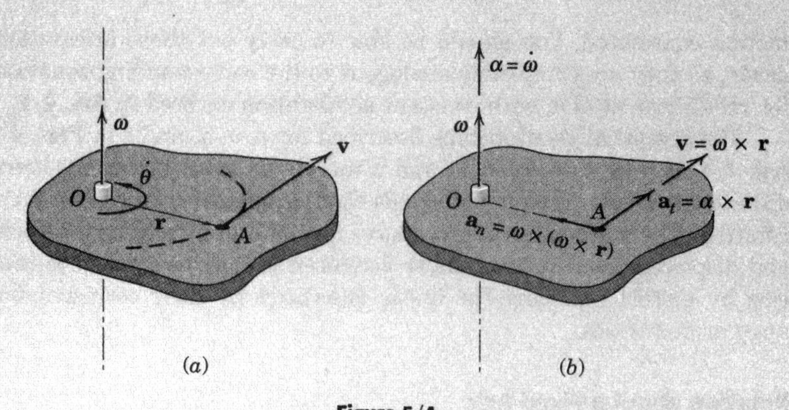

Figure 5/4

Here $\boldsymbol{\alpha} = \dot{\boldsymbol{\omega}}$ stands for the angular acceleration of the body. Thus, the vector equivalents to Eqs. 5/2 are

$$\mathbf{v} = \boldsymbol{\omega} \times \mathbf{r}$$
$$\mathbf{a}_n = \boldsymbol{\omega} \times (\boldsymbol{\omega} \times \mathbf{r}) \qquad (5/3)$$
$$\mathbf{a}_t = \boldsymbol{\alpha} \times \mathbf{r}$$

and are shown in Fig. 5/4*b*.

For three-dimensional motion of a rigid body, the angular-velocity vector $\boldsymbol{\omega}$ may change direction as well as magnitude, and in this case, the angular acceleration, which is the time derivative of angular velocity, $\boldsymbol{\alpha} = \dot{\boldsymbol{\omega}}$, will no longer be in the same direction as $\boldsymbol{\omega}$.

Sample Problem 5/1

A flywheel rotating freely at 1800 rev/min clockwise is subjected to a variable counterclockwise torque which is first applied at time $t = 0$. The torque produces a counterclockwise angular acceleration $\alpha = 4t$ rad/s^2, where t is the time in seconds during which the torque is applied. Determine (*a*) the time required for the flywheel to reduce its clockwise angular speed to 900 rev/min, (*b*) the time required for the flywheel to reverse its direction of rotation, and (*c*) the total number of revolutions, clockwise plus counterclockwise, turned by the flywheel during the first 14 seconds of torque application.

Solution. The counterclockwise direction will be taken arbitrarily as positive.

(a) Since α is a known function of the time, we may integrate it to obtain angular velocity. With the initial angular velocity of $-1800(2\pi)/60 = -60\pi$ rad/s, we have ①

$$[d\omega = \alpha\, dt] \qquad \int_{-60\pi}^{\omega} d\omega = \int_0^t 4t\, dt \qquad \omega = -60\pi + 2t^2$$

Substituting the clockwise angular speed of 900 rev/min or $\omega = -900(2\pi)/60 = -30\pi$ rad/s gives

$$-30\pi = -60\pi + 2t^2 \qquad t^2 = 15\pi \qquad t = 6.86 \text{ s} \qquad \textit{Ans.}$$

(b) The flywheel changes direction when its angular velocity is momentarily zero. Thus,

$$0 = -60\pi + 2t^2 \qquad t^2 = 30\pi \qquad t = 9.71 \text{ s} \qquad \textit{Ans.}$$

(c) The total number of revolutions through which the flywheel turns during 14 seconds is the number of clockwise turns N_1 during the first 9.71 seconds, plus the number of counterclockwise turns N_2 during the remainder of the interval. Integrating the expression for ω in terms of t gives us the angular displacement in radians. Thus, for the first interval

$$[d\theta = \omega\, dt] \qquad \int_0^{\theta_1} d\theta = \int_0^{9.71} (-60\pi + 2t^2)\, dt$$

② $$\theta_1 = [-60\pi t + \tfrac{2}{3}t^3]_0^{9.71} = -1220 \text{ rad}$$

or $N_1 = 1220/2\pi = 194.2$ revolutions clockwise.

For the second interval

$$\int_0^{\theta_2} d\theta = \int_{9.71}^{14} (-60\pi + 2t^2)\, dt$$

③ $$\theta_2 = [-60\pi t + \tfrac{2}{3}t^3]_{9.71}^{14} = 410 \text{ rad}$$

or $N_2 = 410/2\pi = 65.3$ revolutions counterclockwise. Thus, the total number of revolutions turned during the 14 seconds is

$$N = N_1 + N_2 = 194.2 + 65.3 = 259 \text{ rev} \qquad \textit{Ans.}$$

We have plotted ω versus t and we see that θ_1 is represented by the negative area and θ_2 by the positive area. If we had integrated over the entire interval in one step, we would have obtained $|\theta_2| - |\theta_1|$.

Helpful Hints

① We must be very careful to be consistent with our algebraic signs. The lower limit is the negative (clockwise) value of the initial angular velocity. Also we must convert revolutions to radians since α is in radian units.

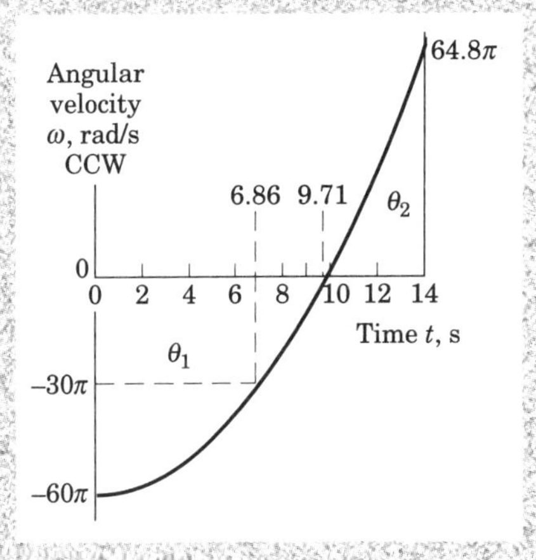

② Again note that the minus sign signifies clockwise in this problem.

③ We could have converted the original expression for α into the units of rev/s^2, in which case our integrals would have come out directly in revolutions.

Sample Problem 5/2

The pinion A of the hoist motor drives gear B, which is attached to the hoisting drum. The load L is lifted from its rest position and acquires an upward velocity of 2 m/s in a vertical rise of 0.8 m with constant acceleration. As the load passes this position, compute (*a*) the acceleration of point C on the cable in contact with the drum and (*b*) the angular velocity and angular acceleration of the pinion A.

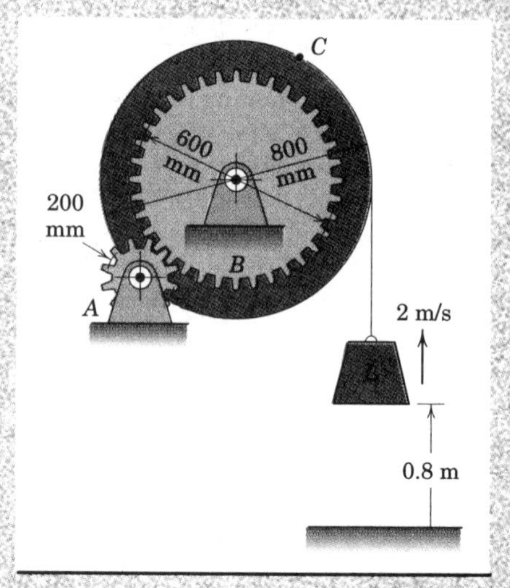

Solution. ***(a)*** If the cable does not slip on the drum, the vertical velocity and acceleration of the load L are, of necessity, the same as the tangential velocity v and tangential acceleration a_t of point C. For the rectilinear motion of L with constant acceleration, the n- and t-components of the acceleration of C become

$[v^2 = 2as]$ $\quad a = a_t = v^2/2s = 2^2/[2(0.8)] = 2.5 \text{ m/s}^2$

① $[a_n = v^2/r]$ $\quad a_n = 2^2/0.400 = 10 \text{ m/s}^2$

$[a = \sqrt{a_n^2 + a_t^2}]$ $\quad a_C = \sqrt{(10)^2 + (2.5)^2} = 10.31 \text{ m/s}^2$ *Ans.*

(b) The angular motion of gear A is determined from the angular motion of gear B by the velocity v_1 and tangential acceleration a_1 of their common point of contact. First, the angular motion of gear B is determined from the motion of point C on the attached drum. Thus,

$[v = r\omega]$ $\quad \omega_B = v/r = 2/0.400 = 5 \text{ rad/s}$

$[a_t = r\alpha]$ $\quad \alpha_B = a_t/r = 2.5/0.400 = 6.25 \text{ rad/s}^2$

Then from $v_1 = r_A\omega_A = r_B\omega_B$ and $a_1 = r_A\alpha_A = r_B\alpha_B$, we have

$$\omega_A = \frac{r_B}{r_A}\omega_B = \frac{0.300}{0.100}5 = 15 \text{ rad/s CW} \qquad \textit{Ans.}$$

$$\alpha_A = \frac{r_B}{r_A}\alpha_B = \frac{0.300}{0.100}6.25 = 18.75 \text{ rad/s}^2 \text{ CW} \qquad \textit{Ans.}$$

Helpful Hint

① Recognize that a point on the cable changes the direction of its velocity after it contacts the drum and acquires a normal component of acceleration.

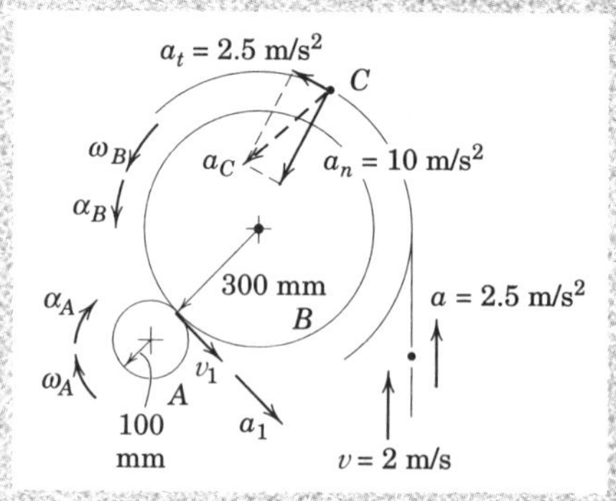

Sample Problem 5/3

The right-angle bar rotates clockwise with an angular velocity which is decreasing at the rate of 4 rad/s². Write the vector expressions for the velocity and acceleration of point A when $\omega = 2$ rad/s.

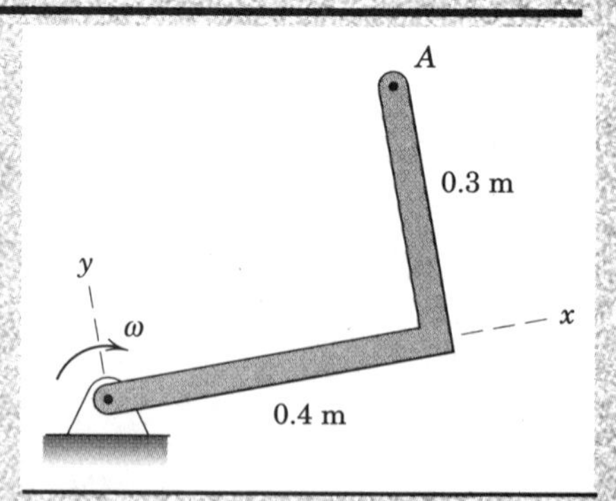

Solution. Using the right-hand rule gives

$$\boldsymbol{\omega} = -2\mathbf{k} \text{ rad/s} \quad \text{and} \quad \boldsymbol{\alpha} = +4\mathbf{k} \text{ rad/s}^2$$

The velocity and acceleration of A become

$[\mathbf{v} = \boldsymbol{\omega} \times \mathbf{r}]$ $\quad \mathbf{v} = -2\mathbf{k} \times (0.4\mathbf{i} + 0.3\mathbf{j}) = 0.6\mathbf{i} - 0.8\mathbf{j} \text{ m/s}$ *Ans.*

$[\mathbf{a}_n = \boldsymbol{\omega} \times (\boldsymbol{\omega} \times \mathbf{r})]$ $\quad \mathbf{a}_n = -2\mathbf{k} \times (0.6\mathbf{i} - 0.8\mathbf{j}) = -1.6\mathbf{i} - 1.2\mathbf{j} \text{ m/s}^2$

$[\mathbf{a}_t = \boldsymbol{\alpha} \times \mathbf{r}]$ $\quad \mathbf{a}_t = 4\mathbf{k} \times (0.4\mathbf{i} + 0.3\mathbf{j}) = -1.2\mathbf{i} + 1.6\mathbf{j} \text{ m/s}^2$

$[\mathbf{a} = \mathbf{a}_n + \mathbf{a}_t]$ $\quad \mathbf{a} = -2.8\mathbf{i} + 0.4\mathbf{j} \text{ m/s}^2$ *Ans.*

The magnitudes of $\mathbf{v}$ and $\mathbf{a}$ are

$$v = \sqrt{0.6^2 + 0.8^2} = 1 \text{ m/s} \quad \text{and} \quad a = \sqrt{2.8^2 + 0.4^2} = 2.83 \text{ m/s}^2$$

PROBLEMS

Introductory Problems

5/1 A torque applied to a flywheel causes it to accelerate uniformly from a speed of 200 rev/min to a speed of 800 rev/min in 4 seconds. Determine the number of revolutions N through which the wheel turns during this interval. (*Suggestion:* Use revolutions and minutes for units in your calculations.)

Ans. $N = 33.3$ rev

5/2 The rectangular plate is rotating about its corner axis through O with a constant angular velocity $\omega = 10$ rad/s. Determine the magnitudes of the velocity $\mathbf{v}$ and acceleration $\mathbf{a}$ of the corner A by (*a*) using the scalar relations and (*b*) using the vector relations.

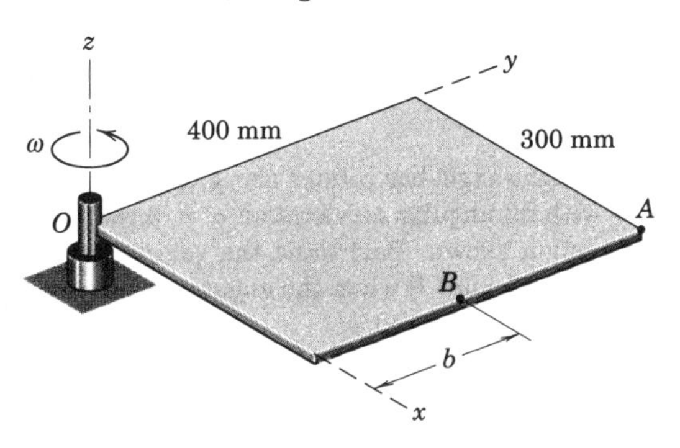

Problem 5/2

5/3 The T-shaped body rotates about a horizontal axis through point O. At the instant represented, its angular velocity is $\omega = 3$ rad/s and its angular acceleration is $\alpha = 14$ rad/s^2 in the directions indicated. Determine the velocity and acceleration of (*a*) point A and (*b*) point B. Express your results in terms of components along the n- and t-axes shown.

Ans. (*a*) $\mathbf{v}_A = 1.2\mathbf{e}_t$ m/s

$\mathbf{a}_A = -5.6\mathbf{e}_t + 3.6\mathbf{e}_n$ m/s^2

(*b*) $\mathbf{v}_B = 1.2\mathbf{e}_t + 0.3\mathbf{e}_n$ m/s

$\mathbf{a}_B = -6.5\mathbf{e}_t + 2.2\mathbf{e}_n$ m/s^2

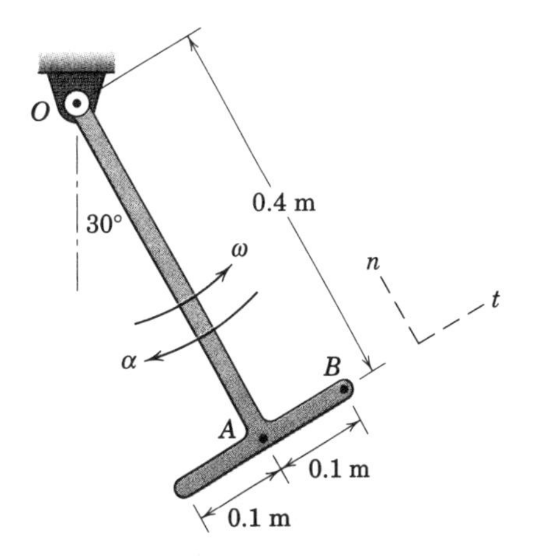

Problem 5/3

5/4 The small cart is released from rest in position 1 and requires 0.638 seconds to reach position 2 at the bottom of the path, where its center G has a velocity of 4.33 m/s. Determine the angular velocity ω of line AB in position 2 and the average angular velocity ω_{av} of AB during the interval.

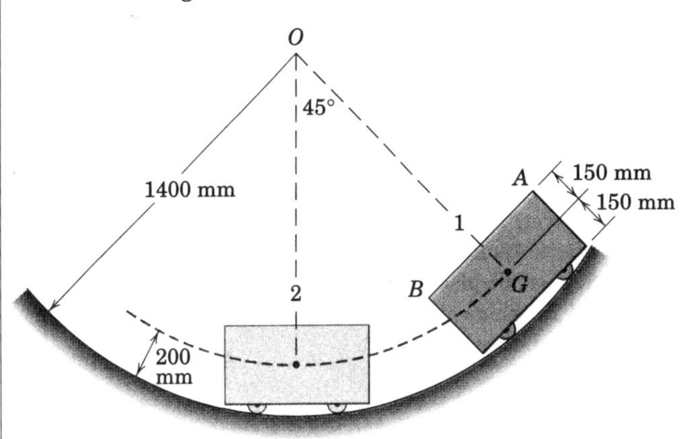

Problem 5/4

5/5 The flywheel has a diameter of 600 mm and rotates with increasing speed about its z-axis shaft. When point P on the rim crosses the y-axis with $\theta = 90°$, it has an acceleration given by $\mathbf{a} = -1.8\mathbf{i} - 4.8\mathbf{j}$ m/s^2. For this instant, determine the angular velocity ω and the angular acceleration α of the flywheel.

Ans. $\alpha = 6$ rad/s^2, $\omega = 4$ rad/s

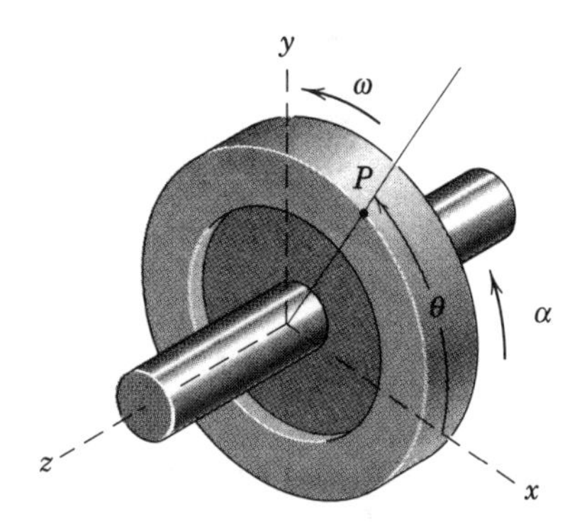

Problem 5/5

5/6 If the acceleration of point P on the rim of the flywheel of Prob. 5/5 is $\mathbf{a} = -3.02\mathbf{i} - 1.624\mathbf{j}$ m/s^2 when $\theta = 60°$, determine the angular velocity ω and angular acceleration α of the 600-mm-diameter flywheel for this position.

5/7 The angular position of a radial line in a rotating disk is given by the clockwise angle $\theta = 2t^3 - 3t^2 + 4$, where θ is in radians and t is in seconds. Calculate the angular displacement $\Delta\theta$ of the disk during the interval in which its angular acceleration increases from 42 rad/s^2 to 66 rad/s^2.

Ans. $\Delta\theta = 244$ rad

5/8 The circular disk rotates about its center O in the direction shown. At a certain instant point P on the rim has an acceleration given by $\mathbf{a} = -3\mathbf{i} - 4\mathbf{j}$ m/s^2. For this instant determine the angular velocity $\boldsymbol{\omega}$ and angular acceleration $\boldsymbol{\alpha}$ of the disk.

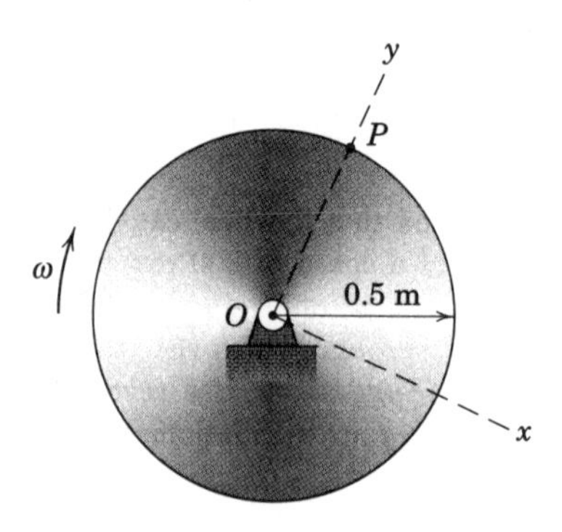

Problem 5/8

5/9 If the rectangular plate of Prob. 5/2 starts from rest and point B has an initial acceleration of 5.5 m/s^2, determine the distance b if the plate reaches an angular speed of 300 rev/min in 2 seconds with a constant angular acceleration.

Ans. $b = 180.6$ mm

5/10 The right-angle bar rotates about the z-axis through O with an angular acceleration $\alpha = 3$ rad/s^2 in the direction shown. Determine the velocity and acceleration of point P when the angular velocity reaches the value $\omega = 2$ rad/s.

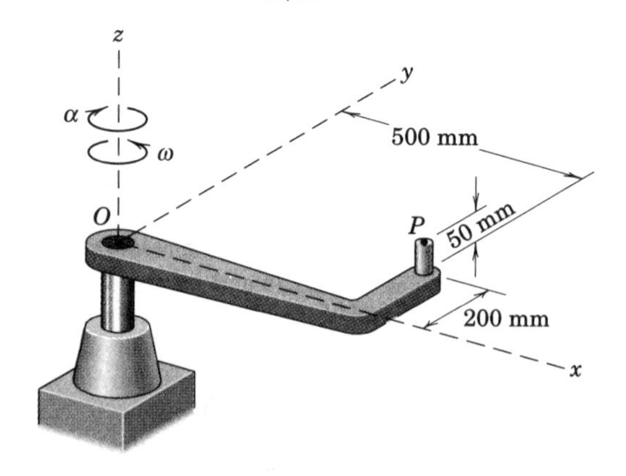

Problem 5/10

Representative Problems

5/11 Experimental data for a rotating control element reveal the plotted relation between angular velocity and the angular coordinate θ as shown. Approximate the angular acceleration α of the element when $\theta = 6$ rad.

Ans. $\alpha = 3.95$ rad/s^2

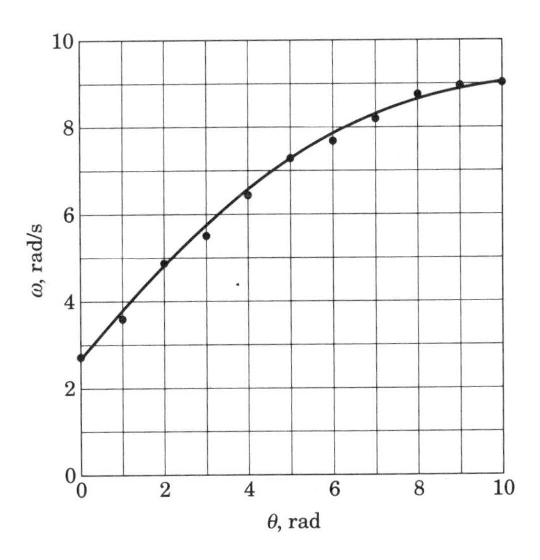

Problem 5/11

5/12 The angular position of a radial line on a rotating disk is given by $\theta = (-1 + 1.5t)e^{-0.5t}$, where θ is in radians and t is in seconds. Plot the angular position, angular velocity, and angular acceleration versus time for the first 20 seconds of motion. Determine the time at which the acceleration is zero.

5/13 The rectangular plate rotates clockwise about its fixed bearing at O. If edge BC has a constant angular velocity of 6 rad/s, determine the vector expressions for the velocity and acceleration of point A using the coordinates given.

Ans. $\mathbf{v}_A = 1.68\mathbf{i} - 1.8\mathbf{j}$ m/s
$\mathbf{a}_A = -10.8\mathbf{i} - 10.08\mathbf{j}$ m/s^2

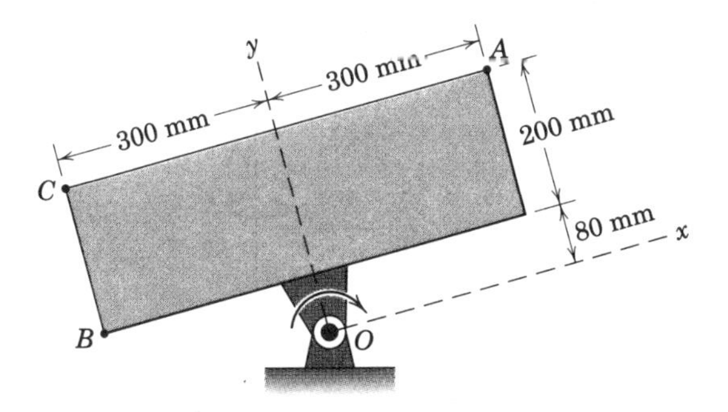

Problem 5/13

5/14 The mass center G of the car has a velocity of 60 km/h at position A and 1.52 s later at B has a velocity of 80 km/h. The radius of curvature of the road at B is 60 m. Calculate the angular velocity ω of the car at B and the average angular velocity ω_{av} of the car between A and B.

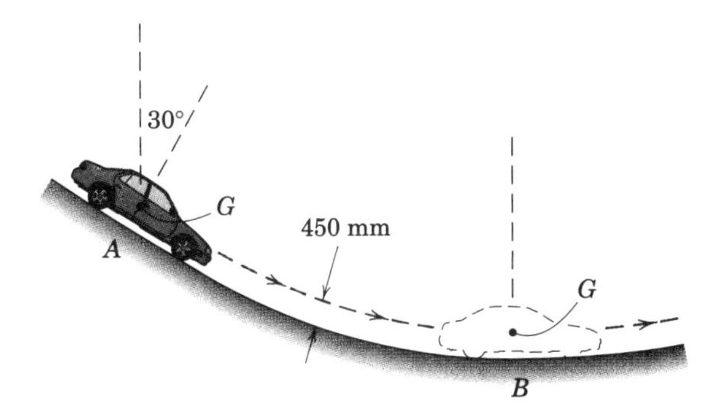

Problem 5/14

5/15 The belt-driven pulley and attached disk are rotating with increasing angular velocity. At a certain instant the speed v of the belt is 1.5 m/s, and the total acceleration of point A is 75 m/s^2. For this instant determine (*a*) the angular acceleration α of the pulley and disk, (*b*) the total acceleration of point B, and (*c*) the acceleration of point C on the belt.

Ans. (*a*) $\alpha = 300$ rad/s^2, (*b*) $a_B = 37.5$ m/s^2
(*c*) $a_C = 22.5$ m/s^2

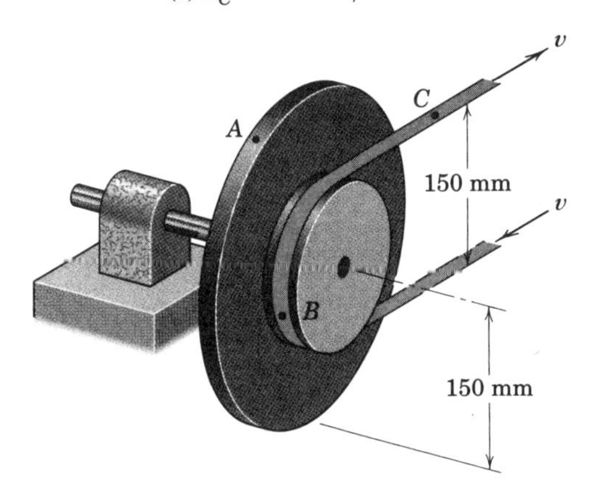

Problem 5/15

5/16 Magnetic tape is being fed over and around the light pulleys mounted in a computer. If the speed v of the tape is constant and if the magnitude of the acceleration of point A on the tape is 4/3 times that of point B, calculate the radius r of the smaller pulley.

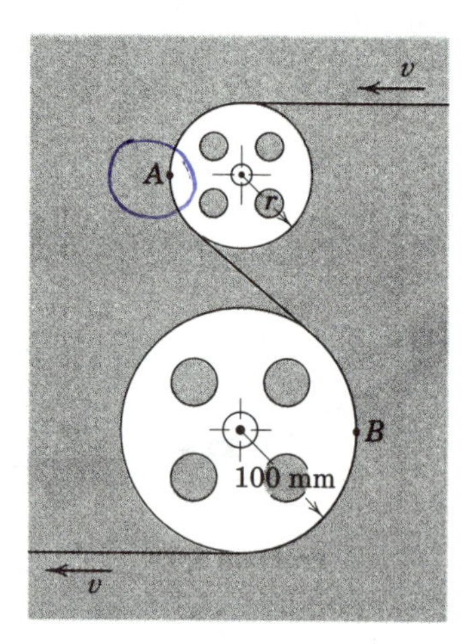

Problem 5/16

5/17 The circular disk rotates about its center O. For the instant represented, the velocity of A is $\mathbf{v}_A = 200\mathbf{j}$ mm/s and the tangential acceleration of B is $(\mathbf{a}_B)_t = 150\mathbf{i}$ mm/s^2. Write the vector expressions for the angular velocity $\boldsymbol{\omega}$ and angular acceleration $\boldsymbol{\alpha}$ of the disk. Use these results to write the vector expression for the acceleration of point C.

Ans. $\boldsymbol{\omega} = 2\mathbf{k}$ rad/s, $\boldsymbol{\alpha} = -\frac{3}{2}\mathbf{k}$ rad/s^2

$\mathbf{a}_C = 25\sqrt{2}(-11\mathbf{i} + 5\mathbf{j})$ mm/s^2

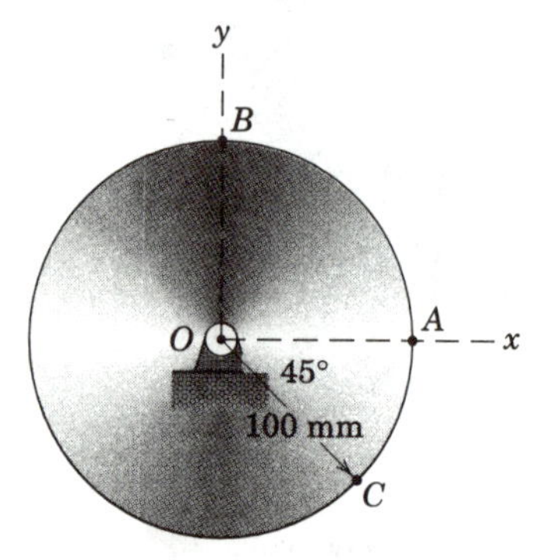

Problem 5/17

5/18 The solid cylinder rotates about its z-axis. At the instant represented, point P on the rim has a velocity whose x-component is -1.28 m/s, and $\theta = 20°$. Determine the angular velocity $\boldsymbol{\omega}$ of line AB on the face of the cylinder. Does the element line BC have an angular velocity?

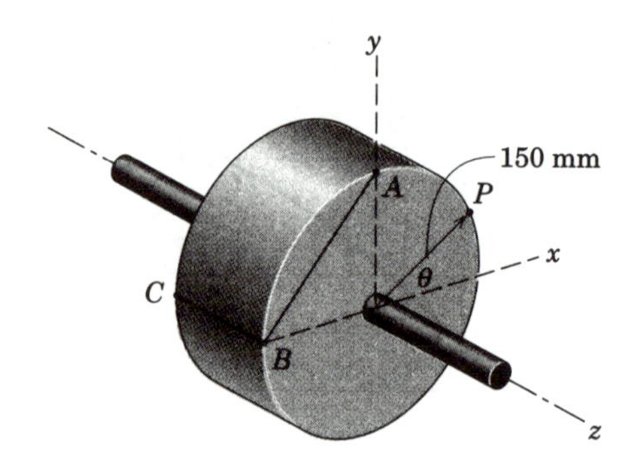

Problem 5/18

5/19 A gear rotating at a clockwise speed of 200 rev/min is subjected to a torque which gives it a clockwise angular acceleration α which varies with the time as shown. Find the speed N of the gear when $t = 5$ s.

Ans. $N = 272$ rev/min

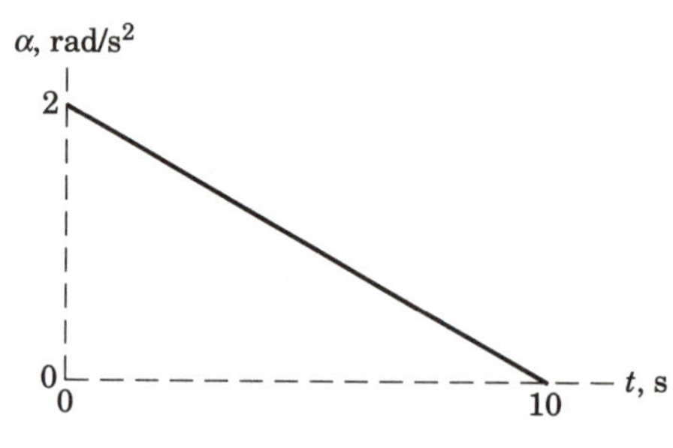

Problem 5/19

5/20 The rotating arm starts from rest and acquires a rotational speed $N = 600$ rev/min in 2 seconds with constant angular acceleration. Find the time t after starting before the acceleration vector of end P makes an angle of 45° with the arm OP.

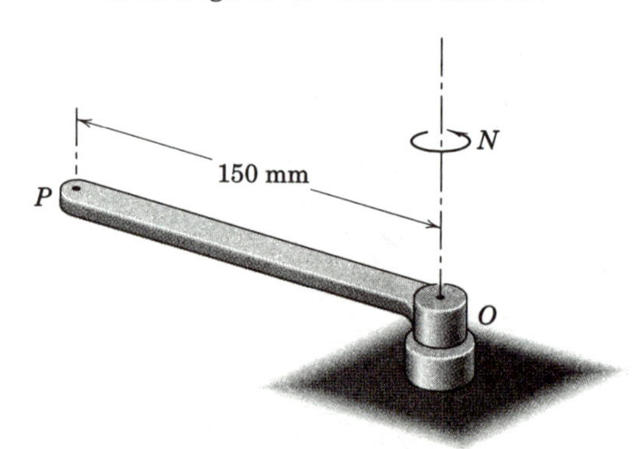

Problem 5/20

5/21 The two V-belt pulleys form an integral unit and rotate about the fixed axis at O. At a certain instant, point A on the belt of the smaller pulley has a velocity $v_A = 1.5$ m/s, and point B on the belt of the larger pulley has an acceleration $a_B = 45$ m/s^2 as shown. For this instant determine the magnitude of the acceleration $\mathbf{a}_C$ of point C and sketch the vector in your solution.

Ans. $a_C = 149.6$ m/s^2

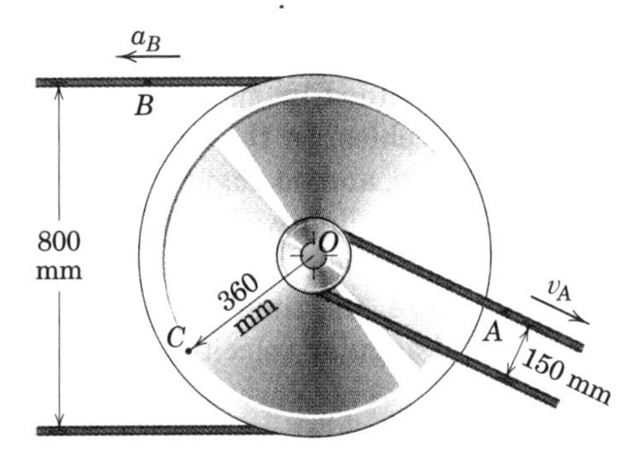

Problem 5/21

5/22 A clockwise variable torque is applied to a flywheel at time $t = 0$ causing its clockwise angular acceleration to decrease linearly with angular displacement θ during 20 revolutions of the wheel as shown. If the clockwise speed of the flywheel was 300 rev/min at $t = 0$, determine its speed N after turning the 20 revolutions. (*Suggestion:* Use units of revolutions instead of radians.)

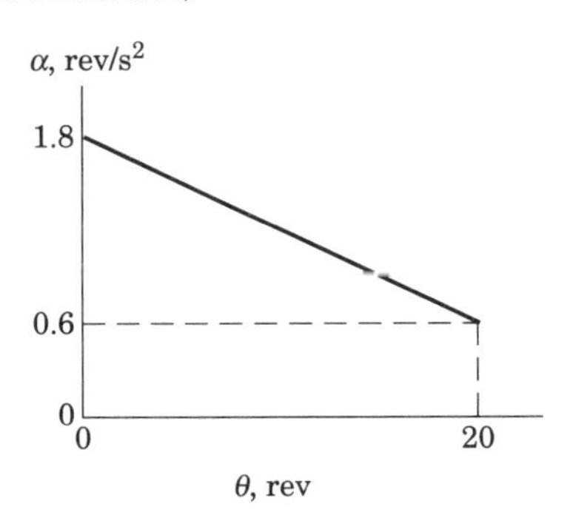

Problem 5/22

5/23 The design characteristics of a gear-reduction unit are under review. Gear B is rotating clockwise with a speed of 300 rev/min when a torque is applied to gear A at time $t = 2$ s to give gear A a counterclockwise acceleration α which varies with time for a duration of 4 seconds as shown. Determine the speed N_B of gear B when $t = 6$ s.

Ans. $N_B = 415$ rev/min

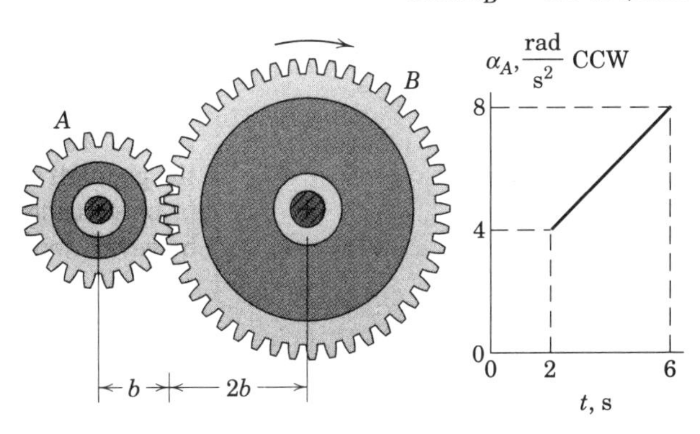

Problem 5/23

▸**5/24** A V-belt speed-reduction drive is shown where pulley A drives the two integral pulleys B which in turn drive pulley C. If A starts from rest at time $t = 0$ and is given a constant angular acceleration α_1, derive expressions for the angular velocity of C and the magnitude of the acceleration of a point P on the belt, both at time t.

Ans. $\omega_C = \left(\frac{r_1}{r_2}\right)^2 \alpha_1 t$

$$a_P = \frac{r_1^{\,2}}{r_2}\alpha_1 \sqrt{1 + \left(\frac{r_1}{r_2}\right)^4 \alpha_1^{\,2} t^4}$$

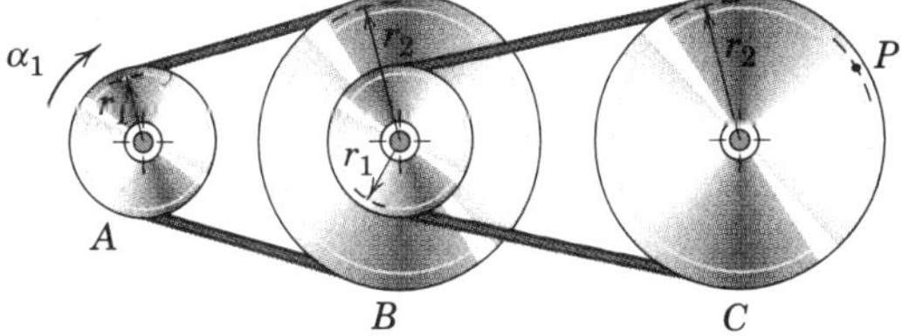

Problem 5/24

5/3 Absolute Motion

We now develop the approach of absolute-motion analysis to describe the plane kinematics of rigid bodies. In this approach, we make use of the geometric relations which define the configuration of the body involved and then proceed to take the time derivatives of the defining geometric relations to obtain velocities and accelerations.

In Art. 2/9 of Chapter 2 on particle kinematics, we introduced the application of absolute-motion analysis for the constrained motion of connected particles. For the pulley configurations treated, the relevant velocities and accelerations were determined by successive differentiation of the lengths of the connecting cables. In this earlier treatment, the geometric relations were quite simple, and no angular quantities had to be considered. Now that we will be dealing with rigid-body motion, however, we find that our defining geometric relations include both linear and angular variables and, therefore, the time derivatives of these quantities will involve both linear and angular velocities and linear and angular accelerations.

In absolute-motion analysis, it is essential that we be consistent with the mathematics of the description. For example, if the angular position of a moving line in the plane of motion is specified by its counterclockwise angle θ measured from some convenient fixed reference axis, then the positive sense for both angular velocity $\dot{\theta}$ and angular acceleration $\ddot{\theta}$ will also be counterclockwise. A negative sign for either quantity will, of course, indicate a clockwise angular motion. The defining relations for linear motion, Eqs. 2/1, 2/2, and 2/3, and the relations involving angular motion, Eqs. 5/1 and 5/2 or 5/3, will find repeated use in the motion analysis and should be mastered.

The absolute-motion approach to rigid-body kinematics is quite straightforward, provided the configuration lends itself to a geometric description which is not overly complex. If the geometric configuration is awkward or complex, analysis by the principles of relative motion may be preferable. Relative-motion analysis is treated in this chapter beginning with Art. 5/4. The choice between absolute- and relative-motion analyses is best made after experience has been gained with both approaches.

The next three sample problems illustrate the application of absolute-motion analysis to three commonly encountered situations. The kinematics of a rolling wheel, treated in Sample Problem 5/4, is especially important and will be useful in much of the problem work because the rolling wheel in various forms is such a common element in mechanical systems.